Ecological Studies, Vol. 108

Analysis and Synthesis

Edited by

O.L. Lange, Würzburg, FRG
H.A. Mooney, Stanford, USA
H. Remmert, †Marburg, FRG

Ecological Studies

Volumes published since 1989 are listed at the end of this book.

Mary T. Kalin Arroyo   Paul H. Zedler
Marilyn D. Fox
Editors

Ecology and Biogeography of Mediterranean Ecosystems in Chile, California, and Australia

With 66 Illustrations

Springer-Verlag
New York Berlin Heidelberg London Paris
Tokyo Hong Kong Barcelona Budapest

Mary T. Kalin Arroyo
Department of Biology
Faculty of Sciences
University of Chile
Casilla 653, Santiago
Chile

Paul H. Zedler
Department of Biology
San Diego State University
San Diego, CA 92182-0057
USA

Marilyn D. Fox
School of Geography
University of New South Wales
Kensington NSW
2033 Australia

QH
541
.5
.M44
E26
1995

Cover illustration: The fruit, seed, and flower of *Quillaja saponaria* Mol. (Rosaceae), a dominant tree in the sclerophyllous vegetation of Central Chile.

Library of Congress Cataloging-in-Publication Data
Ecology and biogeography of Mediterranean ecosystems in Chile,
    California, and Australia/M.T.K. Arroyo, Paul H. Zedler,
    and Marilyn D. Fox, editors.
        p.   cm. — (Ecological studies; v. 108)
    Includes bibliographical references and index.
    ISBN 0-387-94266-1. — ISBN 3-540-94266-1
    1. Mediterranean-type ecosystems—Chile.   2. Mediterranean-type
ecosystems—California.   3. Mediterranean-type ecosystems—
Australia.   4. Plant ecology—Chile.   5. Plant ecology—California.
6. Plant ecology—Australia.   7. Biogeography—Chile.
8. Biogeography—California.   9. Biogeography—Australia.   I. Kalin
Arroyo, Mary T.   II. Zedler, Paul H.   III. Fox, Marilyn D.
IV. Series.
QH541.5.M44E26   1994
574.5′262—dc20   94-3097

Printed on acid-free paper.

Production coordinated by Chernow Editorial Services, Inc., and managed by Francine McNeill; manufacturing supervised by Jacqui Ashri.
Typeset by Asco Trade Typesetting Ltd., Hong Kong.
Printed and bound by Braun-Brumfield, Inc., Ann Arbor, MI.
Printed in the United States of America.

9 8 7 6 5 4 3 2 1

ISBN 0-387-94266-1 Springer-Verlag New York Berlin Heidelberg
ISBN 3-540-94266-1 Springer-Verlag Berlin Heidelberg New York

Foreword

Comparative studies of the biota in mediterranean-climate areas have been especially interesting since the intriguing similarities between these geographically distant regions were revealed by the early biogeographers. The similarities have been taken as good examples of the power of natural selection to shape organisms in a way well suited to their environment—of common solutions to common ecological factors. Such broad-scale examples of convergent evolution, then, have been viewed as natural experiments on a grand scale, providing a useful key in understanding the evolution of life on earth. The growth of our factual knowledge about these regions has been significantly stimulated by the desire to document convergence, and more recently to measure functional similarity. With the increase in understanding that has resulted, however, has come realization that the differences are as intriguing as the similarities.

This volume presents a diverse collection of papers exploring aspects of the ecology, biogeography, and paleobiology in the three mediterranean-climate regions that lie around the Pacific basin. Chile and California have been subjected to careful comparisons in recent decades, and their matching ecosystems have been shown to exhibit both fascinating physiognomic similarities and overwhelming biological differences resulting from the very different ancestral stocks that gave rise to their biological communities. Much more different are the communities of the mediterranean-climate regions in Australia, which diverge markedly from both American systems and have unique features that considerably illuminate the degree to which convergence has apparently been possible, given their distinct history as well as the phylogenetic constraints.

The diversity of the contributions in this volume reflects the variety of approaches and points of view that are necessary to compare present ecosystem function and to understand how functional roles and historical factors may have interacted to assemble the contemporary biotas. As with any effort of this sort, many more questions are raised than are answered. It is also apparent that major gaps in our knowledge remain and that much work still needs doing. Nonetheless, the comparative method is one of the best, and sometimes the only means by which ecologists, evolutionists, and biogeographers can test the generality of their concepts. For this reason the contributions by the Chilean scientists represented here are particularly valuable, because many of them have been in the forefront of efforts to understand mediterranean-climate systems through collaborative and comparative research. This volume demonstrates that the tradition continues and is represented here in such a way as to provide a point of access to the Chilean literature for interested researchers.

We have entered an exciting era—in which many new techniques of analysis are arising in systematics and the need to view ecosystems in a global context is widely recognized. Forging stronger links between systematics and ecology, and between researchers in different countries will be critical to future comparative studies. Because of the threats to biodiversity throughout the world, and especially in these fragile, unique ecosystems, the need for such studies is great.

Peter H. Raven

Preface

The VI Inter-Congress of the Pacific Science Association, "The Pacific, Bridge or Barrier?" held in Viña del Mar, Chile, in August 1989, included a symposium on mediterranean ecosystems of the Pacific region. Intense interest in the ecosystems of central Chile and earlier concentration on the search for convergence among the mediterranean regions made a symposium stressing similarities and differences between the Pacific mediterranean regions seem timely. This geographic constraint fortuitously brought together biologists working in a novel subset of the world's five mediterranean regions: California, Chile, and Australia. Past research programs and publications had focused on the mediterranean regions in the two American continents. Because of strong physiognomic similarities, and being on the west coasts of contiguous continents, the opportunity for comparison was compelling and the outcome of collaborative research particularly fruitful. Past comparative studies involving southern Australia had been more limited, however, and had invariably linked it with the other southern mediterranean region, South Africa. The prospect of drawing similarities or highlighting differences among the three Pacific regions was a challenging one. Following the successful meeting in Viña del Mar, the organizers were encouraged to publish the proceedings to stimulate new approaches in comparative studies in the three regions. In 1991, arrangements were made with the publisher to assemble an extended compilation of papers including other contributors to flesh out aspects of the convergence debate that the meeting had brought out.

Two decades of intensive study had shown that convergence in mediterranean ecosystems manifests itself only at the broadest levels and that the many similarities are embedded in a matrix of fascinating differences. A host of a posteriori explanations for such differences are found in the concluding paragraphs of the enormous volume of literature this exciting topic has generated. These conclusions, however, often reflect research focused on small study sites and emphasizing local community dynamics and ecosystem functioning, giving little attention to the complex evolutionary history of the organisms or details of past climatic change. If the idea of convergence is to have lasting influence in ecology and biogeography, the many explanations for lack of strong convergence need to be elevated to hypotheses and incorporated into a new paradigm to be tested with the same rigor as the convergence hypothesis itself.

For its ultimate development, any new paradigm requires access to new types of information. This was the general philosophy that guided the original symposium and assembly of this volume. We envisaged interchange among individuals working in mediterranean climates in many fields—paleobotanists, systematists, ecologists, animal physiologists—as a first step in preparing a substrate for further thought. As in the title of the original symposium that led to this volume, we specifically asked authors to attend to the differences as well as similarities among mediterranean ecosystems that were revealed by their approaches.

This goal guarantees strong diversity in the papers. The material draws on literature and unpublished data combining different sets of the three regions. Most papers compare all three regions, and others feature new information on the Chile–California nexus. One focuses exclusively on southern Australia, and another on California; we are also pleased to be able to present papers on the history and present land use of mediterranean vegetation in Chile alone or in combination with the other regions.

In regions where people have had and continue to have great influence in shaping landscapes, their effect on natural succession or in imposing new successional forces is addressed in the chapters by Armesto et al. and Fuentes and Muñoz. The earlier date of European settlement in Chile is shown to affect regeneration of matorral (Armesto et al.), the general trend being to slow successional rates and the persistence of anthropogenic plant formations. The way in which these changes are maintained involves increased frequency and intensity of fires, modified patterns of grazing and browsing, and increased clearance of vegetation, especially on steep slopes (Fuentes and Muñoz). M.D. Fox also raises the matter of anthropogenic introductions adversely affecting, and possibly diluting, biogeographic differences among the three regions. At the same time, unique features of the biota in each region are being lost and replaced by Mediterranean annual weeds and grazing animals originally domesticated in the Old World.

Whereas past convergence studies had a strong plant ecophysiological bias, this volume brings together a blend of disciplines not previously canvassed. The pivotal influence of systematics is emphasized in several papers. The past effect of climate in imposing constraints on animal physiology is addressed by Bozinovic et al. They point out that the convergence hypothesis is extremely difficult to test; recognizing the similarity in physiological traits does not guarantee that they have converged in the past. Because of the physiological significance of body size, ancestral body sizes as discerned in fossil assemblages can offer tests of convergence.

Study of specialist life forms such as fossorial mammals can elucidate patterns of animal communities dependence on structure of vegetation. The prominence of geophytes in California and Chile could explain the diversity among subterranean foragers in those regions, and their slighter importance in Australia may explain why this specialization did not arise there (Cox et al.). Community structure responding to the more general shrubby formation has led to similarity between the small mammal faunas in Australia and Chile (B. Fox). Californian community characteristics differed from the other two, possibly in response to productivity.

A paper linking the mammalian contributions to the vegetation chapters is Louda's, reviewing how seed predation affects plant regeneration. She suggests a conceptual model that combines characteristics in plant life history and reproductive success in a predictably stressful environment.

Contributions here about plant communities include three on reproductive strategies. Keeley explores a rich background of information on germination patterns and concludes that all five mediterranean regions display remarkable convergence. The structure of the seed embryo is a surprising delineator of phylogenetic status, geographic affinity, and habitat requirements of herbaceous dicotyledons. Vivrette finds similarities between southern Australia and California, but significant differences in the Chilean flora. Studying morphology of fruits and dispersal of seeds, Hoffmann and Armesto report important differences between phylogenetically matched taxa and the mediterranean regions of California and Chile. They suggest pivotal disturbance regimes in the evolution of dispersal modes.

The poor nutrient content of the soils in the mediterranean regions has received some attention in the past. Lamont's chapter, though, compares in detail soils from southern Australia and California, and compares the relatively rich Chilean soils to those of the other two regions. Below the soil's surface, Canadell and Zedler's thorough review shows that the root structures of mediterranean shrubs are highly diverse, are still poorly studied, and should not be overlooked in forming hypotheses. Again, disturbance is postulated in the development of lignotubers, a highly characteristic structure among mediterranean plants.

More broadly investigating ecomorphological features, Montenegro and Ginocchio, in accordance with earlier work, contend that most influence exerted by phylogeny on evolution of growth form can be overcome by selective forces in the environment. Similarities in plant growth form have often been allied to vegetation physiognomy. M.D. Fox uses data from the two parts of the southern Australia mediterranean region to consider the separate evolutions of shrublands on the same continent. Species composition, expressed as α- and β-diversity, differs in the two Australian regions, and also between Australia and the other two regions. Zedler continues the investigation of diversity and plant life history in the Californian chaparral. He concludes from studying woody life histories that phylogenetic constraints affect the origin of major life-history types.

Ecological convergence, especially in the mediterranean regions, was first espoused by biogeographers, and biogeography combined with phylogeny is an integral part in chapters of this book. Bernhardt, applying a comparative systematics approach, reviews the floral evolution of a tribe of orchids with Australasian–South American distribution. The group's adaptive radiation has been explosive within mediterranean Australia, but not in the Chilean region.

The biogeographic scale is also espoused by Arroyo et al. in comparing the mediterranean floras of central Chile and California. Increasing the biogeographic scale, they found less climatic matching between California and Chile. Clear differences appear in species and generic richness, life-form distribution, and historical development of the two mediterranean floras, suggesting strong phylogenetic constraints on the evolution of the annual habit in a woodier flora in central Chile. Their results could provide an alternative explanation for some differences previously attributed to human influence, as found in small-scale studies of community structure.

For the first time in the convergence literature, the histories of the two southern regions are given detailed treatment here. Dodson and Kershaw review the history of vegetation in southern Australia for both the pre-Quaternary and Quaternary periods. Interestingly, they find that many higher-order taxa in Australia's mediterranean flora developed in the early to mid-Tertiary, but became prominent only during the late Quaternary. Villagrán confines her treatment to the more recent Quaternary and stresses the repetitive changes during the glacial–interglacial cycles for the distribution of mediterranean vegetation in central Chile.

This collection of papers offers new data, new interpretations, new methodologies, and new investigators to the continuing debate on convergence in mediterranean ecosystems. We are acutely aware that the studies reported in this volume merely sample the information available in the three regions, leaving many obvious gaps. It is also apparent, however, that for most of the topics covered the information necessary for detailed quantitative comparison is lacking or decidedly incomplete. We will be gratified if this work stimulates others to fill in the gaps and to collect or compile data that will eventually allow more meaningful comparisons among regions.

With three editors on three continents, the logistics of producing this publication have sometimes been unnerving. Opportunities to meet, however, prompted by visits for other conferences and for sabbaticals, kept enthusiasm alive. We thank the contributors for their willing support for this project. We especially thank Peter Raven for writing the Foreword and for his encouragement in bringing this volume to fruition. His seminal work on the mediterranean floras has fostered the biogeographic approach to comparative studies in mediterranean regions. We thank the Departamento Técnico de Investigación (DTI), University of Chile, and CONICYT, Chile for providing travel support for M.T. Kalin Arroyo at several stages in this venture. We also thank Springer-Verlag, particularly Rob Garber, for his assistance. P.H. Zedler thanks Wende Rehlaender for exemplary editing, and we all extend our gratitude to Shannon Bliss and Lohengrin Cavieres for their very capable assistance in the final editing stages. The final editorial work on this book was supported in part by FONDECYT Project No. 92-1135.

Finally, we dedicate this book to these colleagues, who have made significant contributions in the ecological convergence debate, working on each of the three continents: Francesco di Castri for his pivotal role in pioneering many of the Chilean studies, Hal Mooney for his extraordinary past contribution and continuing enthusiasm for convergence studies focused on California and Chile, and Ray Specht for his long-term, meticulous documentation of the mediterranean vegetation of southern Australia.

Mary T. Kalin Arroyo
Paul H. Zedler
Marilyn D. Fox

Contents

Contributors

Armesto, Juan J.

Department of Biology, Faculty of Sciences, University of Chile, Casilla 653, Santiago, Chile

Arroyo, Mary T. Kalin

Department of Biology, Faculty of Sciences, University of Chile, Casilla 653, Santiago, Chile

Bernhardt, Peter

Department of Biology, St. Louis University, St. Louis, MO 63103, USA

Bozinovic, Francisco

Department of Biology, Faculty of Sciences, University of Chile, Casilla 653, Santiago, Chile

Canadell, Josep

Ecology Unit, Antonomous University of Barcelona, 08193 Bellaterra, Spain

Contributors

Cavieres, Lohengrin

Department of Biology, Faculty of Sciences, University of Chile, Casilla 653, Santiago, Chile

Contreras, Luis C.

Department of Biology, University of La Serena, Casilla 599, La Serena, Chile

Cox, George W.

Department of Biology, San Diego State University, San Diego, CA 92182-0057, USA

Dodson, John R.

School of Geography, University of New South Wales, Kensington NSW, 2033 Australia

Fox, Barry J.

School of Biological Science, University of New South Wales, Kensington NSW, 2033 Australia

Fox, Marilyn D.

School of Geography, University of New South Wales, Kensington NSW, 2033 Australia

Fuentes, Eduardo R.

Department of Ecology, Faculty of Biological Sciences, Pontifical Catholic University of Chile, Casilla 114-D, Santiago, Chile

Ginocchio, Rosanna

Department of Ecology, Faculty of Biological Sciences, Pontifical Catholic University of Chile, Casilla 114-D, Santiago, Chile

Hoffmann, Alicia J.

Department of Ecology, Faculty of Biological Sciences, Pontifical Catholic University of Chile, Casilla 114-D, Santiago, Chile

Jiménez, Hector E.

Department of Biology, Faculty of
Sciences, University of Chile, Casilla
653, Santiago, Chile

Keeley, Jon E.

Department of Biology, Occidental
College, Los Angeles, CA 90041, USA

Kershaw, A. Peter

Department of Geography and
Environmental Science, Monash
University, Clayton, Victoria, 3168
Australia

Lamont, Byron B.

School of Environmental Biology,
Curtin University of Technology,
Perth, 6001 Australia

Louda, Svaťa M.

School of Biological Sciences,
University of Nebraska, Lincoln, NE
68588, USA

Marticorena, Clodomiro

Department of Botany, Faculty of
Biology and Natural Resources,
University of Concepción, Casilla
2407, Concepción, Chile

Medel, Rodrigo G.

Department of Ecological Sciences,
Faculty of Sciences, University of
Chile, Casilla 653, Santiago, Chile

Milewski, A.V.

Division of Mammals, Department of
Zoology, Field Museum of Natural
History, Chicago, IL 60605-2496, USA

Montenegro, Gloria

Department of Ecology, Faculty of
Biological Sciences, Pontifical Catholic
University of Chile, Casilla 114-D,
Santiago, Chile

Muñoz, Mauricio R.

Department of Ecology, Faculty of
Biological Sciences, Pontifical Catholic
University of Chile, Casilla 114-D,
Santiago, Chile

Muñoz-Schick, Melica

Botany Section, National Museum of Natural History, Casilla 787, Santiago, Chile

Novoa, F. Fernando

Department of Ecological Sciences, Faculty of Science, University of Chile, Casilla 653, Santiago, Chile

Rosenmann, Mario

Department of Ecological Sciences, Faculty of Sciences, University of Chile, Casilla 653, Santiago, Chile

Vidiella, Patricia E.

Department of Biology, Faculty of Sciences, University of Chile, Casilla 653, Santiago, Chile

Villagrán, Carolina M.

Department of Biology, Faculty of Sciences, University of Chile, Casilla 653, Santiago, Chile

Vivrette, Nancy J.

Ransom Seed Laboratory, P.O. Box 300, Carprinteria, CA 93014-0300, USA

Zedler, Paul H.

Department of Biology, San Diego State University, San Diego, CA 92182-0057, USA

I.   History of Mediterranean Floras

1. Quaternary History of the Mediterranean Vegetation of Chile

Carolina M. Villagrán

The climate of the mediterranean region of Chile in a wide sense (30° to 39° S) is determined principally by the Southern Pacific Subtropical Anticyclone (SPSA). This system has an average center at around 30° S and presents a conspicuous annual cycle of latitudinal displacement. The winter rains in the mediterranean zone of Chile are associated with the northern position of the anticyclone during its annual cycle. This placement favors frontal activity associated with the westerly wind belt in the middle latitudes. During summer, displacement of the anticyclone to the south, and contraction in the same direction of the westerly wind belt, produces the dry season in the mediterranean region of Chile (Aceituno 1990).

To the north of 30° S, arid conditions are observed throughout the year as a result of the permanent influence of the Pacific Anticyclone, the desiccating action of the Peruvian cold ocean current, and the rain-shadow effect of the Andes on moisture-laden winds from the east. Only the far northern Andes receive summer rains (100 to 300 mm) of tropical origin (Arroyo et al. 1988; Grosjean et al. 1991). South of 39° S, rains of western origin are received through most of the year, although the mediterranean tendency is still found with one or two dry months, as far south as 42° S in inland areas (di Castri and Hajek 1976).

Climatic anomalies appear in central Chile (to the north of 35° S) that have been related to global phenomena of atmospheric circulation, such as the Southern Oscillation (SO). During the negative phase of the SO, associated with El Niño events (ENSO events), atmospheric pressure is unusually low in the SPSA domain. This situation has been associated with a tendency toward warmer and more humid winters in central Chile (Aceituno 1990).

Corresponding with the transitional character of these two contrasting climatic systems, the vegetation of the mediterranean region of Chile is characterized by conspicuous floristic and physiognomic heterogeneity. In this chapter, the Quaternary history of the mediterranean vegetation is discussed, based on the palynological record. Emphasis is given to the influence of paleoenvironmental changes and climatic fluctuations of the last glacial–interglacial cycle. The present geographic distribution of forest species of central–south Chile is interpreted in relation to the dynamics of paleoenvironmental change documented in the palynological record.

Vegetation Formations

A succession of xeric to mesic vegetation formations occurs in the mediterranean region of Chile along the latitudinal gradient. This sequence accords with an increase in rainfall from north to south. Semiarid formations, xerophytic matorral, and "savannas" with *Acacia caven* and *Prosopis chilensis* in the northern mediterranean sector are succeeded in turn by subtropical broad-leaved forest and matorral formations (sclerophyllous forests) in the central sector, and by deciduous forests of *Nothofagus* farther south. The latter gradually intergrade into temperate evergreen rain forests in the Lake region (Fig. 1.1).

The rain-shadow effects exerted by the two mountain ranges also determine strong east–west variation in the vegetation. At 35° S, sclerophyllous formations are restricted to the Coast range and the Andean foothills, and *Acacia caven* woodland and semiarid matorral predominate in the central depression. South of 35° S, deciduous *Nothofagus* forest occurs in the central depression and on the eastern slopes of the Coast range, and different types of evergreen forests are distributed on the western slopes.

Vestigial communities are found discontinuously on mountain summits and gullies of the Coast range (Fig. 1.2). These include olivillo (*Aextoxicon punctatum*) cloud forests at 30°30′ S, palm stands of *Jubaea chilensis*, isolated populations of deciduous *Nothofagus obliqua* in central Chile, coniferous forests with *Araucaria araucana* and *Fitzroya cupressoides*, and the Magellanic moorlands in the south-central region. An elevated number of endemic and monotypic taxa show narrow and discontinuous distribution ranges in the Coast range in central-south Chile. Examples of strongly endemic genera are *Pitavia*, *Valdivia*, *Jubaea*, *Gomortega*, *Tetilla*, *Lardizabala*, and *Latua*.

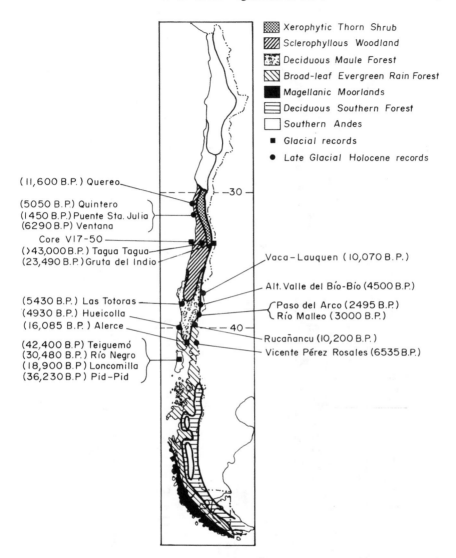

Figure 1.1. Vegetation formations in the mediterranean and southern temperate zone in Chile (Schmithüsen 1956). Palynological records mentioned in text denoted by circles.

Glacial History of Mediterranean Vegetation

Evidence from the temperate Andes in South America shows that during the Quaternary various glacier advances affected practically all the territory to the south of 42°30′ S, the central depression in the Lake region (39° to 42° S), and the Andean slopes to the north of 39° S (Fig. 1.1). During the last glaciation, the maximum temperature depression (6° to 7° C colder) and glacier

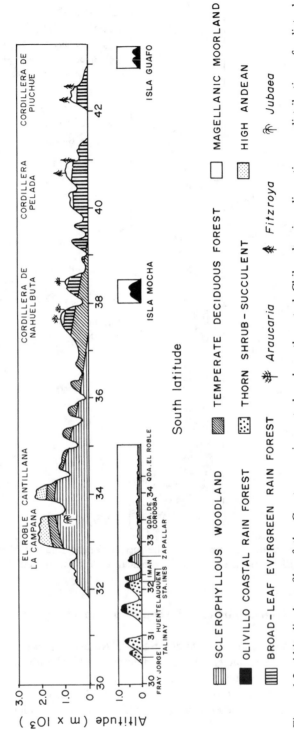

Figure 1.2. Altitudinal profile of the Coast range in central and south-central Chile, showing discontinuous distribution of relictual communities.

advances occurred between 20,000 and 18,000 years B.P., and were in phase with the Northern Hemisphere (Mercer 1976; 1984; Porter 1981; Clapperton 1990; 1991; Denton, unpublished data). A second glacier advance, between 15,000 and 14,500 years B.P., preceded rapid deglaciation that succeeded immediately and without interruption. Unequivocal reversions corresponding to the oscillation of the Younger Dryas period of the Northern Hemisphere have not been registered.

Geomorphologic and paleopedologic investigations (Garleff and Schäbitz 1993; Veit 1993) show that areas lacking morphodynamic activity in the Upper Quaternary were reduced to south of 40° S because of the glacial advance, fluvioglacial processes in the plains of the central depression, and intense vulcanism. These authors have established preliminary altitudinal limits of influence of periglacial processes for the Coast range during the Upper Pleistocene, such as solifluction, indicative of scarce vegetational cover. The inferior altitudinal limit of these processes extends from 1500 m in La Campana (32°55′ S) to 600 m in the Nahuelbuta Range (37°49′ S). This limit is at 300 m in the Sarao Range (40°57′ S), and reaches the base of the Piuchué Range (42°30′ S) farther south (Garleff and Schäbitz 1993).

Palynological Records

The documented glacial history of the vegetation is very scarce for the mediterranean region in Chile. It is restricted to three records at the same latitude (Fig. 1.1). A record for ocean sediments near the coast of Valparaíso (Core V17-50, 34°30′ S, 71°10′ W; Groot and Groot 1966) indicates that the glacial ages were characterized by forests dominated by the hygrophyllous conifer *Podocarpus*, while more diverse broad-leaved forests predominated during the interglacials. The Tagua Tagua record (34°30′ S, 71°10′ W; Heusser 1990a) shows that during the last glacial maximum (25,000 to 14,000 years B.P.) the vegetation in the central depression comprised conifers and *Nothofagus* for trees, Gramineae, Compositae, and Chenopodiaceae for herbs, and aquatic taxa. East of the Andes in the precordillera in Argentina, the record for Gruta del Indio (34°45′ S, 68°22′ W; D'Antoni 1983) indicates glacial vegetation dominated by Gramineae that has been interpreted as a northern extension of the Patagonian steppe.

Palynological evidence from temperate latitudes with mediterranean tendencies (Fig. 1.1), between 39° and 42° S (Heusser 1966; 1984a; 1990b; Villagrán 1985; 1988; 1990), show a glacial vegetation that was dominated by an open forest of *Nothofagus*, conifers, *Drimys*, and Myrtaceae. Cold-temperate taxa from more austral formations such as the Magellanic moorland elements (e.g., *Astelia*, *Dacrydium*) are recorded during the last glacial maximum.

Abundance of subantarctic, cold-temperate elements such as *Nothofagus*, *Prumnopitys*, *Podocarpus*, *Dacrydium*, and *Astelia* in the present mediterranean region of south-central Chile (34° S to 42° S), along with high lake

levels, abundant austral molluscs, microalgae, and other aquatic taxa recorded in Laguna Tagua Tagua in central Chile (Covacevich 1971; Varela
1976; Heusser 1990a) suggest rainy, glacial conditions.

On the island of Chiloé, dominance by conifer and *Nothofagus* trees,
mixed with a mosaic of Magellanic moorland species, suggests that the glacial vegetation would have been equivalent to that found today discontinuously distributed on the mountain summits of the Coast range between 39°
and 43° S. In the central depression of south-central Chile, the singular combination of Gramineae and Compositae pertaining to open communities,
and of trees such as conifers and *Nothofagus*, suggest a physiognomically
homogeneous parklike glacial landscape without an equivalent in the present
Chilean vegetation. This condition was perhaps comparable to the forest–
steppe ecotone on the eastern Andean slopes today. This glacial vegetation is
concordant with an abundant megafauna that became extinct at the begin-

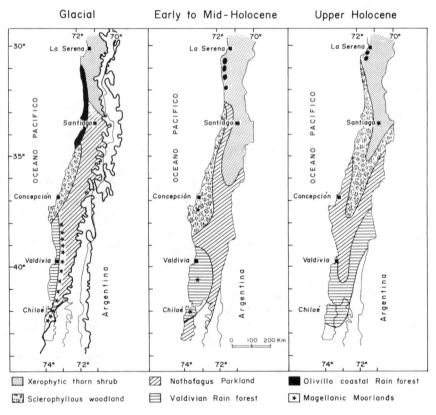

Figure 1.3. Reconstruction of Glacial, Lower to Middle Holocene, and Upper
Holocene vegetation, in accordance with the paleoenvironmental record.

ning of the Holocene (Nuñez et al. 1983). Such glacial vegetation would have occupied the central depression, replacing the present closed-canopy broad-leaved forests that succeed from north to south along the latitudinal gradient. Indicators belonging to the sclerophyllous, deciduous mediterranean and Valdivian forests are lacking. The Coast range, between 35° and 42° S, is proposed as a refugium for such forests during the glacial period (Fig. 1.3).

Reconstruction of the Last Glacial Maximum Climate

According to oceanic temperature reconstructions for the last glacial maximum (CLIMAP 1981), at 18,000 years B.P., oceanic circulation was characterized by: displacement toward the equator and areal contraction of subpolar waters; more pronounced thermal gradients in temperate regions; relatively stable positions for the central gyres in the oceans of the Southern Hemisphere; a rise in divergence and upwelling; and a thermal equator near or slightly deviated north from its present position. These changes suggest an intensification of oceanic, and also probably atmospheric, circulation during the last glacial maximum.

The meridional temperature gradient determines the intensity of the anticyclone and of the westerly wind belt and thus the vigor of the atmospheric circulation (Harrison et al. 1983). Analyzing ocean temperatures from seven cores shows that between 44° and 50° S, the temperature gradient was strongest 18,000 years ago, enabling the intensification (Markgraf et al. 1992) and northern expansion of the westerly wind belt, as proposed by various authors (Heusser 1984a; 1990a; 1990b; Villagrán 1988; 1990; Caviedes 1990). This expansion would explain the presence of glacial records of a cold-temperate subantarctic element to the north of its present geographic distribution, as at low-altitude sites on the island of Chiloé, and of *Nothofagus* and coniferous forests in the central depression in south-central Chile to 34°30′ S.

The pronounced vegetational contrast documented in the glacial records at 34°30′ S, with a grassy steppe to the east of the Andes, a park–steppe ecotone in the central depression of Chile, and a hygrophyllous podocarp forest on the coast, suggest reinforcement of the orographic effects of the Andes and the Coast range, and an even stronger east–west vegetation gradient than is present today. This scenario is consistent with more intense westerlies and greater elevation of the Andes and Coast ranges, reaching 5800 m and 2333 m, between 30° and 35° S, compared with 3400 m and 1533 m, respectively, to the south of 35° S and with more pronounced continental conditions recorded to the south of 32° S on the eastern side of the Andes (Garleff et al. 1991), and would explain the relictual cloud forests in coastal ravines and mountains of north-central Chile that were probably more continuous during the ice ages (Fig. 1.3).

From 27° to 33° S, glacial, periglacial, lacustrine, and pedologic evidence (Veit 1991a; 1991b) indicates colder and drier conditions during the last glacial maximum, with strong morphodynamic activity. The only stable zones were at the altitude of thermal inversions that cause littoral fogs (Veit 1993). There is also now evidence in the northern Andes (24° S) of an arid climate during the glacial maximum, without glacier formation and low lake levels (Grosjean et al. 1991). The present distributions of altiplanic, desert, and mediterranean elements in the high-elevation flora also suggest that desert barriers could have been more pronounced during part of the Pleistocene (Arroyo et al. 1982; Villagrán et al. 1983).

Vigorous anticyclones in the South Pacific and South Atlantic oceans, in positions equivalent to the present winter extremes (Villagrán 1993a) would determine the strong increment in aridity documented in the record in the north of Chile, and the relatively constant position of the center of the arid diagonal during the Upper Quaternary as proposed by Garleff et al. (1991). Considering the stronger atmospheric circulation during the glacial maximum and increased upwelling of cold waters north of 30° S, this scenario is not incompatible with intensified westerlies, but rather implies compression of the present mediterranean climate zone and, in consequence, sharper contrast between the hygrophyllous vegetation in the south and semiarid vegetation in the north.

Late-Glacial Vegetation and Climate

Paleotemperature curves in the Late-Glacial from the Antarctic show a marked tendency toward a rise in temperature from 15,000 B.P., reaching an optimum ca. 10,000 years B.P. Radiocarbon-controlled glacial chronology indicates that deglaciation was rapid and uninterrupted in the Lake region of southern Chile (Mercer 1984), at least from 13,000 years B.P.

The palynological (Villagrán 1991) and fossil-beetle (Ashworth and Hoganson 1987) records from the Lake region show evidence for rapid recolonization of North Patagonian rain forest, beginning 14,000 to 13,000 years B.P. The same arboreal taxa (*Podocarpus*, *Fitzroya*, *Drimys*, Myrtaceae, and *Nothofagus*), present in traces during the glacial maximum, expand and Magellanic moorland species disappear from low-elevation sites. The Valdivian rain-forest species, dominant here today, are still recorded only in traces during the Late-Glacial.

In central Chile the Tagua Tagua record (Heusser 1990a) shows that glacial vegetation persisted in the central depression (34°30′ S) up to 10,000 B.P. In the more northern littoral Quereo (32° S) record, a very diverse semiarid matorral for the Late-Glacial (Villagrán and Varela 1990) with strong representation of aquatic and palustrine taxa suggest humid conditions. The sclerophyllous forest species, today dominant in central Chile, have not yet been recorded; however, many of these species are insect pollinated (Moldenke 1977) and might not accumulate great quantities in the pollen record. The vegetation suggests that humid climate covered a wider geographic area during the Late-Glacial as temperature began to rise.

Geomorphological and soil-development studies carried out in north-central Chile by Veit (1991a; 1991b; 1993) between 27° and 33° S, show that cold and dry phases of the glacial maximum, characterized by strong morphodynamic activity, were succeeded by cold and wet phases in the Late-Glacial (16,000 to 10,000 years B.P.), with development of soils on the coast and in the interior, along with glacial advances in the Andes. Similar climatic conditions prevailed in the northern Chilean Andes (Grosjean et al. 1991) during this time.

This paleoclimate scenario, with glacial advances and a rise in rainfall of eastern origin in the northern Chilean Andes and the Altiplano; glacial advances in the Andes of north-central Chile, with development of paleosoils of temperate–humid climates on the coast and in the interior; and deglaciation in south-central Chile, would be plausible if the Late-Glacial is conceived as a slow relaxation of the glacial mode of circulation, gradual warming of the oceans, and weakening of Atlantic and Pacific anticyclones. The glacier advance, with doubled rainfall of eastern origin in the Altiplano (Kessler 1985), suggests that the Atlantic Convergence Belt was displaced to the south of its current position during the Late-Glacial (Kessler 1991; Villagrán 1993a).

Holocene Vegetational History

Records from central Chile (32° to 35° S) (Fig. 1.4), indicate an abrupt change in the vegetation in the Pleistocene–Holocene transition around 10,000 years ago. Aquatic taxa disappeared and the number of palustrine species, Gramineae, and arboreal traces decreased at Quereo, with a rise in dominance of Compositae and Umbelliferae (Villagrán and Varela 1990). Glacial forest elements disappeared and were substituted for by herbs, mainly Chenopodiaceae/Amaranthaceae in Tagua Tagua (Heusser 1990a). At Gruta del Indio, the steppe glacial vegetation was replaced by allochthonous taxa (D'Antoni 1983). These vegetational changes are consistent with a change from cold-wet to warm-dry conditions during the Pleistocene–Holocene transition.

Evidence from geomorphological and soil studies (Veit 1991a; 1991b; 1993) between 27° and 33° S, shows that during the Holocene there were fluctuating cold-humid and warm-dry phases on the coast of north-central Chile (Fig. 1.4, column 1). The cold-humid phases were accompanied by development of soils and greater snow accumulation in the high mountains; in contrast, the interior climate was dry during all of the Holocene.

The humid versus dry phases of the Early and Middle Holocene, respectively, established by Veit, correspond with equivalent phases in Tagua Tagua (Fig. 1.4, column 4) that were characterized by dominance of Gramineae versus Chenopodiaceae, respectively (Heusser 1990a). Both phases also correspond to fluctuations in the Tagua Tagua lake levels (Varela 1976), with

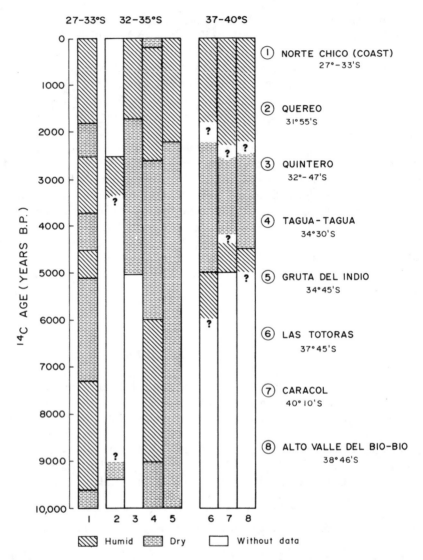

Figure 1.4. Chronological sequence of the humid and dry phases during the last 10,000 years, in accordance with the palynological records discussed in the text and geomorphological-pedological studies by Veit (1993) in north-central Chile.

relatively high levels in the Early Holocene and desiccation in the Middle Holocene.

All the mediterranean-zone records (Quintero, Ventana, and Puente Santa Julia; Fig. 1.1) show that after 5,000 (4,500) B.P. (Fig. 1.4, column 3) an arid phase was followed at around 2,500 years B.P. by a humid phase, with forests and swamps developing on the coast of central Chile (Villagrán 1993b).

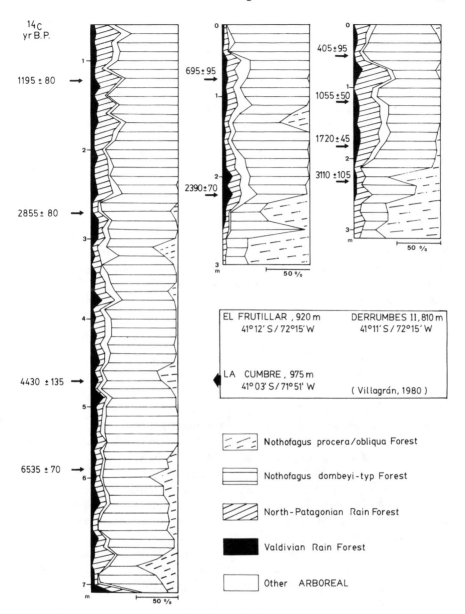

Figure 1.5. Composite diagrams of three records in the Andes from the Lake region that document expansion of deciduous forest element (*Nothofagus procera/obliqua*) during the Middle Holocene. (Adapted from Villagrán 1980.)

The records on the southern border of the mediterranean region in Chile (Fig. 1.4, columns 6 to 8), between 37° and 40° S, show an expansion of cold-temperate forests toward higher sites before 5,000 (4,500) years B.P. seen as *Araucaria* and *Nothofagus dombeyi* forests in the high Bio Bio valley (Rondanelli 1992), and forests with *Nothofagus dombeyi* and *Prumnopitys andina* and Magellanic moorland on the summit of the Nahuelbuta Range at Las Totoras, with expansion of the more thermophyllous forest with Proteaceae and Myrtaceae at El Caracol (Villagrán 1993b). Conversely, the records in the Argentinian Andes, as for Vaca Lauquén at 36°50′S (Markgraf 1987), Paso del Arco at 38°50′S, and Río Malleo at 39°36′S (Heusser 1988), show arid conditions during a large part of the Holocene.

In the Lake region, from the beginning of the Holocene, we see rapid expansion of the more thermophyllous Valdivian rain-forest element (*Eucryphia/Caldcluvia* and *Weinmannia*), and restriction of the Late-Glacial *Nothofagus* type *dombeyi* and conifer forest toward both mountain ranges (Villagrán 1991). During the Early to Middle Holocene the deciduous-forest element (*Nothofagus* type *procera/obliqua*) occupied a dominant position more to the south of its present range, as documented by records at Rucañancu, 39°33′S (Heusser 1984b) and Parque Vicente Pérez Rosales, 41°S (Fig. 1.5) (Villagrán 1980).

Holocene Climate Reconstruction

The glacial evidence already discussed indicates that in the Southern Hemisphere, glacial retreat at high latitudes immediately followed the last glacial advance (15,000 to 14,500 years B.P.); high oceanic temperatures were reached at 11,000 years B.P., with a maximum at 9,400 years B.P., in this way preceding by around 3000 years the maximum in the Northern Hemisphere (Harrison et al. 1983). This asymmetry between the maximum temperatures in the two hemispheres would have facilitated southern displacement of the Pacific Intertropical Convergence Belt during the Early Holocene (Villagrán 1993a). In this way, in the South Pacific during the Early Holocene, ocean temperatures would have been sufficiently high to cause a weakening of the oceanic circulation and a descent in atmospheric pressures in the area influenced by the SPSA. These conditions are analogous to those associated with present ENSO events.

Veit (1991a; 1991b) proposes a major event of very strong precipitation for the coast of north-central Chile between 9600 and 7300 years ago, corresponding to the higher accumulation of snow, higher alluvial cone activity, and intensification of erosion processes in the Andes. These conditions suggest a higher intensity and recurrence of ENSO events during this time. Even though this humid phase is detected in the palynological and lacustrine record of Tagua Tagua in central Chile, during this phase the dominant sclerophyllous forest of today does not appear, and only herbs are recorded. This situation suggests that this humid phase could correspond to a series of more frequent and intense ENSO events than seen today. Such conditions would have favored expansion of flora adapted to episodic rains as seen in the semiarid region today. At the same time these conditions would have restricted the expansion of taxa adapted to more regular seasonal rains, as for many woody mediterranean and cool-temperate forest elements in Chile (Fig. 1.3).

During the Middle Holocene, the relatively drier phases documented in all records from south-central Chile suggest that at this time the SPSA was probably more vigorous and in a more southerly position than today. Again these dry phases seem to have especially affected the northern area of the present mediterranean zone, because forest formation was absent at that time. Present climate and vegetation conditions would have been established during the Upper Holocene (Fig. 1.3).

Biogeographic Effects

Figure 1.6 shows variation in the percentage of forest species throughout Chile by degree of latitude (a total of 178 taxa were considered, including woody species, vines, and flowering plant epiphytes). Maximum richness occurs in a narrow latitudinal band between Maule and Valdivia (36° to 40° S) toward the southern end of the present mediterranean zone in Chile. Between 66% and 77% of species are concentrated in this band, but in sectors immediately to the north and south, only 33% (30° to 35° S) and 40% (40° to 46° S) of the species are seen, respectively.

This same pattern is repeated when we consider groups of woody taxa pertaining to the subantarctic cold-temperate element (e.g., *Nothofagus* and conifers) and the neotropical element requiring warmer temperatures (e.g., Flacourtiaceae and Myrtaceae), separately (Fig. 1.7). If we consider only the endemic species in these groups, the geographical area in which these taxa are concentrated is even more restricted (Fig. 1.7).

This pattern can be related to these events and processes:

1. Glacial advances that have directly affected the development of vegetation in all the territory south of 42° S, the central valley and the Andes in the Lake region (39° to 42° S) and the Andes in central Chile (33° to 39° S).
2. Periglacial processes such as glaciofluvial activity that affected a large part of nonglaciated sectors on the island of Chiloé, and the central depression in the Lake region.
3. Decrease in temperature (6° to 7° C) during the ice ages directly affecting the distributions of taxa and reducing habitats. Processes such as solifluction influenced high sectors (over 1500 m) of the northern Coast range to the base of this same mountain range in Chiloé.

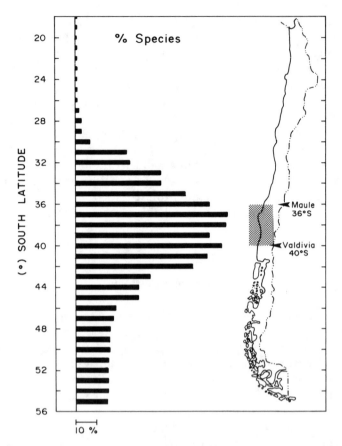

Figure 1.6. Latitudinal variation in the percentage presence of 178 forest species throughout Chile.

4. Change in the rainfall regime was probably the most critical factor for vegetation during glaciation. The intensification and northern expansion of the westerly wind belt, together with more vigorous anticyclones in the South Atlantic and South Pacific and higher upwelling intensity to the north of 30° S, would have provoked a more pronounced rainfall gradient throughout central-south Chile, with consequent compaction of the mediterranean zone.

5. The higher Andes and Coast ranges, to the north of 35° S, together with more vigorous atmospheric circulation, would have reinforced the rain-shadow effect exerted by these mountains, and accentuated the east–west rainfall gradient, provoking marked longitudinal zonation of the vegetation.

6. The interglacials would have been distinguished by unstable oceanic and atmospheric conditions, with maximum temperatures in the Early Holo-

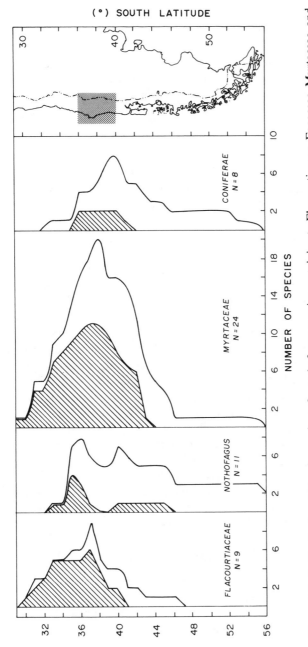

Figure 1.7. Latitudinal variation in number of woody forest species pertaining to Flacourtiaceae, Fagaceae, Myrtaceae, and conifers. Distribution of Chilean endemic species for these families denoted by shading.

cene and glacial readvances during the Late Holocene. The alternating humid and dry phases would have fundamentally affected distribution of species on the northern border of the present mediterranean region, especially to the north of 35° S. It is possible that these phases corresponded to periods of higher climatic oscillation, similar to those associated today with the negative and positive phases respectively of the Southern Oscillation.

Conclusions

Because of its transitional character between the arid and semiarid matorral vegetation in the north, and the temperate rainforest zone in the south, the vegetation in the mediterranean region of Chile (30° to 39° S) experienced profound and repetitive changes during the glacial–interglacial cycles in the Quaternary.

Glacial advances, periglacial effects, and climatic change have particularly affected the northern area in the actual mediterranean zone (30° to 35° S) and the greater part of the temperate zone to the south of 40° S. Stabler climatic conditions prevailed toward the southern end of the actual mediterranean zone, especially on the Coast range, which had better development of soils and more extensive vegetation cover. These stable conditions permitted maintenance of the Chilean forest flora there during the glacial periods. That the area with the highest concentration of species and endemism is between the Maule river and Valdivia (36° to 40° S) is an expression of this condition.

The repetitive climatic fluctuations of lesser scale during the Holocene (10,000 years B.P. to the present) have principally affected distribution of the subtropical element of the sclerophyllous forests in central Chile, which is more adapted to climatic conditions with regular winter rains. The humid postglacial phases registered in central Chile (to the north of 35° S) and in the Norte Chico, probably associated with more frequent and intense ENSO events, have favored expansion and differentiation of semiarid matorral adapted to episodic conditions of precipitation.

Acknowledgments

I acknowledge the significant contributions and comments of Drs. Heinz Veit, Karsten Garleff, and Patricio Aceituno. This study was financed by FONDECYT, Chile Grant 91-0844 and CONICYT/SAREC Project "Climate Changes during Holocene."

References

Aceituno P (1990) Anomalías climáticas en la región Sudamericana durante los extremos de la oscilación austral. Rev Geofís 32:65–78
Arroyo MTK, Villagrán C, Marticorena C, Armesto JJ (1982) Flora y relaciones

biogeográficas en los Andes del norte de Chile (18–19° S). In Veloso A, Bustos E (eds) *El Ambiente Natural y las Poblaciones Humanas de los Andes del Norte Grande de Chile* (Arica, Lat. 18° 28′ S) Vol 1. Rostlac, Montevideo, pp 71–92

Arroyo MTK, Squeo FA, Armesto JJ, Villagrán C (1988) Effects of aridity on plant diversity in the northern Chilean Andes: Results of a natural experiment. Ann Missouri Bot Garden 75:55–78

Ashworth A, Hoganson JW (1987) Coleoptera bioassociations along an elevational gradient in the Lake region of Southern Chile, and comments on the postglacial development of the fauna. Ann Entom Soc Am 80:865–895

Caviedes C (1990) Rainfall variation, snowline depression and vegetational shifts in Chile during the Pleistocene. Clim Change 16:94–114

Clapperton CM (1990) Quaternary glaciations in the Southern Hemisphere: An overview. Quat Sci Rev 9:299–304

Clapperton CM (1991) Glacier fluctuations of the last glacial–interglacial cycle in the Andes of South America. Bamberger Geog Schr 11:183–207

CLIMAP Project Members (1981) Seasonal reconstructions of the earth's surface at the last Glacial maximum. Geol Soc Am Map Chart Ser MC-36:1–18

Covacevich V (1971) Los moluscos pleistocénicos y holocénicos de San Vicente de Tagua Tagua. Unpub Geol Thesis, U Chile, Santiago

D'Antoni HL (1983) Pollen analysis of Gruta del Indio. Quat S Am Ant Penin 1:83–104

di Castri F, Hajek E (1976) *Bioclimatología de Chile*. Ed U Catól Chile, Santiago

Garleff K, Schäbitz F (1993) Tech Rep FONDECYT-91-0844

Garleff K, Schäbitz F, Stingl H, Veit H (1991) Jungquartäre Landschaftentwicklung und Klimageschichte beiderseits der Ariden Diagonale Südamerikas. Bamberger Geog Schr 11:359–394

Groot JJ, Groot CR (1966) Pollen spectra from deep-sea sediments as indicators of climatic changes in Southern South America. Mar Geol 4:467–524

Grosjean M, Messerli B, Schreier H (1991) Seenhochstände, Bodenbildung und Vergletscherung in Altiplano Nordchiles: Ein interdisziplinärer Beitrag zur Klimageschichte der Atacama. Erste Resultate. Bamberger Geog Schr 11:99–108

Harrison SP, Metcalfe SE, Street-Perrott FA, Pittock AB, Roberts CN, Salinger MS (1983) A climate model of the Last Glacial/Interglacial transition based on paleotemperature and palaeohydrological evidence. In Vogel JC (ed) *Late Cainozoic Paleoclimates of the Southern Hemisphere*. AA Balkema, Rotterdam, pp 21–24

Heusser CJ (1966) Late-Pleistocene pollen diagrams from the province of Llanquihue, Southern Chile. Am Phil Soc Proc 110:269–305

Heusser CJ (1984a) Late Quaternary climates of Chile. In Vogel JC (ed) *Late Cainozoic Paleoclimates of the Southern Hemisphere*. AA Balkema, Rotterdam, pp 59–83

Heusser CJ (1984b) Late-Glacial-Holocene climate of the Lake District of Chile. Quat Res 22:77–90

Heusser CJ (1988) Late-Holocene vegetation of the Andean *Araucaria* region, Province of Neuquén, Argentina. Mountain Res Dev 8:53–63

Heusser CJ (1990a) Ice age vegetation and climate of subtropical Chile. Palaeogeog, Palaeoclimat, Palaeoecol 80:107–127

Heusser CJ (1990b) Chilotan piedmont glacier in the Southern Andes during the Last Glacial Maximum. Rev Geol Chile 17:3–18

Kessler VA (1985) Zur Rekonstruktion von spätglazialen Klima und Wasserhaushalt auf dem peruanisch-bolivianischen Altiplano. Zeitschr Gletsch Glazialgeol 21:107–114

Kessler VA (1991) Zur Frage der Änderung der allgemeinen atmosphärischen Zirkulation auf dem Altiplano seit dem Spätglazial. Bamberger Geog Schr 11:351–357

Markgraf V (1987) Paleoenvironmental changes at the northern limit of the sub-antarctic *Nothofagus* forest. Quat Res 28:119–129

Markgraf V, Dodson JR, Kershaw AP, McGlone MS, Nicholls N (1992) Evolution of late Pleistocene and Holocene climates in the circum-South Pacific land areas. Clim Dyn 6:193–211

Mercer JH (1976) Glacial history of southernmost South America. Quat Res 6:125–166

Mercer JH (1984) Late Cainozoic glacial variations in South America south of the equator. In Vogel JC (ed) *Late Cainozoic Paleoclimates of the Southern Hemisphere*. AA Balkema, Rotterdam, pp 45–58

Moldenke AR (1977) Insect–plant relations. In Thrower NJW, Bradbury DE (eds) *Chile–California Mediterranean Scrub Atlas: A Comparative Analysis*. Dowden, Hutchinson & Ross, Stroudsburg, PA, pp 199–217

Núñez L, Varela J, Casamiquela R (1983) *Ocupación Paleoindio en Quereo*. Edic U Norte, Antofagasta, Chile

Porter C (1981) Pleistocene glaciation in the Southern Lake District of Chile. Quat Res 16:263–292

Rondanelli MJ (1992) Historia vegetacional del Holoceno Tardío en la subsecuencia del ecosistema andino Alto Valle del BíoBío, Provincia de Lonquimay, Chile. Estudio paleoecológico basado en el análisis de pólen. Thesis, U Concepción, Concepción, Chile

Schmithüsen J (1956) Die räumliche Ordnung der chilenischen Vegetation. Bonner Geog Abh 17:1–86

Varela J (1976) Geología del Cuaternario de Laguna Tagua Tagua. Actas Primer Congreso Geológico Chileno: D81–113

Veit H (1991a) Jungquartäre Relief-und Bodenentwicklung in der Hochkordillere im Einzugsgebiet des Río Elqui (Nordchile, 30° S). Bamberger Geog Schr 11:81–97

Veit H (1991b) Jungquartäre Landschafts-und Bodenentwicklung im chilenischen Andenvorland zwischen 27 to 33° S. Bonner Geog Abh 85:196–208

Veit H (1993) Tech Rep FONDECYT 91-0844

Villagrán C (1980) Vegetationsgeschichtliche und pflanzensoziologische Untersuchungen im Vicente Pérez Rosales Nationalpark (Chile). Diss Bot 54:1–165

Villagrán C (1985) Análisis palinológico de los cambios vegetacionales durante el Tardiglacial y Postglacial en Chiloé, Chile. Rev Chil Hist Nat 58:57–69

Villagrán C (1988) Expansion of Magellanic moorland during the Late-Pleistocene: Palynological evidence from northern Isla de Chiloé, Chile. Quat Res 30:304–314

Villagrán C (1990) Glacial climates and their effects on the history of the vegetation of Chile: A synthesis based on palynological evidence from Isla de Chiloé. Rev Palaeobot Palynol 65:17–24

Villagrán C (1991) Historia de los bosques templados del sur de Chile durante el Tardiglacial y Postglacial. Rev Chil Hist Nat 64:447–460

Villagrán C (1993a) Una interpretación climática del registro palinológico del último ciclo glacial-postglacial en Sudamérica. Bull Inst Fr Études Andines 22(1):243–258

Villagrán C (1993b) Tech Rep FONDECYT 91-0844, Santiago, Chile

Villagrán C, Varela J (1990) Palynological evidence for increased aridity of the central Chilean coast during the Holocene. Quat Res 34:198–207

Villagrán C, Arroyo MTK, Marticorena C (1983) Efectos de la desertización en la distribución de la flora andina de Chile. Rev Chil Hist Nat 56:137–157

2. Evolution and History of Mediterranean Vegetation Types in Australia

John R. Dodson and A. Peter Kershaw

Mediterranean climates in Australia occur in the southwest and in near coastal South Austalia and adjacent western Victoria. Most of these regions support species-rich sclerophyllous heaths, with or without scattered tree cover, and eucalypt woodland with herbaceous understory. Despite wide taxonomic differences the regions show strong physiological and structural analogues with vegetation in other mediterranean regions.

Here we examine the Tertiary and Quaternary fossil records to tease out the climatic and vegetational patterns of development in the southern Australian mediterranean environments.

Mediterranean Vegetation in Australia

The mediterranean geobotanical region is defined by a seasonal climate with hot summers and cool winters and with a mean annual rainfall of between 300 and 900 mm yr^{-1} falling predominantly in winter (Specht 1969a). The location of areas experiencing a mediterranean climate, together with their vegetation cover, is shown in Figure 2.1.

A characteristic feature of mediterranean landscapes is their species-rich sclerophyllous heath, with or without an open canopy of trees usually composed of eucalypt species. Frequent fires, extensive flowering periods, and

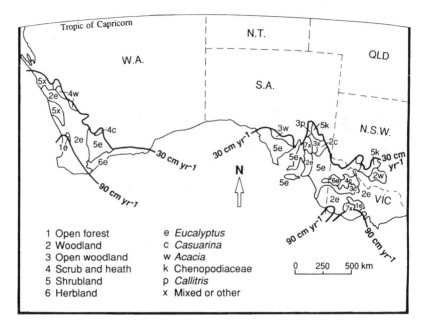

Figure 2.1. Location of the mediterranean climatic zone in Australia, showing the position of isohyets used to define the zone, and the main vegetation types. The vegetation map is based on Groves (1981).

seasonal growth rhythms extending into the dry summer are prominent features of these systems. Specht (1969b) demonstrates that solar radiation and available water control growth rates, but soil fertility is a major factor in controlling distributions, with heaths being restricted mainly to low nutrient status soils.

On relatively fertile and well-drained soils, and where mean annual precipitation exceeds about 400 mm, heath is replaced by *Eucalyptus*- and *Casuarina*-dominated woodland, with an understory of perennial herbs, geophytes, and therophytes (Groves and Specht 1965; Jones 1967; Specht 1969a). Scattered evergreen sclerophyllous shrubs are rare but are increased by firing (Specht et al. 1958). Where precipitation falls below 400 mm yr^{-1}, fertile but solonized soils support mallee eucalypts (multistemmed tall shrubs with lignotubers) with an open grassy understory, and infertile soils support either sclerophyllous heaths or closed scrubs dominated by species of *Leptospermum* or *Melaleuca*. Where mean annual rainfall is within the range 250 to 330 mm yr^{-1} heath disappears and mallee-broombush (*Melaleuca uncinata*) takes over. Fire is an important component in all these formations (Specht 1969a; Specht 1981a; Specht 1981b).

Dominants within mediterranean vegetation are usually drawn from several taxa within the families Casuarinaceae, Epacridaceae, Fabaceae, Myrtaceae, Proteaceae, Rutaceae, one to a few taxa from the families Dilleniaceae, Restionaceae, Rhamnaceae, Santalaceae, Thymelaeaceae, and Xanthorrhoeaceae, and Droseraceae and Cyperaceae can be common in wet heaths. Table 2.1 gives the distribution of major mediterranean genera across other vegetation formations in Australia: based on Francis (1970), Willis (1972) and Harden (1990, 1991) for closed forests; Costermans (1983), Willis (1970, 1972), and Pate and Beard (1984) for forests, open forests, and woodlands; Jessop (1981) for arid lands; Costin et al. (1979) for alpine regions; and Willis (1970, 1972), Costermans (1983), and Harden (1990, 1991) for shrublands and heaths.

Analysis at the generic level is striking because of the many nonspecific taxa within mediterranean ecosystems in Australia. Not surprisingly there is strong divergence with closed forests. Almost no differences appear for vegetation formations with which mediterranean vegetation normally abuts, and even for less-related systems, such as alpine and arid shrublands, resemblances are strong.

Pate and Beard (1984) and Specht (1981a) report that mediterranean systems are very much dominated by species of Australian origin. Advanced flowering plant families predominate, and primitive vascular plants make almost no contribution. These data and the analysis above suggest therefore a scenario of relatively modern development of mediterranean systems in Australia, without time for distinctiveness at the generic level. This makes it difficult to trace the history of mediterranean vegetation, because our generic and family level data cannot be used to detect trends, for the species are recently evolved, and often have similar morphology, including pollen morphology.

Pre-Quaternary Origins

Origins and Pre-Quaternary Development of Components of Mediterranean Vegetation

Because almost no features clearly distinguish mediterranean vegetation from other types of vegetation within the Australian region, any reconstruction is bound to be very general and speculative. Data are available, however, on the history of the flora and vegetation that are now represented in mediterranean regions. Together with independent evidence on changing environments, particularly climate and fire, these data allow some assessment of likely times, causes, and places of origin and development of the vegetation.

The Data Base

To facilitate discussion, the known time ranges of recognizable important taxa in mediterranean areas are shown in Figure 2.2. Pollen abundance data

Table 2.1. Presence (+) and absence (−) of major mediterranean taxa across vegetation types in Australia

Family	Genus	Closed Forest		
		Cool-temperate	Warm-temperate	Tropical-Subtropical
Casuarinaceae	*Casuarina*	−	−	−
Epacridaceae	*Astroloma*	−	−	−
	Brachyloma	−	−	−
	Epacris	−	−	−
	Leucopogon	−	−	−
	Monotoca	−	−	−
	Sprengelia	−	−	−
Fabaceae	*Acacia*	−	+	−
	Bossiaea	−	−	−
	Daviesia	−	−	−
	Dillwynia	−	−	−
	Platylobium	−	−	−
	Pultenaea	−	−	−
Myrtaceae	*Baeckea*	−	−	−
	Callistemon	−	−	−
	Calytrix	−	−	−
	Conospermum	−	−	−
	Eucalyptus	−	−	−
	Kunzea	−	−	−
	Leptospermum	−	−	−
	Melaleuca	−	−	−
Proteaceae	*Banksia*	−	−	−
	Grevillea	−	−	−
	Hakea	−	−	+
	Lomatia	−	−	−
	Persoonia	−	−	−
Rutaceae	*Boronia*	−	−	−
	Correa	−	−	−
	Eriostemon	−	−	−
Dilleniaceae	*Hibbertia*	−	−	−
Droseraceae	*Drosera*	−	−	−
Restionaceae	*Leptocarpus*	−	−	−
	Lepyrodia	−	−	−
	Restio	−	−	−
Rhamnaceae	*Pomaderris*	−	−	−
	Spyridium	−	−	−
Santalaceae	*Exocarpos*	−	−	−
Thymeleaceae	*Pimelea*	−	−	−
Xanthorrhoeaceae	*Lomandra*	−	−	−
	Xanthorrhoea	−	−	−

Forest–Open Forest		Woodland			Shrublands and heaths	
Wet sclerophyll	Dry sclerophyll	Alpine	Dry	Arid	Coastal complex	Lowlands heaths
+	+	−	+	+	+	+
−	+	−	+	−	+	+
+	+	−	+	−	−	+
+	+	+	−	−	+	+
+	+	+	+	+	+	+
+	+	−	+	−	+	+
−	−	−	−	−	+	+
+	+	−	+	+	+	+
+	+	−	+	−	+	+
+	+	−	+	+	+	+
+	+	−	+	−	+	+
+	+	−	+	−	+	+
+	+	−	+	−	+	+
+	+	+	+	+	+	−
+	+	−	+	+	+	+
−	+	−	+	+	+	+
−	+	−	−	−	−	+
+	+	+	+	+	+	+
−	+	+	−	−	+	+
+	+	−	+	+	+	+
+	+	−	+	+	+	+
+	+	−	+	−	+	+
+	+	+	+	+	+	+
+	−	+	+	+	+	+
+	+	−	−	−	+	+
+	+	−	+	−	+	+
+	+	−	+	−	+	+
+	+	−	+	−	+	+
+	+	−	+	+	+	+
+	+	−	+	−	+	+
−	−	+	+	−	+	+
−	+	+	+	−	+	+
−	+	−	−	−	+	+
+	−	−	−	−	+	+
+	+	−	−	−	−	+
−	+	−	−	+	−	+
+	+	+	+	+	+	+
+	+	+	+	+	+	+
+	+	−	+	−	+	+
+	+	−	+	+	+	+

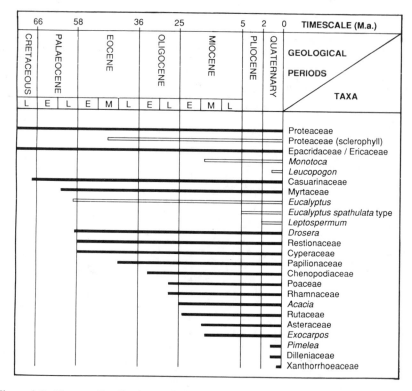

Figure 2.2. Known distribution and time ranges of important taxa in mediterranean areas in Southern Australia. The time scale (Ma) is millions of years ago.

have been collated to provide general histories of the vegetation from sites close to areas with mediterranean vegetation today (Figures 2.3 and 2.4). Figure 2.3 is based mostly on results from the extensive coal measures and overburden in the Latrobe Valley in Victoria, an area bounded by the South-eastern Highlands to the north and the Strzelecki Ranges to the south and lying southeast of the mediterranean region of South Australia. It presently receives mean annual rainfall of about 800 mm. Figure 2.4 is constructed from sequences lying to the north and east of the south mediterranean region that fall along a present precipitation gradient from the very arid interior of the continent to the Southeastern Highlands. Mean annual rainfall is about 100 mm for the Lake Eyre region, 200 mm for the western part of the Murray Basin with the Oakvale-1 core, 400 mm for Jemmalong in the eastern Murray catchment, and 650 mm for Lake George in the South-eastern Highlands.

The Cretaceous pollen spectra from both sequences are older than the period of recognizable angiosperms (Dettman 1989). All taxa relatable to extant forms are from the southern coniferous families Podocarpaceae and

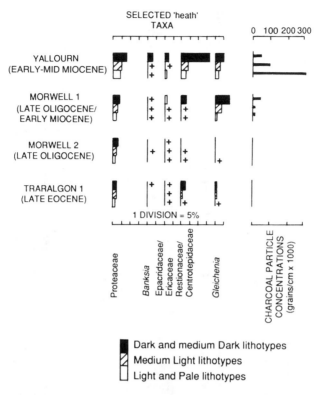

Figure 2.3. The distribution of selected taxa found in the Tertiary brown coal measures in Eastern Victoria. These taxa are important in modern mediterranean vegetation in Australia.

Araucariaceae, of which only one species of *Podocarpus* still survives in mediterranean environments—and this form has very limited distribution in wetter parts of southwestern Western Australia.

Through the latter part of the Cretaceous and Paleocene periods, a substantial number of angiosperm taxa, relatable to present forms, enter the record. These include Proteaceae, Epacridaceae, Casuarinaceae, Myrtaceae, *Drosera*, and Restionaceae, which are conspicuous components of mediterranean systems. Almost all these taxa were widely distributed in southern Gondwanaland and either evolved or underwent a substantial part of their early development in this region (Dettman 1989). Despite the importance of these taxa in mediterranean environments, the general impression of the vegetation, from gross floristic composition and pollen dominants, is one of developing rain forest. The remarkable assemblages from the now-arid Lake Eyre Basin (Sluiter 1991) can be readily compared with extant wetter rainforest community types. The sequence through the early Tertiary is from warm-temperate rain forest dominated by Cunoniaceae, through subtropical

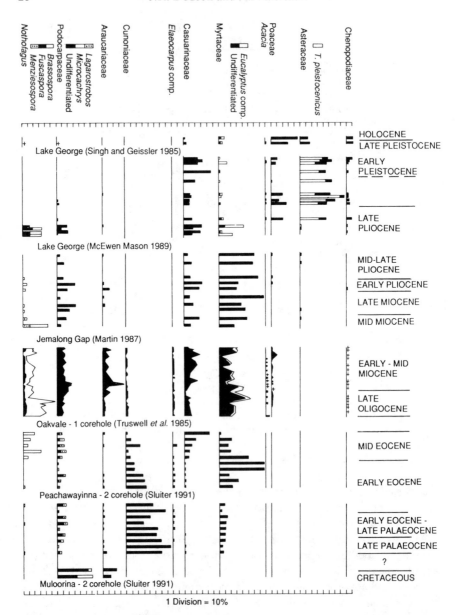

Figure 2.4. Distribution of selected pollen taxa in time, from sites adjacent to the present mediterranean region in Southeastern Australia.

rain forest characterized by high levels of Myrtaceae, and then a return to cooler conditions, as shown by increasing *Nothofagus* (Sluiter 1991). The assemblages from the Latrobe Valley are somewhat different, but again the vegetation was essentially rain forest and the increase in *Nothofagus* from the Mid-Eocene suggests broadly similar vegetation trends.

From these and almost all other sites in Australia from the early Tertiary, it appears that rain forest covered most of, if not the whole continent, with a peak in taxa of tropical affinity in the Early to Mid-Eocene reflecting— regionally—the maximal global Tertiary temperatures (Kemp 1978). After the Mid-Eocene, rain-forest domination continued, with very high *Nothofagus* values extending into the Miocene. Evidence for some opening up of the rain-forest canopy is indicated in the Late Oligocene to earlier Miocene by a substantial increase in the number of herb and sclerophyll woody taxa entering the reocrd, many of which are well represented in mediterranean regions today. Unlike the earlier "Gondwanan element," this latest evolutionary phase is distinguished by cosmopolitan opportunists such as Chenopodiaceae, Poaceae, and Asteraceae, although other taxa, including *Monotoca* and *Exocarpos*, probably evolved from Gondwanan ancestors. A more open rain-forest canopy is also indicated by the development of Araucarian rain forests in the Murray Valley and in sites farther north (Kershaw et al. 1991a). These changes in vegetation were probably caused by a decrease in effective precipitation associated with increasing climatic variability.

Regional differentiation in vegetation became more evident in the latter part of the Miocene under an increasing latitudinal temperature gradient and further reduction in precipitation. The wetter rainforests characterized by *Nothofagus* were much reduced in extent as drier rain forests dominated by Araucariaceae became more coastal and northern in their distribution. In the Murray Basin region, polleniferous sediments were restricted almost totally to the wetter eastern river valleys. Here the high Myrtaceae values are considered to represent a mixed canopy of eucalypts and marginal rain-forest genera dominating a rain forest understory much like the wet sclerophyll or tall open forests in the wetter parts of southern Australia today (Martin 1987). In the Latrobe Valley, precipitation remained sufficiently high to allow a return to dominance of *Nothofagus* under cooler conditions after a minor temperature peak in the Early to Mid-Miocene, although more open vegetation taxa are increasingly represented.

Much of the Pliocene lacks pollen in the Latrobe Valley and also at Lake George, where the illustrated sequence is underlain by palynologically barren sediments extending back into the latest Miocene. These barren phases are consistent with a proposed highly seasonal summer-rainfall climate similar to that in northern Australia today, as deduced from the strongly oxidized sediments in Lake George and other sites in southeastern Australia dated paleomagnetically between about 6 and 2.5 million years ago (Bowler 1982). Some pollen was preserved in river-valley sediments of the Murray catchment and shows continued domination by wet sclerophyll forest with a clear phase of rain-forest expansion in the Early Pliocene. The survival of rain forest through this phase is indicated at Lake George by relatively high *Nothofagus* and Podocarpaceae levels in the basal polleniferous samples.

Bowler (1982) proposed a subsequent switch from a summer to a winter rainfall regime, coinciding approximately with the beginning of major climatic oscillations that typify the Quaternary. This switch is marked in the Lake George and Latrobe Valley sequences by a substantial increase in pollen in plants of open vegetation, particulary Asteraceae, and also *Eucalyptus* in the Latrobe Valley. The high Asteraceae values strongly suggest widespread occurrence of treeless steppe vegetation, particularly in cooler parts of climatic cycles that culminated in the glacial-interglacial oscillations in the later part of the Quaternary. Evidence for rain forest is very limited and more warmth-demanding components have effectively disappeared from the southern part of the continent. Onset of winter rainfall probably was the critical factor in determining the present composition of temperate rain forests, and their restriction to isolated, favorable locations in coastal and mountain areas. With the onset of the Quaternary climates, taxa of the mediterranean vegetation also begin to appear in the pollen record. Although by the end of the Tertiary period the distributions of rain forest and open-canopied vegetation may have been very similar to those of today in southern Australia, clear differences appear in the open-vegetation types. Asteraceae rather than Poaceae dominated treeless vegetation and the understory in sclerophyll communities, and *Casuarina* was generally a more important canopy component than *Eucalyptus*, despite dominance by *Eucalyptus* in some places such as parts of the Latrobe valley. By contrast, the Poaceae, rather than the Asteraceae, expanded in northern Australia and pollen of this family dominates available pollen spectra from the beginning of the Pliocene in the northwest of the continent. Eucalypt pollen values are very low there until the later part of the Pleistocene (Kershaw et al. 1991b; McMinn and Martin 1991).

Development of Heath

The early presence in the geological record of taxa such as Proteaceae, Epacridaceae, Casuarinaceae, Myrtaceae, *Drosera*, and Restionaceae has led to the idea that one major component of mediterranean vegetation—sclerophyll heath—was present in the Late Cretaceous or very early Tertiary (e.g., Westman 1978; Specht 1979). Because conditions appear to have been uniformly wet, status of soil nutrient was considered to have been the critical, causal factor in development of heath, with wet heath differentiating from rain forest on low-fertility soils. Subsequently, dry heath communities, preadapted to unfavorable conditions, evolved in response to increasingly dry and more seasonal conditions in the later part of the Tertiary. Westman (1978) further suggested that the climate would have had to be cool as well as wet to explain the present summer growth pattern in many heath plants living in mediterranean climates where summer moisture stress is not conducive to growth in this season. The evidence for wet and cool conditions from the Paleocene in the Eyre Basin is consistent with proposed early heath origin.

To our knowledge, however, few fossil data support the existence of vegetation of the heath type, judging by either structural or floristic characters, at this time. In fact, within the Proteaceae at east, not until the Mid-Eocene have genera been identified for certain that have representatives in sclerophyll, rather than rain-forest vegetation (Martin 1982).

Somewhat better indication of how wet heath developed may be provided by studying the change in local swamp vegetation within the Latrobe Valley coal seams (see Figure 2.3). In this diagram, average pollen percentages for taxa that could have been components of wet heath are shown for different coal lithotypes, which represent different depositional environments within the four major coal seams. The pollen percentage of heath plants generally increases in relation to true rain-forest taxa through the sequence, indicating that heath vegetation was developing and replacing rain forest within the basin. Percentages of heath plants sufficiently high to indicate a true heath community are not achieved until the time of deposition of the Morwell seam. These occur in the darker lithotypes, which are considered to represent drier raised-bog environments. The clear differentiation of lithotypes in the younger coal seams, together with the first evidence for charcoal within the record, suggest that this heath development may have been related to increased climatic variability, and to fire becoming an important environmental factor (Kershaw et al. 1991a). It is possible, therefore, that wet heath did not occur until the later part of the Tertiary and that climatic variability and associated fire, along with low nutrient status soils, were important to its development and spread.

Development of Open Forests and Woodlands

We find evidence for eucalypt-type pollen as long ago as the Late Eocene, but no positive indication of an opening in the rain-forest canopy until the Oligocene-Early Miocene, when light-demanding dry-land understory taxa such as Chenopodiaceae, Poaceae, and Asteraceae, as well as the dominant woody genus of arid environments, *Acacia*, enter the record (Figure 2.5). Even then, the progressively drier conditions through the Miocene, as indicated by the decline in *Nothofagus* and the disappearance of suitable sites for pollen deposition and preservation from the interior, did not cause development of extensive open forests and woodlands with a herbaceous understory that are so prominent today. Instead, wetter rain forests were replaced by drier facies characterized by Araucariaceae and probably also by sclerophyllous forests dominated by taxa such as Casuarinaceae with limited understories or understories composed mainly of palynologically invisible sclerophyll shrubs.

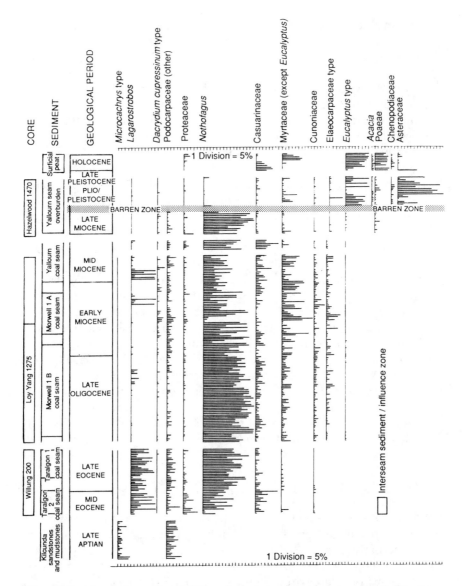

Figure 2.5. Abundance and variability of selected pollen types found in the Tertiary-Quaternary period from the brown coal region in Eastern Victoria.

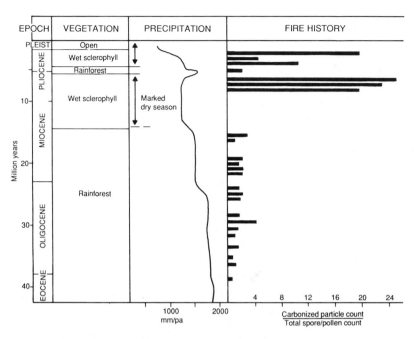

Figure 2.6. An interpretation of the history of vegetation types, precipitation, and fire from the Murray basin region in Southeastern Australia.

In the latter part of the Miocene and Pliocene, pollen assemblages dominated by Myrtaceae in the Murray catchment area are interpreted as wet sclerophyll forest with a rain-forest understory (Martin 1987). These would have required fires for their maintenance (Ashton 1981), and there is certainly a large increase in charcoal in the record at this time (see Figure 2.6). Earlier, charcoal was present in moderate amounts extending back at least to the Late Eocene, suggesting that in drier areas fire may always have been an important component of the environment. It may not necessarily, however, have exercised significant control over community composition prior to the late Tertiary.

An interesting feature in the late Tertiary record is the evolution of a eucalypt with a very distinctive pollen grain that can be related to *Eucalyptus spathulata*, presently restricted to mallee communities within Western Australia (Martin 1989). Its fossil record from the beginning of the Pliocene in both southwestern and southeastern Australia (Bint 1981) until the late Pleistocene (Kate Harle personal communication) demonstrates that it has had much broader distribution in the past within mediterranean and adjacent areas. During the Pliocene it is found in more mesic assemblages than today and nothing indicates whether or not it possessed the mallee growth habit. It does, however, provide the most concrete evidence of early development of a plant, identifiable to a refined taxonomic level, which did become a component in a characteristic mediterranean community.

Quaternary Environments and Vegetation in Southern Australia

The last 1.8 million years of earth history, the Quaternary, was most climatically turbulent. It is notable for the great swings from high latitude and
altitude refrigeration to relative warmth in the interglacials. We attempt here
to generally reconstruct Quaternary climates from global climate and geomorphic patterns in Australia, discuss likely effects on the vegetation in
southern Australia, and then discuss the implications of these changes for
understanding its mediterranean ecosystems.

The records for oxygen isotopes in ocean sediments and loess deposits in
China (e.g., Bradley 1985; Kukla 1987) are the most complete and show the
swings and variability in global climates during the Quaternary. It is now
accepted that these swings were driven by insolation changes associated with
the changing parameters of earth's orbit (Hays et al. 1976). The cycles were
complex, each with its own subtleties and seasonal differences caused by
forcing of the orbital parameters. These variations became noticeably more
pronounced in the last 600,000 years, and one of the most powerful climatic
swings in the whole Quaternary record may have been associated with the
last great cycle, which occupied the time between 100,000 and 10,000 years
ago.

The problem with Australia's Quaternary environmental and vegetational
record is its incompleteness. The pattern in Australia was not one of pronounced glaciation, although minor advances did occur (Galloway 1986).
Early Quaternary evidence of climate is sparse. Long records of climate and
vegetation, extending from the present to more than 100,000 years ago, are
known only from eastern Australia (Singh and Geissler 1985; Kershaw 1986;
Colhoun 1988). Several additional records, however, provide evidence in the
lead up to the last glacial maximum, about 18,000 radiocarbon years ago,
and through to the present. These data, fossil animal evidence (e.g., Hope et
at. 1977; Hope 1982), and geomorphic evidence from, for example, lake
levels (e.g., Bowler 1982; Harrison 1989), linear dunes fields (e.g., Ash and
Wasson 1983; Bowler and Wasson 1984), and sea-level change (e.g., Jennings
1971; Thom and Chappell 1975) confirm that glacial-maximum climate was
conspicuously dry and sea level was about 120 m below the present. In the
southeast, Tasmania was a broad promontory.

In the late Tertiary Australia was extensively forested but sclerophyll and
open-land taxa were well established. Records from the late Quaternary
show great changes in vegetation. Major taxa in the present and penultimate
interglacials are similar, though some regional extinctions occurred. Lacking
detailed vegetation data, spatially and temporally, we can only speculate on
the development of the present patterns. If we assume that the great climatic
oscillation in the last glacial–interglacial cycle was a repeat of many earlier
ones, although possibly more intense, then the more recent data can be used
as a guide for speculating on the longer sequence. One can surmise that
changes in species ranges and adaptations paralleled the continuous shifts in

climate, and were most conspicuous when these were most telling. Because interglacial conditions existed only during about 10% of the last 600,000 years, and Australia is relatively flat, the vast expanses in the interior have continually favored arid taxa. Only coastal regions, now drowned by the relatively high sea level and the dissected country of the ranges, would have provided habitat for lowland humid taxa.

Quaternary vegetation records from southeastern Australia abound; however, long records are few and the southwest is poorly served. Pollen diagrams from southeastern South Australia, within or adjacent to the present mediterranean climate region, show a pattern of eucalypt woodland or forest with both heathland and grassland replaced by open steppelike grassland, with arid and cool elements from about 25,000 until after 15,000 B.P. We have no detailed records of forest with or without substantial mediterranean elements from anywhere in southeastern Australia at last time of glacial maximum, although Kenyon (personal communication) has the first glacial-maximum record of forest, from lowland eastern Victoria. The establishment of present patterns was well under way by the Holocene and later only minor adjustments resulted from mid-Holocene warming and late Holocene deterioration in climate (e.g., Bowler et al. 1976; Macphail and Hope 1984; Dodson 1987).

The data clearly show a great climatic shift away from mediterranean climates in southeastern Australia, for perhaps 10,000 years around glacial maximum time. This timing raises questions about the form in which many plant formations survived. Without data on the geographic expansion of climate during the peak of the glacial climate as it was realized in Australia, and without knowledge of the distribution patterns of key taxa throughout the period, it is not clear whether Australia's mediterranean vegetation migrated some hundreds of kilometers, to wherever mediterranean climate patterns remained or developed, or whether they survived in suitable refugia, possibly in species collections unlike those of the present. There is evidence of disharmonious faunas from around glacial maximum time (Hope et al. 1977).

Within and adjacent to the eastern portion of Australia's mediterranean ecosystems a number of pollen diagrams have records spanning part or all of the period surrounding the last glacial maximum. These show the destruction and reassembly of mediterranean vegetation, especially as it relates to *Eucalyptus*- and *Casuarina*-dominated formations. Carefully examining these records shows, however, that *Banksia, Melaleuca, Leptospermum, Gonocarpus*, Rutaceae, Epacridaceae (including *Monotoca*), *Pimelea, Amperea, Exocarpos*, and even *Pomaderris* are present in the records at Lake Leake and Wyrie Swamp in South Australia (Dodson 1975; 1977) and from Lake Wangoom, Lake Terang, Tower Hill, and Lake Bullenmerri in western Victoria (Dodson 1979; Edney 1987; D'Costa et al. 1989; Kershaw et al. 1991b). Dodson (1983) pointed out that many of these taxa tend to have low pollen representation, suggesting then that they survived full glacial conditions within the region. This survival may account for the rapid recovery of mediterranean systems after the glacial period's aridity waned.

Pollen records from outside the region, at Lake George (New South Wales), Cave Bay, Tullabardine, Newall Creek, and Darwin Crater (Tasmania) suggest patterns similar to those elsewhere in southeastern Australia (Hope 1978; Singh and Geissler 1985; Colhoun and van de Geer 1986; Colhoun 1988; van de Geer et al. 1989). In northern New South Wales, at Ulungra Springs and Cuddie Springs, full glacial conditions also supported some mediterranean elements (Dodson and Wright 1989; Dodson et al. 1993). Additional sites need investigation in areas now below sea level and sites from the southwest of the continent. Dodson (1989) recently pointed out that for much of the Quaternary the southwest of Australia was a humid island of habitat well isolated from that in the east. As a result the mediterranean regions in the west and east would have had differences in historical development.

The data indicate a strong likelihood that many of the major genera of mediterranean vegetation survived the last glacial maximum in lowland regions not far from their present distributions. Considering this status and knowing that mediterranean vegetation is difficult to differentiate at the generic level, and that all sedimentary records examined in Australia show a strong charcoal signal for at least the last 40,000 years (e.g., Singh et al. 1981), then we find that the uniqueness fo mediterranean vegetation begins to fade. Could it be therefore that the major adaptatons to mediterranean climate are expressed at the species level? Better taxonomic resolution, perhaps involving macrofossil evidence, is needed to test this suggestion if we are to achieve a temporal view of development.

Conclusions

The flora of the mediterranean climatic region in southern Australia is diverse and composed of Gondwanan and cosmopolitan families. Many of the genera and most of the species are endemic to Australia. Amount and seasonality of precipitation, status of soil nutrients, and fire regime are important controls on function, but apparently not distribution, of the region's characteristic taxa.

Many taxa of higher order in Australia's mediterranean vegetation developed in the early to mid-Tertiary, when most of Australia was still covered with rain forest. Important shrub and tree taxa became abundant in the Pliocene at a time when a winter rainfall regime developed over southern Australia. It is probable that characteristic herbaceous taxa became prominent elements only during the climatic oscillations in the later Quaternary. In the glacials, these switched from comparative aridity and coolness, and possibly reduced fire regime, to warm and humid conditions like those in the present interglacial. Such climatic swings must have promoted considerable sorting and redistribution of species.

The strong taxonomic relationship of Australia's mediterranean flora to that of many other vegetation formations, however, is undoutedly tied to its development during the late Tertiary and Quaternary climatic transformations. The ancient and generally nutrient-poor soils of Australia have been a relative constant in the development of the flora, but an increasingly obvious fire component has undoubtedly been a contributing factor.

References

Ash JE, Wasson RJ (1983) Vegetation and sand mobility in the Australian desert dunefield. Zeitschr Geomorph, Supp Band 45:7–25

Ashton DH (1981) Fire in tall open forests (west sclerophyll forests). In Gill AM, Groves RH, Noble IR (eds) *Fire and the Australian Biota*. Aust Acad Sci. Canberra, pp 339–366

Bint AN (1981) An early Pliocene pollen assemblage from Lake Tay, southwestern Australia and its phytogeographic implications. Aust J Bot 29:277–291

Bowler JM (1982) Aridity in the late Tertiary and Quaternary of Australia. In Barker WR, Greenslade PJM (eds) *Evolution of the Flora and Fauna of Arid Australia*. Peacock Publications, Adelaide, pp. 35–45

Bowler JM, Hope GS, Jennings JN, Singh GS, Walker D (1976) Late Quaternary climates of Australia and New Guinea. Quat Res 6:359–394

Bowler JM, Wasson RJ (1984) Glacial age environments of inland Australia. In Vogel JC (ed) *Late Cainozoic Palaeoclimates of the Southern Hemisphere*. Balkema, Rotterdem

Bradley, RS (1985) *Quaternary Palaeoclimatology: Methods of Palaeoclimatic Reconstruction*. Allen and Unwin, Boston

Colhoun EA (compiler) (1988) Cainozoic vegetation of Tasmania. Special Paper, Dep Geog, U Newcastle, Australia

Colhoun EA, van de Geer G (1986) Holocene to Middle Last Glaciation Vegetation History at Tullabardine Dam, Western Tasmania. Proc Royal Soc London, B 229:177–207

Costermans L (1983) *Native Trees and Shrubs of South-Eastern Australia*. Rigby, Adelaide

Costin AB, Gray M, Totterdell CJ, Wimbush DJ (1979) *Kosciusko Alpine Flora*. CSIRO, Melbourne

D'Costa DM, Edney P, Kershaw AP, De Deckker P (1989) Late Quaternary palaeoecology of Tower Hill, Victoria, Australia. J Biogeog 16:461–482

Dettman ME (1989) Antarctica: Cretaceous cradle of austral temperate rainforests. In Crame JA (ed) *Origins and Evolution of the Antarctic Biota*. Geol Soc Spec Pub No. 47, pp 89–105

Dodson JR (1975) Vegetation history and water fluctuations at Lake Leake, southeastern South Australia. II. 50,000 to 10,000 B.P. Aust J Bot 23:123–150

Dodson JR (1977) Late Quaternary palaeoecology of Wyrie Swamp, southeastern South Australia. Quat Res 8:97–114

Dodson JR (1979) Late Pleistocene vegetation and environments near Lake Bullen-merri, western Victoria. Aust J Ecol 4:419–428

Dodson JR (1983) Modern pollen rain in southeastern New South Wales, Australia. Rev Palaeobot Palynol 38:249–268

Dodson JR (1987) Mire development and environmental change, Barrington Tops, New South Wales, Australia. Quat Res 27:73–81

Dodson JR (1989) Late Pleistocene vegetation and environmental shifts in Australia and their bearing on faunal extinctions. J Arch Sci 16:207–217

Dodson JR, Fullagar R, Furby JH, Prosser I (1993) Humans and megafauna in a late Pleistocene arid environment from Cuddie Springs, New South Wales. Arch Oceania 28:94–99

Dodson JR, Wright RVS (1989) Humid to arid to subhumid vegetation shift on Pilliga Sandstone, Ulungra Springs, New South Wales. Quat Res 32:182–192

Edney P (1987) Late Quaternary vegetation and environments from Lake Wangoom, western Victoria. Unpub M.A. Thesis, Monash U

Francis WD (1970) *Australian Rain-Forest Trees.* Australian Government Publishing Service, Canberra

Galloway RW (1986) Australian snow fields past and present. In Barlow BA (ed) *Flora and Fauna of Alpine Australasia: Ages and Origins.* CSIRO, Melbourne

Groves RH (1981) *Australian Vegetation.* Cambridge U Press, Cambridge

Groves RH, Specht RL (1965) Growth of heath vegetation. I. Annual growth curves of two heath ecosystems in Australia. Aust J Bot 13:261–280

Harden G (ed) (1990) *Flora of New South Wales.* Vol. 1. U New South Wales Press, Sydney

Harden G (ed) (1991) *Flora of New South Wales.* Vol. 2. U New South Wales Press, Sydney

Harrison SP (1989) Lake level records from Australia and New Guinea. UNGI Rapport (U Uppsala) 72:1–142

Hays JD, Imbrie J, Shackleton NJ (1976) Variations in the earth's orbit: pacemaker of the ice ages. Science 194:1121–1132

Hope GS (1978) The Pleistocene and Holocene vegetational history of Hunter Island, north-western Tasmania. Aust J Bot 26:493–514

Hope JH (1982) Late Cainozoic vertebrate faunas and the development of aridity in Australia. In Barker WR, Greenslade PJM (eds) *Evolution of the Flora and Fauna of Arid Australia.* Peacock Publications, Adelaide

Hope JH, Lampert RJ, Edmondson E, Smith MJ, van Tets GF (1977) Late Pleistocene faunal remains from the Seton rock shelter, Kangaroo Island, South Australia. J Biog 4:363–385

Jennings JN (1971) Sea level changes and land links. In Mulvaney DJ, Golson J (eds) *Aboriginal Man and Environment in Australia.* ANU Press, Canberra

Jessop J (ed) (1981) *Flora of Central Australia.* Reed, Sydney

Jones R (1967) Productivity and water use efficiency of a Victorian heathland. Unpub PhD Thesis, U Melbourne

Kemp EM (1978) Tertiary climatic evolution and vegetation history in the southeast Indian Ocean region. Palaeogeog, Palaeoclim, Palaeoecol 24:169–208

Kershaw AP (1986) Climatic change and Aboriginal burning in north-east Australia during the last two glacial/interglacial cycles. Nature 122:47–49

Kershaw AP, Bolger P, Sluiter IR, Baird J, Whitelaw M (1991a) The origin and evolution of brown coal lithotypes in the Latrobe Valley, Victoria, Australia. Int J Coal Geol 18:233–249

Kershaw AP, D'Costa DM, McEwen Mason JRC, Wagstaff BE (1991b) Palynological evidence for Quaternary vegetation and environments of mainland southeastern Australia. Quat Sci Rev 10:391–404

Kukla G (1987) Loess stratigraphy in central China and correlation with an extended oxygen isotope stage scale. Quat Sci Rev 6:191–220

Macphail MK, Hope GS (1984) Late Holocene mire development in montane southeastern Australia: a sensitive climate indicator. Search 15:344–349

Martin ARH (1982) Proteaceae and the early differentiation of the central Australian flora. In Barker WR, Greenslade PJM (eds) *Evolution of the Flora and Fauna of Arid Australia*. Peacock Publications, Adelaide, pp 77–83

Martin HA (1987) The Cainozoic history of the vegetation and climate of the Lachlan River region, New South Wales. Proc Linnean Soc NSW 117:45–51

Martin HA (1989) Evolution of mallee and its environment. In Noble JC, Bradstock RA (eds) *Mediterranean Landscapes in Australia*. CSIRO, Melbourne, pp 83–92

Martin HA (1994) Australian Tertiary phytogeography: evidence from palynology. In Hill RS (ed) Australian Vegetation History: Cretaceous to Recent. Cambridge U Press: in press

McEwan Mason JRC (1989) The palaeomagnetics and palynology of late Cainozoic cored sediments, from Lake George New South Wales, Southeastern Australia. Unpub PhD thesis, Monash U, Melbourne

McMinn A, Martin HA (1991) Late Cainozoic pollen history from Site 765, Eastern Indian Ocean. Scientific Reports of the Ocean Drilling Programme, Leg 123, in press

Pate JS, Beard JS (1984) *Kwongan: Plant Life of the Sandplain*. U Western Australia Press, Nedlands, 284 pp

Singh G, Geissler EA (1985) Late Cainozoic history of vegetation, fire, lake levels and climate at Lake George, New South Wales. Phil Trans Royal Soc London, Series B, 311:379–447

Singh G, Kershaw AP, Clark RL (1981) Quaternary vegetation and fire history of Australia. In Gill AM, Groves RH, Noble IR (eds) *Fire and the Australian biota*. Aust Acad Sci, Canberra, pp 23–54

Sluiter IR (1991) Early Tertiary vegetation and climates, Lake Eyre region, northeastern South Australia. In Williams MAJ, De Deckker P, Kershaw AP (eds) Geol Soc Aust Spec Pub No. 18, pp 99–118

Specht RL (1969a) A comparison of the sclerophyllous vegetation characteristic of Mediterranean type climates in France, California, and southern Australia. I. Structure, morphology, and succession. Aust J Bot 17:277–292

Specht RL (1969b) A comparison of the sclerophyllous vegetation characteristic of Mediterranean type climates in France, California, and southern Australia. II. Dry matter, energy, and nutrient accumulation. Aust J Bot 17:293–309

Specht RL (1979) Heathlands and related shrublands of the world. In Specht RL (ed) *Heathlands and Related Shrublands*. A. Lange and Springer, Berlin, pp 1–18

Specht RL (1981a) Mallee ecosystems in southern Australia. In di Castri F, Goodall DW, Specht RL (eds) *Mediterranean-Type Shrublands*. Elsevier, Amsterdam

Specht RL (1981b) Primary production in mediterranean-climate ecosystems regenerating after fire. In di Castri F, Goodall DW, Specht RL (eds) *Mediterranean-Type Shrublands*. Elsevier, Amsterdam

Specht RL, Rayson P, Jackman ME (1958) Dark Island heath (Ninety-Mile Plain, South Australia). VI. Pyric succession: changes in composition, coverage, dry weight, and mineral nutrient status. Aust J Bot 6:59–88

Thom BG, Chappell JMA (1975) Holocene sea levels relative to Australia. Search 6:90–93

Truswell EM, Sluiter IR, Harris WK (1985) Palynology of the Oligo-Miocene sequence in the Oakvale-1 corehole, Western Murray Basin, South Australia. Bur Min Res J Aust Geol Geophys 9:267–295

van de Geer G, Fitzsimons SJ, Colhoun EA (1989) Holocene to middle last glaciation

vegetation history at Newall Creek, western Tasmania. New Phytol 111:549–558
Westman WE (1978) Evidence for the distinct evolutionary histories of canopy and
    understorey in the *Eucalyptus* forest-heath alliance of Australia. J Biogeog 5:365–
    376
Willis JH (1970) *A Handbook to Plants in Victoria.* Vol. 1. Cambridge U Press,
    Cambridge
Willis JH (1972) *A handbook to Plants in Victoria.* Vol. 2. Melbourne U Press,
    Melbourne

II. Comparative Biogeography

3. Convergence in the Mediterranean Floras in Central Chile and California: Insights from Comparative Biogeography

Mary T. Kalin Arroyo, Lohengrin Cavieres,
Clodomiro Marticorena, and Melica Muñoz-Schick

A growing body of empirical evidence shows that mediterranean climate areas, with their long dry summers and winter rainfall, support high species richness, a rich array of life forms, and high levels of endemism (Goldblatt 1978; Raven and Axelrod 1978; Hopper 1979; 1992; Naveh and Whittaker 1979; Bond and Goldblatt 1984; Lamont et al. 1984; Pate et al. 1984; Pignatti and Pignatti 1985; Westoby 1988; Cowling et al. 1989; 1992; Greuter 1991; Keeley 1991; Arroyo and Uslar 1993; Arroyo et al. 1993a). These features have been variously ascribed to the peculiar mode of origin of mediterranean floras (Raven and Axelrod 1978; Raven 1988), the transitional position of mediterranean vegetation between cool-temperate and dry-tropical types (di Castri 1990), strong temporal fluctuations in precipitation, determining spatial and temporal patchiness (Zedler 1990), and low-nutrient soils (Westoby 1988).

As knowledge has accumulated, however, considerable variation in biotic and abiotic characteristics in climate areas of mediterranean type has been revealed (Cody and Mooney 1978; di Castri 1981; 1990). Di Castri, attempting to systematize such variation, grouped the five main mediterranean climate regions into three complexes comprising, respectively, South Africa–Australia, California–central Chile, and the Mediterranean Basin in a "crossroads" position, based on considering tectonic structure, status of soil nutrients, climatic determinants, proneness to fire, and biotic characteristics. Nevertheless, even within these clusters, as exemplified by careful studies such as Milewski's (1979) comparing the climates in the South Africa–Australia complex, differences can be considerable.

California and central Chile are considered the closest-knit of the three mediterranean clusters (Mooney et al. 1977; di Castri 1981; 1990). Here we ask if consistent trends appear in distribution of life forms, richness of species and genera, and patterns of generic differentiation at a biogeographic scale in central Chile and California. To interpret our results, we also consider the extent to which climates in central Chile and California differ at this large scale and whether the histories of the mediterranean floras in central Chile and California have been parallel. These questions are all logical correlates of early efforts in central Chile and California, where the notion was tested that similar climates in different parts of the world acting on organisms with proximally independent phylogenetic histories will result in structurally and functionally similar ecosystems (Mooney et al. 1970; Aschmann 1973; Parsons 1976; Mooney and Cody 1977; Mooney et al. 1977; Thrower and Bradbury 1977; Cody and Mooney 1978).

The importance of focusing on details of climatic similarity and the history of floras when comparing diversity and life-form distribution at a biogeographic scale for the mediterranean areas in central Chile and California should not be underestimated. A striking result of the studies on small climatically matched sites in central Chile and California was the lack of convergence in community structure (Mooney et al. 1977). The differences seen are now generally attributed to more extensive human disturbance and absence of fire as a natural factor in central Chile (Mooney and Cody 1977; Fuentes et al. 1986; Aschmann 1990; Fuentes 1990). Both California and central Chile have been impacted by humans (Mooney et al. 1977; Axelrod 1989; Minnich 1989), but the influence has been more pervasive in Chile. Fire is considered an unprecedented perturbation in the Chilean landscape (Armesto et al. this volume) and has not been associated with the evolution of special life-history traits (Muñoz and Fuentes 1989) as in California, where it occurs naturally. The resident biota at any pair of local, climatically equivalent sites, however, will comprise species adapted to a wider range of climatic conditions than at those sites. If local sites are drawn from two larger areas that are on the average less than identical in climate than the climatically matched sites themselves, strong convergence on climatically matched sites will not necessarily be expected.

Second, complete convergence may be constrained where biogeographic history imposes different phylogenetic constraints (Kochmer and Handel 1986; Schluter 1986; Primack 1987; Ricklefs 1987). To detect effects of these kinds, intimate knowledge of history is essential. Moreover, comparisons must be made at a scale that ensures full expression of the long-term evolutionary signal; the biogeographic scale as proposed here seems particularly appropriate.

Regional Climatic Analogs and Floristic Base

To undertake the biogeographic comparisons, we delineated areas of analogous climate in central Chile and California. We began by selecting a large area of mediterranean climate in western North America for which floristic information was readily available: the California Floristic Province within California (Hickman 1993). As originally defined (Howell 1957), the latter consists of the state of California excluding the Great Basin and Desert provinces and extends beyond California, taking in the coastal portion of Baja California to around 30° N, as well as a small area in the extreme south of Oregon. So as not to combine different taxonomic concepts within a floristic list, we have considered only the portion of the California Floristic Province *within the confines of the state of California* (see Hickman 1993 for details). For clarity this area is referred to as the *California FP* throughout this chapter.

The California FP includes the semiarid, subhumid, and a portion of the humid type of mediterranean climate (di Castri 1973). A regional climatic analog in central Chile for the California FP was determined by interpolation from the latitudinal extensions of these same types of mediterranean climate in central Chile, given by di Castri (1973), and consisting of the area from around 31° to 31°30′ S to 37°15′ S (Fig. 3.1). To simplify obtaining floristic data, we have delimited the Chilean climatic analog at the nearest provincial boundaries, consisting respectively of the northern limit of the province of Choapa in the Coquimbo (IV) Region and the southern limits of the provinces of Concepción and Ñuble, Bíobío (VIII) Región (Fig. 3.1). It should be obvious to the reader that the Chilean climatic analog, just as in California, does not correspond to a natural vegetation province. A more natural boundary in central Chile would be found farther south.

A complete list of native plant species occurring in the California FP was compiled by revising the coded distributional data provided for each species in the Jepson Manual (Hickman 1993). To derive a species–area curve for the California FP, additional data for smaller areas within the California FP were taken from published county floras and checklists, excluding in advance adventive species and hybrids.

Exploration of the Chilean flora commenced early in the eighteenth century with important expeditions by the French botanist Feuillée in 1709 and 1711 (Feuillée 1714–25) and Ruiz and Pavón of Spain in 1777 and 1778 (Ruiz and Pavón 1794; Alvarez 1954). The first authoritative work by a Chilean naturalist, Juan Ignacio Molina, was published toward the end of 18th century (Molina 1782; Gunckel 1929). This early phase culminated in pioneer works by Gay (1845–1854) and Reiche (1896–1911). These floras, remarkable for their time, are incomplete for taxonomic and geographic coverage, however, and are now totally out of date nomenclaturally. Although no complete modern floristic work has been published, many recent monographic studies have enabled publication of comprehensive checklists

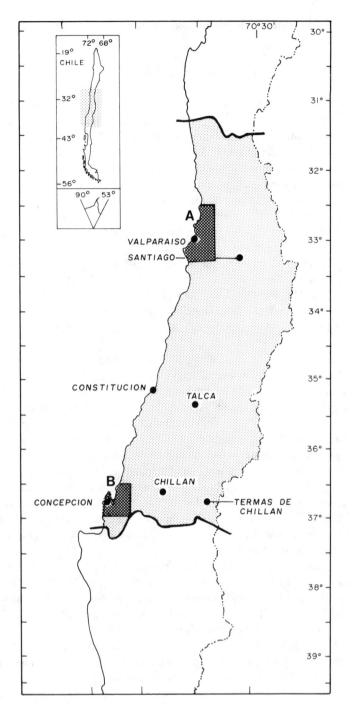

Figure 3.1. Area of central Chile used as a climatic analog of the California FP. Also shown are the Concepción and Valparaíso areas referred to in Table 3.10.

(Marticorena and Quezada 1985; Marticorena 1990) and a tree flora (Rodríguez et al. 1983). Additionally, an electronic specimen-based data base of collections in the major national herbaria (SGO and CONC) and of specimens (including types) cited in the taxonomic literature, indexed by latitude and longitude, has recently been compiled. The floristic data for central Chile and the winter rainfall area as a whole were gathered by screening the Chilean data base for species in the countrywide checklists (with some hitherto unpublished additions and modifications) occurring within the two areas, respectively. Additional checklists for the Coast range region, the Andean cordillera and for smaller areas within the Chilean analog were similarly generated to assess floristic similarity between the dominant physiographic units and to generate a species–area curve for Chile.

Climate

Details of the climates in California and central Chile can be found in Barbour and Major (1988) and in di Castri and Hajek (1976), respectively, and a complete comparison of central Chile and California climate was made by Mooney et al. (1977). Summarizing and extending this comparison, the key forcing factors are: (1) the position of the Southern Hemisphere high at 40° S (Aceituno 1988), farther equatorward of its position in western North America at 42° N (Axelrod 1973); (2) the significantly steeper and higher Andes, with many peaks rising to well over 5000 m, located, on average, about 100 km from the Pacific Ocean as opposed to about 200 km for the Sierra Nevada; (3) the higher, transversely drained central depression in Chile, 130 to 600 m elevation, in contrast to the lower, longitudinally drained Central Valley in California, lying mostly below 100 m; (4) the Humboldt current colder than the California current; (5) lack of tropical influence in central Chile.

1. The first factor, in central Chile, leads to a much steeper rainfall gradient (Fig. 3.2a) and relatively more rainfall in the summer months in relation to any given total amount of rainfall at the more southerly locations. The latter results in lower summer aridity for any annual value of aridity in central Chile (Fig. 3.2b). Emphasis on summer precipitation in defining the mediterranean climate (di Castri, 1973) effectively leads to "anomalies." First, the southern limit of the Chilean analog receives a similar amount of summer rainfall but far less total rainfall than the northern border of California. Conversely, areas of low summer rainfall, still corresponding to a mediterranean climate by definition (di Castri 1973), persist much farther north in western North America for the same amounts of annual rainfall received by locations such as Valdivia (ca. 40° S), considered to correspond to the seasonal rain-forest zone (Alaback 1991; Arroyo et al. 1993b).

2. Effects of this and factors (3) and (4) combined are readily appreciated

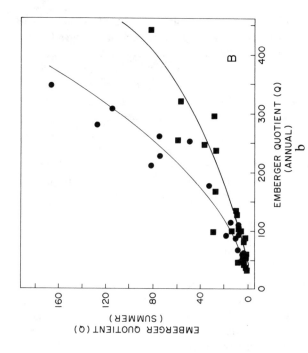

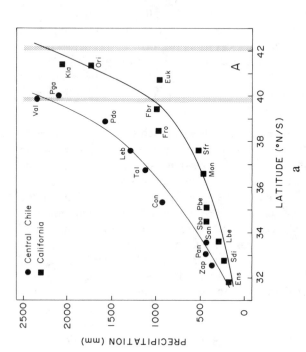

Figure 3.2. (a) Mean annual precipitation for selected coastal locations at different latitudes in central Chile and California. Central Chile: Zap—Zapallar; Pan—Punta Angeles; San—San Antonio; Con—Constitución; Tal—Talcahuano; Leb—Lebu; Pdo—Puerto Domínguez; Val—Valdivia; Pga—Punta Galera. California: Ens—Ensenada; Sdi—San Diego; Lbe—Long Beach; Sba—Santa Barbara; Pbe—Pismo Beach; Mon—Monterey; Sfr—San Francisco; Fro—Fort Ross; Fbr—Fort Bragg; Euk—Eureka; Ori—Orick Prairie Creek; Kla—Klamath. Climatic data, central Chile: Hajek and di Castri (1975); di Castri and Hajek (1976); Santibañez and Uribe (1990). Climatic data: California—Baja California: California: Climatology of the United States No. 81, California 1951–80 (NOAA), Secretaría de Agricultura y Ganadería, Dirección General de Geografía y Meteorología, México. (b) Summer aridity plotted against yearly aridity for coastal locations in central Chile and the California–Oregon area. Aridity expressed as Emberger's Pluviometric Quotient (Q) (Emberger 1955). Q decreases with increasing aridity.

by comparing annual and summer aridity for coastal areas (Table 3.1). Emberger's (1955) Pluviometric Coefficient (Q) for the summer months and the total year are fairly similar at the semiarid extremes of the two mediterranean areas. At the humid extreme the summer Q values for Punta Lavapie (central Chile) and Klamath (northern California) are identical (82). The corresponding annual Q values for these localities are 212 and 443, respectively, with Klamath close to Valdivia (435) in Chile (di Castri and Hajek 1976). The regional annual average for central Chilean coastal areas is, nevertheless, just slightly lower than in the California FP, but the summer average is slightly higher (Table 3.1). Similarity in the regional annual values, in spite of wide differences in the upper extreme values, accrues from a relatively larger extension of drier mediterranean climate in southern California FP that compensates for very high rainfall on the northern border of California when an average value is calculated.

The higher central depression in Chile, though very arid at the northern extreme, is regionally much less arid (Table 3.1) than the central valley in California on account of its higher elevation and strong inland penetration of maritime air masses through several Coast range corridors. Although many climatic stations are at high elevations in the California FP (NOAA 1982), data are very scant in the Andean range in central Chile. Moreover, expected adiabatic effects on precipitation confound interpretation of the few data available. Palomar Mt. (33°21′ N; 1690 m—$Q = 90$) in southern California and San José de Maipo (33°29′ S; 1060 m—$Q = 86$) in central Chile show a similar degree of aridity. Juncal (32°52′ S; 2250 m—$Q = 45$) in central Chile, however, is more arid than Cuyamaca (32°59′ N; 1417 m—$Q = 90$) in southern California, but Sewell (34°06′ S; 2134 m— $Q = 187$) in central Chile is less arid than Lake Arrowhead (34°15′ N; 1586 m —$Q = 126$). No cordilleran climatic stations are southward of Sewell in central Chile. With no discernible trend in the inner cordilleras, similar annual and lower Summer aridity in coastal areas, and lower aridity in the central depression, it is probably safe to assume that, overall, the Chilean analog is somewhat less arid than the California FP.

Table 3.1. Comparison of aridity (annual and summer values) for coastal and central valley locations in central Chile and California. Values are Emberger's (1955) Pluviometric Quotient (Q)

	Central Chile		California FP	
	Regional mean	Range	Regional mean	Range
Coastal (annual)	157	64–261	165	50–443
Coastal (summer)	37	4–82	24	3–82
Central valley (annual)	86	7–141	45	15–113
Central valley (summer)	18	7–46	6	2–25

[a] High values indicate low aridity. Summer values are for June, July and August in the California FP and December, January and February in central Chile, normalized for direct comparison with annual values. Regional means are averages of representative stations per each degree of latitude.

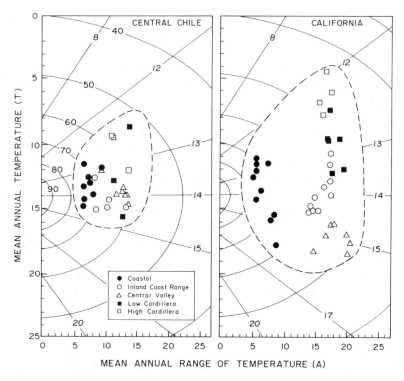

Figure 3.3. Mean annual temperature and annual temperature range for locations in central Chile and California, plotted on a Bailey (1964) nomogram. Arcs are equability lines, per Bailey (1964). Data are representative stations (where available) at each degree of latitude. See Fig. 3.2 for sources of climatic data.

3. Factors (2) and (3) combined lead to greater thermal uniformity, stronger equability, and generally cooler conditions in central Chile (Fig. 3.3). Thermal conditions converge most at coastal locations, diverging progressively inland, with a hiatus in temperature conditions already evident in the inner Coast range. Another evident difference are the lower temperatures on the northern coast of central Chile in relation to southern California because of the colder Humboldt current. This effect leads to a lower difference in mean annual temperature comparing the arid and humid extremes of the Chilean analog (Fig. 3.3).

4. The last difference (5) emphasizes that the two mediterranean areas are embedded in contrasting regional climatic scenarios (cf. Arroyo et al. 1988). In Chile, sparse winter rainfall continues as the only source of precipitation to as far north as 20° S in the coastal desert, and to ca. 25° S in the inner Andean highlands (Arroyo et al. 1988; Grojean et al. 1991; Rundel et al. 1991). North of 25° S, precipitation is received principally during the summer season from a tropical source. Strictly defined, climate in much of the Chil-

ean deserts is of the hyperarid mediterranean type (di Castri and Hajek 1976; Nahal 1981). Only some highland locations between 22° and 24° S receive precipitation from both sources (Grojean et al. 1991).

In western North America, westward-moving storm cells originating in the Gulf of Mexico impose a pattern of summer rainfall as far as 32° N, with the Sonoran Desert area bordering on the southern edge of the mediterranean area receiving between 40% and 70% summer precipitation (MacMahon 1988). This important difference in the summer rainfall regime accounts for the contrasting wildfire regimes in central Chile and California. Vegetation in central Chile is shielded from summer lightning ignitions because the mediterranean area lies well outside the influence of the intertropical convergence, whereas fire propagated by lightning is a natural phenomenon for many mediterranean vegetation types in California (Mooney 1977).

5. Finally, as a result of the first (1) and second (2) features combined, the area of mediterranean climate in central Chile is less than half the corresponding area in the California FP (details in Table 3.10).

A multivariate analysis (Decorana) (Fig. 3.4) performed on climatic data for coastal and central valley locations from 30° to 45° S in central Chile and 32° to 45° N in the California–Oregon region highlights differences in climate between the two Pacific coasts and shows that a straightforward climatic match at a regional level is lacking. Clearly visible at this level of detail are two distinctly varying climatic matrices. Points of climatic encounter can always be found, but they tend to be drawn from different relative geographic positions in the two climatic matrices.

Floristics and Vegetation

Comparisons of matorral and chaparral, or sclerophyllous scrublands, formed the basis for the original convergence studies in central Chile and California (Parsons and Moldenke 1975; Keeley and Johnson 1977; Mooney et al. 1977; Cody and Mooney 1978). These types of vegetation were selected for study because their biota are taxonomically distinct and much subsequent work on the ecological dynamics of mediterranean vegetation has been concentrated here (e.g., Montenegro et al. 1985; Fuentes et al. 1986). Nevertheless, many other types of vegetation are found within the broader climatic analogs considered here. It is interesting, then, to consider the overall floristic affinity of the two native mediterranean floras and the ecogeographic correspondence of other individual types of vegetation in the two regions.

Floristics

The tree floras in central Chile and the California FP, echoing earlier findings for the matorral and chaparral, are taxonomically distinct, sharing no

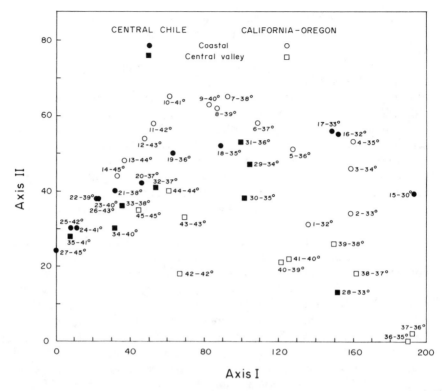

Figure 3.4. Multivariate analysis (DECORANA) of climate for coastal and central-valley locations in Chile and California-Oregon. Latitude is indicated for each location. Variables analyzed were mean annual precipitation; mean summer precipitation; mean annual temperature; mean temperature in the warmest and coolest month, respectively; annual and summer aridity (de Martonne Index: $A = P/(T + 10)$, where P = mean annual or summer precipitation (mm) and T = mean annual or summer temperature (°C)—di Castri and Hajek 1976). Coastal locations: California-Oregon: (1) San Diego, (2) Laguna Beach, (3) Santa Barbara, (4) Pismo Beach, (5) Monterey, (6) Half Moon Bay, (7) Fort Ross, (8) Fort Bragg, (9) Eureka, (10) Klamath, (11) Brooking, (12) Readsport, (13) Newport, (14) Seaside. Coastal locations: Chile: (15) La Serena, (16) Zapallar, (17) Valparaíso, (18) Constitución, (19) Talcahuano, (20) Lebu, (21) Puerto Domínguez, (22) Valdivia, (23) Punta Galera, (24) Maullín, (25) Moro Lobos, (26) Quellón, (27) Puerto Aisén. Central Valley locations: Chile: (28) Colina, (29) San Fernando, (30) Talca, (31) Chillán, (32) Los Angeles, (33) Temuco, (34) Osorno, (35) Puerto Montt. Central-valley locations: California-Oregon: (36) Bakersfield, (37) Kettleman, (38) Stockton, (39) Sacramento, (40) Orland, (41) Red Bluff, (42) Ashland, (43) Roseburg, (44) Corvallis, (45) Oregon City. Axis I showed highest correlation with mean annual precipitation ($r = 0.90$); Axis II with mean summer temperature ($r = 0.67$).

Table 3.2. Summary statistics of the mediterranean floras of central Chile and the California FP. Shared genera are those contributing species to the life form in the indicated geographic region, that are also found in the second region. Percentage species contributed is the percentage of total species in that life form contributed by the shared genera. Life-form identity in the second region is not a requirement to be a shared genus

					Shared genera		
Region	Number genera	Number species	% species	Species/ genus	N	%	Species contributed (%)
Central Chile[a]							
Annuals	151	378	15.8	2.5	99	65.6	65.6
Perennial herbs	355	1516	63.4	4.3	159	44.8	50.3
Shrubs[b]	144	426	17.8	2.9	35	24.3	33.3
Trees	47	70	2.9	1.5	2	4.3	2.9
Total flora	591	2395		4.0	233	39.4	48.4
California FP							
Annuals	273	1279	30.2	4.7	111	40.7	59.3
Perennial herbs	503	2384	56.2	4.7	176	35.0	48.2
Shrubs[b]	148	467	11.0	3.2	29	19.6	26.8
Trees	43	110	2.6	2.6	3	7.0	11.8
Total flora	806	4240		5.3	233	28.9	48.2

[a] Life-forms uncertain in 5 species. [b] Includes woody vines and woody epiphytes. Life-form spectrum: central Chile vs. California FP: $G = 200.00$, $df = 3$, $p < 0.001$.

species and only *Salix* and *Prosopis* at the generic level (Table 3.2). Additionally, *Rhamnus*, represented by shrubs in central Chile, forms trees and shrubs in California. Central Chile has proportionately far fewer gymnosperms (6 species) than the California FP (54 species), with only *Austrocedrus chilensis* (Cupressaceae) showing any dominance (Schlegel 1962). *Araucaria araucana* (Araucariaceae) forms pure stands at higher elevations on the Cordillera Nahuelbuta just south of the boundary of the Chilean climatic analog, an area that continues to have a mediterranean climate, but with less pronounced summer drought than within the California FP. Both general mediterranean areas have appreciable numbers of endemic tree genera: western North America has *Sequoia*, *Sequoiadendron*, *Lyonothamnus*, and *Umbellularia* (Raven and Axelrod 1978); Chile has *Gomortega*, *Jubaea*, *Legrandia*, and *Pitavia*, and central Chile also has the endemic family Gomortegaceae.

Among shrub components are substantial numbers of shared genera contributing heavily to the species pool (Table 3.2). These range from such widespread genera as *Senecio*, *Baccharis*, *Gaultheria*, *Lycium*, *Ribes*, and *Rubus*, in which the Chilean and California species have been independently derived, to amphitropical disjuncts such as *Krameria*, *Viguiera*, *Frankenia*, *Larrea*, and *Encelia*. The total contribution of the amphitropically disjunct genera, however, is disproportionately low. The largest genera of shrubs in central Chile (*Adesmia*, 38 spp. and *Haplopappus*, 32 spp.) and the California FP (*Arctostaphylos*, 57 spp. and *Ceanothus*, 41 spp.) are unshared, although shared *Baccharis* (30 spp.) and *Berberis* (20 spp.) are especially speciose in Chile.

The number of shared genera is even higher in the herbaceous floras, with the common herbaceous genera contributing very strongly to the species pool (Table 3.2). The tendency is most exaggerated in annuals where 66% (99) of native genera with annuals in Chile are also found in California, and conversely 41% (111) of native genera with annuals in California are found in central Chile (Table 3.2). Probably about half the shared herbaceous genera in central Chile and California correspond to amphitropical disjunctions (Raven 1963), but as in the shrub life form, these contribute disproportionately to the species total. Overall, then, and as expected, the floristic distinctiveness seen in the woody flora of the matorral and chaparral is maintained at the larger biogeographic scale considered here. This distinctiveness becomes progressively less evident when herbaceous groups are considered.

Vegetation

Table 3.3 shows the main vegetation types in central Chile and those in California occupying corresponding ecogeographic positions. In central

Table 3.3. Main zonal vegetation types in central Chile and their ecogeographic analogs in the California FP. Vegetation types in central Chile modified after Schmithüsen (1956); for California, as in Barbour and Major (1988)

Central Chile	California FP
Relic coastal rain forest	Relic redwood forest
Evergreen *Nothofagus* forest	No close equivalent
Deciduous *Nothofagus* forest	Oak woodland
	Montane and subalpine forests (Transverse and Peninsula ranges)
	Montane and subalpine forests (Klamath Mts.)
	Lower montane Sierran forest
	Upper montane Sierran forest
Sclerophyllous forest	Mixed evergreen forest
Matorral	Chaparral
	Northern coastal scrub
Northern coastal matorral	Coastal sage scrub
Coastal succulent scrub	Coastal sage succulent scrub
Inland succulent scrub	No close equivalent
Montane sclerophyllous forest (*Kageneckia angustifolia*)	Montane chaparral
Montane coniferous forest (*Austrocedrus chilensis*)	Subalpine Sierran forest
No equivalent	Valley grassland
No equivalent	Coastal prairie
Alpine	Alpine

Chile a very abundant vegetation type is deciduous *Nothofagus* forest (with
N. obliqua, N. glauca, N. alessandrii, N. alpina, N. antarctica in various combinations; Donoso 1987). Such deciduous forest occurs in the extreme south
in the central depression and then northward along the Coast range and
Andes. Deciduous *Nothofagus* forest occupies the position not only of oak
woodland around the perimeter of the Central Valley in California, which
it somewhat resembles physiognomically, but also of many gymnosperm-
dominated Sierran and southern Californian mountainous forests and Coast
range forests (Table 3.3). Coastal rainforest, dominated by *Aextoxicon
punctatum*, found in isolated patches in central Chile (Pérez and Villagrán
1985), recalls the more southern relic stands of redwood forest in northern
California. The Chilean forests, however, are vinier and richer in angiosperm
tree species. Chilean matorral, with such sclerophyllous species as *Lithrea
caustica, Peumus boldus, Cryptocarya alba, Quillaja saponaria*, and *Colliguaja*
spp., is found on the inner slopes of the Coast range and Andean slopes
north of 36° S and in coastal locations from around 32° S southward, and is
closely matched to chaparral (Mooney et al. 1977; Hanes 1988; Keeley and
Keeley 1988). On the wetter Pacific slopes of the Coast range, and in damp
valleys in the Andes, matorral often intergrades imperceptibly into dense,
tall sclerophyllous forest with such species as *Beilschmiedia* spp., *Dasyphyllum excelsum, L. caustica, Kageneckia oblonga, P. boldus, C. alba, Azara*
spp., *Myrceugenia* spp., *Escallonia* spp., *Persea lingue*, of which many are the
dominants in matorral. Such sclerophyllous forest seems to correspond eco-
geographically, and physiognomically is very similar to mixed evergreen forest in California, also a sclerophyllous formation. At the northern extreme of
the mediterranean zone in central Chile, the typical matorral shrubs (*L.
caustica, P. boldus, C. alba*) become scattered with appearance of such
species as *Bahia ambrosioides, Fuchsia lycioides*, and *Adesmia* spp. forming a
softer scrub, similar to coastal sage scrub in California (Mooney et al. 1977).
This zone in turn, as in California, acquires more succulents, northward, as
aridity increases. An inland version of succulent scrub with *Puya* and colum-
nar Cactaceae appears on steep north-facing slopes in the Andes, and in the
Coast range to around 33° to 34° S in central Chile. This type of vegetation
seems to have no close equivalent in the California FP. The latter is probably
an artefact of lack of a high mountain range well inland in southern California where vegetation of this type would be expected. On the other hand,
Californian northern coastal scrub has no equivalent in south-central Chile.
Coastal matorral in Chile generally includes the same dominants as the typical inland matorral communities do.

   North of 34° S in the Andes, from around 1500 m upward, deciduous
Nothofagus forest is replaced by montane sclerophyllous forest dominated
by *Kageneckia angustifolia*, with a closest ecogeographic equivalent in montane chaparral on the Transverse and Peninsula ranges in southern California, but quite different physiognomically in its more woodland aspect (Arroyo and Uslar 1993). Above the deciduous *Nothofagus* and *Kageneckia*

belts, discontinuous patches of *Austrocedrus chilensis* may occur (Schlegel 1962), occupying an ecogeographic position similar to that of the subalpine Sierran forests. Toward the southern end of the Chilean climatic analog, evergreen *N. dombeyi* and deciduous *N. pumilio* may appear close to tree line. Both geographical areas show extensive alpine areas above tree line.

Native grasslands, as seen in the Central Valley (Heady 1988) and northern coasts of California, are conspicuously absent in central Chile. Although the central depression in Chile today is highly altered and frequently invaded by secondary *Acacia caven* and Mediterranean weeds (Fox 1990; Fuentes et al. 1990; Ovalle et al. 1990), herbarium records and loci of native vegetation attest to original sclerophyllous forest and deciduous woodlands in its southern extreme, and matorral and *Prosopis chilensis* woodlands in its northern extreme. Finally, vernal pools can be found scattered throughout central Chile. Nowhere to be found, however, are the very extensive networks of vernal pools seen in the central valley of California and in southern California.

This comparison above all shows that the California FP has a greater variety of vegetation types than central Chile, and that one vegetation type in Chile may occupy the place of several in California. This last situation may partially reflect different concepts for dividing the vegetation landscape in the two mediterranean areas. Lower latitudinal and altitudinal differentiation of vegetation in central Chile is probably also an important factor, however. In central Chile, distinction between rain forest and sclerophyllous elements is often ambiguous, with so-called rain forest elements frequently coexisting with the more typical sclerophyllous trees in coastal forest in the woodland canopy. Remarkably, some rain forest species found in coastal sites in central Chile around 35° S (e.g., *Nothofagus antarctica, Campsidium valdivicum*) extend as far south as the southern, nonseasonal rain forest zone (Donoso 1987; Arroyo et al. 1993b). Along the latitudinal gradient the turnover from rain forest to sclerophyllous trees is typically very gradual (Villagrán and Armesto 1980) and broad ecotones over local aridity gradients are the norm (San Martín et al. 1984). Only on steep north- and south-facing slopes does spatial segregation of rain forest and sclerophyllous elements become clearly evident (Armesto and Martínez 1978; Pérez and Villagrán 1985; Arroyo et al. 1993a,b).

Analogously, along any altitudinal gradient, in the semiarid extreme of the mediterranean area in central Chile, unlike the fairly clear succession through coastal sage scrub, chaparral and gymnosperm-dominated forests seen in California, we find a steady progression through matorral, sclerophyllous forest, and upland deciduous or montane sclerophyllous forest (Hoffmann and Hoffmann 1982). Axelrod et al. (1991) suggest that gradual altitudinal turnover of vegetation in central Chile reflects equable climate. With a more oceanic climate and lower extremes in temperature, species can extend their altitudinal ranges more widely, and thus are likely to be found in limited quantities well beyond their typical ranges. Lower latitudinal differentiation in the main types of vegetation in central Chile, in turn, can be related to a shallower temperature gradient along the coast resulting from the strong influence of the colder Humboldt current at the northern end of the mediterranean area (cf. Fig. 3.3).

An important correlate of low vegetation turnover in central Chile is high floristic affinity between the Coast and Andean mountain ranges (Table 3.4). Comparative data on trees for central Chile and California suggest that such floristic affinity is higher in central Chile. In the central Chilean analog, 97% of tree species occur in the Coast range and 86% in the Andean range (Table 3.5) with relatively few species being limited or practically so to the Coast range (e.g., *Gomortega keule*, *Pitavia punctata*, *Nothofagus alessandrii*, *Beilschmiedia spp.*, *Jubaea chilensis*). In the California FP, 95% of forest trees occur on the Coast range and mountains in southwestern California (based on Griffin and Critchfield 1976). Only 65%, however, reach the Sierra Nevada. Many of the proportionately more numerous tree species limited to the Coast range in California have very limited distributions (e.g., *Abies bracteata*, *A. grandis*, *Alnus rubra*, and several species of both *Cupressus* and *Pinus*), a feature also characterizing a number of endemic species in the Chilean coast range, but in much lower degree.

The wider distribution of Chilean trees across the vegetation landscape might also reflect weaker internal biogeographic barriers and steeper altitudinal gradients. The higher central depression, with evidence of forest or woodland in the recent past, undoubtedly would have facilitated easier interchange of tree species between the two mountain ranges than the lower unforested Central Valley in California. Stronger altitudinal gradients in central Chile, on the other hand, would have favored faster temperature and precipitation tracking during any past period of climate change (Arroyo et al. 1993a). Compounded over historical time, and especially in the Pleistocene and Holocene, when both areas experienced floristic upheavals (Heusser 1983; 1989; 1991; Heusser et al. 1985; Villagrán 1990; this volume; Villagrán and Varela 1990), such effects might have led to easier reexpansion of Chilean tree species out of coastal refugia during periods of climatic amelioration.

Table 3.4. Floristic similarity for the Coast and Andean ranges in central Chile according to life form: Predominantly alpine species were discarded

| Life form | Number of species | | | |
	Coast range	Andean range	Common	Floristic similarity[a]
Annuals	251	241	183	0.744
Perennial herbs	844	810	565	0.683
Shrubs[b]	286	272	201	0.720
Trees	68	60	58	0.906

[a] Similarity coefficient: $S = 2a/(b + c)$, where a = number of species common to the two mountain ranges, b = total species in Coast range, c = total species in Andean range.
[b] Includes woody vines and woody epiphytes.

Table 3.5. Genera and number of species per genus of trees in the mediterranean flora in central Chile and their distribution in the Coast Range and Andean Range

Genus	Family	Total trees	Coast range	Andean range
Angiosperms				
Aextoxicon	Aextoxicaceae	1	1	1
Lithrea	Anacardiaceae	1	1	1
Schinus	Anacardiaceae	3	3	3
Pseudopanax	Araliaceae	1	1	1
Trichocereus	Cactaceae	2	2	1
Maytenus	Celestraceae	2	2	2
Dasyphyllum	Asteraceae	2	2	1
Caldcluvia	Cunoniaceae	1	1	1
Weinmannia	Cunoniaceae	1	1	1
Aristotelia	Elaeocarpaceae	1	1	1
Crinodendron	Elaeocarpaceae	1	1	1
Eucryphia	Eucryphiaceae	2	1	2
Nothofagus	Fagaceae	8	8	7
Azara	Flacourtiaceae	2	2	2
Gomortega	Gomortegaceae	1	1	0
Citronella	Icacinaceae	1	1	1
Beilschmiedia	Lauraceae	2	2	0
Cryptocarya	Lauraceae	1	1	1
Persea	Lauraceae	2	2	2
Corynabutilon	Malvaceae	1	1	0
Acacia	Mimosaceae	1	1	1
Prosopis	Mimosaceae	1	1	1
Laurelia	Monimiaceae	1	1	1
Peumus	Monimiaceae	1	1	1
Amomyrtus	Myrtaceae	1	1	1
Blepharocalyx	Myrtaceae	1	1	1
Legrandia	Myrtaceae	1	0	1
Luma	Myrtaceae	2	2	2
Myrceugenia	Myrtaceae	3	3	3
Tepualia	Myrtaceae	1	1	1
Jubaea	Palmae	1	1	0
Adesmia	Papilionaceae	2	2	1
Sophora	Papilionaceae	1	1	1
Embothrium	Proteaceae	1	1	1
Gevuina	Proteaceae	1	1	1
Lomatia	Proteaceae	3	3	3
Kageneckia	Rosaceae	2	2	2
Quillaja	Rosaceae	1	1	1
Pitavia	Rutaceae	1	1	0
Salix	Salicaceae	1	1	1
Escallonia	Saxifragaceae	1	1	1
Rhaphithamnus	Verbenaceae	1	1	1
Drimys	Winteraceae	1	1	1
Gymnosperms				
Austrocedrus	Cupressaceae	1	1	1
Podocarpus	Podocarpaceae	1	1	1
Prumnopitys	Podocarpaceae	1	1	1
Saxegothaea	Podocarpaceae	1	1	1
Totals		70	68	60

Comparative History of the Mediterranean Floras

California

In California the dominant woody elements present today can be traced to the Arcto-Tertiary and Madro-Tertiary geofloras (Axelrod 1973; Raven and Axelrod 1978; Axelrod 1988; 1989; 1992). The humid coastal forests in northern California are the remains of rich Paleocene and Eocene forests that dominated the Pacific coastal strip and inland (Raven and Axelrod 1978). These forests (Arcto-Tertiary Geoflora) had numerous taxa whose nearest allies are now found only in summer rainfall areas in eastern North America or in eastern Asia. Inland, south of about latitude 44° N, the wet forests gave way to gymnosperm-dominated communities with *Abies*, *Picea*, *Pinus*, and *Pseudosuga*, showing direct affinity with the modern Sierra Nevada and Rocky Mountain forests. Development of a summer-dry climate in western North America in the Miocene at around 15 Ma (Axelrod 1992) led to restriction of the earlier Tertiary subtropical forests to equable coastal or cool upland locations with regional extinction in many genera (e.g., *Abies*, *Larix*, *Picea* spp.).

Woody chaparral elements in the California FP, although distributed into northern California and southern Oregon, have their centers of distribution in San Diego-Baja California (Epling and Lewis 1942), with many evident disjunctions today to northern Mexico and Arizona (Raven and Axelrod 1978). Chaparral elements are considered to be derived from the Madro-Tertiary Geoflora (Axelrod 1973; 1992), which the fossil record shows to have dominated in continental locations in western North America with summer rainfall, well before a mediterranean climate developed in California, and is still represented there by vicariant species (Axelrod 1989). The oldest known records of the chaparral alliance are from a continental location in Colorado-Utah in the Middle Eocene. Here chaparral shrubs were considered to have formed part of the forest understory, occurring in pure stands only on drier sites (Axelrod 1989). Sclerophyllous elements were also present in the Eocene at more coastal locations south of the mediterranean zone in Baja California (Raven and Axelrod 1978). In inland southern California and part of the Sonora–Mohave desert region, following a trend for westward migration, the dominant vegetation by the Miocene was sclerophyllous, with live oaks, madrone, and other taxa present. Chaparral as a distinctive type of vegetation did not, however, appear in central California until 3.5 Ma (Axelrod 1989).

Central Chile

Compared with California, the Tertiary fossil record for the Chilean mediterranean area is fragmentary. Evidence suggests, though, that sclerophyllous elements have appeared along a trajectory that might have been quite different from that in California (cf. Solbrig et al. 1977b). As indicated above, many dominant matorral elements also appear as trees in sclerophyllous forest, and unlike in Californian chaparral, the biogeographic center of matorral species probably is in a more mesic part of the gradient. This location provides a salient empirical clue for understanding the comparative biogeography of matorral and chaparral.

Subtropical assemblages covered middle and some far-southern latitudes in southern South America from Paleocene to Miocene with little east–west differentiation across the continent at that stage, judging by the similarity of fossil floras from Lota and Coronel on the coast of Chile (37° S), and Río Pichileufú (41° S) across the Andes (Berry 1922; 1938; Romero 1986). Petrified wood samples resembling the rain forest elements, *Aextoxicon*, *Myrceugenia*, and *Laurelia*, all found north of about 43° S today, and of the now endemic mediterranean genus *Gomortega*, have been described as far south as 51° S in Patagonia (Nishida et al. 1988a,b) for the Late Oligocene–Early Miocene, indicating significant latitudinal contraction of vegetation belts in post-Miocene times.

Recently, Troncoso (1991) described an important Miocene fossil-leaf flora assemblage from the central Chilean coast at Matanzas (33°57′ S) including a rich array of modern temperate rain forest elements or their close relatives, such as *Araucaria*, *Caldcluvia*, *Nothofagus*, *Weinmannia*, *Podocarpus*, *Mitraria*, *Boquila*, *Ovidia*, and *Saxegothopsis*, several extinct woody rain forest taxa of warmer climates (e.g., *Phoebe*), and numerous fern species. Unequivocal evidence of typical sclerophyllous elements is lacking. The assemblage is interpreted as warm rain forest (Troncoso 1991) with affinity to the modern Yungas, Atlantic, and Paraná vegetation provinces of tropical–subtropical eastern South America (Cabrera and Willink 1980), indicating a situation parallel to that seen for rain forest elements in coastal California in the Miocene, although the presence of such genera as *Caldcluvia* and *Ovidia* could indicate cooler conditions. Most woody rain forest elements with clear-cut modern affinities in the Matanzas' flora are presently restricted to more southerly locations. This phase is reflected in the many trees and woody vines in mediterranean latitudes today that are more typically encountered in the southern rain forests (e.g., *Aextoxicon*, *Campsidium*, *Pseudopanax*, *Caldcluvia*, *Lapageria*, *Mitraria*, *Weinmannia*, *Eucryphia*, *Myrceugenia*, *Nothofagus antarctica*, *Laurelia*, *Tepualia*, *Rhaphithamnus*, *Gevuina*, *Podocarpus*, *Prumnopitys*, *Saxegothaea*) (Table 3.5) (see also Arroyo et al. 1993b).

Several Miocene records of *Nothofagus* are available in central Chile (e.g., Nishida 1981; Pons and Vicente 1985; Troncoso 1991). Climatic interpretation is problematic, however, because of the many climates and habitats inhabited by the genus (mediterranean to subantarctic), and presence of strong latitudinal clinal variation in several species (Donoso 1987). The Miocene wood record from an inland location, now above tree limit in the Andean cordillera, at 32°35′ S (Pons and Vicente 1985), with affinity to *N. obliqua* (Pons and Vicente 1985), is a case in point, as is Nishida's (1981) fossil *N. pseudobliquum* from Isla Quiriquina (36°30′ S) for the Miocene. Modern *N. obliqua* occurs from 32° to 41° S through a succession of seasonal rain forest (>2000 mm precipitation) to semiarid mediterranean habitats (ca. 400 mm precipitation) and exhibits notorious latitudinal clinal variation in nut size (Donoso 1979), demographic mode (Casassa 1985), leaf size, and texture (Rodríguez et al. 1983). The *Nothofagus* fossil records could be taken as evidence of some incipient development of sclerophyllous vegetation in central Chile by Miocene times. Considering the range of leaf variation shown in *N. obliqua* today, however, the Miocene material could equally well have come from a seasonal rain forest habitat, with strong development of sclerophylly occurring at a later date.

Considering more closely the present distributions of important genera with sclerophyllous species in the central Chilean flora (e.g., *Azara, Colliguaja, Cryptocarya, Escallonia, Dasyphyllum, Myrceugenia, Kageneckia, Lithrea, Maytenus, Peumus, Quillaja, Schinus*) is revealing. If central Chilean sclerophyllous elements had arisen through the western migration of inland continental vegetation as in California, that source and a concentration of their closest relatives today would be expected in the Chaco vegetation lying immediately to the east of Monte desert (Sarmiento 1972; Cabrera and Willink 1980). Semidesert conditions appeared earliest in the South American subtropics in the Chaco and Monte areas (Mares et al. 1985). Subsequently, however, the Monte underwent further aridification in the Pliocene and Pleistocene (Solbrig et al. 1977a), and although it shares strong floristic ties with the Chaco (Sarmiento 1972), modern imprints of a more widespread sclerophyllous vegetation would be unexpected in the Monte, and in fact are not there today (Sarmiento 1972). Indeed it is the poorly represented shrub genera in the Chilean matorral, such as *Krameria, Larrea, Encelia*, which are directly linked to the modern Monte vegetation. Only a limited number of sclerophyllous elements (e.g., *Lithrea, Schinus, Maytenus*), however, occur in the true Chaco (Sarmiento 1972; Cabrera and Willink 1980; Ramella and Spichiger 1989), although the marginal mediterranean elements, *Acacia* and *Prosopis*, are present in dry Chaco (Sarmiento 1972). Fossil material of *Lithrea* and *Schinus* have recently been recorded for the Pliocene in the Mesopotamian flora of Argentina (Anzotegui and Lutz 1987). These authors stress, however, that the fossil species find their closest allies in the *modern* Chaco species in these genera.

A very typical pattern for some genera with sclerophyllous species is seen in *Myrceugenia, Dasyphyllum, Escallonia, Crinodendron*, and *Azara*. These genera, in addition to occurring in sclerophyllous vegetation in central Chile, have closely related species in the southern rain forests of Chile as well as disjunct members in tropical or subtropical vegetation types in southeastern Brazil, northeastern Argentina, and along the southern margin of the seasonal Planalto. This distribution pattern clearly mirrors the present affinities of the ancient Miocene rain forests of the Chilean coast (Troncoso 1991). Some intervening stations in southern Bolivia and north-central Argentina are typical (Arroyo et al. 1993a,b). *Quillaja* and *Colliguaja* show the typical southeastern Brazilian disjunction pattern, but lack rain forest species in southern Chile. Deviating somewhat from this more typical pattern is *Kageneckia*, with two strongly sclerophyllous species in central Chile and a second disjunct vicariant species in thorn scrub east of the Andes in Peru, Bolivia, and northwestern Argentina on the northern border of the Chaco (which is closely related to central Chilean upland *K. angustifolia*). Wood tentatively referred to *Kageneckia/Quillaja* from the Paleocene of Chile Chico (44° S) (Nishida et al. 1990; Ohsawa and Nishida 1990) opens the possibility that *Kageneckia* might be a genus that developed fairly early, either east of the Andes or over the present site of the Andean axis.

Considering that coastal central Chile was still covered with dense rain forest in the upper Miocene (Troncoso 1991), it is unlikely that typical lowland sclerophyllous taxa appeared much before the Pliocene. By then the Andes had already gained height (Arroyo et al. 1988), and would have been a formidable barrier for passage of lowland species from the east, unlike California, where access has always been fairly easy from continental areas in the south. Although clearly based on very fragmentary evidence, we suggest the hypothesis that, from the timing of events and the stronger biogeographic barriers in South America, many typical sclerophyllous elements of the central Chilean flora, unlike with California, evolved directly in situ from ancient rain-forest stock.

Any modification of this hypothesis would come from better knowledge of the vegetation history of the Atacama desert and adjacent areas immediately east of the Andes. The chronology of the Atacama desert, although still poorly understood, suggests that dry climates appeared only in the middle Miocene (cf. Rundel et al. 1991). Some drier elements found in sclerophyllous vegetation today thus could have evolved at an earlier period over the present location of the Atacama Desert, only to migrate south to the present area of the mediterranean region as even drier climates appeared in the Pliocene. Disjunct populations of *Colliguaja odorifera* (Marticorena and Quezada 1991) and *Proustia cuneifolia* (Rundel et al. 1991) in the fog zones on Cerro Moreno (20° S) and Paposo (25° S), respectively could be interpreted as support for this hypothesis. Such disjunctions could also have formed, however, during some wetter period in the Pleistocene (cf. Graf 1992).

Life Forms and the Annual Habit

Mediterranean-climate vegetation is probably more diverse in life forms than any other kind. This diversity is accounted for by maintenance of ancient woody groups, evolution of many shorter-lived herbaceous species responding to opening up of the vegetation in the late Tertiary, and the summer-dry climate per se that has selected for drought-tolerant (e.g., geophytes) and

Table 3.6. Incidence of the woody and herbaceous habits in central Chile and the California FP: Data from Table 3.2

Region	Genera	Species	Percentage species	Spp./genus
Central Chile[a]				
Herbaceous	437	1894	79.2	4.3
Woody	180	496	20.7	2.8
California FP				
Herbaceous	671	3663	86.4	5.5
Woody	171	577	13.6	3.4

[a] Life forms uncertain in five species; Central Chile vs. California FP: Woody vs. herbaceous: $G = 56.02$, $df = 1$, $p < 0.001$.

drought-evading (e.g., annuals) life-history strategies. The ancient woody groups display numerous interesting morphological adaptations such as sprouting capacity (Zedler and Zammit 1989) and lignotuber development (Montenegro et al. 1983), which further increase the life form diversity.

Raunkiaer (1934) initially suggested that mediterranean climates should have high percentages of annuals that survive the long, dry summer season as seeds. Consistent with the latter are the 51% annuals in Israel (Eig 1931/32) and 27% annuals in the entire California Floristic Province (including its extensions into southern Oregon and northern Baja California) (Raven and Axelrod 1978). Danin and Orshan (1990) illustrated, moreover, that the annual habit increases in frequency with decreasing precipitation up to a critical point in Israel. A similar situation was reported by Arroyo et al. (1988) for another arid area, the high-elevation desert in northern Chile. The mediterranean areas in Australia and South Africa, however, feature relatively few annuals and thus deviate strongly from theoretical expectation. Annuals are poorly represented in the fire-prone Australian mallee (M. Fox, personal communication). Hopper (1979) suggests 7% annuals for the southwest of Western Australia, including mediterranean vegetation and some humid forest. In fire-adapted mediterranean vegetation in the South African Cape Province only 6.4% of the flora reportedly is annual (Bond and Goldblatt 1984).

The life-form spectra of the vascular plant floras in central Chile and the California FP are shown in Table 3.2. (suffrutescent species were placed in the perennial herb category and woody vines in the shrub category; arborescent cacti were grouped with trees.) Finer life form division proved difficult because of variability in the bibliographic information. The life form spectra prove to be statistically different (Table 3.2). A fundamental difference between the two floras is a higher contribution of woody taxa in central Chile (Table 3.6). Trees are represented in similar proportions in the two regions, but the general category of woodiness is probably more reliable for comparison because of strong habit variation in sclerophyllous vegetation.

The Annual Habit

Another profound difference is seen in contribution of annual species (Tables 3.2, 3.7). Chile (16%) has proportionately far fewer native annual species than the California FP (30%). Indeed, the frequency of annuals in the Chilean mediterranean flora is remarkably similar to that in the entire flora, including the northern deserts and the southern temperate rain forests (Arroyo et al. 1990). The difference revealed for central Chile and the California FP clearly cannot simply be the result of a few genera with outstanding capacity to produce annuals in California because the same result is echoed at the generic level (Table 3.7). That is, proportionately more native genera in the California FP contain annuals than in central Chile.

That genera with native annuals are proportionately fewer in central Chile is consistent with an overall reduced tendency for lineages in general to evolve annuals in the central Chilean flora. It is generally agreed that the evolutionary pathway to the annual habit must involve a prior, short-lived perennial herb intermediary. Consequently, in strongly woody floras there will be constraints on the evolution of the annual habit. We have just shown that woodiness is proportionately overrepresented in the central Chilean flora. Such excess could result from historical effects such as less extinction, as proposed earlier, or intrinsic selection for the woody habit per se for some ecological reason, or both.

Separation of these last confounding effects can be achieved by assessing the frequency of the annual habit in the herbaceous flora, discarding all genera that include only woody species. When this procedure is followed, the trend for fewer native annuals in the central Chile flora continues to persist

Table 3.7. Incidence of the annual habit in the mediterranean floras of Chile and California: Adopted from Arroyo et al. (1990) for species occurring in mediterranean area. N = annuals

Geographical area	Genera			Species		
	Total	N	%	Total	N	%
Central Chile						
Dicotyledons	432	134	31.0	1807	351	19.4
Monocotyledons	127	16	12.6	520	26	5.0
Gymnosperms	5	0	0	6	0	0
Ferns	27	1	3.7	57	1	1.8
Total	591	151	25.5	2390[a]	378	15.8
California FP						
Dicotyledons	627	239	38.1	3307	1187	35.9
Monocotyledons	136	34	25.0	788	92	11.7
Gymnosperms	14	0	0	54	0	0
Ferns	29	0	0	91	0	0
Total	806	273	33.9	4240	1279	30.2

[a] Life forms of five species additional in central Chile uncertain; $G = 176.49$, $df = 1$, $p < 0.001$ (Total flora-species); $G = 11.27$, $df = 1$, $p < 0.001$ (Total flora–genera).

Table 3.8. Incidence of the annual habit among herbaceous genera and species in the mediterranean floras of Chile and California

	Herbaceous genera			Herbaceous species		
Geographic area	Total	With annuals	% with annuals	Total	Annuals	% annuals
Central Chile	437	151	34.6	1894	378	20.0
California FP	671	273	40.7	3663	1279	34.9

[a] $G = 4.23$, $df = 1$, $p < 0.05$ (genera); $G = 139.18$, $df = 1$, $p < 0.001$ (species).

both at the species and generic levels (Table 3.8). This result indicates that the propensity for fewer annuals in the mediterranean flora of central Chile is a deeply ingrained feature of that flora and not merely a frequency-dependent effect of a strongly woody flora.

The earlier climatic analysis (p. 49) suggested that central Chile is regionally less arid than the California FP. Fewer annuals in the flora of central Chile at the regional level thus could reflect this lower degree of aridity. Richerson and Lum (1980) provided data for the sum total of annuals (native and introduced) occurring in California counties. Comparing the frequency of the annual habit in some Coast-range counties in California and 1° areas of the Coast-range area in central Chile with a standard aridity index (Emberger's Pluviometric Quotient, 1955) shows that annuals are consistently less frequent in Chile than in the California FP when aridity is controlled for (Fig. 3.5a). The relation also holds (not shown graphically) when annuals are considered as a percentage of the herbaceous flora. This relation suggests that absolute aridity is unlikely to be the critical factor behind the widely different frequencies of the annual habit in the two mediterranean regions. The hiatus in annual frequencies would probably be even more dramatic if the comparison could have been made for native species only. Higher levels of disturbance throughout central Chile have been shown to be associated with proportionately more introduced annuals locally than in California (Mooney et al. 1977).

Because the amount of summer rainfall for any total amount of rainfall tends to be higher in central Chile, a given level of aridity as measured by the Emberger quotient could be misleading for the stressful period of the year. One way of getting around this problem is to consider an aridity index based only on the summer months. This choice is clearly artificial for true availability of water, but because annuals respond mostly to lack of water in the surface layers of the soil during the drier period, it probably provides a better measure of aridity as perceived by annuals in an evolutionary sense. When a measure of summer aridity (Fig. 3.5b) is considered, some convergence appears for the wettest areas (areas of relatively low summer aridity)—that is, in the humid extremes of the two mediterranean areas. Nevertheless, the hiatus is maintained at the arid extremes (Fig. 3.5b), where as much as a 10%

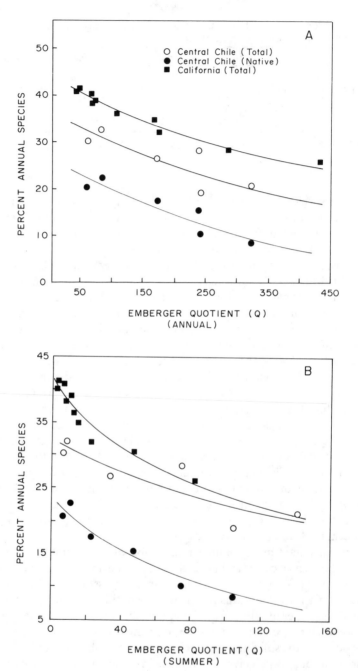

Figure 3.5. Incidence of annual habit in the Coast range areas of central Chile and California according to degree of annual aridity (A) and summer aridity (B), as expressed by Emberger's quotient (Q) (Emberger 1955). Q decreases with increasing aridity. Data are for 1-degree intervals in Chile and for selected counties in California per degree of latitude, as reported in Richerson and Lum (1980).

difference in the contribution of annuals persists. This difference remains even though introduced annuals are probably contributing more locally to the annual count in Chile than in California.

That annuals continue to be much less frequent in the semiarid extreme of central Chile, even when summer aridity is controlled for, suggests that some other factor besides macroclimate must be controlling diversification of annuals in central Chile and California. One possibility is clearly the fire regimen. The Californian chaparral has a large number of fire-adapted annuals that use fire-stimulated germination (Keeley 1991) as a cue to recognize availability of open, noncompetitive sites. One study suggests that the non–fire-prone herbaceous vegetation in central Chile fails to respond markedly to artificially set fires (Keeley and Johnson 1977). With fire as a recurrent form of disturbance, as in California, the annual habit would be highly favored. Strong differences in the frequency of the annual habit in the semiarid extremes in the two mediterranean areas would be consistent with this hypothesis in that fire tends to be commoner in southern California.

To take this issue into a more conjectural domain, we argued that if fire or some other exogenous factor has been important in selecting for the annual habit in the California FP, shared genera should be less likely to include annual species in central Chile. Remarkably, the incidence of the annual habit is statistically indistinguishable among shared genera in central Chile and California (Table 3.9). Of the respectively 99 and 111 shared genera occurring in central Chile and California that have annuals in those regions, 82 are the same genera in both regions, indicating that shared genera have similar incidence of the annual habit because many genera have common behavior. This impressive list of genera are represented by native annuals in both regions: *Amaranthus* (Amaranthaceae), *Agoseris*, *Bidens*, *Blennosperma*, *Conyza*, *Eclipta*, *Gnaphalium*, *Helenium*, *Lasthenia*, *Madia*, *Microseris*, *Psilocarphus*, *Xanthium* (Asteraceae), *Bowlesia*, *Daucus* (Apiaceae), *Amsinckia*, *Cryptantha*, *Pectocarya*, *Plagiobothrys* (Boraginaceae), *Descurainia*, *Lepidium*, *Rorippa* (Brassicaceae), *Downingia*, *Triodanis* (Campanulaceae), *Minuartia*, *Sagina*, *Stellaria* (Caryophyllaceae), *Atriplex*, *Chenopodium* (Chenopodiaceae), *Crassula* (Crassulaceae), *Cyperus*, *Scirpus* (Cyperaceae), *Euphorbia* (Euphorbiaceae), *Centaurium*, *Cicendia*, *Gentiana*, *Gentianella* (Gentianaceae), *Geranium* (Geraniaceae), *Alopecurus*, *Aristida*, *Bromus*,

Table 3.9. Incidence of annual habit in shared and unshared genera in the mediterranean floras of central Chile and California

Geographic area	Shared genera	With annuals	Unshared genera	With annuals
Central Chile	233	99 (42.5%)	358	52 (14.5%)
California FP	233	111 (47.6%)	573	162 (28.3%)

$G = 1.24m$, $df = 1$, NS (shared genera); $G = 24.67$, $df = 1$, $p < 0.001$ (unshared genera); $G = 57.30$, $df = 1$, $p < 0.001$ (Chile—shared vs. unshared genera); $G = 26.95$, $df = 1$, $p < 0.001$ (California—shared vs. unshared genera).

Deschampsia, Eragrostis, Muhlenbergia, Phalaris, Vulpia (Poaceae), *Phacelia* (Hydrophyllaceae), *Hypericum* (Hypericaceae), *Juncus* (Juncaceae), *Lilaea* (Lilaeaceae), *Camissonia, Clarkia, Gayophytum, Oenothera* (Onagraceae), *Argemone* (Papaveraceae), *Astragalus, Lotus, Lupinus, Trifolium, Vicia* (Papilionaceae), *Plantago* (Plantaginaceae), *Collomia, Gilia, Linanthus, Navarretia, Polemonium* (Polemoniaceae), *Lastarriaea, Oxytheca* (Polygonaceae), *Calandrinia, Montia* (Portulacaceae), *Myosurus, Ranunculus* (Ranunculaceae), *Aphanes* (Rosaceae), *Galium* (Rubiaceae), *Limosella, Lindernia, Mimulus, Orthocarpus* (Scrophulariaceae), *Nicotiana, Solanum* (Solanaceae), *Parietaria* (Urticaceae), and *Plectritis* (Valerianaceae).

Unlike the situation seen in shared genera, the sets of unshared genera differ significantly for presence of annuals, with strong statistical over-representation of the annual habit in the California FP (Table 3.9). The unpredicted behavior of shared genera in the two floras, because of the hypothesis entertained, could be interpreted as evidence for strong phyletic control of evolution of the annual habit, with shared genera showing a similar propensity to evolve annuals because of expression of a deeply ingrained phylogenetic trend, regardless of differences in the fire regime or other exogenous factors. The shared genera, however, also show a statistically higher tendency to have annuals than either set of unshared genera in the two regions (Table 3.9). This tendency is undoubtedly a consequence of the propensity to be annual, which increases the probability of being a shared genus via the amphitropical dispersal pattern. This last situation unfortunately, confounds testing the phylogenetic constraints hypothesis—the dispersal filter, in favoring the annual habit, has led to the annual habit already being fixed in many shared genera in at least one geographic area. Such fixation would tend to impede detecting a similar propensity to evolve annuals locally under two sets of environmental conditions, as we had hoped to be able to test.

This difficulty in no way negates that some form of phylogenetic determinism is operating in the shared genera—it only makes such determinism more difficult to prove. If indeed phylogenetic determinism has contributed in the shared genera, then we must also entertain the corollary that the different propensities to evolve annuals in the unshared genera have a strong phylogenetic basis. Because unshared genera in central Chile exhibit a much lower tendency to evolve annuals, it is plausible, but certainly cannot be proved at this stage, that there have been much stronger phylogenetic constraints on the evolution of the annual habit in central Chile than in the California FP.

Taking this last possibility into account leads to an alternative hypothesis worthy of exploration for the overall higher frequency of native annuals in the flora of the California FP. This trend might first reflect a more distal effect for genera in California to be more predisposed to evolve annuals in general. Fire and other factors could have further stimulated the evolution of annuals in California, but hierarchically speaking, these last factors might have been less important than intrinsic phylogenetic predisposition.

Overall Life-Form Spectrum

The distribution of life forms other than annuals along the rainfall gradient has not yet been studied in detail. Figure 3.6, however, shows a principal-components analysis of the overall life form spectrum (percentage of trees, shrubs, perennial herbs, and annual herbs, respectively) for areas of the Coast and inner cordilleras in central Chile and California. Unlike the strong overlap previously shown for climate (Fig. 3.4), the life form spectra tend to form two partially overlapping clusters. Close matches are sometimes seen where climate is evidently very different, as between Klamath County on the northern border of California (41° N) and the Andean range at 33° S. Klamath County is one of the wettest areas in the California FP, Klamath itself receiving 2066 mm rainfall. The Andes at latitude 33° S, typified by San José de Maipo, 1060 m altitude, are in the semiarid area of the mediterra-

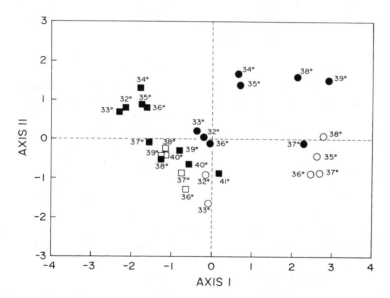

Figure 3.6. Principal-component analysis of life forms (including native and introduced species) for locations at different latitudes in the general mediterranean areas in central Chile and the California FP. Number with each data point indicates latitude. ● Central Chile-Coast range: ○ central Chile-Andean range; ■ California-Coast range; □ California-Sierra Nevada. Data for Chile are for 1 degree intervals corresponding to the latitudes indicated. Data for California from Richerson and Lum (1980): Coast range: Del Norte County (41° N); western Humboldt County (40° N); western Mendocino County (39° N); western Sonoma County (38° N); San Mateo-San Francisco Counties (37° N); western Monterey County (36° N); western San Obispo County (35° N); western Santa Barbara County (34° N); Orange County (33° N); western San Diego County (32° N). Sierra Nevada: Tehama County (40° N); Butte County (39° N); El Dorado County (38° N); East Tulare County (37° N); Mid Kern County (36° N).

nean region, and receive only 623 mm precipitation annually. Clearly, little overall convergence appears in the life form spectra in the floras of central Chile and California.

Patterns of Diversity

Species and Generic Richness

Some mediterranean regions have proven to be exceptions to the widely documented trend for a general increase in species richness from polar to tropical latitudes. Foremost among these is the Cape Region in South Africa, with more than 8000 species, which is claimed to compare with some tropical regions for species density (Goldblatt 1978; Major 1988; Cowling et al. 1989; 1992) and the Southwest Botanical Province in Western Australia, with an estimated 8000 species (Hopper 1992). Greuter (1991) recently estimated 24,000 + / − 600 vascular plant species for the entire, much larger Mediterranean Checklist area (countries bordering the Mediterranean Sea, plus Portugal, Bulgaria, the Crimea, and Jordan), based on an extrapolation from 44.5% species covered to date in the Med-Checklist. Raven and Axelrod (1978) gave the figure of 4452 vascular plant species for the entire California Floristic Province. Historical explanations for these trends include coexistence of paleoendemics and neoendemics, with explosive speciation in some herbaceous and woody groups following the opening up of the woody matrix upon development of a seasonally dry climate (Raven and Axelrod 1978).

Consistent with the latter, in the California FP as defined here, eight genera have more than 50 species: *Juncus* (51), *Arctostaphylos* (57), *Astragalus* (57), *Mimulus* (57), *Lupinus* (60), *Phacelia* (70), *Erigeron* (82), and *Carex* (126). The much smaller central Chilean analog has three genera with more than 50 species: *Calceolaria* (58), *Adesmia* (82), and *Senecio* (109). Important taxonomically isolated woody monotypic genera in central Chile include *Gomortega*, *Pitavia*, *Legrandia*, and *Peumus*, and in California, *Lyonothamnus*, *Umbellularia*, and *Sequoiadendron* are noteworthy. In California such monotypic annual genera as *Blepharizonia* and *Holozonia* stand out, as do *Agallis* and *Microphyes* in Chile.

The factors determining the size of any regional species pool nevertheless are complex, and involve more than history. Alpha, beta, and gamma diversity also impinge on the final number of species in a region (Cowling et al. 1992). Central Chile and California are areas of high physiographic diversity, central Chile having more extreme relief with steeper altitudinal gradients than California. Although physiographic diversity is more extreme in Central Chile, east−west climatic differentiation was shown to be less pronounced, and climate generally more equable. Thus any expected increase in beta diversity from higher physiographic diversity in central Chile could be counterbalanced by shorter climatic gradients. The importance of beta diversity in the California flora is clear from the work of Richerson and Lum (1980). These authors showed that species richness was highest in mountain areas with high topographic and climatic diversity. Unfortunately, however, no appropriate data in California and central Chile are available at this stage for comparing beta and other higher-order diversity.

Although many local areas in central Chile are still poorly collected, species accumulation curves approach an asymptote at the regional level considered in this paper (Maldonado et al. 1993), suggesting that any estimate of total flora for central Chile will be fairly reliable. Our survey suggests that the native vascular flora in central Chile for an area of 104,000 km² conservatively has some 2395 species (2332 angiosperms, 6 gymnosperms, and 57

Table 3.10. Number of native species, genera, and the species/genus ratio for different-sized areas in Chile and the California FP

Source	Area (km²)	Number species	Number genera	Species/ genus ratio	
Chile					
La Plata valley	10	249	163	1.53	(1)
Marga Marga valley	450	457	279	1.64	(2)
Concepción area	2100	672	346	1.94	(3)
Valparaíso area	3300	799	356	2.24	(3)
Santiago valley	4000	654	304	2.15	(4)
Valparaíso and Metrop. Regions	31,794	1627	488	3.33	(3)
Coast range	38,131	1514	515	2.94	(3)
Andean range	38,500	1777	483	3.68	(3)
Central Chile	103,740	2395	591	4.05	(3)
Winter rainfall area	294,608	3385	681	4.97	(3)
California					
San Bruno Mts.	12	398	243	1.64	(5)
Tiburon Peninsula	15	370	NA	NA	(6)
San Francisco Co.	117	640	NA	NA	(6)
Mt. Diablo Co.	146	525	NA	NA	(6)
Santa Monica Mts.	828	642	336	1.91	(7)
Marin Co.	1369	1004	413	2.43	(8)
Orange Co.[a]	2024	568	296	1.92	(9)
Santa Cruz Mts.	2590	1246	460	2.71	(10)
Santa Barbara region	7085	1390	450	3.09	(11)
San Luis de Obispo Co.	8570	1287	NA	NA	(12)
Monterey Co.	8650	1224	433	2.83	(13)
San Diego Co.	11,030	1516	551	2.75	(14)
Kern Co.	21,200	1463	468	3.13	(15)
Coast range	63,458	2525	NA	NA	(6)
California FP	278,000	4240	806	5.26	(16)

[a] Pteridophytes not included; see Figure 3.1 for location of Concepción and Valparaíso areas.
Sources: (1) Schlegel (1966); (2) Jaffuel and Pirion (1921), and additional herbarium records; (3) Unpublished checklists; (4) Navas (1973; 1976; 1979); (5) McClintock et al. (1990); (6) Johnson et al. (1968); (7) Raven et al. (1986); (8) Howell (1970); (9) Boughley (1968); (10) Thomas (1961); (11) Smith (1976); (12) Hoover (1970); (13) Howitt and Howell (1964); (14) Beauchamp (1986); (15) Twisselmann (1967); (16) Hickman (1993).

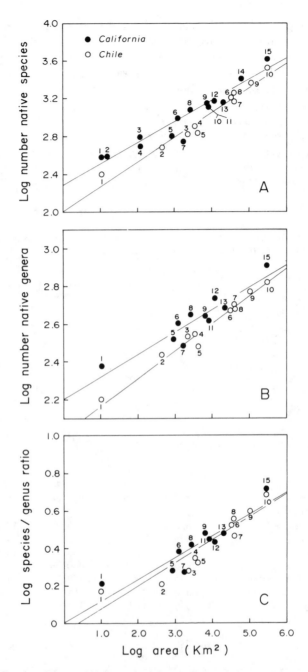

Figure 3.7. Species richness (A), generic richness (B), and the species/genus ratio (C) versus area for central Chile and the California FP. *Chile*: (1) Quebrada de la Plata, (2) Valle de Marga Marga, (3) Concepción area, (4) Valparaíso area, (5) Santiago valley, (6) Valparaíso and Metropolitan Regions, (7) Coast range, (8) Andean range,

pteridophytes) in 591 genera, compared with 4240 native vascular plant species (4095 species of angiosperms, 54 gymnosperms, 91 pteridophytes) in 806 genera for an area of 278,000 km² in the California FP (Table 3.10). Important climatic differences were seen for central Chile and the California FP. But it is also true that these two areas reputedly have the most similar climates in the world. How then, do they compare for species and generic richness?

Table 3.10 assembles data for native species, native genera, and the native species/native genus ratio for different-sized areas in the two mediterranean areas. Some local effects in Chile are interesting. The entire Coast range, which is smaller in area than the inner Andean range and has negligible alpine vegetation, has many more genera and only 263 fewer species than the entire Andean range. The Santiago valley has fewer species and genera than the much smaller Concepción and Valparaíso areas. These differences could be interpreted as evidence for higher richness in the Coast range than inland in Chile. The Coast range as defined here, however, refers to a general geographic region including areas to sea level. The inner Andean range begins at a higher elevation. Likewise, the Santiago valley is situated at an average elevation of around 600 m, whereas the Valparaíso and Concepción areas are directly adjacent to the Pacific coast. Because altitudinal declines in species richness are to be expected, more detailed data would be needed to conclude that species richness is intrinsically higher in the Coast range.

More species were expected in the combined Valparaiso and Metropolitan Regions traversing the Coast range, Central Valley, and Andean range, than in the more uniform entire Coast range of similar area. Richness is very similar, however (Table 3.10). The earlier findings of high floristic affinity

---

Figure 3.7 (*cont.*)
(9) central Chile, (10) Winter rainfall area. *California FP*: (1) San Bruno Mts. (2) Tiburon Peninsula (3) San Francisco County, (4) Mt. Diablo County, (5) Santa Monica Mts., (6) Marin County, (7) Orange County, (8) Santa Cruz Mts., (9) Santa Barbara region, (10) San Luis de Obispo County, (11) Monterey County, (12) San Diego County, (13) Kern County, (14) Coast range, (15) California FP. See Table 3.10 for sources of data. Regression lines: *Species richness*: Chile: Log $S = 2.013 + 0.262 \log A$ ($F = 158.66$, $df = 8$, $p < 0.001$); California: Log $S = 2.268 + 0.223 \log A$ ($F = 107.33$, $df = 13$, $p < 0.001$); *Generic richness*: Chile: Log $S = 2.062 + 0.138$ Log A ($F = 312.72$, $df = 8$, $p < 0.001$); California: Log $S = 2.215 + 0.116 \log A$ ($F = 37.00$, $df = 8$, $p < 0.001$); *Species/genus ratio*: Chile: Log $S = -0.050 + 0.123 \log A$ ($F = 49.79$, $df = 8$, $p < 0.001$; California: Log $S = 0.012 + 0.115 \log A$ ($F = 36.25$; $df = 8$, $p < 0.001$). Slopes not significantly different in all cases (Species richness: $Fs = 1.569$; Generic richness: $Fs = 1.272$, Species/genus: $Fs = 0.105$). Intercepts significantly different for species richness ($Fs = 8.174$) and generic richness ($Fs = 9.675$). Intercepts for species/genus not significantly different ($Fs = 1.035$) (Sokal and Rohlf 1981, Chapter 14).

between the major physiographic units in central Chile are consistent with this finding.

Figure 3.7 shows log-log plots of the double-logarithmic form of the power function (Williamson 1988), $\text{Log } S = k + z \text{ Log } A$ for central Chile and the California FP, where S = species or genus number, or the species/genus ratio; A = area in km^2, and k, z are constants. The three parameters and area were strongly linearly correlated under this model. Slopes of the curves for species richness, generic richness, and the species/genus ratio are not significantly different in comparing central Chile and California (Fig. 3.7). The intercepts for species and generic richness were significantly different, but not for the species/genus ratio. These results indicate that central Chile is less rich than the California FP for both species and number of genera, but nevertheless, on average, the overall pattern of species accumulation per genus has been similar.

Earlier we showed that woody species and genera are proportionately overrepresented in central Chile compared with the California FP, and annuals are underrepresented. Because intrinsic species and generic richness are lower in central Chile, annuals clearly have to be absolutely less abundant in Chile. With lower intrinsic species richness, though, a smaller absolute-sized woody flora becomes possible. Considering the regression curves allows us to detect whether Chile has more woody species and genera on an absolute scale.

Figure 3.8 shows the observed and the expected number of woody species and genera in areas ranging from 2100 km^2 to 103,740 km^2 in central Chile. The expected numbers are based on the Chilean regression equations and the percentage contribution of woody species and genera in the total flora. Also shown are the numbers of woody species and genera for areas of identical size in the California FP, based on the corresponding regressions and percentage contribution of woodiness in that flora. The expected numbers of woody genera and species for Chile are very similar to the observed numbers, indicating that extrapolations based on the proportional representation of the woody and herbaceous habit in the entire flora are valid. For the range of areas considered, it can be appreciated that woodiness is always more strongly represented in central Chile (Fig. 3.8). For example, an area in the California FP the size of central Chile can be expected to include 331 woody species and 133 genera with woody species, as opposed to 496 species and 180 genera observed in central Chile. A much smaller area, equivalent to the size of the Valparaíso coast area (3300 km^2) in California, can be expected to have 154 woody species in 89 genera, compared to an observed 170 species in 99 genera in central Chile. Thus, interestingly, although total species and generic richness is intrinsically lower in central Chile, woody species and genera still far outnumber those in California, considering equal areas.

Several factors may be proposed to explain lower overall richness in central Chile. Woody taxa in the Chile mediterranean area, and in general, are typically dioecious or genetically self-incompatible (Peralta et al. 1992; Arroyo

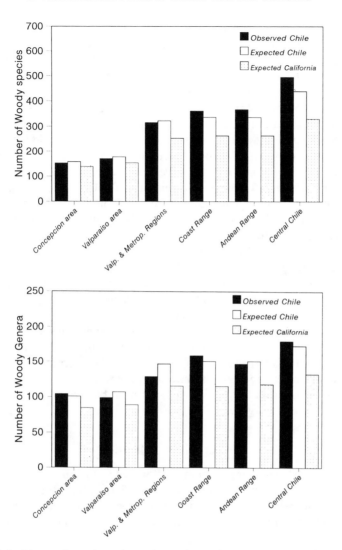

Figure 3.8. Observed number of woody species and genera in areas of various sizes in central Chile compared with the expected numbers for equivalent-sized areas in the California FP based on regression equations in Fig. 3.7. Also given are the expected numbers for central Chile based on the Chilean regression lines. For area size see Table 3.10.

and Uslar 1993), and can be expected to have low speciation rates. This feature is corroborated by the generally lower species/genus ratios in relation to herbaceous genera in both mediterranean areas (Table 3.2). Thus, with a generally woodier flora, fewer species might be expected from the start. It is tempting to relate lower overall richness in central Chile to the greater bio-geographic isolation of central Chile. Central Chile is strongly isolated to the east by the Andes and to the north by the absolute Atacama Desert (Arroyo et al. 1988). The isolation argument, however, is clearly untenable for woodiness where the numbers of woody species and genera are higher than in the California FP. Isolation might, however, have affected the potential size of the herbaceous flora. The species/genus ratio for the woody flora is lower than the overall species/genus ratio in central Chile, yet this last ratio does not differ significantly in the two regions. This situation indicates that herbaceous genera in central Chile, on the average, must be contributing more species than in the California FP. Direct support for this conclusion is seen in the very similar species/genus ratio for perennial herbs in central Chile (4.3) in relation to California (4.7), in circumstances that central Chile comprises a much smaller area. Possibly fewer herbaceous genera have been able to enter Chile because of stronger biogeographic barriers, with those genera present accumulating greater numbers of species individually. In summary, lower overall species richness in central Chile would seem to be predominantly an effect of life-history constraints on speciation rates in woody taxa and far fewer faster-speciating herbaceous genera in the flora. Overall lower generic richness probably reflects the stronger biogeographic barriers, as they affect herbaceous genera.

Generic Diversification

One may also ask whether genera in the floras of California and central Chile have shown similar diversification patterns. That is, have the sets of genera in each region behaved similarly for their speciation patterns? One way of assessing this behavior is to determine how the taxonomic information is

Table 3.11. Taxonomic diversity for genera in the mediterranean floras in central Chile and California, measured by $H' = -\Sigma p_i \ln p_i$, where p_i is the proportion of the total species in the ith genus. H'_{max} is maximum diversity (equal numbers of species per genus throughout). Original data are those in Table 3.2

| | Central Chile | | California FP | |
Life form	H'	H'/H'_{max}	H'	H'/H'_{max}
Annuals	4.645	0.926	4.911	0.875
Perennial herbs	5.161	0.879	5.475	0.880
Shrubs	4.204	0.846	4.197	0.840
Trees	3.672	0.954	3.273	0.870
Total flora	5.554	0.870	5.886	0.879

distributed among the component genera or plant families. We resort to the Shannon information index, $H' = -\Sigma p_i \ln p_i$, where p_i is the proportion of species in the ith genus, which allows a first approximation (Table 3.11). All other things being equal, a high H' value indicates that species are distributed equitably among the component genera, or that genera have accumulated species at relatively similar rates. In contrast, a low H' value indicates that some genera have contributed outstanding numbers of species. Relative taxonomic diversity, calculated as observed H' divided by maximum diversity (maximum entropy *sensu* Pignatti and Pignatti 1985) can be compared for floras of different sizes. In spite of the very different absolute sizes of the central Chilean and Californian floras, relative H' shows strong convergence for the perennial herb and shrub life forms. Species of annuals and trees, nevertheless, are more equitably distributed among their respective genera in central Chile compared with the California FP. This distribution indicates that fewer genera of trees and annuals in central Chile are making exaggerated contributions to the respective species pools than in the California FP. Relative diversity considering the total floras, however, is very similar, indicating that overall patterns of generic diversification have been remarkably similar in the two floras, in spite of the richness differences seen earlier.

Regional Versus Local Species Richness

Ricklefs (1987) suggest that regional richness and alpha diversity may be coupled. Bond (1983) also drew attention to the size of the regional species pool in determining alpha diversity of fynbos communities. Because historical processes impinge on the size of the regional species pool, local species richness on climatically (and topographically) similar sites need not converge. The converse notion is that higher alpha diversity could be an indicator of high regional species richness. To date, in this paper, we have seen that combined historical, climatic, and perhaps life-history constraints, have precipitated lower overall species richness in central Chile in relation to the California FP. Yet woody species richness is intrinsically higher in central Chile.

Limited published data on species richness at small scales are available for the mediterranean area in central Chile (e.g., Mooney et al. 1977; Montenegro in Specht 1988). Interpreting available data, moreover, is somewhat troublesome, in that the data given for 0.1-ha plots are often the total values from independent 100 m^2 plots (e.g., Mooney et al. 1977), in which case true alpha diversity is probably overestimated. Nevertheless, these data tend to suggest that both overall species richness and woody species richness are higher at the alpha level in central Chile than in California. Because central Chile has more woody taxa at both the biogeographical and local scales, there seems to be some support for Ricklef's (1987) hypothesis. The lack of correspondence for total species richness, on the other hand, may reflect many more introduced herbaceous species in Chile, artificially increasing the total species number. Unfortunately, this hypothesis cannot be explored further at this time, for the proportion of introduced species in the Chilean plots is not given.

An associated question is whether the species–area curves generated from biogeographic data predict the number of species found on local vegetation plots. The Californian curve predicts that an area of 0.1 ha would have around 40 species of plants. This prediction compares very favorably with an average of 36 species given in Specht (1988). The Chilean curve, however, predicts only around 17 native species in total, and 4 woody species for 0.1 ha. These last numbers underestimate grossly local alpha diversity in central Chile. Even if the introduced species were removed from the small plots, there can be little doubt that a greater total and many more woody species would remain. Thus, although the sizes of the regional woody species pool in central Chile and California correctly predict more woody species at an alpha level in central Chile, the regression equation for Chile fails to generate the numbers of species expected at the alpha level.

Discussion

Unlike the small, climatically matched sites considered originally by Mooney et al. (1977), the larger mediterranean analogs considered here were seen to be climatically equivalent only at the broadest level. The primary forcing factors of climate, operating singly or in combination, lead to differences in the slopes of the rainfall gradients, degree of aridity, and equability, with a consequent nonsystematic pattern of climatic overlap. We cannot overstate the relevance of the latter. The organisms in small climatically matched sites on different continents, especially if they are long-lived and slow-evolving, will represent responses to a much broader array of climatic conditions than experienced in those sites. If local sites are drawn from two larger areas, less than identical in climate on the average than the climatically matched sites themselves, as for Chile and California, strong convergence on small climatically matched sites will not necessarily be expected. Central Chile is regionally less arid than the California FP. Interestingly, three of four sclerophyllous Chilean species on the convergent evolution site have much greater levels of water-tissue stress than their Californian counterparts (Dunn 1975), suggesting modal species adaptation to overall cooler biogeographic conditions. The former situation brings up an interesting dilemma! Because it will always be difficult to find on different continents, areas of identical climate big enough to be the evolutionary domain of many species, a rigorous test of the convergence hypothesis becomes almost impossible.

The overrepresentation of woodiness in central Chile could merely be the result of less extinction due to easier migration over the Chilean physical landscape. Lower regional aridity might also come into play, however. The difference in aridity nevertheless is not too great, and if woodiness does indeed reflect lower aridity, the increase in Chile is remarkable, and suggests that small differences in climate can profoundly affect composition of life forms. A more probable explanation for a more strongly woody flora in central Chile relates to equable climate. Such climate might have favored maintenance of more older woody tropical lineages in general in the Chilean mediterranean flora as climatic conditions changed in the upper Tertiary (Axelrod et al. 1991) because of less stringent selection for cold hardiness in the Southern Hemisphere. Indeed, Mooney (1977) shows that native trees in central Chile are more sensitive to frost than their Californian counterparts. Strong woodiness also characterizes the higher-elevation deserts in northern Chile (Arroyo et al. 1988), where annual temperature range is low, suggesting a regional trend.

The higher proportion of angiosperm trees in the mediterranean of central Chile and in the southern rain forest flora compared with western North America (Arroyo et al. 1993b) is probably another distal manifestation of equable climate (Axelrod et al. 1991). Northern Hemisphere conifers show greater winter hardiness than angiosperms (Sakai et al. 1981). Winter hardiness is low in Southern Hemisphere angiosperms (Sakai et al. 1981). Over evolutionary time, because of differences in continentality determined by the contrasting land/ocean ratios in North and South America, these ecophysiological differences have led to hemispheric differences in the distribution of angiosperm and gymnosperm trees, with greater retention of angiosperms in the equable southern climates (Sakai et al. 1981; Axelrod et al. 1991).

A coherent explanation for the lower frequency of annuals in the mediterranean flora in central Chile has still to be formulated. Some evidence was provided for fire being an important selective factor, at least in the semiarid sectors. Evidently, the more strongly woody flora in central Chile has somehow impeded the evolution of annuals. Detailed studies might eventually show fewer short-lived perennial herb lineages in central Chile than in California, as a result of the overall stronger tendency for longevity, and this trend alone would impede the evolution of annuals. The hypothesis that evolutionary constraints have been involved needs to be tested with rigorous phylogenetic analysis.

Zedler (1990) pointed to another potential selective advantage of the annual habit in mediterranean climates. He suggested that temporal variation in rainfall might be more relevant than the absolute amount of precipitation. If rainfall were consistently low, stress-tolerant adaptations in perennial woody plants could evolve and be utilized with success. If interannual variation in rainfall is high, however, the annual habit will become advantageous, with populations expanding in years of good rainfall and surviving in the soil seed bank in years with limited precipitation. The Miller et al. (1977) comparison of long-term climatic records for Valparaíso and San Diego showed that the California location exhibited greater year-to-year variability in precipitation. This kind of selection pressure, as Zedler (1990) points out, should be especially important in seasonally wet habitats such as vernal pools, which are especially common in California. The greater abundance of vernal pools there would further select for the annual habit, which is already evident considering that some important genera of annuals in California (e.g., *Limnanthes, Eryngium, Downingia, Lasthenia*) (Holland and Jain 1988; Jokerst 1990; Stone 1990) are practically restricted to this habitat. Although Chile has vernal pools, it is presently unknown how many Chilean native annuals are restricted to this habitat. A complex web of factors probably underlies differences in the frequency of annuals in the two mediterranean floras.

Looking more broadly at mediterranean floras, it is indeed noteworthy that the incidence of the annual habit in the fire-prone Cape flora in South Africa is as low as (e.g., 6% for New Zealand) and lower than (e.g., 7.3% for Japan) in some floras in which an extended dry season is lacking (figures as in Raven and Axelrod 1978). Appreciable amounts of summer rainfall in mediterranean areas in South Africa (di Castri 1990) might contribute to this unexpected situation. On the other hand, the South African mediterranean flora, in an equable coastal climate, also has an excess of woody lineages, as in central Chile. Clearly, the diversity of biotic and abiotic conditions in mediterranean floras provides an extraordinary setting for teasing apart the relative roles of phylogenetic versus ecological factors in the evolution of annuals.

The comparisons of species and generic richness effected in this paper are still very crude, in the sense that no control is included for physiographic diversity among the data sets. But considering that fairly consistent trends have appeared, it is probably safe to assume that differences in regional species and generic richness in central Chile and the California FP are real. Undoubtedly, the most significant finding was that central Chile has intrinsically more woody species, even though total species richness is lower than in the California FP. This discovery has led us to question earlier conclusions of the convergent evolution program that more woody species and lower dominance concentration among woody species in central Chile is a result of human disturbance (Mooney et al. 1977). Such high richness in woody species could simply be the outcome of the now demonstrated, more intrinsically woody flora in central Chile. Perhaps some other differences between the central Chilean and Californian floras, presently attributed to human disturbance, will turn out to reflect real intrinsic differences that evaded detection in earlier studies because of trends being parallel to the expectations of human disturbance. Coexistence of many woody species locally might be facilitated by the higher generic richness of the woody Chilean flora, which could be expected to enhance niche differentiation among co-occurring species. The relative roles of disturbance and intrinsic properties of the Chilean flora as explanations for differences in community structure on small climatically matched plots need to be studied further. It would be appropriate here to address alpha diversity and dominance concentration across a gradient of human disturbance.

Finally, we return to the notion that sclerophyllous elements in central Chile developed in situ from ancient rain forest stock rather than by migration from a continental source, as proposed for California. This hypothesis, though based on very slim evidence, would explain many features of the woody Chilean mediterranean flora—the low floristic affinity of the sclerophyllous forests in central Chile and the Chaco, the common occurrence of closely related species in the mediterranean and southern rain forest zones, notable clinal variation in widespread species like *Nothofagus obliqua*, and, above all, the many transcontinental generic disjunctions. In a sense, the present disjunctions in many sclerophyllous genera in central Chile and in eastern South America can be seen to mimic the transcontinental and intercontinental disjunctions seen today in the Arcto-Tertiary elements in the California flora. Whereas the woody mediterranean flora in California has arisen through the assembly of two distinct biogeographic elements (Arcto-Tertiary and Madro-Tertiary Geofloras), most woody elements in central Chile (from mesic to more xeric) seem to have arisen gradually from a common rain forest stock analogous to the Arcto-Tertiary Geoflora, with one biogeographic pattern predominating in present distributions. The proposed contrasting modes of origin could have many ecological repercussions. Mooney et al. (1977) showed that chaparral species concentrate vegetative growth as spring begins, whereas in Chilean matorral species, it is more protracted, carrying through spring and summer. Also, as mentioned earlier, evergreen species in central Chile are more frost sensitive than Californian species (Mooney 1977). Mooney (1977) related the latter factor to higher climate equability in Chile. Perhaps these differences reflect a history in more continental climates for the Californian chaparral. Critical to testing historical hypotheses is better knowledge of the Tertiary fossil record in southern South America. A close look at the Atacama Desert, inland areas in central Chile, and the present area of the Monte desert is of paramount importance.

Acknowledgments

Research supported by FONDECYT Grant No. 88-1177, John D. and Catherine T. MacArthur Foundation Grant No. 90-9929, and WWF-AID Biodiversity Support Program Grant No. I-7506. Lohengrin Cavieres gratefully acknowledges doctoral support from the Andrew W. Mellon Foundation. Final manuscript preparation was supported by FONDECYT Grant No. 92-1135.

References

Aceituno P (1988) On the functioning of the southern oscillation in the South American sector. Part I: Surface climate. Mon Wea Rev 116:505–524

Alaback PB (1991) Comparison of temperate rainforests of the Americas. Rev Chil Hist Nat 64:399–412

Alvarez E (1954) Algunos aspectos de la obra de Ruiz and Pavón. Anal Inst Bot Cavanilles 12:5–111

Anzotegui LM, Lutz AI (1987) Paleocomunidades vegetales del terciario superior ("Formación Ituzaigó") de la Mesopotamia argentina. Rev Asoc Cien Nat Lit, Santa Fe, Argentina 18:105–228

Armesto JJ, Martínez JA (1978) Relations between vegetation structure and slope aspect in the mediterranean region of Chile. J Ecol 66:881–889

Arroyo MTK, Uslar P (1993) Breeding systems in a temperate mediterranean-type climate montane sclerophyllous forest in central Chile. Bot J Linn Soc 111:83–102

Arroyo MTK, Squeo F, Armesto JJ, Villagrán C (1988) Effects of aridity on plant diversity in the northern Chilean Andes: Results of a natural experiment. Ann Missouri Bot Garden 75:55–78

Arroyo MTK, Marticorena C, Muñoz M (1990) A checklist of the native annual flora of continental Chile. Gayana, Botánica 47:119–135

Arroyo MTK, Armesto J, Squeo F, Gutiérrez J (1993a) Global change: Flora and vegetation of Chile. In Mooney H, Fuentes E, Kronberg B (eds) *Earth System Response to Global Change: Contrasts between North and South America*. Academic Press, New York, pp 239–263

Arroyo MTK, Riveros M, Peñaloza A, Cavieres L, Faggi AM (1993b) Phytogeographic relationships and regional richness patterns of the cool temperate rainforest flora of southern South America. In Lawford R, Alaback P, Fuentes ER (eds) *High Latitude Rain Forests and Associated Ecosystems of the West Coast of the Americas: Climate, Hydrology, Ecology and Conservation*. Springer-Verlag, Tiergartenstrasse, Heidelberg, Germany (accepted)

Aschmann H (1973) Distribution and peculiarity of mediterranean ecosystems. In di Castri F, Mooney HA (eds) *Mediterranean-Type Ecosystems: Origin and Structure*. Springer-Verlag, Berlin, pp 11–19

Aschmann H (1990) Human impact on the biota of mediterranean-climate regions of Chile and California. In Groves RH, di Castri F (eds) *Biogeography of Mediterranean Invasions*. Cambridge Press, Cambridge, pp 33–41

Axelrod DI (1973) History of the mediterranean ecosystem in California. In di Castri F, Mooney HA (eds) *Mediterranean-Type Ecosystems: Origin and Structure*. Springer-Verlag, Berlin, pp 225–277

Axelrod DI (1988) Outline history of California vegetation. In Barbour MG, Major J (eds) *Terrestrial Vegetation of California*. California Native Plant Society. Spec Pub No. 9, USA, pp 139–193

Axelrod DI (1989) Age and origin of chaparral. In Keeley SC (ed) *The California Chaparral: Paradigms Reexamined*. Sci Ser No. 34, Nat Hist Mus LA County, Los Angeles, CA, pp 7–19

Axelrod DI (1992) Miocene floristic change at 15 Ma, Nevada to Washington, U.S.A. In Ventkatachala BS, Dilcher DL, Maheshwari HK (eds) *Essays in Evolutionary Plant Biology*. Birbal Sahni Inst of Palaeobot, Luchnow, pp 234–239

Axelrod DI, Arroyo MTK, Raven P (1991) Historical development of temperate vegetation in the Americas. Rev Chil Hist Nat 64:413–446

Bailey HP (1964) Toward a unified concept of the temperate climate, Geog Rev 54:516–545

Barbour MG, Major J (eds) (1988) *Terrestrial Vegetation of California*. Calif Native Plant Soc, Spec Pub No. 9, USA

Beauchamp RM (1986) *A Flora of San Diego County, California*. Sweetwater River Press, National City, CA

Berry EW (1922) The flora of the Concepción Arauco coal measures of Chile. Johns Hopkins U Stud Geol 4:73–142

Berry EW (1938) Tertiary flora from the Río Pichileufú, Argentina. Spec Pap Geol Soc Am No. 12

Bond P, Goldblatt P (1984) Plants of the Cape flora, a descriptive catalogue. J S African Bot Supp Vol 13:1–455

Bond WJ (1983) On alpha diversity and richness of the Cape Flora: a study in southern Cape Fynbos. In Kruger FJ, Mitchell DT, Jarvis JUM (eds) *Mediterranean-Type Ecosystems: The Role of Nutrients*. Springer-Verlag, Berlin, pp 225–243

Boughley AS (1968) A checklist of Orange County flowering plants. Museum of

Systematic Biology, U California, Irvine, Res Ser No. 1, pp 1–89

Cabrera AL, Willink A (1980) *Biogeografía de America Latina*, 2nd Edition, Org Am States, Washington

Casassa I (1985) Estudio demográfico y florístico de los bosques de *Nothofagus obliqua* (Mirb.) Oerst en Chile central. M.S. Thesis, U Chile, Santiago

Cody ML, Mooney HA (1978) Convergence versus nonconvergence in mediterranean ecosystems. Ann Rev Ecol System 9:265–321

Cowling RM, Gibbs Russell GE, Hoffman MT, Hilton-Taylor C (1989) Patterns of plant species diversity in southern Africa. In Huntley BJ (ed) *Biotic Diversity in South Africa: Concepts and Conservation*. Oxford U Press, Cape Town, pp 19–50

Cowling RM, Holmes PM, Rebelo AG (1992) Plant diversity and endemism. In Cowling RM (ed) *The Ecology of Fynbos: Nutrients, Fire and Diversity*. Oxford U Press, Cape Town, pp 62–112

Danin A, Orshan G (1990) The distribution of Raunkiaer life forms in Israel in relation to the environment. J Veg Sci 1:41–48

di Castri F (1973) Climatographical comparisons between Chile and the western coast of North America. In di Castri F, Mooney HA (eds) *Mediterranean-Type Ecosystems: Origin and Structure*. Springer-Verlag, Berlin, pp 21–36

di Castri F (1981) Mediterranean-type shrublands of the world. In di Castri F, Goodall DW, Specht RL (eds) *Mediterranean-Type Shrublands*. Elsevier, Amsterdam, pp 1–52

di Castri F (1990) An ecological overview of the five regions of the world with mediterranean climate. In Groves RH, di Castri F (eds) *Biogeography of Mediterranean Invasions*. Cambridge U Press, Cambridge, pp 3–15

di Castri F, Hajek ER (1976) *Bioclimatología de Chile*. Dirección de Investigación, Vicerectoría Académica, U Catól Chile, Santiago

Donoso C (1979) Genecological differentiation in *Nothofagus obliqua* (Mirb) Oerst. in Chile. For Ecol Man 2:53–66

Donoso C (1987) Variación natural en especies de *Nothofagus* en Chile. Bosque 8:85–97

Dunn EL (1975) Environmental stresses and inherent limitations affecting CO_2 exchange in evergreen sclerophylls in mediterranean climates. In Gates DM, Schmerl RB (eds) *Perspectives of Biophysical Ecology*. Springer-Verlag, Berlin, pp 159–181

Eig A (1931/1932) Les élements et les groupes phytogéographiques auxiliaires dans la flore palestinienne. 2 parts. Feddes Repert Specierum Nov Regni Veg Beih 63(1): 1–201; 63(2):1–120

Emberger L (1955). Une classification biogéographique des climats. Ann Biol 31: 249–255

Epling C, Lewis H (1942) The centers of distribution of the chaparral and coastal sage. Amer Midl Nat 27:445–462

Feuillée LE (1714–25) Journal des observations physiques, mathématiques et botaniques. Faites par l'ordre du Roy sur les côtes orientales de l'Amérique meridionale, & dans les Indes Occidentales, depuis l'année 1707 jusques en 1712, 3 vols, Paris

Fox M (1990) Mediterranean weeds: exchanges of invasive plants between the five Mediterranean regions of the world. In di Castri F, Hansen AJ, Deussche M (eds) *Biological Invasions in Europe and the Mediterranean Basin*. Kluwer Academic Publishers, Dordrecht, pp 179–200

Fuentes ER (1990) Central Chile: how do introduced plants and animals fit into the landscape? In Groves RH, di Castri F (eds) *Biogeography of Mediterranean Invasions*. Cambridge U Press, Cambridge, pp 43–49

Fuentes ER, Hoffmann AJ, Poiani A, Alliende, MC (1986) Vegetation change in large clearings: patterns in the Chilean matorral. Oecologia 68:358–366

Fuentes ER, Avilés R, Segura A (1990) The natural vegetation of a heavily man-transformed landscape: the savanna of central Chile. Interciencia 15:293–295

Gay C (1845–1854) *Historia Física y Política de Chile.* 15 vols., Maulde y Renou, Paris

Goldblatt P (1978) An analysis of the flora of Southern Africa: its characteristics, relationships and origins. Ann Missouri Botan Garden 65:369–436

Graf K (1992) Pollendiagramme aus den Anden. Physische Geographie, Vol. 34, Geograph Inst Univ Zürich, Zürich

Greuter W (1991) Botanical diversity, endemism, rarity, and extinction in the mediterranean area: An analysis based on the published volumes of Med-Checklist. Bot Chron 10:63–79

Griffin JR, Critchfield WB (1976) The distribution of forest trees in California. USDA For Serv, Res Paper PSW-82/1972

Grosjean M, Messerli B, Schreier H (1991) Seehochstände, Bodenbildung und Vergletscherung im Altiplano Nordchiles: Ein interdisziplinärer Beitrag zur Klimageschichte der Atacama. Erste Resultate. In Garleff K, Stingl H (eds) *Südamerika Geomorphologie und Paläoökologie im jüngeren Quartär,* Bamberger Geog Schr, Vol. 11, Bamberg, pp 99–108

Gunckel H (1929) Don Juan Ignacio Molina. Su vida, sus obras y su importancia científica. Rev Univ (Santiago) 14:195–216, 14:320–341

Hajek ER, di Castri F (1975) *Bioclimatología de Chile.* Dirección de Investigación, Vicerectoría Académica, U Catól Chile, Santiago

Hanes TL (1988) Chaparral. In Barbour MG, Major J (eds) *Terrestrial Vegetation of California.* Calif Native Plant Soc. Spec Pub No 9, USA, pp 417–490

Heady HF (1988) Valley grassland. In Barbour G, Major J (eds) *Terrestrial Vegetation of California.* Calif Native Plant Soc Spec Pub Number 9, USA, pp 491–514

Heusser CJ (1983) Quaternary pollen record from Laguna de Tagua Tagua, Chile. Science 219:1429–1432

Heusser CJ (1989) Southern westerlies during the last glacial maximum. Quat Res 31:423–425

Heusser CJ (1991) Biogeographic evidence for late Pleistocene paleoclimate of Chile. In Garleff K, Stingl H (eds) *Südamerika Geomorphologie und Paläoökologie im jüngeren Quartär.* Bamberger Geograph Schr, Vol. 11, Bamberg, pp 257–270

Heusser CJ, Heusser LE, Peteet DM (1985) Late-Quaternary climatic change on the American North Pacific coast. Nature 315:485–487

Hickman JC (ed) (1993) *The Jepson Manual: Higher Plants of California.* U California Press, California

Hoffmann, AJ, Hoffmann AE (1982) Altitudinal ranges of phanerophytes and chamaephytes in central Chile. Vegetation 48:151–163

Holland RF, Jain SK (1988) Vernal pools. In Barbour MG, Major J (eds) *Terrestrial Vegetation of California.* Calif Native Plant Soc. Spec Pub No 9, USA, pp 515–533

Hoover RF (1970) *The Vascular Plants of San Luis Obispo County, California.* U California Press, Berkeley, Los Angeles

Hopper SD (1979) Biogeographical aspects of speciation in the south-west Australian flora. Ann Rev Ecol System 10:399–422

Hopper SD (1992) Patterns of plant diversity at the population and species level in south-west Australian mediterranean ecosystems. In Hobbs J (ed) *Biodiversity in Mediterranean Ecosystems in Australia.* Surrey Beatty & Sons Pty Limited, Norton, Australia, pp 27–46

Howell JT (1957) The California flora and its province. Leaflets in Western Botany 8:133–138

Howell JT (1970) *Marin Flora.* U California Press, Los Angeles

Howitt BF, Howell JH (1964) The vascular plants of Monterey County, California. Wasmann J Bio 22:1–184

Jaffuel F, Pirion A (1921) Plantas fanerógamas del Valle Marga-Marga. Rev Chil Hist Nat 25:350–405

Johnson MP, Mason LG, Raven PH (1968) Ecological parameters and plant species diversity. Am Nat 102:297–306

Jokerst JD (1990) Floristic analysis of volcanic mudflow vernal pools. In Ikeda DH, Schlising RA (eds) *Vernal Pool Plants: Their Habitat and Biology*. Stud Herbar, California State U, Chico, No. 8, U Foundation, Chico, pp 1–29

Keeley JE (1991) Seed germination and life history syndromes in the California chaparral. Bot Rev 57:81–116

Keeley JE, Johnson AW (1977) A comparison of the pattern of herb and shrub growth in comparable sites in Chile and California. Am Mid Nat 97:120–132

Keeley JE, Keeley SC (1988) Chaparral. In Barbour MG, Billings WD (eds) *North American Terrestrial Vegetation*. Cambridge U Press, New York, pp 165–207

Kochmer JP, Handel SN (1986) Constraints and competition in the evolution of flowering phenology. Ecol Mono 56:303–325

Lamont BB, Hopkins AJM, Hnatiuk RJ (1984) The flora: Composition, diversity and origins. In Pate JS, Beard JS (eds) *Kwongan: Plant Life of the Sandplain*. U West Aust Press, Nedlands, pp 27–50

MacMahon JA (1988) Warm deserts. In Barbour MG, Billing WD (eds) *North American Terrestrial Vegetation*. Cambridge U Press, Cambridge, pp 231–264

Major M (1988) Endemism: a botanical perspective. In Myers AA, Giller PS (eds) *Analytical Biogeography*. Chapman and Hall, New York, pp 118–146

Maldonado S, Arroyo MTK, Marticorena C, Muñoz M, León P (1993) Utilidad de las bases de datos para estudios en biodiversidad: evaluación preliminar de parámetros en las Asteraceae de Chile central (30 to 40° S). Anales del Instituto de Biología, U Nac Auto México, Ser Bot: in press

Mares MA, Morello J, Goldstein G (1985) The Monte desert and other subtropical semi-arid biomes of Argentina, with comment on their relation to North American arid areas. Evenari et al. (eds) *Hot Deserts and Arid Shrublands*, Elsevier Science Publishers, Amsterdam, pp 203–237

Marticorena C (1990) Contribución a la estadística de la flora vascular de Chile. Gayana, Botánica 47:85–113

Marticorena C, Quezada M (1985) Catálogo de la flora vascular de Chile. Gayana, Botánica 42:1–157

Marticorena C, Quezada M (1991) Adiciones y notas a la flora de Chile. Gayana, Botánica 48:121–126

McClintock E, Reeberg P, Knight W (1990) A Flora of the San Bruno Mountains, San Mateo County California. Calif Nat Plant Soc, Spec Pub No 8, USA

Milewski AV (1979) A climatic basis for the study of convergence of vegetation structure in mediterranean Australia and South Africa. J Biogeog 6:293–299

Miller PC, Bradbury DE, Hajek E, LaMarche V, Thrower NJW (1977) Past and present environment. In Mooney HA (ed) *Convergent Evolution in Chile and California: Mediterranean Climate Ecosystems*. Dowden, Hutchinson & Ross, Stroudsburg, PA, pp 27–72

Minnich RH (1989) Chaparral fire history in San Diego County and adjacent northern Baja California: An evaluation of natural fire regimes and the effects of suppression management. In Keeley SC (ed) *The California Chaparral: Paradigms Reexamined*. Sci Ser No 34, Nat Hist Mus LA County, pp 37–47

Molina JI (1782) *Saggio Sulla Storia Naturale del Chili*. Tomasso d'Aquino, Bologna

Montenegro G, Avila G, Schatte P (1983) Presence and development of lignotubers in shrubs of the Chilean matorral. Can J Bot 61:1804–1808

Montenegro G, Serey I, Gómez M (1985) Growth forms of arid and semi-arid bioclimatic zone in Chile through the monocharacter approach. Med Amb 7:21–30

Mooney HA (1977) Frost sensitivity and resprouting behavior of analogous shrubs of California and Chile. Madroño 24:74–78

Mooney HA, Cody ML (1977) Summary and conclusions. In Mooney HA (ed) *Convergent Evolution in Chile and California: Mediterranean Climate Ecosystems*.

Dowden, Hutchinson and Ross, Stroudsburg, PA, pp 193–199

Mooney HA, Dunn L, Shropshire F, Song L (1970) Vegetation comparisons between the mediterranean climatic areas of California and Chile. Flora 159:480–496

Mooney HA, Kummerow J, Johnson AW, Parsons DJ, Keeley S, Hoffmann A, Hays RI, Giliberto J, Chu C (1977) The producers—their resources and adaptive responses. In Mooney HA (ed) *Convergent Evolution in Chile and California: Mediterranean Climate Ecosystems.* Dowden, Hutchinson and Ross, Stroudsburg, PA, pp 85–153

Muñoz MR, Fuentes ER (1989) Does the fire induce shrub germination in the Chilean matorral? Oikos 56:177–181

Nahal I (1981) The mediterranean climate from a biological viewpoint. In di Castri F, Goodall DW, Specht RL (eds) *Mediterranean-Type Shrublands.* Elsevier Scientific Publishing Company, Amsterdam, pp 63–86

Navas LE (1973) *Flora de la Cuenca de Santiago de Chile.* Tomo I. Edic U Chile. Editorial Andrés Bello

Navas LE (1976) *Flora de la Cuenca de Santiago de Chile.* Tomo II. Edic U Chile. Editorial Andrés Bello

Navas LE (1979) *Flora de la Cuenca de Santiago de Chile.* Tomo III. Edic U Chile. Editorial Andrés Bello

Naveh Z, Whittaker RH (1979) Structural and floristic diversity of shrublands and woodlands in Northern Israel and other mediterranean areas. Vegetatio 41:171–190

Nishida M (1981) Petrified woods from the Tertiary of Quiriquina Island (a preliminary report). In Nishida M (ed) *A Report of the Paleobotanical Survey to Southern Chile.* Chiba U, Chiba, pp 38–40

Nishida M, Nishida H, Ohsawa T (1988a) Preliminary notes on the petrified woods from the Tertiary of Cerro Dorotea, Ultima Esperanza, Chile. In Nishida M (ed) *A Report of the Botanical Survey to Bolivia and Southern Chile* (1986–1987). Chiba U, Chiba, pp 16–25

Nishida M, Nishida H, Nasa T (1988b) Anatomy and affinities of the petrified plants from the Tertiary of Chile V. Bot Mag, Tokyo 101:293–309

Nishida M, Ohsawa T, Nishida H (1990) Anatomy and affinities of the petrified plants from the Tertiary of Chile (VI). Bot Mag, Tokyo 103:255–268

NOAA (1982) Monthly normals of temperature, precipitation, and heating and cooling degree days 1951–80, California. Climatography of the United States No. 81 (by state). National Oceanic and Atmospheric Administration, Environmental Data and Information Service, National Climatic Center, Asheville, NC, USA

Ohsawa T, Nishida M (1990) Preliminary notes on petrified dicotyledonous woods from central Patagonia, XI Region, Chile. In Nishida M (ed) *A Report of the Paleobotanical Survey to Patagonia, Chile* (1989). Chiba U, Chiba, pp 19–21

Ovalle C, Aronson J, Del Pozo A, Avendaño J (1990). The espinal: agroforestry systems of the mediterranean-type climate region of Chile. Agrofor Sys 10:213–239

Parsons DJ (1976) Vegetation structure in the mediterranean climate scrub communities of California and Chile. J Ecol 64:435–447

Parsons DJ, Moldenke AR (1975) Convergence in vegetation structure along analogous climatic gradients in California and Chile. Ecology 56:950–957

Pate JS, Dixon KW, Orshan G (1984) Growth and life form characteristics of kwongan species. In Pate JS, Beard JS (eds) *Kwongan: Plant Life of the Sandplain.* U West Aust Press, Nedlands, pp 84–100

Peralta I, Rodríguez J, Arroyo, MTK (1992) Breeding systems and aspects of pollination in *Acacia caven* (Mol.) Mol. (Leguminosae: Mimosoideae) in the mediterranean type climate zone of central Chile. Bot Jahrb 114:297–314

Pérez C, Villagrán C (1985) Distribución de abundancias de especies de bosques relictos de la zona mediterránea de Chile. Rev Chil Hist Nat 58:157–170

Pignatti E, Pignatti S (1985) Mediterranean type vegetation of SW Australia, Chile and the Mediterranean Basin, a comparison. Ann Bot 43:227–243

Pons D, Vicente JC (1985) Découverte d'un bois fossile de Fagaceae dans la formation Farellones (Miocene) des Andes d'Aconcagua (Chili): Importance paléobotanique et signification paléobotanique et signifaction paléo-orographique. 110 Congrès National des Sociétés Savantes, Montpellier

Primack RB (1987) Relationship among flowers, fruits, and seeds. Ann Rev Ecol System 18:409–430

Ramella L, Spichiger R (1989) Interpretación preliminar del medio físico y de la vegetación del Chaco boreal. Contribución al estudio de la flora y de la vegetación del Chaco. I. Candollea 44:639–680

Raunkiaer C (1934) Life Forms of Plants and Statistical Plant Geography. Calderon Press, Oxford

Raven PH (1963) Amphitropical relationships in the floras of North and South America. Quar Rev Bio 38:151–177

Raven PH (1988) The California flora. In Barbour MG, Major J. (eds) Terrestrial Vegetation of California. Calif Native Plant Soc. Spec Pub No 9, USA, pp 109–138

Raven PH, Axelrod AI (1978) Origin and relationships of the California flora. U Calif Pub Bot 72:1–134

Raven PH, Thompson HJ, Prigge BA (1986) Flora of the Santa Monica Mountains, California. Southern California Botanists, Spec Pub No 2

Reiche KF (1886–1911) Flora de Chile, 6 vols., Editorial Cervantes, Santiago

Richerson PJ, Lum K (1980) Patterns of plant species diversity in California: Relation to weather and topography. Am Nat 116:504–536

Ricklefs RE (1987) Community diversity: relative roles of local and regional processes. Science 235:167–171

Rodríguez R, Matthei O, Quezada N (1983) Flora Arbórea de Chile. Edit U Concepción, Concepción

Romero E (1986) Paleogene phytogeography and climatology of South America. Ann Missouri Bot Garden 73:449–461

Ruiz H, Pavón J (1794) Florae peruvianae, et chilensis prodromus, sive novorum generum plantarum peruvianarum, et chilensium descriptiones, et icones, Madrid

Rundel PW, Dillon MO, Palma B, Mooney HA, Gulmon SL, Ehleringer JR (1991) The phytogeography and ecology of the coastal Atacama and Peruvian Deserts. Aliso 13:1–49

Sakai A, Paton DM, Wardle P (1981) Freezing resistance of trees of the south temperate zone, especially subalpine species of Australasia. Ecology 62:563–570

San Martín J, Figueroa H, Ramírez C (1984) Fitosociología de los bosques de ruil (Nothofagus alessandri Espinosa) en Chile central. Rev Chil Hist Nat 57:171–200

Santibañez F, Uribe JM (1990) Atlas Agroclimático de Chile: Regiones V y Metropolitana. U Chile, Fac Cien Agr For, Santiago

Sarmiento G (1972) Ecological and floristic convergences between seasonal plant formations of tropical and subtropical South America. J Ecol 60:367–410

Schlegel F (1962) Hallazgo de un bosque de cipreses cordilleranos en la Provincia de Aconcagua. Bol U Chile 32:43–46

Schlegel F (1966) Pflanzensoziologische und floristische Untersuchungen über Hartlaubgehöze im La Plata-Tal bei Santiago de Chile. Bericht der Oberhessischen Gesellschaft für Natur- und Heilkunde zu Giessen, Neue Folge, Nat Abt 34:183–204

Schluter D (1986) Tests for similarity and convergence of finch communities. Ecology 67:1073–1085

Schmithüsen LJ (1956) Die ramuliche Ordnung der Chilenischen Vegetation. Bonn Geogr Abh 17:1–86

Smith CF (1976) A Flora of the Santa Barbara Region, California. S Barbara Mus Nat Hist, Santa Barbara

Sokal R, Rohlf FJ (1981) *Biometry*, 2nd Edition, WH Freeman and Company, New York

Solbrig OT, Blair WF, Enders FA, Hulse AC, Hunt JH, Mares MA, Neff J, Otte D, Simpson BB, Tomoff CS (1977a) The biota: The dependent variable. In Orians GH, Solbrig OT (eds) *Convergent Evolution in Warm Deserts*. Dowden, Hutchinson & Ross, Stroudsburg, PA, pp 50–66

Solbrig OT, Cody ML, Fuentes ER, Glanz W, Hunt JH, Moldenke AR (1977b) The origin of the biota. In Mooney HA (ed) *Convergent Evolution in Chile and California: Mediterranean Climate Ecosystems*. Dowden, Hutchinson and Ross, Stroudsburg, PA, pp 13–26

Specht RL (1988) (ed) *Mediterranean-Type Ecosystems*. Kluwer Academic Publishers, Dordrecht

Stone RD (1990) California's endemic vernal pool plants: Some factors influencing their rarity and endangerment. In Ikeda DH, Schlising RA (eds) *Vernal Pool Plants: Their Habitat and Biology*. Stud Herb, California State U, Chico, No. 8, University Foundation, Chico, pp 89–107

Thomas JH (1961) *Flora of the Santa Cruz Mountains of California*. Stanford U Press, Stanford

Thrower NJW, Bradbury DE (eds) (1977) *Chile-California Mediterranean Scrub Atlas: A Comparative Analysis*. Dowden, Hutchinson and Ross, Stroudsburg, PA

Troncoso A (1991) Paleomegaflora de la formación Navidad, Miembro Navidad (Mioceno), en el área de Matanzas, Chile central occidental. Bol Mus Nac Hist Nat, Chile 42:131–168

Twisselmann EC (1967) A flora of Kern County, California. Wasmann J Biol 25:1–395

Villagrán C (1990) Glacial climates and their effects on the history of the vegetation of Chile: A synthesis based on palynological evidence from Isla de Chiloé. Rev Palaeobot Palynol 65:17–24

Villagrán C, Armesto JJ (1980) Relaciones florísticas entre las comunidades relictuales del Norte Chico y la Zona Central con el bosque del sur de Chile. Bol Mus Nac Hist Nat (Chile) 37:87–101

Villagrán C, Varela J (1990) Palynological evidence for increased aridity on the central Chilean coast during the Holocene. Quat Res 34:198–207

Westoby M (1988) Comparing Australian ecosystems to those elsewhere. BioScience 38:549–556

Williamson M (1988) Relationship of species number to area, distance and other variables. In Myers AA, Giller PS (eds) *Analytical Biogeography*. Chapman and Hall, London, pp 92–115

Zedler PH (1990) Life histories of vernal pool vascular plants. In Ikeda D, Schlising RA (eds) *Vernal Pool Plants: Their Habitat and Biology*. Stud Herb Calif State U, Chico, pp 123–146

Zedler PH, Zammit CA (1989) A population-based critique of concepts of change in the chaparral. In Keeley SC (ed) *The California Chaparral: Paradigms Reexamined*. Sci Ser No 34, Nat Hist Mus LA County, pp 73–83

4. Plant Life History and Dynamic Specialization in the Chaparral/Coastal Sage Shrub Flora in Southern California

Paul H. Zedler

The goal of ecologists to explain why an ecosystem includes the number of species that it does, and why different landscapes have different numbers of species has been reaffirmed by the recent focus on biodiversity and its role in ecosystem function. Old and new discussions of diversity debate the extent to which each taxonomic entity can be considered to fulfill a unique role in the landscape (Schulze and Mooney 1993). Past studies make it clear that both spatial (i.e., habitat-related) and temporal (i.e., succession-related) aspects of specialization must be considered in any comprehensive explanation of biodiversity. Many ecologists consider it axiomatic that in a large and diverse landscape (the physical attributes of which we can take to define a fixed playing field) any taxonomic entity (crudely, species) is able to survive and reproduce in only part of the landscape (e.g., Whittaker 1975). More generally, they consider that for each species, in some part of the landscape growth and reproduction are at a maximum, and in other areas performance is less successful (Cody and Mooney 1978). Concepts of niche diversification arising from the competitive-exclusion principle argue for species to differentiate so as to divide up the landscape (MacArthur 1972). It is theorized, and also verified in many situations, that the performance of a plant species across the landscape will be strongly influenced by other species, including species of higher animals and microbes, acting as mutualists, predators in the broad sense, and competitors. Thus there are theoretical and empirical demonstrations of entities that will be referred to here as habitat specialists.

It has also been demonstrated that the distribution of species across a landscape will depend on the frequency, intensity, and spatial pattern of disturbances—events that cause widespread mortality or damage to plants. With the degree of occupancy of habitats varying in response to disturbance, specialization for different temporal segments of the resource gradient becomes theoretically possible (Stewart and Levin 1973, Comins and Noble 1985). That is, there can be dynamic specialists—species whose existence depends on being able to exploit opportunities that exist for a limited time. The existence of such specialists can be considered axiomatic for what is sometimes called "classical" succession theory. These dynamic specialists need not disperse. Some species are able to persist through multiple generations on a site because of an ability to tolerate or recover from environmental and other variation. The second possibility is for species to be temporal specialists that suffer local extinction regularly but migrate between disturbance patches because of high dispersability.

For this discussion, species of the first type are referred to as "resident," emphasizing that over an interval equivalent to a few generations neither extinction nor wide dispersal to establish new populations is likely. Among residents we may further distinguish between species that spend on average a significant period between generations as dormant propagules and those that are continuously present in the vegetative state, though they may also have dormant propagules. The former may be called "ephemeral" even though they have life spans of many years, and the latter "persistent." We also see that some species are dominants, which because of biomass and size dominate the plant community. Others are subdominants, species that "fill in"—occurring beneath or between or upon dominant species. The alternative life history that lies at the other end of the persistence of occupancy gradient from the resident species I refer to as "itinerant," to emphasize that the survival of such species in the vegetation is a consequence of movement between patches. Traditionally, itinerant species have often been named "pioneer" or "fugitive" species (Louda, this volume), but both labels unnecessarily emphasize only one of a number of possible roles for species with high dispersability.

From these assumptions, the expectation is that a highly evolved species assemblage should tend to have the number of dominant taxa determined by the diversity of habitat types and the number of itinerant taxa determined primarily by types and patterns of disturbance. Subdominant diversity in turn would be expected to depend on the number of dominants and the variety of subsidiary niches, or subhabitat types. An important special case are the "regeneration niches" (Grubb 1977). These could either be resource patches that the dominants fail to utilize (e.g., gaps) or microhabitats that the presence of the dominants creates or enhances (e.g., epiphytes). These observations must be tempered, however, by acknowledging that empirical demonstrations of habitat diversity as a cause rather than a correlate of the number of species, are lacking (Brown 1988).

Generalists can be only a negative influence on biodiversity measured by numbers of taxa, particularly if it they are dominant species. A dominant able to grow in many types of habitat or to fulfill many roles in succession after disturbance must decrease the diversity of species in proportion to its success as a generalist. The circumstances under which the gains accruing to generalists would be great enough to permit them to exclude habitat specialists are not obvious and probably are complex, involving such factors as the amounts and direction of gene flow (important for the spread of rare valuable generalist traits) and the degree of contrast between subhabitat types.

Mediterranean Landscapes

The mediterranean landscapes in the Pacific region exemplify the assortment of species by substrate. Distribution patterns of moisture conditions and water regime are particularly marked (Specht 1981, Steward and Webber 1981), but the chemical aspects of substrates also exert both subtle (e.g., Rehlaender 1992) and strong influences on performance and distribution of plants (e.g., Kruckeberg 1984). Generalist species also occur over wide geographic ranges on a variety of substrates, and attempts to correlate species with substrates sometimes reveal no simple pattern or strong fidelity (Hopkins and Griffin 1984).

The long dry season and resulting high frequency of extensive fires insures recurrent disturbance. This pattern might be expected to create opportunities for itinerant species, and therefore a flora in which such species had a greater than usual role. But this expectation is not borne out by studies of the dynamics of mediterranean systems, a seeming contradiction that requires explanation.

Purpose and Study Area

Purpose

In this review I consider the woody plant flora in a diverse Mediterranean-climate region—San Diego County, California—with the goal of understanding plant life-history specialization. It is prompted by desire to understand how extensively specialization can be attributed to dynamic roles. In other words, the question is how many species appear to be successional specialists of the most extreme kind—species that move about the landscape tracking disturbance patches or habitat patches at the appropriate stage of invasibility. To accomplish this, I review the major life-history patterns of the plants and also consider other aspects of specialization.

It would be ideal to consider all plants and all animals and microbes that significantly affect them. Instead, I follow the common expedient of limiting the analysis to one plant type, woody plants, with only incidental comments about other life forms. Such a class of functionally similar plants is often called a "guild" (Simberloff and Dayan 1991), but that is a loaded term when applied to plants and I scrupulously avoid it. The survey is further limited by considering only the coastal to lower montane shrub-dominated portion of San Diego County.

The Study Area

San Diego County is the southernmost part of coastal California. It has a total area of 10,890 km^2, and ranges in elevation from sea level to 1991 m. The coastal climate is mediterranean, with cool wet winters and dry hot summers. The annual precipitation increases with elevation from 25 cm at the coast, to 36 to 48 cm in the higher parts of the mountains. It then drops rapidly in the easternmost part of the county in the rain shadow of the mountains toward a minimum of 2 to 3 cm at the lowest parts of the desert. The mediterranean-climate drought is ameliorated in the mountains by summer convective storms and in the region as a whole by occasional tropical storms strong enough to carry into southern California. The environment in this part of California from an ecological perspective is described in detail in Miller et al. (1977).

The large-scale vegetation pattern from the coast to the peak of the Peninsular mountain chain is from coastal scrub with many drought and summer deciduous shrubs to evergreen chaparral, to oak, mixed oak-conifer, and conifer forests at the highest elevations. From the mountains to the desert there is a steep gradient in roughly reverse order of physiognomic types down to the desert floor, where a biogeographically distinct assemblage of typically Sonoran desert shrublands is found. Grasslands, mostly dominated by introduced annuals, occur locally throughout the gradient. Riparian woodlands also occur throughout the gradient.

In this study, I consider only the shrub-dominated western portion of the county with large areas of chaparral and coastal sage scrub. I exclude the montane forests, the complex shrub communities transitional to the desert, and the Sonoran desert shrublands in the rain shadow of the mountains. The coastal estuaries, which would add a number of halophytic shrubs (e.g., *Salicornia* spp.) are also excluded.

The area considered is approximately 7000 km^2, the elevation range approximately 0 to 2000 m, and the rainfall range 8 cm to 120 cm (Fig 4.1.). The major community types dominated by woody plants usually recognized within this zone are (1) chaparral—dense shrublands dominated by evergreen, mostly sclerophyllous shrubs; (2) coastal sage scrub—lower, more open shrublands dominated by summer- and drought-deciduous shrubs, but also with some evergreen species; (3) oak woodland—savannas to closed forests dominated by *Quercus* spp.; and (4) riparian—forest, woodlands, and shrublands associated with the larger drainages and subject to flooding. Grasslands also occur in the present vegetation, and presumably did so in the

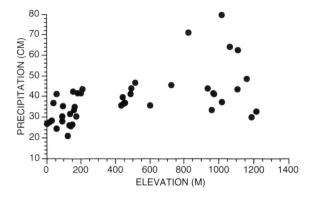

Figure 4.1. Precipitation as a function of elevation for stations in San Diego County, CA. Data from Close et al., 1970.

past, though the extent and distribution of the primeval grasslands is debated, partly because invasion by introduced annual grasses from mediterranean regions has altered the ecological relations of the grasslands. One possibility is that annual grasses and frequent human ignitions have allowed grassland to expand at the expense of shrublands (Zedler et al. 1983). Present grasslands usually occur in conjunction with oak woodland and mostly are relatively small patches. For brevity, I refer to the whole area as the "chaparral landscape," even though these other types of vegetation occur within it.

The Flora

For this study, only the woody vegetation is considered. A woody species is one that has significant above-ground secondary thickening to form a definite woody base persisting for more than two years. This includes trees, shrubs, and subshrubs. Although some monocots do not quite meet this strict definition (e.g., *Yucca* spp.) they are included because their growth habitat is essentially shrublike. Species, genus, and family are the only taxonomic entities considered. Although species are chosen as the lowest taxonomic units this does not dismiss the importance of taxonomic and genetic diversity within species. Many species in the chaparral region are known or suspected to vary substantially over their geographic range. Many subspecies have been described. But the lack of uniformity in taxonomic effort makes it difficult to deal with any level below that of species, and studies that relate genetic diversity to ecological performance are very few. The species lists and species statistics are based on Beauchamp (1986).

The county has approximately 1516 native vascular plant species and another 464 nonnative species (Beauchamp 1986). The native species are distributed among 551 genera. For this study, woody species that occur in

Table 4.1. Woody genera in the chaparral landscape with 3 or more species ranked in order of species number with cumulative percentages of the total woody plant flora of 201 species. Data from Beauchamp 1986

Genus	No. species	Cumulative % of total flora
Ceanothus	14	7.0
Ericameria	7	10.4
Arctostaphylos	7	13.9
Baccharis	6	16.9
Mahonia	6	19.9
Salix	6	22.9
Artemisia	5	25.4
Opuntia	5	27.9
Quercus	5	30.3
Salvia	5	32.8
Diplacus	5	35.3
Ribes	4	37.3
Rhamnus	4	39.3
Solanum	4	41.3
Lonicera	3	42.8
Lupinus	3	44.3
Garrya	3	45.8
Eriogonum	3	47.3
Clematis	3	48.8
Keckiella	3	50.2

the chaparral zone as described above were selected from this total flora. Because the floras and vegetation in the desert and the chaparral are relatively distinct, in most instances deciding whether to place a taxon in the chaparral category was clear. The woody plant flora thus described consists of 201 species in 101 genera and 43 families, with an average of 2 species per genus. Many genera are monotypic within the region, and a minority of the genera contribute a majority of the species (Table 4.1). The largest family is the Asteraceae, with 39 species. The shrub life form is the most common, with subshrubs and trees next most abundant (Table 4.2).

Table 4.2. Breakdown of the woody flora of the chaparral landscape in San Diego County

Growth form	Number	%
Tree	17	8
Shrub	128	64
Vine	8	4
Subshrub	37	18
Succulent	7	3
"Yuccoid"	4	2
Total	201	100

Assessing Diversity of Habitat

Southern California is topographically, edaphically, and climatically complex. To understand how well the diversity of species matches that of habitats, one would like to know the length of the gradients, the correlation among the major gradients, and the area of habitat at each point along the gradients, as well as the spatial arrangement of habitat patches. Detailed data of this kind are not available, but a reasonable approximation of at least minimal levels of diversity can be derived from information on climate and soils.

Total precipitation, mostly as rain, tends to increase with elevation in the chaparral landscape (Fig. 4.1). The variance within any zone is substantial, however, mostly relating to the distance of a station from the coast and its location relative to mountain masses.

Soils also vary substantially across the county. A detailed soil survey has been published for San Diego County, including the entire area of the chaparral landscape portion of the county (Bowman 1973). The basic map unit is the soils series, a grouping of soils similar in the number, kind, sequence, and properties of the soil horizons (Birkeland 1984). Because the soil-series concept was developed with the aim of assessing the capacity of soils for agriculture and other uses, it is also a useful first approximation to the edaphic component of habitat diversity. In unglaciated mountainous regions the soils series also strongly correspond to bedrock geology.

We can get an impression about the diversity of subhabitats in San Diego County by calculating the combinations of topographic, climatic, and edaphic conditions. A conservative estimate is that the area being considered has 52 named soil series. This figure omits the generic categories that were mapped, including, for example, "rough broken land," "loamy alluvial land," and "terrace escarpments." Furthermore, most series are broken into two to seven or more phases based on variations in slope, surface texture, rockiness, or soil depth. Each series occurs over a range of elevations, but none spans the entire elevational gradient. Taking three broad elevational bands of 500 m each, we find 42 soils series in the lowest band, 30 in the intermediate band, and 10 at the highest elevations (Table 4.3). With this division of the elevation gradient it may be said that at least 82 edaphic/climate habitat types are present in the chaparral region.

Many factors besides soil series and the position on the elevation/precipitation/temperature gradient will be influential. Among these are slope position and the site's relation to the adjacent topography. For example, on a ridge, topsoil depth is generally less and plant canopies are exposed to full sun for the entire day, whereas on the lower parts of the adjacent slope, soil depth usually is greater and shading is present for part of the day. The aspect, the direction in which the slope faces, is also a factor and all possible aspects will be included within many of the soil series. Microtopography is also important. A minor drainage a fraction of a meter deep will be sufficient to provide a distinctly different water regime, with plants in the drainageway

Table 4.3. Number of soil series within each elevation zone, with the minimum, maximum, and average precipitation for stations within each zone. Number of stations gives the sample size for the precipitation data. Climatic data from Close et al. (1970) and soils data from Bowman (1973)

Elevation band (m)	Annual precipitation (cm)			Number of stations	Number of soil series
	Maximum	Average	Minimum		
0–500	43.7	33.2	20.6	30	42
500–1000	71	45	33	11	30
1000–1500	79	50	30	8	10
Total elevation zone × soil series combinations =					82

receiving more water in small storms and deeper penetration by water in larger storms. Considering all these possibilities, it is very conservative (and admittedly arbitrary) to assume that on average at least two phases related to aspect and two related to slope position or microtopography are found within each of the 82 soil/climate units. Thus, conservatively, there are 328 distinctly different subhabitats within the chaparral landscape in San Diego County.

Species Number Versus Habitat Diversity

Comparing the 328 distinct habitat types to the 201 species of woody plants found in the county suggests that it has fewer species than distinctly different habitats to be occupied. Thus the problem is not to explain why there are so many species of plants but rather to consider why the number of species is so low. The estimate of habitat diversity is of course crude, but because the estimate is conservative, it does not seem likely that refinements would substantially change the qualitative conclusion reached here—that the evolution of habitat specialists in the chaparral either does not proceed rapidly enough to exploit all the distinct habitats available or that forces act against habitat specialization, so that species will always be fewer than habitat types. In other words, habitat heterogeneity appears to be sufficient that the number of woody species could easily be accounted for by the number of habitat types to be occupied. Thus each woody species could hypothetically have a refuge—some part of the landscape where it was competitively superior and from which it could not be displaced by competition from other species.

Because there are more edaphic–climatic niches than species, and because the landscape does not have huge blank areas, it follows that many species must be habitat generalists that occur over a range of distinctly different habitat types. Evidence is considerable, however, for habitat specialization, for both soil type and elevation/climate regions.

Elevational and Substrate Specialization in the Genus *Ceanothus*

The specialization of the woody flora along the elevational gradient may be illustrated with the genus *Ceanothus* (Rhamnaceae, Fig. 4.2). Species in this genus are found throughout the chaparral/coastal sage scrub and into the forest vegetation of the higher mountains. The genus is divided into two sections differing markedly in morphology. The section Cerastes appears to be more derived and drought-adapted because of its stomatal crypts, absence of a lignotuber, and other traits that seem to fit it better to dry conditions. In contrast, the section Euceanothus is closer in appearance to other genera of the Rhamnaceae that occur in wetter climates (McMinn 1939, Nobs 1963).

The genus as a whole shows a pattern suggesting specialization to different portions of the climatic gradient. The species ranges form the classical sequence with the mid-distributions tending to be displaced but with considerable overlap between the species. Specialization also appears for different substrates. *C. foliosus* occurs primarily on mafic gabbro substrates. *C. ophiochilus* and *C. otayensis* also are both narrow edaphic endemics restricted to pyroxenite-rich rock and acidic, phosphorus-poor metavolcanic rock respectively at opposite ends (north and south respectively) of the county. It is intriguing that some instances of greatest overlap in elevational ranges occur between species that do not overlap edaphically. *C. cyaneus*, whose range is completely encompassed by three other species, is not known to be substrate restricted, but is a rare local endemic, possibly of hybrid origin, restricted to a few sites (Frazier 1993). Other evidence of soil climate related subdivision of the landscape is that the two sprouting species do not overlap significantly

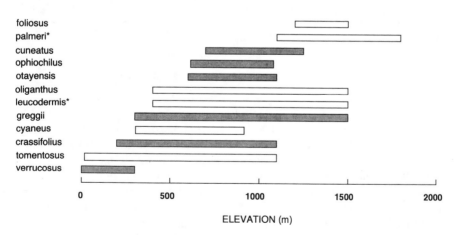

Figure 4.2. Distribution of species of the genus *Ceanothus* (Rhamnaceae) in San Diego County across the elevational gradient. Shaded bars indicate species of the section Cerastes, open bars species of section Euceanothus. Species marked with an asterisk produce burls and sprout after fire, all others are normally killed by fire.

in elevational range, and each section of the genus has members across the elevational gradient.

In this realized pattern of specialization it is clear that most differentiation has occurred among habitats—soils and climate—and little among life-histories because the majority of the species are nonsprouters fulfilling approximately the same dynamic role in the communities in which they occur. The nonsprouters of the Cerastes section tend to have greater longevity, and thus are less prone to apparent extinction—that is, to persist only in the soil seed bank, between fires—than the nonsprouters of the Euceanothus section. But this difference is qualitative only.

Thus radiation in this genus has apparently been stimulated by opportunities or circumstances created by geological and climatic history rather than by aspects of the disturbance regime.

Edaphic Endemics

Other species and genera show strong patterns of substrate restriction by rock type in San Diego County. In this region of mostly acidic plutonic and metamorphic rocks the calcicoles and calcifuge distinction is not important in woody plants. Halophyte substrate specialists are present (Zedler 1982), but are not considered here. Unlike many areas of California, San Diego has no significant areas of serpentine rocks, and therefore no serpentine endemics. As with *Ceanothus* discussed above, however, striking patterns of substrate restriction are associated with mafic gabbros and acidic metamorphosed volcanic rocks. Fourteen of 201 species (7%) show such marked restriction either to gabbro (e.g., *Salvia clevelandii*), gabbro and metavolcanic rock (*Cupressus forbesii*), or to metavolcanic substrates (*Arctostaphylos otayensis*). Thus the 7% may be taken as a minimal estimate of specialization to substrate, in the woody flora as a whole. In contrast, 3 of 12 (25%) of the *Ceanothus* species are substrate restricted.

Life-History Attributes of the Woody Plants

Each species in the region may be thought of as a unique solution to the problems of surviving in a mediterranean-climate region. Because of the importance of fire as a disturbance factor, we might expect that the flora would show unequivocal evidence of fire adaptation. Though some species have traits that can reasonably be explained only by selection by fire, many have no obvious fire adaptations, and others have traits which may be fire selected but which can be explained in other ways (Zedler and Zammit 1989).

Sprouting is often cited as evidence of fire adaptation (Vogl 1970), and it is a common trait in the woody plants of San Diego County. Of 116 species that can be characterized with some assurance, 56% are sprouters, 9% are weak or variable sprouters, and 28% rarely or never sprout after fire. More than a third of these nonsprouters are subshrubs like *Gutierrezia* spp. (Asteraceae). Two species (2%) endure fire, meaning that the bulk of the above-ground tissues come through the fire without damage. The best example in this category is *Yucca whipplei* (Agavaceae), which typically suffers singed leaves, but not death of the single growing point.

It is debatable if this high incidence of sprouting can be attributed to fire selection. Though resprouting after fire is critical to the population ecology in many species, it may be an ancestral trait that proved to be useful in the present mediterranean climate as Wells (1969) suggests. Fire seems more likely to be the cause of the speciation in nonsprouting *Ceanothus* and *Arctostaphylos* (Wells 1969). In San Diego County, for example, the non-sprouting species in these two genera account for about 8% of the woody flora. Wells argues that this process may have been fostered by the short generation times that are a consequence of a mediterranean fire regime. Thus it may be that fire has not diversified or enhanced dynamic roles, but fostered habitat specialization. Some support for this notion comes from the fact mentioned above that two of the rarest species of *Ceanothus* in the county (*C. ophiochilus, C. otayensis*) are nonsprouting habitat specialists. But the general ancestral presence of sprouting does not mean that no aspect of sprouting behavior will have been affected by fire. Canadell and Zedler (this volume) argue that some aspects of the resprouting mechanism, and specifically the occurrence of the lignotuber, may have been at least partly selected by fire.

It is even less clear that fire has had any major effect on the dispersal characteristics of the woody plants. A classification of plants by dispersal mechanisms reveals that a significant portion, 28%, have no obvious means of dispersal (Table 4.4). Almost as many, 22%, are wind dispersed, and thereby possess the potential for invading disturbed areas in large numbers. Forty-two percent are animal dispersed, of which half have traits that suggest birds as the major vectors, though studies of *Rhus integrifolia* by Lloret and Zedler (1991) revealed that in this species, which has prominently displayed red fleshy fruits, most of the fruits fall to the ground and are subsequently taken by small mammals. Unlike Australia, in California shrublands ant dispersal is minimal. *Dendromecon rigida* is one of the few species in the chaparral with elaiosomes and ant dispersal (Bullock 1989).

Table 4.4. Most probable dispersal mechanisms of the woody flora of San Diego County, based on 159 species that could be classified with reasonable confidence

Mechanism	No. species	Percentage
Wind	35	22
Bird	32	20
Other animal	32	20
Ballistic	15	10
None apparent	45	28

A significant portion of the woody flora, (9%), all species of *Ceanothus*, have ballistic dispersal. Their explosive capsules distribute seeds with greater local uniformly than most dispersal mechanisms, but over a limited distance. The seeds of this genus are relatively large (3.1 and 5.6 mg for the two species studied by Frazier (1993)) and they are taken by seed predators, which probably are agents of some secondary dispersal (Mills and Kummerow 1989). It is questionable if ballistic dispersal was selected by fire, but it seems well suited to species with seeds stimulated to germinate by fire. For such a species essentially the entire local environment is available for establishment, and there is little advantage in hazarding long-distance dispersal. Because establishment is limited to a brief postfire window, there is also the possibility of significant sib competition, which ballistic dispersal may minimize because it ensures more even distribution of seeds around a parent plant. It is also a desirable dispersal mechanism for species that frequently occur on sloping sites where the uphill component of the ballistic dispersal counteracts a tendency for net seed movement to be downhill from the parent plant.

The high rate of germination of seeds in burned areas characteristic of *Ceanothus* and a few other species is not, however, the rule in the woody vegetation. Of 89 species that could be characterized with confidence, at least 51% rarely or never establish seedlings in burned areas. Furthermore, many of the 49% of species that do establish in burns do so in low numbers or are quite variable in success of establishment (e.g., *Arctostaphylos glandulosa*). It is difficult to account for this pattern by assuming strong fire selection on life-history traits. The postfire environment provides an ideal opportunity for seedling establishment, yet most species fail to exploit it.

The other side of this pattern is the capacity of species to establish seedlings in situations other than burns. Of 106 species that could be classified, 72% are known to germinate and survive without fire stimulus in unburned vegetation. Thus many species that are regularly found in burns also establish seedlings in other situations. For example, *Salvia mellifera*, a dominant species in coastal sage scrub and low-elevation chaparral that germinates after fire also establishes seedlings in unburned situations, especially in small gaps.

Life Histories of the Woody Plants—Dynamic Roles

The capacity of a species to persist in habitat already occupied and to invade new habitats that become available is a function of the combination of traits like those identified in the preceding section. Following the logic outlined in the introduction, the dynamic roles played by the woody species in the chaparral landscape can be classified by three basic features: (1) whether they tend to be resident or itinerant, (2) the circumstances under which they are most likely to establish new individuals, and (3) their longevity. For the resident/itinerant breakdown, it is necessary to specify a spatial scale, for at the scale of the entire county all species considered here are resident, and at the scale of 0.01 m² all are itinerant. For the present discussion 1 to 10 ha is taken as the area. Thus a resident species is one for which the probability of local extinction in a 1 to 10-ha area is very low and an itinerant species one for which the probability of extinction is very high. Long-lived is defined as > 100 years. This span is chosen not only because it is a round number but also because it is greater than the average maximum return times of fire under current conditions, which the data of Minnich (1989) show to be approximately 70 to 72 years. Long-lived then means capable of surviving in large numbers to the next fire.

The basis for this classification is partly the literature, but mostly personal observations and unpublished data. Because detailed studies have not been made of all species, the calculations are based on 150 species. Six resident strategies were identified among species in this group.

1. *Resident persistent long-lived species whose establishment is independent of large-scale disturbance.* ($N = 66, 44\%$) These are species that could be said to provide the continuity to the vegetation. They are able to tolerate the variations in climate and disturbances either by being able to resist the negative effects of the disturbance (e.g., dense semisucculent leaves of *Yucca whipplei* protecting against fire) or by vegetative recovery after the disturbance (resprouting of *Platanus occidentalis* (Platanaceae) after fire or flood or *Cercocarpus betuloides* (Rosaceae) after fire). Though recruitment may be episodic because of variations in rainfall and seed production, it does not depend on major disturbances such as fire or flood. Thus populations are in general made up of genets of different ages, though often the population of aboveground ramets may be nearly even-aged, especially in shrubs that resprout after fire. Species in this group are mostly also able to recruit new stems by basal sprouting in periods between fires, and thus are able to maintain healthy canopies even when intervals between fires are very long (Zedler 1977, Keeley 1992a,b).

Many of the species in this group (e.g., spp. in the genera *Clematis*, *Juglans*, *Lonicera*, *Rhamnus*, *Ribes, and Sambucus*) have affinities with the Arcto-Tertiary flora in eastern North America, Europe, and eastern Asia. It may be that these species retain traits developed in more mesophytic habitats where fire was not a major factor. In keeping with this interpretation, many of the species are most abundant in relatively moister habitats such as north-facing slopes (Steward and Webber 1981).

2. *Resident persistent dominant and canopy species, intermediate to long-lived, which respond favorably to fire, but do not require it for establishment.* ($N = 14, 9\%$) Species in this group can tolerate and even increase after fire, but they do not require it for establishment. This group includes most of the summer-deciduous species of the coastal sage scrub—*Artemisia californica*, *Encelia californica, Viguieria laciniata*, and *Salvia* spp., and the evergreen but only moderately sclerophyllous *Eriogonum fasciculatum*.

These species can sprout after fire, but are sensitive to fire intensity, so that sprouting can be significant or negligible depending on the circumstances (Westman et al. 1981). Longevity for most of these species appears to be limited. Older individuals often fragment, causing the shrub literally to fall apart.

These species are classified as resident because when they are present in vegetation not strongly influenced by human disturbance, they appear to persist indefinitely. But because of their capacity to germinate without fire or other special disturbance, they are also capable of invading roadsides, abandoned fields, and other artificially disturbed areas, where they can be present as the initial shrub-dominated stage of what appears to be a human-induced secondary succession. *Eriogonum fasciculatum* is known for this ability (Kirkpatrick and Hutchinson 1980) and at least partly because of human disturbance it is one of the most abundant shrubs in the study area (Steward and Webber 1981). This propensity for spreading into disturbed areas has fueled a debate as to the original extent of shrublands dominated by these species. This once-esoteric question has assumed practical importance because a threatened bird, the California gnatcatcher, is associated with areas dominated by *Artemisia californica*.

3. *Resident mostly persistent moderate- to long-lived species whose establishment appears to be keyed to fire.* This set of 32 species is a large proportion of the shrub flora (21% of species that could be classified). It includes many of the species that are dominant in the chaparral, and the two large genera *Arctostaphylos* and *Ceanothus*. (Excluding these two genera reduces the number of species to eleven.) These species are the most likely to have evolved their seed germination and vegetative recovery traits under selection by fire (Zedler and Zammit 1989). Most of the species in this group have limited dispersal, partly explaining why they are resident species for which movement among patches is mostly a small-scale phenomenon. The ballistic dispersal of *Ceanothus* is discussed above. *Arctostaphylos* spp. produce thinly fleshed berries that are eaten in large numbers by coyotes and widely distributed in their feces. The limited flesh on the berries and the seeming undigested state in which they are passed suggests that they are taken as filler or perhaps the canid equivalent of junk food.

4. *Resident ephemeral or persistent smaller, short-lived shrubs with fire-stimulated establishment, but the ability to respond to other disturbances.* ($N = 10$ spp., 7%) The species in this group are small shrubs or subshrubs that are short-lived but germinate after fire. Among these is *Lotus scoparius*, a species which is often dominant in burns three to five years after the fire, but which dies out rapidly and is of only incidental occurrence in long-unburned chaparral, though it can persist in coastal sage scrub (DeSimone and Burk, 1992). Thus *L. scoparius* could be considered a dynamic specialist in chaparral not seriously disrupted by human beings, because it establishes after disturbance but virtually disappears as the longer-lived resident species assume dominance and the canopy closes. This role could be said to be

successional, but it is not in any sense a pioneer, nor is it itinerant because the fruits have no special dispersal mechanism, and the species maintains a large dormant seed bank (Zammit and Zedler 1988).

In contrast, *Eriophyllum confertiflorum*, also placed in this group, is a suffrutescent shrub that has fire-stimulated seeds but also persists in gaps in moderate abundance in old chaparral. Most of these species also will invade along roadsides and other disturbed areas such as lots scraped to prepare them for construction. As with the coastal sage scrub species, these species have a natural dynamic role, and another associated with human disturbance.

5. *Resident persistent short-lived subcanopy or smaller gap species, establishment with or without fire or other pronounced disturbance.* (*N* = 10 spp., 7%) This set of species is differentiated from those above because they are not markedly stimulated by fire and partly for this reason are present in gaps or below open canopies of larger shrubs in long-unburned vegetation. Like the preceding group they profit from human disturbance and are often more abundant on stand margins and roadsides than in the undisturbed vegetation. Most of these species are not vigorous postfire sprouters or seeders, and populations are often reduced after fire. *Hazardia* (*Happlopappus*) *squarrosa*, however, does sprout well after fire and probably achieves larger size and maximum growth in the early postfire years. It persists well, however, and is a characteristic species of chaparral (Louda 1982).

6. *Itinerant* (widely dispersed) species. (*N* = 18 spp, 12%) Included here are all species that have high seed production, sufficient dispersal capacity, and germination characteristics appropriate for exploiting short periods of habitat suitability and locating small gaps remote from the dispersing parent. This group can be further subdivided into two subgroups.

6a. *Resident short- to long-lived riparian species dependent on flooding.* (*N* = 13 spp., 9%) Species of this type exploit the habitat conditions produced by flooding. All six species of *Salix* and both species of *Populus* found in the county are included here. Because these genera are important to riparian ecosystems, their life histories have been described in detail and are essentially similar over broad geographic ranges (e.g., Johnson 1994). These species occur along the larger drainages in which ground water is present throughout the growing season. Reestablishment generally follows large storm events that deposit alluvium that provides the substrate on which the seedlings germinate. Their seeds are highly dispersible by wind and water (Fowells 1965, McBride and Strahan 1984), but because they occupy essentially linear habitats, most of the dispersal is along the drainage course. Under exceptional circumstances willows will seed into any moist habitat, including chaparral sites, but they do not persist without a reliable source of water at depth (Zedler and Scheid 1988). Despite this potential for invasion, the willows and cottonwoods will usually be resident at linear scales more than 50 m in drainage length. Their capacity for invasion is most clearly displayed where there has been human interference. It is common, for example, to find willows growing at points of leakage along irrigation lines in urban lawns.

The life-history of these species is such that they could form the pioneer phase in a riparian succession (i.e., itinerant) because they require open conditions for establishment and have low shade tolerance. But in fact within the landscape most populations appear to be stable at the chosen scale. Willow populations fluctuate somewhat in extent and greatly in structure and stature, but do not normally disappear from areas with significant populations except when there is human interference.

Baccharis sarothroides (Asteraceae), a shrub in this group, may have been a true itinerant in the original landscape, but its distribution has increased so greatly with human disturbance as to make it difficult to determine precisely what its primeval dynamics might have been. *B. sarothroides* is now the most conspicuous shrub invader of denuded or severely disturbed areas. The minute fruits are produced in large numbers, carried long distances, and germinate without fire or other specialized stimuli. Large stands of *B. sarothroides* occur on roadsides and other disturbed upland areas, and the species is the first shrub to invade in anthropogenic successional sequences. Its capacity to locate even small habitat patches is attested to by its occurrence in neglected flower beds and pavement cracks in urban areas. Excluding sites subjected to human disturbance, its primary habitat is at the upper margins of drainages where occasional severe flooding has an effect similar to human removal of vegetation. In the original landscape it is most likely to have functioned as a true dynamic specialist, dispersing to disturbed patches caused by flooding and yielding either to longer-lived upland shrubs and trees on drier sites or to taller willows, cottonwoods, or sycamores in wetter places.

6b. *Widely dispersed short-lived species not requiring massive disturbance for establishment.* ($N = 5$ spp., 3%) This small group includes the only species that can be said to function as itinerant elements of the coastal sage scrub or the chaparral. All these are in the Asteraceae, and possess the traits often found in members of the family of small wind-dispersed seeds and short life span. *Ericameria parryi*, for example, is a shrub that is generally between 1 and 2 m tall when mature. It is usually found as scattered individuals, and on this basis I have assumed that its occurrence is a consequence of dispersal into small disturbance patches caused by death of individual shrubs or perhaps left after fire because of hot spots or other causes of failure by resprouts and seedlings. At a large scale (ca. 1 km^2), however, the species would presumably be resident.

Life Histories of Woodland Dominants

The previous discussion, except for the part dealing with riparian systems, has considered examples mainly from the shrub communities that are most prevalent in the landscape. But woodlands are also a characteristic, if less abundant, vegetation type in southern California. Excluding the lower fringes of the montane coniferous forest, three woodland types are scattered in the chaparral/coastal scrub matrix in San Diego County—oak woodland, coastal (Torrey) pine woodland, and interior cypress woodland. These woodlands share the characteristic that each has a scattered to dense canopy of small to medium-size trees rising well above the height of the shrubs that generally grow in association with the trees.

The cypress woodlands, which in San Diego County are dominated either by *Cupressus forbesii* or *C. arizonica* ssp. *stephensonii* (Cupressaceae), occur on phosphorus-deficient soils (Zedler unpublished, and Rehlaender 1992). Though the cypresses slowly overtop associated shrubs and suppress them, the close spacing of the trees and their short stature makes crown fire at some time inevitable. The cypresses are thin-barked and, unlike some other mediterranean species in the Cupressaceae (one species of *Actinostrobus* in Western Australia and *Widdringtonia spp.* in South Africa), they are incapable of resprouting. Mortality is thus essentially 100% in any fire, and reestablishment is from seed stored in the serotinous cones of the canopy in the first year or two after fire. Thus the population dynamics of the *Cupressus* is essentially the same as that of any of the other nonsprouting seeders such as species of *Arctostaphylos* or *Ceanothus*. The main distinction is that the cypresses have no significant soil seed bank and are long-lived. Therefore, unlike some species of *Ceanothus*, they do not slowly disappear from a stand, but rather gradually increase in dominance even when an area is long unburned. These species are substrate specialists and resident in the strongest sense of these words, and though populations may shift boundaries slightly and fluctuate in density, the cypress woodlands are clearly very stable elements in the landscape unless humans upset the pattern of fire recurrence (Zedler 1981).

The dynamics of oak woodlands are very different. Two species of oaks regularly reach tree size, *Quercus agrifolia* and *Q. engelmannii*. Both of these attain ages in excess of 200 years, and thus would be expected to live through one or more fires during their life span. But by virtue of thick bark and a large dormant bud bank, even after intense fires that kill all exposed buds, epicormic sprouting will occur in the larger branches so that the tree can rapidly rebuild its crown (Plumb 1980). Germination and establishment occur primarily in unburned areas, and seedlings and saplings have been shown to be sensitive to fire (Lawson 1993). In broad outline the life history of these tree-form oaks is similar to that of the sprouting nonseeders, species such as scrub oak (*Quercus dumosa*) and *Heteromeles arbutifolia* that survive fire but do not exploit fires for seedling establishment. The primary difference is that their crowns rarely are completely killed by fire.

The ecology of Torrey pine (*Pinus torreyana*) is somewhat intermediate between the two examples just discussed. Torrey-pine woodland occupies a small area of broken topography along the immediate coast. Though Torrey pine trees reach diameters and heights equal to that of the larger oaks, they do not have the fire resistance of oak trees. Experimental burns (Scheidlinger and Zedler unpublished) show that Torrey pine is sensitive to fire. It seems probable that the stands regenerated after fire because of the partial serotiny. But given its coastal location, it is also likely that fires were rare and that some, perhaps most regeneration occurred between fires (McMaster and Zedler 1981). Seedling establishment under present circumstances in unburned stands is low, but fairly widely distributed in the stand (McMaster 1980).

The tree-form species in these woodlands are all characterized by a resident persistent strategy. The dynamics of the tree populations are relatively simple, with little evidence that successional processes operate at any but the smallest local scale, for example, in the increased survival of oak seedlings that has been attributed to the nurse-plant effect of co-occurring shrubs (Callaway 1992). No tree-size species of upland sites function as itinerant or even ephemeral resident specialists.

Specialization in the Herbaceous Flora

Excluding the exotic species that are now a prominent part of California vegetation (Heady 1988), the herbaceous plants in the chaparral region are overwhelmingly resident species. After fire, though soils are invasible by suitably adapted herbs, the generality that most of the native cover arises from resprouts or from the seed bank present before the fire also applies to the native herbs (Hanes 1971). Some of these postfire species are able to persist under the canopies of the shrubs or in gaps between canopies, and others are capable of persisting for many decades in the soil. One example among many is *Phacelia brachyloba* (Hydrophyllaceae), an annual that is abundant in higher-elevation chaparral after fire but only accidental in unburned chaparral. A dense population was observed in a chaparral burn in which the age of the chaparral before the fire was more than 95 years. This age suggests that the seeds can survive in the soil for at least a century (Zedler and DeSimone, unpublished). *P. brachyloba* and the large *Phacelia* genus generally (25 species in San Diego County) have no special dispersal mechanism, and except for chance movement and possibly some local secondary dispersal by ants, most seeds will come to rest within a short distance of the parent plant.

At the other end of the life-history scale are species that are the herbaceous equivalent of long-lived dominant residents. One conspicuous plant in this category is *Marah macrocarpus* (Cucurbitaceae), a species with a large subterranean tuber that germinates in mature chaparral, grows into the canopy of shrubs in winter and spring, and is dormant in the summer/autumn drought. Because it is dormant at the time that most fires occur and has a deeply placed storage organ and growing point, it has little risk of being killed by fire, and probably benefits from the release of mineralized nutrients.

For a time after fire on sites where it is abundant, *M. macrocarpus* is the most prominent herbaceous plant, its stems running over the ground or climbing burned snags and flowering abundantly. Perennials like *M. macrocarpa* seem to benefit from fire but not to require it either for seedling establishment or vegetative survival.

Although these limited examples obviously fall short of proof, I assert as a hypothesis that as with the woody plants, analyzing the life histories of the herbs would reveal true itinerants to be a relatively small portion of the population, and local extinction and invasion events at larger scales as rare for them as for the woody plants.

Discussion

Habitat Specialization and Generalization in the Woody Flora

It has been argued that 201 woody species are probably far fewer than the number that could coexist given the range of habitat diversity in the landscape. Most of the woody plants are resident-persistent, especially in relation to fire. Human disturbance not considered, most species persist on a site for several to many generations. Dispersal between habitat patches for the bulk of species is therefore slow and uncertain—a trickling rather than a flood. One might therefore expect that species should be evolving toward habitat specialization. That this evolution is in fact apparently far from complete suggests that past climatic fluctuations and habitat changes and the resulting shifts by the populations in the landscape have been extensive enough to offset periods of site-specific selection and favor more general adaptations. *Ceanothus* and *Arctostaphylos* are the exceptions that prove the rule. In these genera, radiation in recent times (but probably mostly predating establishment of the current mediterranean climate, Axelrod 1973) toward edaphic and climatic specialization demonstrates the possibilities for habitat specialization that most lineages have failed to realize.

The most prominent shrub generalist in the chaparral region is *Adenostoma fasciculatum* (Rosaceae), which has both an extensive range (Hanes 1965) and the ability to dominate large areas, often forming nearly monospecific stands. It is notable that *Adenostoma* is one of the relatively few chaparral species that combines an ability to sprout after fire with abundant production of seeds that germinate after fire and to some degree in periods without fire (Keeley and Zedler 1978). It has been reported that *A. fasciculatum* has a limited life span and that old stands deteriorate (Hanes 1971, 1988). We have not been able to confirm this for San Diego County. Healthy populations of *A. fasciculatum*, producing large numbers of seeds and having a large seed bank can be found in stands approaching 100 years since the last fire (Zammit and Zedler 1988).

The existence of a species that is locally dominant but also widely distributed in a complex landscape is difficult to reconcile with an evolutionary model that postulates significant coevolution taking place at a local scale along a unidimensional coenocline. *A. fasciculatum* now, and presumably at least since the early Pleistocene, has coexisted with diverse species, yet maintains approximately the same life-history characteristics throughout its range. It seems *A. fasciculatum* expanded from some ancestral habitat carrying its essential life-history characters, maintaining them despite local differences in site characteristics and biotic interactions.

This argument for the conservatism of life-history traits, and therefore of the dynamic roles of species in the ecosystem, is supported by the pattern of radiation along the elevational gradient and for edaphic diversity shown by *Ceanothus* (Fig. 4.2) and *Arctostaphylos*. Within these genera, radiation has been very unequal within the two major life-history groups (sprouting seeders, nonsprouting seeders) with the nonsprouters producing many more species (Wells 1969). Thus, speciation in these genera has been fostered more by diverse habitat than by opportunities to develop specialized dynamic roles.

Considering the success of *A. fasciculatum*, speciation may have been "unnecessary" for *Ceanothus* to succeed, and it may be only an artifact of the population structure and aspects of the breeding behavior. This view is given some support by the studies of Frazier (1993), who found that hybrids, even if between *Ceanothus* species with strikingly different life histories, seem not to be ecologically disadvantaged, at least in the zone of overlap and hybridization.

Dynamic Specialization in the Woody Flora

Dynamic specialization in the mediterranean region of California appears to be in a primitive state relative to theoretical expectations from models that postulate differentiation along temporal gradients. Although in a typical chaparral postfire successional sequence changes occur in abundance, invasion, and local extinction of species, they have little significance except in unusual circumstances (Cooper 1922, Zedler et al. 1983). This lack of a postdisturbance seral stage and of the widely dispersed early and midsuccessional itinerants that would be required to produce it has been seen in other fire-prone vegetation types (Whelan 1986). Successional specialization, to the extent that it exists, is revealed mostly by species that are stimulated to germinate after fire but ultimately decline and remain on the site as a dormant seed bank in the manner of the subshrub *Lotus scoparius*. Hanes (1971) calls these systems "autosuccessional" for fire, which is the only important natural disturbance. Expressing this in the terminology of this chapter, natural succession in the southern California chaparral zone is dominated by resident-persistent woody species and mostly subshrub or herbaceous resident-ephemeral species.

Therefore, if we were to follow the extreme recommendation by Zedler (1981), who argued that succession is best defined by extinction and invasion, we would be obliged to conclude that in normal circumstances succession (excluding human-induced successional sequences) does not operate in most of the chaparral region. This is not to say, of course, that there is no change. It could be that studies at the scale of 1 to 10 ha could find little evidence of directional change, yet landscape-level studies could reveal a shifting mosaic, with species populations and the community boundaries that they define always moving, as has been postulated for a Santa Barbara County chaparral–woodland–grassland landscape (Callaway and Davis 1993).

Taxonomic Distribution of Life-History Types

Considering the distribution of life-history types with respect to taxonomic groups suggests that the evolution of life history novelties may be severely constrained, thus severely limiting possibilities for the appearance of dynamic specialties. It is apparent that life-history types are linked to levels as high as the family. The Lamiaceae, for example, has five genera in the woody flora and eleven species. Ten of these, including the coastal sage scrub dominants *Salvia mellifera* and *S. apiana*, have life histories of type 2—that is, they are resident species that recover from fire by seeding and some sprouting, but they are also capable of recruiting without fire. One species, *Monardella linoides*, a species of washes and alluvial areas subject to periodic flooding, is classed with life history 6b. The Rhamnaceae has two genera (*Adolphia* and *Rhamnus*) in addition to *Ceanothus*, which was discussed above. *Adolphia* is similar to *Ceanothus* in being responsive to fire. In contrast, all three species of *Rhamnus*, though vigorous sprouters, establish seedlings in unburned chaparral but rarely in recent burns (Horton and Kraebel 1945, Keeley 1992a, 1992b).

The Asteraceae have a high proportion of wind-dispersed seeds, and several species (e.g., *Baccharis pilularis*) are among the few chaparral/sage scrub species that are capable of invading burns and other disturbances as well as grasslands (McBride 1974), and thus are somewhat itinerant. All the species are short- to moderate-lived. None plays the role of a long-lived resident-dominant comparable to the fire-resilient sprouters *Quercus dumosa*, *Malosma laurina*, or *Heteromeles arbutifolia*. Thus inability to achieve or unlikelihood to evolve great longevity appears to be a phylogenetic constraint shared among many species in the Asteraceae. This correlation between life history and phylogeny is hardly surprising, for some traits impose constraints and offer possibilities that limit the range of ecological roles that a species can fulfill. The strong correlation of life-histories among related species suggests that the origin of major life-history types and therefore of dynamic specialties is more likely to be a macroevolutionary event than the outcome of microevolutionary adjustments occurring because of selection under the influence of a habitat and set of co-occurring species. If so, then the assembly of a regional ecosystem is probably severely constrained by the set of genera and other high taxonomic entities that are available.

The lack of successional specialists may also be attributed to the scale, frequency, and stochastic properties of disturbance. The maximum opportunity for the itinerant life history presumably appears when resource-rich but competitor-poor patches of habitat are large and frequent and predictable. Small patches will be invaded by nearby species even if these are poorly dispersed. Infrequent occurrence of patches will make it difficult for the itinerant to maintain its population without a refuge from which it may invade, so that it cannot be considered an itinerant but an opportunistic resident, such as the originally riparian *Baccharis sarothroides* discussed earlier. Fires in mediterranean regions clearly create sufficiently large disturbance patches, but these patches are not devoid of competitors. If fires are very frequent, selection pressure is strong on the species affected to develop mechanisms for surviving fire, one of which, sprouting, most woody lineages already have to some degree. Thus in a hypothetical landscape gradually subjected to increasingly frequent, large fires, there are possibilities for perfecting both itinerancy and residency. In this race, the residents clearly have had the advantage. For most species fewer changes are probably required to survive fire than to develop a successful itinerant mode, and thus resident strategies can keep pace with increasing frequency of fire. But because fire inevitably opens sites for germination, even if seeds and sprouts of residents survive, an itinerant strategy can still succeed if sufficient propagules reach the new burn. To do so, large populations of the itinerant would need to be adjacent to the fire. But both the size of mediterranean fires and stochastic variation in fire occurrence make a pure itinerant strategy very risky, and significant progress toward a fire-patch—exploiting itinerant strategy is statistically improbable. If the seed-producing populations are small, they will colonize only a small proportion of the burn during the favorable period. For populations to be large, fires must be timed so that unburned populations peak in seed production at the time of most likely fire occurrence. If the fire fails to occur, however, large-scale local extinction will result. Conversely, residents strategies deal with stochastic fire behavior by persistent seed banks and great vegetative longevity, or both. Thus a number of factors work against the perfection of fire-exploiting itinerant strategies.

Implications for Comparison among
the Pacific Mediterranean Regions

The conclusions from this survey of woody plant-life histories do not necessarily refute those reached for convergence between Mediterranean regions (e.g., Cody and Mooney 1978). It is true that if phylogenetic constraints limit the development of life histories optimal with respect to similar climates and disturbance regimes, then convergence between regions with respect to dynamic roles could hardly be expected. Arguments made here suggest that if the life histories of the taxa available in the floras were different to begin with, they would probably end up different. The same may not be true for morphology and physiology, however. Physical factors, such as the frequency of killing frost, or the depth to which rooting must extend to allow survival of an evergreen through an extended drought, can be expected to impose absolute constraints on morphology and physiology, and hence some convergence may be expected. But it is difficult to imagine life-history constraints relevant to dynamic roles operating with the same absolute effect. It may be hypothetically true, for example, that species able to exploit the postfire situation to establish seedlings will fare better on average than those which do not (Whelan 1986). But for species that are able to persist by other means, it may be rare for selection pressures to be intense enough over large enough areas long enough to overcome all but the most trivial barriers to developing the necessary adaptations. Thus in Chile, and in California where a large proportion of species in the flora sprout but do not establish seedlings in burns, the rate of convergence toward the more conspicuously fire-adapted floras of Australia may be negligible despite the presence of fire as an important factor in all regions.

If phylogenetic constraints limit the evolution of life-history types, the possibilities for new dynamic roles and especially itinerant roles in disturbance-facilitated succession would also be limited. Regional floras may be assembled by selecting life-history types from those species that are physiologically suited to the prevailing conditions. It appears that adaptation to habitat (climate, soils) is primary, and dynamic adaptations secondary. Therefore, unless there were some reason for taxonomic groups with similar life histories to be similarly situated to exploit the opportunities offered by a mediterranean landscape, one would not expect assemblages of species in three disparate areas to have the same proportions of dynamic specialists, or even necessarily to have the vegetation dominated by species with similar life histories.

Because of the primacy of habitat specialization, one would expect to find that dynamic strategies that tended to "hold ground" by whatever means would be favored, as discussed above. Species that are resident by means of resprouting and dormant seeds are present in all regions, but the proportions of species that adopt these basic strategies and their variants differ. It seems certain that a portion of this difference is due to chance, but equally certain that some is attributable to differences in past selective regimes, and particularly to past disturbance regimes. Probably, however, most of the difference will be attributable to differences in the sets of species on which the mediterranean climate has acted rather than to the details of respective selective regimes in the late Pleistocene, including fire (Axelrod 1989).

Finally, the limitations on the means of comparison among regions must be considered. Like many other studies, the concepts of landscape exploitation addressed in this chapter are tied to species as the fundamental unit. In fact, one species could have in its collective local populations as much or more genetic variation in dynamic and habitat specialties as a set of species, and a genetically homogeneous species as much ecologically important genetic information controlling, for example, organ plasticity, as either of these. Mediterranean areas offer interesting possibilities for studies that will explore such questions. For example, Arroyo et al. (this volume) point out that in Chile the woody flora consists mainly of ancient taxa with one species per genus. This assemblage may be the repository of as much or more genetic diversity as the more species-rich woody assemblages in Australia and California.

Acknowledgments

This paper was completed while in residence at the Biology Department, University of Wollongong. Professor Rob Whelan, his staff, and his associates are thanked for their hospitality. Research that has contributed to the paper has been supported by the National Science Foundation, the California Department of Transportation, the California Department of Forestry (including the Integrated Hardwood Management Program managed by the University of California, Berkeley), the California Department of Parks and Recreation, and the Department of Defense through the U.S. Navy and the U.S. Air Force. The largest debt is owed to my students, who have provided energy, ideas, and insights.

References

Axelrod DI (1973) History of the mediterranean ecosystems in California. In Di Castri F, Mooney HA (eds) *Mediterranean Ecosystems: Origin and Structure.* Springer-Verlag, New York, pp 225–277

Axelrod DI (1989) Age and origin of chaparral. In Keeley SC (ed) *The California Chaparral. Paradigms Revisited.* Nat Hist Mus LA County, Los Angeles, pp 7–19

Beauchamp RM (1986) *A Flora of San Diego County, California.* Sweetwater River Press, National City, CA

Birkeland PW (1984) *Soils and Geomorphology.* Oxford U Press, New York

Bowman RH (1973) Soil survey. San Diego area, California. Part 1. USDA Soil Cons Ser For Serv, Washington, D.C

Brown JH, (1988) Species diversity. In Myers AA, Giller PS (eds) *Analytical Biogeography.* Chapman and Hall, London, pp 57–89

Bullock SH (1989) Life history and seed dispersal of the short-lived chaparral shrub *Dendromecon rigida* (Papaveraceae). Am J Bot 76:1506–1517

Callaway RM (1992) Effect of shrubs on recruitment of *Quercus douglassii* and *Quercus lobata* in California. Ecology 73:2118–2128

Callaway RM, Davis FW (1993) Vegetation dynamics, fire, and the physical environment in Coastal Central California. Ecology 74:1567–1578

Close DH, Gilbert DE, Peterson GD (1970) *Climates of San Diego County.* U Calif Agric Ext Serv

Cody ML, Mooney HA (1978) Convergence versus nonconvergence in Mediterranean ecosystems. Ann Rev Ecol Syst 9:265–321

Comins HN, Noble IR (1985) Dispersal, variability, and transient niches: Species coexistence in a uniformly variable environment. Am Nat 126:706–723

Cooper WS (1922) *The Broad-Leaf Sclerophyll Vegetation of California*. Carnegie Inst Washington, Washington, DC

DeSimone SA, Burk JA (1992) Local variation in floristics and distributional factors in Californian coastal sage scrub. Madroño 39:170–188

Fowells HA (1965) Silvics of forest trees of the United States. US DA, Washington, DC

Frazier CK (1993) An ecological study of hybridization between chaparral shrubs of contrasting life histories. M.S. Thesis, San Diego State U, San Diego, CA

Grubb PJ (1977) The maintenance of species richness in plant communities: the importance of the regeneration niche. Bio Rev 52:107–145

Hanes TL (1965) Ecological studies on two closely related chaparral shrubs in Southern California. Ecol Mono 35:213–235

Hanes TL (1971) Succession after fire in the chaparral of Southern California. Ecol Mono 41:27–52

Hanes TL (1988) California chaparral. In Barbour MG, Major J (eds) *Terrestrial Vegetation of California*. California Native Plant Society, Davis, CA, pp 417–469

Heady HF (1988) Valley grassland. In Barbour MG, Major J (eds) *Terrestrial Vegetation of California*, California Native Plant Society, Davis, CA, pp 491–514

Hopkins AJM, Griffin EA (1984) Floristic patterns. In Pate JS, Beard JS (eds) *Kwongan: Plant Life of the Sandplain*. U West Aust Press., Nedlands, Western Australia, pp 69–83

Horton JS, Kraebel CJ (1955) Development of vegetation after fire in the chamise chaparral of southern California. Ecology 36:244–262

Johnson C (1994) Woodland expansion in the Platte River, Nebraska: Patterns and causes. Ecol Mono 64:45–84

Keeley JE (1992a) Demographic structure of California chaparral in the long-term absence of fire. J Veg Sci 3:79–90

Keeley JE (1992b) Recruitment of seedlings and vegetative sprouts in unburned chaparral. Ecology 73:1194–1208

Keeley JE, Zedler PH (1978) Reproduction of chaparral shrubs ofter fire: A comparison of the sprouting and seeding strategies. Am Mid Nat 99:142–161

Kirkpatrick JB, Hutchinson CF (1980) The environmental relationships of Californian coastal sage scrub and some of its component communities and species. J Biegeog 7:23–38

Kruckeberg AR (1984) *California Serpentines: Flora, Vegetation, Geology, Soils, and Management Problems*. U California Press, Berkeley, CA

Lawson, DM (1993) The effects of fire on stand structure of mixed *Quercus agrifolia* and *Q. engelmannii* woodlands. M.S. Thesis, San Diego State U, San Diego, CA

Lloret F, Zedler PH (1991) Recruitment pattern of *Rhus integrifolia* populations in periods between fire in chaparral. J Veg Sci 2:217–230

Louda SM (1982) Distribution ecology: Variation in plant recruitment over a gradient in relation to insect seed predation. Ecol Mono 52:25–41

MacArthur RH (1972) *Geographical Ecology*. Harper and Row, New York, NY

McBride JR (1974) Plant succession in the Berkeley Hills, California. Madroño 22:317–329

McBride JR, Strahan J (1984) Establishment and survival of woody riparian species on gravel bars of an intermittent stream. Am Mid Nat 112:235–245

McMaster GS (1981) Patterns of reproduction in Torrey pine (*Pinus torreyana*). M.S. Thesis. San Diego State U, San Diego, CA

McMaster G, Zedler PH (1981) Delayed seed dispersal in *Pinus torreyana* (Torrey Pine). Oecologia 51:62–66

McMinn HE (1939) *An Illustrated Manual of California Shrubs*. U California Press, Berkeley, CA

Miller PC, Bradbury DE, Hajek E, LaMarche V, Thrower NJW (1977) Past and present environment. In Mooney HA (ed) *Convergent Evolution in Chile and California*. Dowden, Hutchinson, and Ross, Stroudsburg, PA, pp 27–72

Mills JN, Kummerow J (1989) Herbivores, seed predators, and chaparral succession. In Keeley SC (Ed) *The California Chaparral. Paradigms Revisited*. Nat Hist Mus LA County, Los Angeles, pp 49–55

Minnich RA (1989) Chaparral fire history in San Diego County and adjacent northern Baja California: An evaluation of natural fire regime and the effects of suppression management. In Keeley SC (ed) *The California chaparral. Paradigms Revisited*. Nat Hist Mus LA County, Los Angeles, pp 37–47

Nobs ME (1963) Experimental studies on species relationships in *Ceanothus*. Pub No 623. Carnegie Inst Washington, DC

Plumb T (1980) Response of oaks to fire. In Plumb T (ed) *Ecology, Management, and Utilization of California Oaks*. USDA For Serv. Pacific Southwest For Range Exp Sta. Berkeley, CA, pp 202–215.

Rehlaender WE (1992) Nutrient status of the chaparral plant–soil system during stand development after fire: The effect of stand age and substrate type. M.S. Thesis. San Diego State U, San Diego, CA

Schulze ED, Mooney HA (1993) Biodiversity and ecosystem function. Springer-Verlag, Berlin

Simberloff D, Dayan T (1991) The guild concept and the structure of ecological communities. Ann Rev Ecol System 22:115–143

Specht RL (1981) Heathlands. In Groves RH (ed) *Australian Vegetation*. Cambridge U Press, Cambridge, pp 253–275

Steward D, Webber PJ (1981) The plant communities and their environment. In Miller PC (ed) *Resource Use by Chaparral and Matorral*. Springer-Verlag, New York, pp 43–68

Stewart FM, Levin BR (1973) Partitioning of resources and the outcome of interspecific competition: A model and some general considerations. Am Nat 107:171–198

Vogl RJ (1970) Fire and plant succession. In Sym on role of fire in intermountain west. Interm Fire Council, Missoula, Montana, pp. 65–75

Wells PV (1969) The relation between mode of reproduction and extent of speciation in woody genera of the California chaparral. Evolution 23:264–267

Westman WE, O'Leary JF, Malanson GP (1981) The effects of fire intensity, aspect and substrate on post-fire growth of Californian coastal sage scrub. In Margaris NS, Mooney HA (eds) *Components of Productivity of Mediterranean Regions: Basic and Applied Aspects*. Dr. W. Junk, The Hague, pp 151–179

Whelan RJ (1986) Seed dispersal in relation to fire. In Murray DR (ed) *Seed Dispersal*. Academic Press, Sydney, pp 237–271

Whittaker RH (1975) *Communities and Ecosystems*, 2nd Edition. Macmillan Publishing Company, New York, NY

Zammit C, Zedler PH (1988) The influence of dominant shrubs, fire, and time since fire on soil seed banks in mixed chaparral. Vegetatio 75:175–187

Zedler JB (1982) The ecology of southern California coastal salt marshes: A community profile. US Fish and Wildlife Serv, Bio Serv Program. FWS/OBS-31/54, Washington, DC

Zedler PH (1977) Life history attributes and the fire cycle: a case study in Chaparral dominated by *Cupressus forbesii*. In Mooney HA, Conrad CE (eds) *Proc Sym Env Cons Fire and Fuel Man Mediter Ecosy*. US DA, For Serv, Washington, DC, pp 451–458

Zedler PH (1981) Vegetation change in chaparral and desert communities of San

Diego County. California. In West DC, Shugart HH, Botkin DB (eds) *Forest Succession*, Springer-Verlag, New York, pp 406–430

Zedler PH, Gautier CR, McMaster GS (1983) Vegetation change in response to extreme events: The effect of a short interval between fires in California chaparral and coastal scrub. Ecology 64:809–818

Zedler PH, Scheid GA (1988) Invasion of *Carpobrotus* and *Salix* after fire in a coastal chaparral site in Santa Barbara County, California. Madroño 35:196–201

Zedler PH, Zammit CA (1989) A population-based critique of concepts of change in the chaparral. In Keeley SC (ed) *The California Chaparral. Paradigm Revisited.* Nat Hist Mus LA County, Los Angeles, pp 73–83

5. Biogeography and Floral Evolution in the Geoblasteae (Orchidaceae)

Peter Bernhardt

Can terrestrial orchids be a convincing case study for the evolutionary ecology of mediterranean vegetation within the Pacific rim? The family, Orchidaceae, has a tropical center of diversity and 67 to 73% of its species are epiphytes (Gentry and Dodson 1987). It shows considerable secondary diversity, however, outside tropical regions, especially in the smaller, monandrous subfamilies: Spiranthoideae, Neottioideae, and Orchidoideae (*sensu* Burns-Balogh and Funk 1986). Mediterranean regions appear to be centers of speciation for the subfamilies Neottioideae and Orchidoideae (Dressler 1981; Dafni and Bernhardt 1990). The depauperate orchid flora of southern California (see Luer 1975) must be treated as an exception, not the rule, compared to mediterranean Australia, Chile, South Africa and southern Europe.

The Tribe Geoblasteae (Neottioideae; *sensu* Burns-Balogh and Funk 1986), is of particular interest to Pacific biogeography for its disjunct distribution in South America and Australasia. Burns-Balogh and Funk (1986) defined this tribe by a suite of interconnected characters emphasizing column structure (e.g., degree of fusion and morphology of staminodes, stylar neck, rostellum) and the organization of the pollinarium (e.g., number of pollinia, attachment to viscidium, viscidium consistency).

The Tribe Geoblasteae is subdivided into two subtribes. These two taxa have been defined by a combination of vegetative and geographic characters that overlap (see Dressler 1981). The Chloraeinae have fleshy, fascicled roots, the leaves are spirally arranged or whorled, and this subtribe is confined to South America and New Caledonia. In contrast, the Caladeniinae tend to produce root-stem tuberoids (*sensu* Dressler 1981) and an individual shoot rarely produces more than one basal leaf subtending each peduncle or inflorescence. This subtribe, absent from the Western Henisphere, has an Australasian center of diversity. The placement of "intermediate" genera within the two subtribes varies because some taxonomists weigh morphological and geographical characters differently (see review by Ackerman and Williams 1981).

Floristic studies have made it abundantly clear that both subtribes have undergone intensive speciation within mediterranean regions on both sides of the Pacific (Correa 1956; 1969; Nicholls 1969). Although no final consensus has been reached about the position of some genera within the two subtribes, the suggestion is strong that Chloraeinae and Caladeniinae represent a grade as much as they do a clade. Even when these two subtribes have suffered the more traditional submerging within the large, rather synthetic tribe Diurideae, most orchidologists treat Chloraeinae and Caladeniinae as sister taxa. That is, they appear more closely allied to each other than to Diuridinae, Prasophyllinae, and so on (*sensu* Dressler 1981 and see Ackerman and Williams 1981).

In this chapter we combine taxonomic and phytogeographic data with aspects of comparative morphology and pollination ecology. This combination could provide new considerations of how adaptive radiation and phylogenetic constraints may account for the origin, evolution, and migration of Geoblasteae with a mediterranean emphasis.

Information Base

Modern revisions have been selected to represent each orchid flora to clarify and standardize nomenclature. Correa (1956; 1968; 1969 and 1970) represents the Geoblasteae in South America following the author's examination of the original collection by Correa and the collection of Geoblasteae within the herbarium of the Department of Biology, Universidad Concepción, Chile (CONC). Australasian Geoblasteae is represented by Halle (1977) for New Caledonia, Johns and Molloy (1985) for New Zealand, and Clements (1982) for Australia and Tasmania. Any taxonomic conflicts between Johns and Molloy (1985) and Clements (1982) defers to the latter authority.

Specimens derived from private, spirit collections of the author and dried and/or spirit collections from AD, BAB, CANB, HB, MO, NSW, and PERTH. Dried material was rehydrated by soaking specimens in 10% NaOH for 48 hours to soften and clear material. The rehydrated specimen was then washed in distilled water and taken through three half-hour changes of increasing concentrations of ethanol before storing in 70% ethanol. Spirit collections of whole flowers represent preservation in FAA or specimens were killed in 3/1 ethanol/glacial acetic acid for 2 to 24 hours before storing in 70% ethanol.

All genera in the Geoblasteae were examined for the presence of derived vegetative and reproductive characters. If one species in the genus lacked an advanced character the *entire* genus was treated as lacking the derived state. A character is regarded as derived based on comparison with states in out-group tribes (e.g., Neottieae, Pterostylideae; *sensu* Burns-Balogh and Funk 1986), evolutionary trends suggested by Dressler (1981); and Stebbins's (1974) treatment of reversible versus irreversible trends.

Diversity Versus Distribution

A total of 22 genera and about 170 species comprise the Geoblasteae (Table 5.1). There are six genera (*Bipinnula, Chloraea, Codonorchis, Gavilea, Geoblasta,* and *Megastylis*) within the Chloraeinae. The 16 remaining genera and 91 species are treated as Caladeniinae.

Table 5.1. Phytogeography and species' diversity of the Geoblasteae

Genus	Southwestern Australia	E. Australian Archipelago	South America
Adenochilus	0	2	0
Aporostylis	0	1	0
Arthrochilus	0	4	0
Bipinnula	0	0	10
Burnettia	0	1	0
Caladenia	46	29	0
Caleana	1	1	0
Chiloglottis	0	5	0
Chloraea	0	0	47
Codonorchis	0	0	1
Drakaea	3	0	0
Elythranthera	2	0	0
Eriochilus	2	1	0
Gavilea	0	0	13
Geoblasta	0	0	1
Glossodia	1	1	0
Leporella	1	1	0
Lyperanthus	3	3	0
Megastylis	0	7	0
Paracaleana	1	1	0
Rimacola	0	1	0
Spiculaea	1	0	0
Total genera	10	14	5
Total endemic genera	3	7	5
Number of species	61	58	72

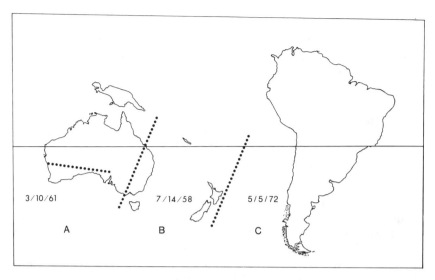

Figure 5.1. Comparative distribution and diversity of the Geoblasteae. Schematic diagram of the land masses showing their relative positions to the Tropic of Capricorn (solid black line). Regions of diversity are isolated by dotted lines. (A) southwestern Australia. (B) eastern Australian archipelago. (C) South America. Three sets of numbers compare relative diversity: number of endemic genera, total of genera, total number of species.

The distribution of the tribe may be subdivided into three regions based on geographic/environmental isolation, comparative diversity, and proposed morphological alliances of endemic genera (Table 5.1; Fig. 5.1). Southwestern Australia is exclusively mediterranean with a California type of mosaic of climatic and edaphic parameters (James and Hopper 1981). Conversely, the eastern Australian archipelago is the most geographically fragmented region, incorporating tropical montane through temperate sclerophyll habitats (including at least one mediterranean region; see Fox, this volume). Within the archipelago the tribe is distributed from north to south with endemic genera found in such extreme habitats as *Melaleuca squarrosa* swamps (*Burnettia*) and the ledges of sandstone clifs (*Rimacola*; see Nicholls 1969). South America is a region similar to the eastern Australasian archipelago in its range of climatic and edaphic zones, but there insularity of the Geoblasteae depends on topographical barriers and desertification instead of isolation by sea (see Correa 1969).

Patterns of endemism and species diversity appear to depend on a combination of land-mass dimensions, spatial isolation, and latitude. The Australian continent has far more species than the smaller islands in the archipelago (Table 5.2). Tasmania has more species of Geoblasteae than any of the smaller islands, but it has been separated from the Australian mainland for

Table 5.2. The comparative diversity of the Geoblasteae through the southern Pacific basin

Land mass	Genera	Species	Endemic genera	Monotypic genera
Australia	14	89	4 (6)[a]	5
Tasmania	9	32	0 (6)[a]	3
New Zealand	6	9	1	1
New Caledonia	2	8	1	0
South America	5	76	5	2

[a] *Arthrochilus, Burnettia, Caleana, Eriochilus, Glossodia, Leporella* are endemic to Australia *and* Tasmania.

less than 400,000 years. Tasmania shares ten genera with the Australian continent, compared to the one genus Australia shares with New Caledonia and five with New Zealand. Not surprisingly, South America has the most endemic genera and shares no species with Australasia (Table 5.2).

Megastylis (New Caledonia) is the only genus in the tribe endemic north of the Tropic of Capricorn (Fig. 5.1), with some species distributed through tropical lowlands (starting only 500 m above sea level; Halle 1977). *Arthrochilus* has its center of diversity north of the Tropic of Capricorn and a few species found southward into temperate New South Wales (Jones 1988). Otherwise, the number of species in this tribe drops precipitously approaching this latitude until the tribe is represented only by a few plastic and polymorphic species (see below). This predominantly temperate distribution adds to the argument that the Geoblasteae originated in a warm, temperate region in the southern Pacific. Their limited ability to radiate within the tropics and cold, subtemperate zones (see Correa 1969; George 1981) suggests a phylogenetic constraint similar to the distribution of the allied tribe, Thelymitreae (Burns-Balogh and Bernhardt 1988).

Radiation Within Mediterranean Areas

Conservatively applying the checklist of Orchidaceae provided by Marticorena and Quezada (1985) indicates that 42 to 44 species of Geoblasteae show continuous or partial distribution through mediterranean South America. But 61 species of Geoblasteae are found in the mediterranean southwest of Australia (Table 5.1). Thus about 56% of South American Geoblasteae and 62% of Australasian species occur in broadly mediterranean environments. This striking, shared trend goes even further when we notice that *Gavilea* and Australasian *Drakaea*, *Elythranthera*, and *Spiculaea* are almost endemic to mediterranean zones (Correa 1968; Clements 1982) on opposite sides of the Pacific. Therefore, both of these mediterranean regions appear to be species sinks (sensu Stebbins 1974) and centers of evolution at the generic level for some taxa in the same tribe.

Still, however, no morphological, floristic, or paleontological evidence suggests that either mediterranean region is the ancestral site of origin for this tribe. Notice, for example, that the eastern Australasian archipelago has more genera than southwestern Australia (and more than all of South America) plus a marginally higher number of endemic genera (Table 5.1), even though this region holds fewer species. Although macroevolution within this tribe has occurred independently in two isolated mediterranean regions, diversity remains very skewed. For example, Chile is the center of diversity for only two genera, *Chloraea* and *Gavilea*, but their combined number of species accounts for 45% of the Chloraeinae (Marticorena and Quezada 1985; Marticorena 1990).

Shared Trends Within the Geoblasteae

Seven trends are recurrent throughout the Geoblasteae. The phytogeographic trend is dicussed above and is summarized thus: most genera occur south of the Tropic of Capricorn and north of Antarctica, with mediterranean regions serving as centers of speciation, but with few taxa in the true deserts (Fig. 5.1, Table 5.2).

Ecological plasticity comprises the second trend. No species is pan-Pacific but a few species in each subtribe are widely distributed regardless of topography, climate, or soil. For example, *Caladenia carnea* occurs from New Caledonia south through Tasmania and westward into southwestern Australia (Halle 1977; Clements 1982; Johns and Molloy 1985; Jones 1988).

The third trend refers to shared peaks in flowering seasons from early spring to early summer. Of course, tropical taxa (Halle 1977) and species restricted to alpine–subalpine zones (e.g., *Adenochilus* and *Lyperanthus* in New Zealand; Johns and Molloy 1985) are exceptions. Examining sheets at CONC, we find that Chloraeinae with subantarctic–Patagonian distributions may delay flowering until February (austral summer).

The fourth trend indicates that some species of Geoblasteae self-pollinate via mechanical autogamy when pollinators are lacking. Self-pollination is reported for some Chilean Chloraeinae at higher altitudes where appropriate insects are infrequent or sluggish (Arroyo and Squeo 1990; Catling 1990) and in Caladeniinae from many habitats in which vernal flowering may actually start in late winter (e.g., *Chiloglottis*; Beardsell and Bernhardt 1982; Catling 1990; Dafni and Bernhardt 1990). In fact, self-pollination where no dependable pollinators are available is a fail-safe mechanism shared by many vernal flowering herbs regardless of family (Schemske et al. 1978).

In the fifth trend the apices of sepals or entire sepals and lateral petals tend to become long and narrow or filiform. These perianth segments are often ornamented with papillae, hammer glands (e.g., *Chloraea*, *Caladenia*), or vasculated appendages (e.g., *Bipinnula*; Fig. 5.2). These elaborate structures may function as perfume glands (Holman 1976; Vogel 1990).

The labellum of a flower in the Geoblasteae may be sessile (e.g., *Geoblasta*, *Glossodia*) or connected to the base of the column, or column foot, via a claw (Fig. 5.3). Both subtribes show a sixth trend in which the labellum

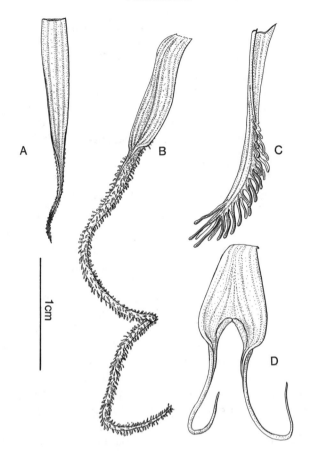

Figure 5.2. Lateral sepals of Geoblasteae and Pterostylideae. (A) *Rimacola elliptica* with acute, papillate apex. (B) *Caladenia filamentosa* with stalked trichomes. (C) *Bipinnula gibberti* with vasculated appendages (each appendage has one vascular trace). (D) *Pterostylis angusta*, two lateral sepals fused basally with apices antenniform.

lamina and its claw are connected by a thin, flexible hinge (e.g., *Bipinnula*, *Caladenia*, *Codonorchis*) or a hinge connects the labellum to the column base (*Spiculaea*) and/or its foot (*Arthrochilus* and see Fig. 5.3).

Finally, the pollen grains are arranged in tetrads in almost all genera in this tribe (Ackerman and Williams 1981). In *Caladenia* and *Chloraea*, strap-like bands connect grains. These bands are the only features unique to the shared trends within the Geoblasteae. All six preceding trends parallel those found in other orchid tribes and appear to be somewhat recurrent within the family (Dressler 1981; Vogel 1990). Such trends appear to be more indicative of limits to variation within most orchid genera than a proscribed pattern of

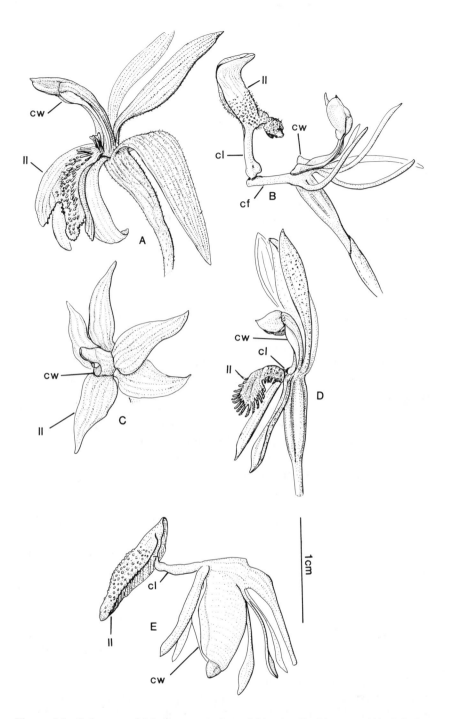

Figure 5.3. Column and labellum variation within the Geoblasteae. (A) *Caladenia sericea*. (B) *Drakaea glyptodon*. (C) *Gavilea kingii* (notice the fused bases of the column wings, cw, forming a short tube). (D) *Leporella fimbriata*. (E) *Paracaleana nigrita*: cf = column foot, cl = labellum claw, cw = column wings, 11 = labellum lamina.

adaptive radiation, for the Geoblasteae have spread from some center of origin (Burns-Balogh and Bernhardt 1988; Dafni and Bernhardt 1990).

Putative Origin and Dispersal

Dressler (1981) proposed that the Chloraeinae represent a subtribe that "remained" in South America while other orchid groups migrated to Australasia via a coastal, Antarctic corridor. Thus the vagile seeds of ancestral Geoblasteae first moved through the southern Pacific within the temperate, *Nothofagus* woodlands and forests. This vegetation extended across southern South America, Australia, and the once-temperate coastal strip of Antarctica as late as 7 to 10 million years ago (Thorne 1986; Carlquist 1987; Christophel and Greenwood 1989).

The antithesis of this theory places the origin of the tribe somewhere within nascent Australasia because the eastern half of this region remains richest in both genera of Geoblasteae and allied neottioids (*sensu* Lavarack 1976). The tribe would have originated in the tropical "shadow" of Laurasia (e.g., modern *Megastylis*), moving southeast into temperate Antarctica–South America and also westward into what would become mediterranean Australia.

After examining all genera in this tribe I support the original theory of Dressler (1981). An obvious correlation between phytogeography and relict taxa forms when we compare the distribution of ten advanced characters throughout the southern Pacific (Table 5.3). South American genera appear to be least advanced. Genera of mediterranean Australia are most advanced and those throughout the eastern Australasian archipelago seem intermediate or show a modest increase in advanced characters compared to South America taxa. If we accept a northeastern, Australasian origin for the Geoblasteae, we are left with the uncomfortable coincidence that southwestern Australia has become the repository for the most advanced genera but South America is the refugium for the most primitive.

A far more parsimonious scenario would treat temperate South America as the center of origin for the tribe that has undergone a secondary but explosive radiation upon invading Australasia. The Caladeniinae would be derived from the Chloraeinae based on this proposal and the morphological evidence. Each advanced character (Table 5.3) may be represented by at least one species of Chloraeinae in South America. The advanced character is simply not consistent throughout *every* species in the same South American genus. For example, some *Bipinnula* spp. have a ligulate, labellum claw and only one basal leaf, but such species occupy only one of two sections in this genus (Izaguirre de Artucio 1973).

Small genera of Geoblasteae endemic to either mediterranean region seem to have evolved from larger, invasive genera. For example, the coalescence of column wings and their adnation to the base of the labellum most probably

Table 5.3. The correlation between morphological trends and the distribution of the Geoblasteae

Advanced character state	Southwestern Australia	E. Australian Archipelago	South America
1. Tuberoid(s) present	10/10[a]	11/14	2/5
2. Leaf solitary and basal	10/10	10/14	0/5
3. Single, terminal flower *or* reduced receme (2 to 4 flower buds)	5/10	9/14	2/5
4. Nectaries absent	10/10	14/14	3/5
5. Ligulate-terete labellum claw and/or elongated column foot interconnected by flexible hinge	4/10	3/14	0/5
6. Labellum lamina entire *or* trilobate state suppressed	7/10	10/14	2/5
7. Lamina "teeth" or serrations replace terete marginal appendages	4/10	3/14	1/5
8. Central lamina appendages few (< 12) but at least one appendage has 1 to 3 vascular traces	6/10	4/14	1/5
9. Sigmoid column emergent and elongate	2/10	1/14	0/5
10. Column wings broad-continuous or widely and abruptly flared	6/10	4/14	1/5

[a] Number of genera in which all species have the character/total number of genera in region (Fig. 5.1).

accounts for the derivation of *Gavilea* from *Chloraea* (Fig. 5.3). The elongated column foot and suppression of the trilobate lamina and its vasculated appendages divides *Drakaea s.s.* from the *Caladenia barbarossa* complex within *Caladenia s.s.* Orchidologists appear to define mediterranean genera of Geoblasteae on comparatively minor changes in the suite of floral characters.

Whatever the direction of migration, the monotypic genus *Codonorchis*, may provide a fascinating model for the exchange of Geoblasteae between continents, coupled with a transitional morphology. The distribution of this taxon in South America is subantarctic as well as submediterranean, for its range extends as far south as Tierra del Fuego and the Falklands (Correa 1970). Examining labels at CONC shows that it is both a herb of *Nothofagus* woodlands and it invades light gaps in closed forests where older trees have fallen. Although placed within the Chloraeinae, *Codonorchis* produces a tuberoid (Fig. 5.4). The labellum is trilobate with a simple claw hinge but has two basal appendages (calli), each with one, vascular trace (the same feature as in *Caladenia*, Section Eucaladenia). Atypically, the pollen grains

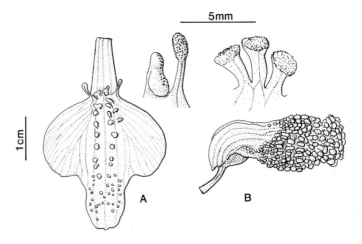

Figure 5.4. Putative mimesis in the labella of *Codonorchis* and *Bipinnula*. (A) Trilobate labellum of *Codonorchis lessonii* featuring appendages with swollen, yellow, papillate apices (only the two basal, stalked appendages are vasculated). (B) Labellum of *Bipinnula gibberti* with a terete claw and labellum lobes suppressed. Warty appendages are hairy with marginal warts darkly pigmented and vasculated.

are monads (Ackerman and Williams 1981). Leaves form a basal rosette instead of a spiral up the stem, which is far more typical for Chloraeinae. Understandably, some taxonomists prefer to place this genus in its own subtribe (Brieger 1974–1975; S. Hopper, personal communication). *Codonorchis*, however, is probably of greater use to the evolutionary ecologist as a model of incipient radiation based on exaptive morphology.

Macroevolutionary Constraints on Generic Diversity

The species of Geoblasteae in Australia are 22% more numerous than those in South America, and Australasia contains 77% of the tribe's genera. Indeed, the mediterranean southwest in Australia has almost a third more species than mediterranean South America. Are we merely looking at a history of lumping in South America and splitting in Australasia, or have subtle selective pressures exerted different trends on opposite sides of the Pacific over the ages? Does this history imply that the Caladeniinae have greater macroevolutionary potential than the Chloraeinae? I argue that the orchidologists' bias has been influenced by real evolutionary constraints.

Because Australasia is an archipelago spread over a wide edaphic, climatic area, insular evolution (*sensu* Carlquist 1974), coupled with allopatry and genetic drift (Gentry and Dodson 1987; Burns-Balogh and Bernhardt 1988) may be invoked. Simple geographic isolation has probably led to the evolution of *Megastylis*, *Aporostylis*, *Drakaea*, and *Elythranthera*. Edaphic or topographic isolation or both have contributed to the isolation and evolution of *Adenochilus*, *Burnettia*, *Rimacola*, and *Spiculaea*. Although the Geoblasteae in South America lacked the opportunity for isolation on large island chains it is obvious that their opportunities for fragmentation caused by topographic and/or climatic changes were equal to or greater than those of taxa in Australasia. Correa (1969) split *Chloraea* into three "groups," observing that morphology correlated partly with distributions interrupted by the Andes, Amazonian basin, and deserts. Izaguirre de Artucio (1973) refrained from dividing *Bipinnula* into two genera but noted that the two sections show degrees of discontinuity in both morphology and distribution. It is not enough to argue that modern taxonomists of the Chloraeinae are too few and too conservative. Correa (1968) "rehabilitated" *Geoblasta*, cleaving the monotypic genus from *Chloraea s.s.* by a dichotomy of vegetative and floral characters.

Australian orchidologists have been criticized for repeatedly subdividing taxa. But only two naturally occurring intergeneric hybrids (both within the Geoblasteae) have been documented for the entire terrestrial orchid flora in continental Australia; *Caladenia deformis X Glossodia major* and *Caladenia sericea X Elythranthera brunonis*. In fact, the *Caladenia X Elythranthera* is a rare, recurrent F1 confined to the mediterranean southwest. This status implies a taxonomic approach far more conservative in Australia than in the classic treatment of the European genera in the Mediterranean basin. Within the orchid flora in the Mediterranean basin, intergeneric hybrids have been recorded between 16 genera, and F1s frequently show high fertility (Dafni and Bernhardt 1990).

Therefore, the discrepancy in the number of genera of Geoblasteae, on opposite sides of the Pacific, probably reflects differences in floral presentation and pollination mechanisms leading to different emphases in reproductive isolation. Specifically, the largest genera in South America (*Bipinnula*, *Chloraea*, and *Gavilea*) take in *all* the nectariferous species in the tribe. Information remains scant, but these nectariferous taxa offer similar pollination mechanisms. The insect lands on the labellum and probes for nectar by placing its head or mouthparts or both into a shallow chamber or down a short tube-floral throat. With the dorsal portion of the insect's head or thorax positioned under the receptive column the forager carries off the pollinarium upon backing out of the flower (Gumprecht 1980).

Conversely, all the Australasian genera display syndromes of food (pseudanthery) or sexual (pseudocopulation) mimesis (Dafni and Bernhardt 1990). Within *Caladenia*, for example, species may show a gradation between dummy anthers and dummy females on labellum lamina (Stoutamire 1982; Bates 1985). Different scents, pigmentation, and labellum sculptures stimulate and deceive different insect families and genders. The flowers of Australasian Geoblasteae also show sharp divergences in respective degrees of resupination (Gann and Carlquist 1985), length and elasticity of the labellum claw, and angle, length and width of the column wings that hold the deceived

Table 5.4. Distribution and endemism of *Pterostylis* spp. through the
southern Pacific basin.

Land mass or region	Total species	Total endemics
Antarctica	0	0
Southwestern Australia	16	6
East coast Australia	52	33
Tasmania	26	1
New Zealand	19	8
New Caledonia	3	1
South America	0	0

insect during transference of pollinia (Fig. 5.3). Even though the pseudo-
copulatory flowers of Australasian Geoblasteae are pollinated exclusively by
male Hymenoptera, different orchid genera are pollinated by different genera
of wasps within the family Tiphiidae (Stoutamire 1981; Dafni and Bernhardt
1990), one genus of ponerine ant (Peakall 1989), and a genus of the wasplike
sawflies (Bates 1989).

This overwhelming trend toward pollination-by-deceit within Australasia
must be contrasted with the Chloraeinae. Floral mimesis occurs in South
America but seems limited to nectarless *Codonorchis* (Gumprecht 1980),
Geoblasta (Correa 1968), and the Uruguayan *Bipinnula* spp. (Izaguirre de
Artucio 1973). With its yellow papillose calli (Fig. 5.4), *Codonorchis* may
be a pseudantherous flower like the *Caladenia* spp. in section Eucaladenia
(Dafni and Bernhardt 1990). *Geoblasta* and the Uruguayan *Bipinnula* spp.
have dark colors, thickened plates, and the hairy "warts" of pseudocopula-
tory flowers (Fig. 5.4) like Australasian *Chiloglottis* (Dafni and Bernhardt
1990). It is also possible that some *Chloraea* spp. grade toward pseudocopulation based on the "blackened" sculptures on their labella and sepal
apices. Comparative rates of nectar secretion have not been recorded in
Chloraea or *Bipinnula* spp., and so such observations remain speculative.
Mimetic flowers, though, are obviously underrepresented in South American
Geoblasteae compared to Australasia. Unlike mediterranean Australia, the
mediterranean coast of South America appears to be particularly depauperate in Geoblasteae with mimetic flowers.

If Geoblasteae have migrated from South America to Australasia, the
dominance of pollination-by-deceit in Australasia may merely represent the
canalizing of ancestral stock. That is, nectariferous taxa might never have
arrived in Australasia if the Australasian taxa had been derived from a common ancestor with *Codonorchis* or *Geoblasta* or both.

Opportunities for exploiting pollinators in Australasia differ from those in
South America. A large proportion of the Caladeniinae are pollinated by
male tiphiid wasps. The female wasps are wingless soil dwellers, but the
flying males take nectar and are deceived by both food and sexual mimics
(Stoutamire 1981). The subfamily Thynninae covers the majority of "orchid

wasps," but South America has only about 200 species, whereas Australia has nearly 800 species (L.S. Kimsey, personal communication). It is obvious that the adaptive radiation of pollination-by-deceit in Australian Geoblasteae has depended greatly on explosive speciation among their dominant pollinators.

Leporella has wide distribution in southern Australia from temperate east to mediterranean west. It is treated as a monotypic genus (Table 5.1, Fig. 5.3D) and is pollinated exclusively by males of the bulldog ant, *Myrmercia urens*. The broad distribution but low diversity of this genus suggests a standard pattern of macroevolution via character displacement (Futuyma 1986). The ancestors of modern *Leporella* were probably inadequate competitors for male tiphiid pollinators. A similar pattern may reflect evolution of the widespread but submediterranean monotypic genus *Caleana*, which appears to be pollinated exclusively by male sawflies in the genus *Pteryogophorous* (Bates 1989).

Therefore, it is most likely that the treatment of generic diversity in the Geoblasteae reflects the taxonomist's intellectual response to the relative scale of morphological elaboration as reflected by the pollination syndrome of the flowers. Nectariferous taxa are more likely to be lumped than mimetic taxa, for column-labellum morphology and sculpturing of mimics is far more exaggerated than the rather monotonous pouched, spurred, or tubular blossoms of nectar producers. This habit appears to be indicative of classification throughout the Orchidaceae regardless of tribe or subfamily. Note the parallelisms inherent in the nectariferous, moth-pollinated genera *Angraecum*, *Neofinetia*, and *Anesiella* and their incorrect initial lumping within *Angraecum s.l.* (see Dressler 1981). Separation of the nectariferous genera, *Dactylorrhiza*, *Gymnadenia*, and *Platanthera* is still subject to debate, because they form intergeneric hybrids in the field and show overlapping floral and vegetative characters (Dressler 1981; Dafni and Bernhardt 1990).

Evolution of the Pterostylideae

Finally, unlike the Geoblasteae in South America, this tribe has spread so extensively through Australasia that it may have produced a second tribe following further character displacement of pollination mechanisms. The Pterostylidae consists of a single genus unique to Australasia (Table 5.4), but the majority of species flower from midwinter to midspring (Dafni and Bernhardt 1990). All species are nectarless but the major pollinators appear to be fungus gnats (Diptera; Mycetophilidae) active during the cool and wet months when sporocarps flourish. Pollination syndromes described so far suggest pseudocopulation, in which male gnats land on a labellum lamina that is ornamented and colored to resemble a female gnat. The labellum of most *Pterostylis* spp. appears to be irritable. That is, if the gnat lands on the lamina, the insect's pressure appears to trigger an electrochemical response,

causing the lamina to flip upward into the column, "pressing" the gnat between labellum and rostellum for 30 to 90 seconds (Jones 1981; 1988). In
fresh flowers, the labellum resets itself spontaneously but slowly, often taking up to eight or nine hours in the field, as in *P. curta* (Bernhardt, work in
progress) before it returns to its original position.

This irritable mechanism is not to be confused with the system of floral
cantilevers that exploit the insect's attempt to fly off with the dummy female,
as in wasp-pollinated *Drakaea, Spiculaea,* some *Caladenia* spp., and some
other genera of Geoblasteae. The labellum of *Pterostylis* spp. moves independently when touched, as in Australian *Caleana* and *Paracaleana*. In
Caleana major the touched lamina is slammed into the broadened column so
violently that it makes an audible click (Bernhardt, personal observation).
Some *Pterostylis* spp., however, lack the irritable labellum (Jones 1988) and
the actual mode of pollinia dispersal remains undefined in these species,
though self-pollination seems unlikely.

In *Pterostylis* the slender, often elongated, yet extravagantly winged
column is coupled with a labellum featuring a hinged, ligulate claw and
a nontrilobate lamina that often bears a dark, insectiform, callous-*cum*-
appendage. This character seems indicative of the advanced Caladeniinae
(Table 5.3; Figs. 5.3, 5.5). *Pterostylis* has also retained the trend toward
elongated-filiform apices of the perianth segments (Fig. 5.2). In contrast to
the Caladeniinae the lateral petals and dorsal sepal of *Pterostylis* spp. are
broad enough to overlap and form a galea (helmet) and the lateral sepals
coalesce (Figs. 5.2, 5.5). This action produces an inflated chamber that gnats
enter and leave via a sinus composed of the galea and the semirelaxed labellum that is positioned above the lateral sepals (Fig. 5.5; Woolcock and
Woolcock 1984).

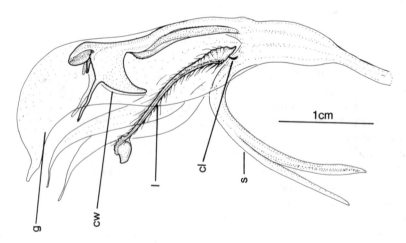

Figure 5.5. Cleared flower of *Pterostylis plummata:* cl = claw, cw = column wings,
g = galea, l = labellum lamina, s = fused, lateral sepals.

Column-labellum morphology allies Pterostylidae with Caladeniinae, but the monad pollen and leaf organization of *Pterostylis* is far closer to the Chloraeinae and *Codonorchis* in particular. The biogeography of *Pterostylis* indicates a southerly but eastern temperate origin in Australasia (Table 5.4). Speciation of *Pterostylis* remains distinct from the Geoblasteae in one important detail. Specifically, *Pterostylis* has not undergone significant speciation within mediterranean regions. Note, for example, that both endemism and the number of *Pterostylis* spp. are many times higher in eastern Australia than in the mediterranean southwest (Table 5.4). Consequently, the Pterostylidae are isolated from the Geoblasteae in their respective patterns of distribution, floral phenology, and their exploitation of pollinators despite their prospective common origin. This isolation is further evidence that the Australasian Geoblasteae could not have originated in the mediterranean west.

Conclusions

The Tribe Geoblasteae consists of more than 170 species divided unequally between South America and Australasia. Although the Geoblasteae are distributed from tropical sites through temperate, coastal ecosystems, those mediterranean regions, south of the Tropic of Capricorn, remain centers of speciation for this tribe on both sides of the Pacific. Continental aridity has fragmented distribution of the tribe in Australasia and South America. Genera within the Geoblasteae share up to seven ecological, phenological, and morphological "trends" regardless of respective phytogeography.

An analysis of inflorescence structure and floral anatomy (emphasizing labellum and column characters) strongly suggests that genera endemic to South America may be less advanced than the genera of Australasia. It is probable that the tribe originated in South America and migrated to Australasia via a temperate Antarctic corridor. Therefore, although the adaptive radiation of the Geoblasteae within mediterranean Australia has been explosive, and reflects elaborate floral modification, it represents a secondary invasion. Although mediterranean regions in the southern Pacific encourage adaptive radiation and speciation within the Geoblasteae, the evidence does not suggest that this tribe originated in a mediterranean ecosystem in South America.

Macroevolutionary distinctions within the Geoblasteae are reflected taxonomically with five endemic genera in South America but seventeen genera endemic to Australasia. Furthermore, the number of species of Geoblasteae in mediterranean Australia is almost 33% higher than in mediterranean South America. These differences cannot be assigned exclusively to the bias of taxonomists or the common factors influencing rates of insular evolution. It is more likely that disjunctions in generic diversity are caused by phylogenetic constraints related to the history of pollination mechanisms and the comparative diversity of pollinating Hymenoptera in the southern Pacific.

Finally, the Australasian tribe, Pterostylideae, is probably derived from ancestral Geoblasteae. The pterostylids evolved via character displacement in temperate habitats and secondary and inferior migration to mediterranean habitats.

Acknowledgments

A Beaumont Fellowship was provided by St. Louis University to support transportation and fieldwork in Chile. The author thanks Mr. Ron Heberle of Albany, Western Australia, and Mr. John Kelly of St. Arnaud, Victoria, for their collections and donations. I also thank A.G. and Barbara Wells for allowing me to examine their color slides depicting the pollination of Western Australian orchids. I am most grateful to Dr. Pamela Burns-Balogh for her review of the manuscript. The Missouri Botanical Garden has supported this research for the past seven years, providing lab space, equipment, storage space, and continuing maintenance of specimens on loan. The Royal Botanic Garden of Sydney provided open access to living specimens, herbarium sheets, and spirit collections from 1990 to 1992 under the authority of Dr. Peter Weston, keeper of orchids.

References

Ackerman JD, Williams NH (1981) Pollen morphology of the Chloraeinae (Orchidaceae: Diurideae) and related subtribes. Am J Bot 68:1392–1402

Arroyo MTK, Squeo F (1990) Relationship between plant breeding systems and pollination. In Kawano, S. (ed) *Biological Approaches and Evolutionary in Plants.* Academic Press, San Diego, CA, pp 205–227

Bates R (1985) Colorful Caladenias. Aust Orchid Rev 50:6–11

Bates R (1989) Observations on the pollination of *Caleana major* R.Br. by male sawflies (*Pterygophorus* sp.). Orchadian, September: 208–210

Beardsell D, Bernhardt P (1982) Pollination biology of Australian terrestrial orchids. In Williams, EG, Knox, RB, Gilbert, JH, Bernhardt, P. (eds) *Pollination '82,* U of Melbourne Press, Parkville, Victoria, Australia, pp 166–183

Brieger FG (1974–1975) III Unterfamilie: Neottioideae. Band infferung 5–6. In Brieger, FG, Maatsch, R., Senghas, K (eds) *Die Orchideen: ihre Beschreibung, Kultur, und Zuchtung.* Paul Parey, Berlin, Germany

Burns-Balogh P, Bernhardt, P (1988) Floral evolution and phylogeny in the tribe Thelymitreae (Orchidaceae: Neottioideae). Pl Syst Evol 158:19–47

Burns-Balogh P, Funk VA (1986) A phylogenetic analysis of the Orchidaceae. Smithsonian Contr Bot, No. 61. Smithsonian Inst Press, Washington, DC, USA

Carlquist SJ (1974) *Island Biology.* Columbia U Press, NY

Carlquist SJ (1987) Pliocene *Nothofagus* wood from the transantarctic mountains. El Aliso 11:571–583

Catling PM (1990) Auto-pollination in the Orchidaceae. In Arditti, J. (ed) *Orchid Biology: Reviews and Perspectives.* Timber Press, Portland, OR, pp 123–158

Christophel DC, Greenwood DR (1989) Changes in climate and vegetation in Australia during the Tertiary. Rev Palaeobot Palyn 58:95–109

Clements M (1982) *Preliminary Checklist of Australian Orchidaceae.* Canberra Publishing and Printing, Canberra, Australia

Correa MN (1956) Las especies argentinas del genero *Gavilea.* Bol Soc Argentina Bo 6:73–86

Correa MN (1968) Rehabilitación del genero *Geoblasta* Barb. Rodr Rev del Mus de la Plata, Ser. 9, 11 (Botanica No. 54):69–74

Correa MN (1969) *Chloraea* genero sudamericano del genero de Orchidaceae. Darwinia (Buenos Aires) 15:374–500

Correa MN (1970) Orchidaceae. In *Flora Patagonica.* I.N.T.A. Collection. Vol 8, parte 2, pp 188–299

Dafni A, Bernhardt P (1990) Pollination of terrestrial orchids of southern Australia and the Mediterranean region; Systematic, ecological and evolutionary implications. In Hecht MK, Wallace B., Macintyre RJ (eds) *Evolutionary Biology*, Plenum Publishing, NY pp 193–252

Dressler R (1981) *The Orchids: Natural History and Classification.* Harvard U Press. Cambridge, MA

Futuyma, DJ (1986) *Evolutionary Biology.* Sinauer Associates, MA

Gann W, Carlquist S (1985) Orchid flower position and its significance. Orchid Dig 49:99–101

Gentry AH, Dodson CH (1987) Diversity and biogeography of neotropical vascular epiphytes. Ann Missouri Bot Gard 74:205–233

George AS (1981) Orchidaceae. In Boyland DE, Carolin RC, George AS, Jessop JP, Maconochie JR (eds) *Flora of Central Australia.* Reed Pty Ltd., Sydney, N.S.W., Australia, pp 515–516

Gumprecht R (1980) Blossom-structure and pollination mechanisms in endemic orchids of South America. Medio Ambiente 4:99–102

Halle N (1977) *Flore de la Nouvelle Calédonie. 8 Orchidacées.* Musée National de l'Histoire Naturelle, Paris, France

Holman R (1976) The chemical composition of fragrances of some orchids. In Szmat HH, Wemple J (eds) *First Symposium on the Scientific Aspects of Orchids.* Chemistry Dept., U Detroit, MI, USA

Izaguirre de Artucio HP (1973). Las especies uruguayas de *Bipinnula* (Orchidaceae). Bol Soc Argentina Bot 15:261–276

James SH, Hopper SD (1981) Speciation in the Australian flora. In Pate JS, McComb AJ (eds) *The Biology of Australian Plants.* U Western Australia Press, Nedlands, W.A., Australia, pp 362–382

Johns J, Molloy B (1985) *Native Orchids of New Zealand.* Reed, Wellington, New Zealand

Jones DL (1981) The pollination of selected Australian orchids. In Lawler L, Kerr RD (eds) *Proceedings of the Orchid Symposium; 13th International Botanical Congress, 1981.* Orchid Society of New South Wales, Harbour Press, Sydney, Australia, pp 40–43

Jones DL (1988) *Native Orchids of Australia.* Reed Books Pty. Ltd., Frenchs Forest, N.S.W., Australia

Lavarack PS (1976) The taxonomic affinities of the Australasian Neottioideae. Taxon 25:289–296

Luer CA (1975) *The Native Orchids of the United States and Canada, Excluding Florida.* New York Botanical Garden, NY, USA

Marticorena C (1990) Contribución a la estadistica de la flora vascular de Chile. Gayana, Bot 85:85–113

Marticorena C, Quezada M (1985) Catálogo de la flora vascular de Chile. Gayana Bot 41:1–157

Nicholls WH (1969) *Orchids of Australia.* Nelson, Melbourne, Australia

Peakall R (1989) The unique pollination of *Leporella fimbriata* (Orchidaceae): pollination by pseudocopulating male ants (*Myrmecia urens*, Formicidae). Pl Syst Evol 167:137–148

Schemske DW Wilson MF, Melampy MN, Miller LJ, Werner L, Best LB (1978) Flowering ecology of some spring, woodland herbs. Ecology 59:351–366

Stebbins GL (1974) *Flowering Plants: Evolution Above the Species Level*. Harvard U Press, Cambridge, MA, USA

Stoutamire WP (1981) Pollination studies in Australian terrestrial orchids. *Natl Geog Soc Res Rep* 13:591–598

Stoutamire WP (1982) Wasp pollinated species of *Caladenia* (Orchidaceae) in south-western Australia. Aust J Bot 31:383–394

Thorne RF (1986) Antarctic elements in Australasian rainforests. Telopea 2, pp 611–617

Vogel S 1990. *The Role of Scent Glands in Pollination*. Amerind Publishing Pvt. Ltd., New Delhi, India

Woolcock C, Woolcock D (1984) *Australian Terrestrial Orchids*. Nelson, Melbourne, Australia

III. Vegetation Structure and Soils

6. Australian Mediterranean Vegetation: Intra- and Intercontinental Comparisons

Marilyn D. Fox

On either side of the Pacific Ocean are three regions having the winter rainfall–summer drought characteristic of the Mediterranean Basin. This type of climate is incorporated in Köppen's Csa and Csb categories (Köppen 1923). These regions are the Baja California and California region in North America, the central coastal region in Chile, and the southern portion of Australia. The relative sizes of the climatically matched regions are: 0.5, 0.1, and 0.2×10^6 km^2 for California, Chile, and Australia respectively (Dick 1975). The vegetation in these three regions is characterized by broadleaf sclerophyllous shrublands: respectively the chaparral in California, the matorral in central Chile, and the mallee and kwongan (Pate and Beard 1984) in southern Australia. All three regions, however, have a range of vegetation formations from forest to low shrubland.

Southern Australia offers a special case for comparative studies in mediterranean ecosystems. Unlike these other two regions sharing the winter-dominant rainfall climate, the southern parts of Australia are stretched longitudinally and comprise two discrete regions separated by extensive desert (See Fig. 2.1). Of these, the southwest of the continent comprises an arc of mediterranean communities bounded on one side by the extensive desert and grading to high-rainfall forests in the far southwest. By comparison, the eastern portion is more irregularly shaped, bounded to its north by the arid zone and grading eastward into other semiarid shrublands, woodlands, and eventually to humid temperate forests. The two regions had been distinct biogeographic regions long before the onset of the mediterranean climate, perhaps only 2.5 Ma (Bowler 1982), and so represent two independently derived ecosystems from phylogenetically related stock.

Previous comparisons of mediterranean regions have concentrated on those from connected continents (California and Chile) (e.g., Parsons and Moldenke 1975, Parsons 1976, Mooney 1977), on continents with similar evolutionary histories (Australia and South Africa) (e.g., Milewski 1979), or on combinations of these compared to the Mediterranean Basin (e.g., Aschmann 1973). In this chapter I first compare the climate and vegetation of the two Australian regions, and then compare these with those of California and central Chile. It is necessary first to consider the distribution of the climatic types and to distinguish the vegetation types associated with them in Australia.

Mediterranean Climate in Southern Australia

The mediterranean climate has two strong components that affect ecosystem functioning. These are the winter dominance of rainfall and the summer onset of drought. The summer climate is therefore particularly stressful and the temperature regime is critical to most life forms. For this reason Köppen (1923) distinguished two subtypes of his Cs mediterranean climate type with *a* and *b* suffixes referring to the hottest mean monthly temperature ($>22°C$ and $<22°C$ respectively). Whereas southern California and central Chile feature the Csa climate, the southwestern part of Western Australia has both the Csa and Csb types, and only the Csb type is found east of the Great Australian Bight. These cross-continental differences are significant for the phenology of the plants in each part (Bell and Stephens 1984). The climate in southern Australia features winter-dominant rainfall that is usually of low intensity and spread over several rain days. It may also have occasional significant summer falls, however, which tend to be from thunderstorms of higher intensity but shorter duration. The mediterranean ecosystems in southern Australia also extend into the semiarid climate (BSk) region (Specht 1981a); significantly, these regions almost meet at the head of the Great Australian Bight.

Another important distinction within the western mediterranean region in southern Australia, coincident with the Csa and Csb delineation, is that the far-western portion, roughly equivalent to the littoral sand plain, has very marked seasonal rainfall with winter : summer rainfall $>3:1$ (Parkinson 1986). The rest of southern Australia experiences less extreme winter dominance (W : S $> 1.3:1$ but less than 3:1). The distinction between Csa and Csb therefore lies in the severity of both the summer drought and the temperatures prevailing during that drought.

The Australian mediterranean regions are also distinguished by the variability in their rainfall (Saunders and Hobbs 1992). Historical data indicate an overall reduction in winter rainfall of about 4% in southwestern Australia during this century (Pittock 1988). Predictions for future climate change are in the same direction, but at an accelerated rate (Greenwood and Boardman 1990), and would lead to the southward spread of the desert and significant reduction in the area of mediterranean climate in southern Australia by the mid-twenty-first century. This trend has considerable significance for the conservation of the unique communities occupying the mediterranean regions in southern Australia which may be endangered by a change in climate.

Vegetation Physiognomy

The vegetation of mediterranean Australia comprises a mosaic of short and tall shrublands grading up to woodlands (Specht 1981a, Fox 1982); the distribution of these is illustrated in Figure 2.1. All these communities are characterized by scleromorphic plants and the genera *Eucalyptus*, *Acacia*, and *Casuarina* are prominent. Westman (1978) proposed that the eucalypts may have been a more recent addition to many older sclerophyll communities and may respond differently to environmental factors in the same communities. The variety of plant communities associated with the mediterranean climate in southern Australia may then represent one synusium of scleromorphic shrubs or grasses responding to climatic and edaphic patterns, and another overlying eucalypt synusium where conditions are suitable. Certainly the eucalypts and other members of the Myrtaceae have a distinctive flowering period, peaking in early summer, compared to the late-winter peak for the heath family, Epacridaceae (Bell and Stephens 1984). Specht (1973) attributes the summer growth rhythym of the mediterranean flora in Australia to its presumed tropical origin (Specht and Rayson 1957), although the herbaceous flora more closely approximates the typical mediterranean pattern.

Parts of southern Australia have a short, species-rich shrubland with a diverse array of plants from the Epacridaceae, Restionaceae, and Fabaceae, but with few or no eucalypts, which is referred to as heathland or kwongan. These communities are principally related to nutrient-poor soils (Specht and Moll 1983), and they also occur in wetter lowland regions near the coast around Australia in a range of climate types (Beadle 1981). Their occurrence in southern Australia partly coincides with the mediterranean climate, but even there they extend into zones of both higher and lower rainfall. Similar communities are found farther north in eastern Australia, as on the southeast coast of Queensland (25° latitude, 1600 mm rainfall) and in the vicinity of Sydney (33° latitude, 1200 mm rainfall). The sand-plain regions in southwestern Western Australia support a very rich array of communities of the heathland type and within the mediterranean (Csa) rainfall zone (although extending beyond it to coastal, higher-rainfall areas to the south and more arid regions to the north and west). The flora of these communities is extremely rich and shows a high level of endemism (Hopkins and Griffin 1984).

Shrublands or scrubland with a dominant stratum of multistemmed eucalypts, referred to as mallee, are more typical of the mediterranean (Csb and

BSk) climate. Although some limited stands occur elsewhere, usually associated with skeletal soils, the bulk of the mallee distribution coincides with
the mediterranean climate zone. The mallee communities mapped by Carnahan (Carnahan and Deveson 1990) occur between the 180- and 750-mm
isohyets, and are most common and dominate between 250 and 460 mm
(Parsons 1981). The evidence from outlier stands of mallee is that the mallee
has expanded and contracted in the past in response to changing climate
(Beadle 1981). Although primarily associated with the winter-rainfall zone
(despite having a summer growth rhythm), some species occur in areas with
uniform rainfall, and some in areas with summer-dominant rains.

Mallee soils are coarse, calcareous, and well drained, with little surface
flow and few drainage channels in mallee landscapes. The structure of the
mallee stands is quite variable, depending partly on soil fertility, fire history,
and constituent species. Some stands feature single-stemmed, relatively tall
(>10 m) individuals that grade into woodland formation; some comprise
multistemmed plants only 1 to 3 m tall. The predominant form through
much of the mallee range is of tall shrublands (3 to 8 m) with multiple stems,
each supporting a relatively sparse, clumped canopy (Fox 1990b).

Recent revisions of the eucalypt subgenera have elucidated more biogeographic patterns in the distributions of the mallee species and proliferation of
species descriptions (see, e.g., Hill 1989, 1990; Hopper 1990, 1992). Hopper
(1990) indicates the rapidity of change in this field for the flora in southern
Western Australia: Aplin (1977) listed 165 species of mallee eucalypts, Green
(1985) listed 185 species, Brooker and Kleinig (1989) raised the tally to about
300 eucalypt species (of which about 225 are mallees); and additional new
species have been identified since then. Hill (pers. comm.) estimates the number of mallee eucalypt species associated with the mediterranean regions
conservatively at 300, most from the subgenus Symphyomyrtus. Neither of
the other mediterranean regions considered here has such numerical dominance by one genus (or group of closely related subgenera).

The understory associated with mallee stands depends on rainfall and soil
type. I recognize five types of understory, combining work of Specht (1972)
and Beadle (1981) with my own descriptions (Fox 1990a):

1. On slightly more fertile but saline soils, the understory features halophytic
   shrubs in the families Chenopodiaceae and Zygophyllaceae; as these
   grade to more saline or more waterlogged conditions, the mallee drops
   out and low-halophytic shrublands predominate.
2. On infertile soils with intermediate rainfall, a dense thicket of *Melaleuca*
   and/or *Casuarina* forms the understory—this is the "mallee-broombush"
   of Specht (1972); in places, mallee eucalypts overtop the thicket, but elsewhere the eucalypts may be missing. This arrangement is particularly well
   developed in the eastern mallee sector, in western Victoria and parts of
   South Australia.
3. On infertile soils in wetter areas a rich shrub layer of xeromorphs (from
   the families Proteaceae and Myrtaceae) forms the understory of "mallee

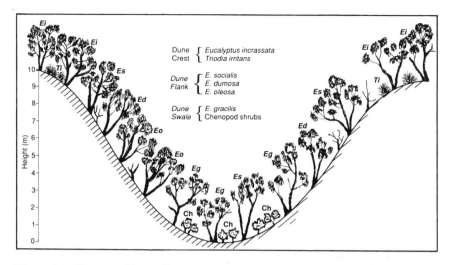

Figure 6.1. Topographic gradient in western New South Wales showing the species replacement of mallee eucalypts across a dune–swale sequence.

heath," which may be ecotonal; this combination is most common nearer the southern coastal regions.
4. On infertile soils in the driest areas, the understory is of herbaceous, hummock grasses (*Triodia* spp. and *Plectrachne* spp.); as rainfall declines the mallee drops out and the hummock grasslands continue into the arid interior of the continent.
5. When mallee remains unburned for a long time and much woody litter accumulates, the understory may comprise only annuals native and introduced. In highly disturbed sites introduced species predominate.

Within a landscape unit one will also find large-scale vegetation patterning, and both structure and floristic composition can vary on a topographic scale of tens of meters (Fox 1990a). In swales between dunes, conditions may be quite saline and the chenopod understory may predominate; on the flanks of the dunes the understory is heathy, and on the crest it may be dominated by *Triodia*. Similarly, there are topographically related shifts in mallee species dominance, and the same catena from swale to dune crest may involve five or more eucalypts. This combination is illustrated in Figure 6.1. Similar local shifts in composition of mallee communities have been recognized in the western region (Beard 1990). At a broader landscape scale we also find topographic and edaphic control of mallee and other communities (Fox 1991).

Intracontinental Comparisons

The part of southern Australia supporting mallee and the other types of mediterranean vegetation comprises two geographic regions separated by the arid Nullarbor Plain, which extends to the coast at the head of the Great

Australian Bight. The western sector is larger, with an estimated area of 147×10^3 km^2, and the total area of the eastern sector is less than half that, 67.6×10^3 km^2 (Dick 1975). When the associated BSk type is included (semiarid, winter dominant), the regions are more than doubled, the eastern one extending into central New South Wales (Dick 1975).

The mallee region in Western Australia comprised a more contiguous landscape prior to European occupation. Except for outliers, much of southern Western Australia formed a continuous block of mallee communities grading to kwongan and woodland formations. By comparison, the mallee region east of the Great Australian Bight comprised naturally fragmented patches. Mallee stands on the Eyre and Yorke Peninsulas in South Australia were separated from each other and from more extensive mallee farther east. In turn, the mallee in eastern South Australia, Victoria, and New South Wales was fragmented by the Murray–Darling river systems.

The soils in southern Australia are generally considered to be the poorest in plant nutrients of all the mediterranean regions (Williams and Raupach 1983). A difference between the two regions is apparent, however, and the eastern part of the mediterranean region is relatively more fertile than is the far southwest (Lamont, this volume). Eastern mallee communities commonly occur on aeolian calcareous brown earths, but in the west mallee communities grow on duplex soils with a distinct sandy and lateritic horizon over a clay horizon (Hill 1989). Calcareous soils are preponderant in the eastern mallee and noncalcareous soils in the west (Blackburn and Wright 1989).

Doing (1981) proposed a Mallee Botanical Geographic Region (Em), subdivided into an Eastern Mallee Province (Eme), a Nullarbor Province (Emn), and a Western Mallee Province (Emw). Across this region, the mallee eucalypts are distributed discontinuously; the relative representation can be gained from numbers (now dated) given in Parsons (1981), who listed 108 species, of which 71 occur only in the Western Mallee Province, 16 only in the Eastern Mallee Province, and 21 are shared. The species occurring in both west and east (the most widely distributed are *Eucalyptus dumosa*, *E. foecunda*, *E. gracilis*, *E. incrassata*, *E. oleosa*, and *E. socialis*), show closer relationships to other species in Western Australia and so probably originated there (Burbidge 1947). Recent biogeographic studies suggest that the major development of the mallee habit has been in southwestern Western Australia and recent radiations across inland and coastal dune systems brought a few successful taxa into eastern Australia (Hill 1989). Other taxa exhibit the same or similar biogeographic divisions; for example, Schodde (1981), in reviewing the bird communities in the Australian mallee, recognized six "blocs," five of which (western Australia, Eyre Peninsula, Adelaide-Flinders, Murray, and central New South Wales) correspond to the mediterranean region in southern Australia, and the birds had different regional distributions, with little overlap.

Table 6.1 lists the principal mallee groups recognized by Hill (1989) with the number of species from each eucalypt subgenus, habitat type, and asso-

Table 6.1. Summary of the principal mallee groups recognized by Hill (1989) with habitat type and associated vegetation. Hill also comments on relict mallees and facultative mallees. The notation W and E with the group number indicates western and/or eastern occurrences

	Subgenera	Number of species	Habitat	Associated vegetation
1W	Eudesmia	5	Lateritic sand plain	Kwongan, extremely rich
	Symphyomyrtus	60		
	Monocalyptus	9		
2W & E	Eudesmia	3	Calcareous coastal dunes on	Shrub thicket with *Eucalyptus*,
	Symphyomyrtus	14		
	Dumaria	8	Pleistocene and	*Banksia, Acacia,*
	Monocalyptus	5	Holocene dune fields	or *Melaleuca*
3W	Symphyomyrtus	23	Mallee woodland	*Cratystylis,*
	Dumaria	10		*Melaleuca,* chenopod understory
4W	Symphyomyrtus	12	Siliceous dunes from granite and quartzite	Shrub thicket
	Monocalyptus	4		
5W & E	Corymbia	2	Aeolian desert dunes	Hummock grassland understory *Triodia* or *Plectrachne*
	Eudesmia	4		
	Symphyomyrtus	31		
6W & E	Corymbia	3	Arid-zone rocky ridges	Shrub savannah and hummock grass with *Acacia* in upper stratum
	Blakella	1		
	Eudesmia	2		
	Symphyomyrtus	11		
	Exsertaria	8		
7E	Symphyomyrtus	6	Mallee boxes	Mixed shrubby with *Acacia*
8E	Monocalyptus	14	Mallee ashes, high rainfall	Low scleromorphic shrub understory

ciated vegetation. Groups 1 to 7 occur within the mediterranean climate region, some exclusively in the west or east, some in both.

The subgenus Symphyomyrtus is clearly the dominant group; its constituent species are predominantly mallees, with few obligate monopodial trees. By comparison, most of the other eucalypt subgenera comprise predominantly tree-form with fewer mallee-form species. Facultative mallees are the species which can, on more fertile or deeper soils, produce trees but which, when stressed, have the multistemmed mallee habit. The richest mallee stands are in southwestern Western Australia. Hill (1990) records up to 20 species in a few hectares near Ravensthorpe in Western Australia; the common range in the west is 8 to 10 species (Beard 1981). The mallees in southeastern Australia are less rich, with 3 to 4 sympatric eucalypt species being the norm (Fox 1990a).

Because of the high incidence of ephemeral species (annuals and short-

Table 6.2. The number of genera of higher plants represented in the floras of the heathlands in southwestern (West)
and southeastern (East) Australia (based on Table 11.1,
Specht 1981b): The "Total" column is for all Australian
heathlands (mediterranean and other types of climate).

	West	East	Total
Gymnosperms	3	2	8
Monocotyledons	92	41	164
Dicotyledons	195	118	399
Total	290	161	571

lived perennials) in the understory, richness and composition of species vary from season to season and year to year. Cumulative species richness at a site is greater than any one season's display. For example, mallee near Mungo National Park in western New South Wales has recorded 93 species from a tenth-hectare permanent plot (six observation dates over five years) but the richest collection was 61 species, and the poorest, 23 species. Floristic similarity between subsequent seasons was commonly between 50% and 70% (Fox 1990a).

We find an indication of the relative richness of heathland communities from western and eastern Australia in Specht (1981b), in which he lists the number of genera recorded in the subhumid temperate heathlands. The western communities are consistently richer (Table 6.2). Lamont et al. (1984) report a total of 433 genera for the kwongan in Western Australia; the number of genera for the entire continent has been estimated at 2962.

Comparisons of community diversity and floristic affinity between the western and eastern mallee regions have not been made. Hopkins and Griffin (1984) report data for sets of kwongan sites in the west, however. Mean species richness for all sites ranges from 60 to 77 per tenth hectare, and β-diversity ranges from 0.27 to 0.49 for near sites, to 3.54 for sites separated by more than 30 km (Table 4.1, Hopkins and Griffin 1984). Fox (1990a) reported species richnesses for eastern mallee sites and degree of similarity between seasons. Using the same sites, β-diversity was calculated in the same way as that used by Hopkins and Griffin (1984) [as (total species ÷ mean richness) − 1 (Whittaker 1972)]. These values are plotted as a function of distance between sites in Fig. 6.2 together with the values reported by Hopkins and Griffin (1984) for the western region. The much higher values and greater rate of increase with distance are apparent for the western kwongan. The regression equation for these western sites equates to a species-turnover rate of 7.1 species/100 km. By comparison, the eastern sites yield a rate of 0.84 species/100 km. Interestingly, the two regression lines converge at zero distance, and the intercept for the eastern regression coincides with the value reported for adjacent quadrats by Hopkins and Griffin (1984).

A crude comparison can be made for a set of mallee sites in the Eastern

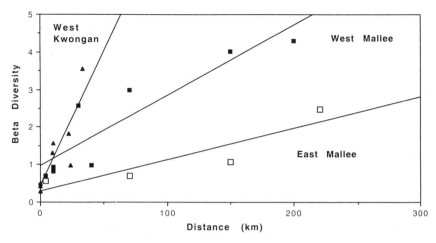

Figure 6.2. Trends in β-diversity for mediterranean shrublands in western and eastern parts of southern Australia. Western values are taken from Table 4.1 in Hopkins and Griffin (1984), eastern values are for sites in western New South Wales (Fox 1990a) calculated in the same way. Regression equations are significant (West $r = 0.8509$, 6 d.f., $p < 0.005$; East $r = 0.9017$, 2 d.f., $p < 0.05$).

Goldfields region in Western Australia (31°50′ S, 122°30′ E). These comparisons lack precision because the data (Newbey et al. 1984) used to calculate β-diversity were collated from releves, hence the areas sampled may not be identical. The trends are so strong (Fig. 6.2), however, that such inconsistencies do not influence the main conclusion drawn. The western mallee sites indicate a species-turnover rate of around 2 species per 100 km, clearly more than the eastern mallee and fewer than the kwongan sites, as can be seen from the slopes of the lines in the graphic presentation (Figure 6.2). The regression equations and significance are given in the caption to Fig. 6.2. Notice that the slopes are significantly different ($p < 0.02$ and $p < 0.05$ for the kwongan–west mallee and west mallee–east mallee) and that the variance explained by each equation is 72% (kwongan), 83% (west mallee), and 81% (east mallee), respectively.

Intercontinental Comparisons

In each region colloquial names have been given to the mediterranean shrublands. In California and Chile the Spanish conquests are remembered in "chaparral" and "matorral," respectively. In Australia the Aboriginal words "mallee" and "kwongan" are used. Figure 6.3 shows the broad distribution of each shrubland in the three regions.

Both central Chile and southern California are characterized by longitudinal bands of vegetation communities, the pattern dictated by the parallel

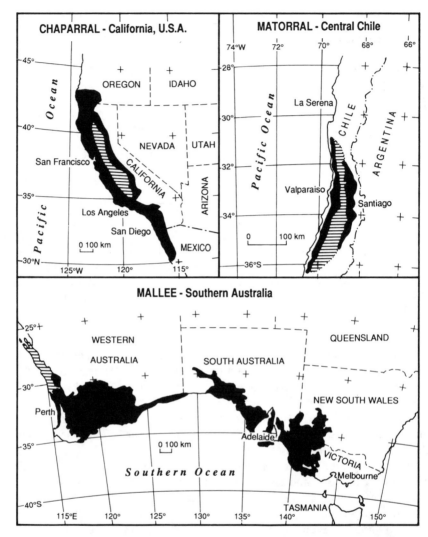

Figure 6.3. The extent of the mediterranean shrublands in California, Chile, and Australia. Based on figures in Thrower and Bradbury (1973, 1977), Hanes (1981), and Carnahan and Deveson (1990). The diagonally hatched area in Western Australia is the region which corresponds to the Csa climate and which supports the kwongan. In both California and Chile the central valleys are hatched. The projections and scales are similar for California and Australia, but the central Chile region is enlarged.

coastal and inland mountain ranges. Both regions have distinctive coastal communities—coastal matorral and coastal scrub in Chile and coastal sage and coastal sage with succulents in California. Inland of this region in Chile is a band of coastal arborescent matorral corresponding to the Coast range, the Central Valley's *Acacia* savanna, and then the sub-Andean matorral. At the latitude of Santiago, this band grades steeply to xeric Andean scrub. The main Californian chaparral zone is immediately adjacent to the coastal communities with woodlands toward its interior. By comparison, no strong directional pattern appears in the occurrence of the mediterranean shrublands in southern Australia. The main mallee regions occur almost from the coastline to the arid shrubland border toward their interior. A narrow littoral dune complex abuts heath with scattered mallee, but not as extensive as the coastal communities in Chile and California. The eastern block of mallee is also fragmented, with occurrences on the Eyre Peninsula, the Yorke Peninsula, and parts of eastern South Australia, western Victoria, and the far southwest in New South Wales (Fig. 6.3).

Table 6.3 summarizes the general environmental constraints in the three regions. Notable is the relatively narrow latitudinal range for the Chilean matorral (5° of latitude); more pronounced is the width of the mallee region (32° of longitude).

Both California and central Chile have the mediterranean climate type recognized by Köppen (1923) as Csa, with a hot, dry summer. By comparison, only part of the western mallee region shares this type of climate; the rest, together with the eastern mallee region, experiences the milder Csb. For

Table 6.3. The broad geographic distributions of the Pacific mediterranean regions and their principal environmental characteristics (Hanes 1981, Rundel 1981, Fox and Fox 1986a)

	California	Central Chile	Southern Australia
Mediterranean shrubland	Chaparral	Matorral	Mallee, kwongan
Latitudinal range	31° to 42°30′ N (12°30′ lat.)	32°30′ to 36°30′ S (4° lat.)	27° to 37°30′ S (10°30′ lat.)
Longitudinal range	124° to 116° W (8° long.)	73° to 70° W (3° long.)	115° to 147° E (32° long.)
Elevational range	10 to 1760 m	10 to 1800 m	10 to 800 m
Rainfall range	300 to 1100 mm	300 to 800 mm	250 to 750 mm
Orogenic system	Young	Young	Old
Topography	Rolling, with high mountains close to coast and inland	Rolling, with high mountains close to coast and inland	Flat, eroded
Volcanic activity	Moderate	Major	Minor
Seismic activity	Major	Major	Minor
Soil phosphorus	Moderately low	High	Extremely low
Soil nitrogen	Low	Moderately low	Extremely low
Fire frequency	Medium	Low	High

this reason it could be argued that it would be more valid to compare the chaparral and matorral with the kwongan. In the past, however, the more widely distributed mallee has been the main choice for study (e.g., di Castri 1981).

It is unfortunate that Charles Darwin (Darwin 1894) did not visit California and record his impressions of the vegetation there. He did though, spend considerable time in central Chile and less in southwestern Australia. His descriptions of the two vegetations are still instructive:

July 23 1834, p. 253
The vegetation near Valparaiso:
The vegetation in consequence [of the summer drought and winter rainfall] is very scanty: except in some deep valleys, there are no trees, and only a little grass and a few low bushes are scattered over the less steep parts of the hills.

There are many very beautiful flowers; and, as in most other dry climates, the plants and shrubs possess *strong and peculiar odours.*

16 August 1834, p. 256
Climbing the Campana or Bell Mountain:
During the ascent I noticed that nothing but bushes grew on the northern slope, whilst on the southern slope there was a bamboo about 15' high.

This asymmetry in community distributions was one of the points investigated in the California–Chile research program (Miller 1981). In that study it was concluded that California showed greater difference between pole-facing and equator-facing slopes than did Chile. Measurements from central Chile at Fundo Santa Laura, however, were from a deep gully where the sheltered microclimate and mesic vegetation would tend to blanket differences. Elsewhere on gentler, more exposed slopes the differences in vegetation are very pronounced, as in the cross-section for La Dormida–Quebrada de Tiltil (Thrower and Bradbury 1977), which illustrates gradation from xeric conditions with cacti such as *Trichocereus chiloensis* on equator-facing slopes and taller, denser, scleromorphic shrublands on the pole-facing slopes. Compared to a similar topographic cross-section in the Australian mallee (Fig. 6.1) dominated by the single genus *Eucalyptus*, the Chilean cross-section features ten genera of small trees and shrubs on the southern aspect alone. No strong aspect effect appears in the Australian mediterranean vegetation, mainly because of the greatly reduced aspect effects on a very subdued landscape.

Southern Australia differs from all the other mediterranean regions by having very flat topography. California and Chile especially have tall mountains inland and parallel to the coast, leading to parallel climate patterning (Aschmann 1973) and corresponding vegetation pattern. Australia has shallow environmental gradients from the more mesic to arid zones. These trends have meant that during periods of climatic change in the past, minor changes in rainfall have influenced broad expanses of country. By comparison, the steep elevational gradients in Chile and California result in narrow vegetation bands.

The "strong and peculiar odours" Darwin mentioned were from aromatic shrubs and herbs. The secondary chemical composition of the Californian and Chilean floras has received some attention (Mabry and Difeo 1973). In Australia the eucalypts are notably rich in essential oils (Markham and Noble 1989), but the other shrubs and herbs mostly have less aromatic foliage than the floras of the other mediterranean regions.

On February 7, 1836, Darwin (1894) described the vegetation near King George's Sound (southwestern Australia) near the town of Albany: "Everywhere we found the soil sandy, and very poor; it supported either a coarse vegetation of thin, low brushwood and wiry grass, or a forest of stunted trees."

This area is within the more mesic climate just to the west of the mediterranean region. Darwin (1894) also commented on the abundance of Casuarinas and grass trees of the genus *Xanthorrhoea*, and on the relative paucity of eucalypts there. His "stunted trees," though, are characteristic of the broader mediterranean region in southern Australia.

Because of the striking physiognomic similarities between California and central Chile, comparative studies have been many. For example, Parsons and Moldenke (1975) compared vegetation structure along analogous climatic gradients in California and Chile. Despite distinct genetic histories, the structure of the vegetations in these areas has converged under similar climatic constraints. Climatically analogous sites on the two continents were more similar in a number of structural attributes than were climatically different sites in each region. Montenegro and Ginocchio (this volume) investigated the structure of the three mediterranean vegetation formations using ecomorphological characters and found evidence of strong convergence, with no significant differences among chaparral, matorral, and mallee.

The species diversity, particularly among plants, of mediterranean ecosystems has been mentioned often. The high α-diversity may be augmented by disturbance factors (Naveh and Whittaker 1979), particularly anthropogenic land-use practices. Bond (1983) ascribes the high diversity of South African fynbos and Australian kwongan to the extremely impoverished soils. Additionally, the β-diversity, contributed to by topographic factors such as slope and aspect, and differences in substrate and climate, accounts for the high species turnover among sites. Hopkins and Griffith (1984) reported a 10% species turnover per 25-km shift among kwongan sites. Cody (1986) reports the differential in species-turnover rates for *Ceanothus* and *Arctostaphylos* along transects parallel to the coast in California compared to the steeper gradients normal to the coast—that is, cutting across the topographic and rainfall gradients. Similar treatments for *Protea* and *Leucadendron* in South Africa produce higher turnover rates for the steeper environmental gradients but lower rates for the gradients of gradual climatic shift. Within the mediterranean region in southwestern western Australia, species turnover rates in *Banksia* and *Acacia* are very gradual over the gentle environmental gradients, but over the longer climatic contours, high turnover levels are reached, reflecting the γ-diversity of the flora (Cody 1986).

Data on species richness on a standard-area basis are still somewhat limited. Of those available at the tenth-hectare scale, however, the western shrubby communities are generally rich, with values ranging to 103 species (Bell and Watson 1988). By comparison, the mallee and shrubby communities in the eastern part of southern Australia are poorer in species. In eastern mallee sites, Fox (1988) reported a range of values from a drought-affected low of 14 to 57. Whittaker et al. (1979) found 39 species, and Specht and Specht (1989) report values of 33 and 36 for mallee-broombush in South Australia.

Species-richness values for Californian sites tend to be lower than for most of the Australian sites. Westman (1981, 1983) reported values for cha-parral, coastal sage scrub, and coastal succulent scrub. Figure 6.4 depicts the range of species-richness values reported for the mediterranean shrublands in Chile, California, and southern Australia. The few data available for Chilean matorral (Montenegro and Teillier 1988) are extremely high. No indication is given as to whether the species included are only natives or include introduced species as well. These data may also represent pooled smaller plots, which would also exaggerate their richness.

Most of the data available for species richness combine the native species and the adventive, the latter species mainly introduced subsequent to European contact. In some places it has been suggested that the introduced species, mainly annuals, are replacing native species removed by grazing pres-

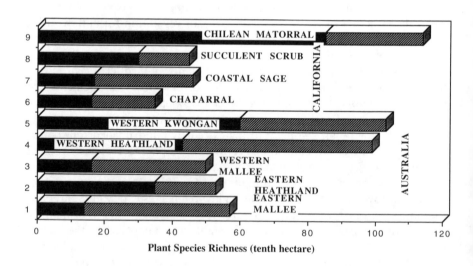

Figure 6.4. Comparison of vascular plant species α-diversity for mediterranean shrublands in southern Australia, California, and Chile. Data for Australia are from Bell and Watson (1988), George et al. (1979), Specht and Specht (1989), and Fox (1988), for California from Westman (1981, 1983), and for Chile from Montenegro and Teillier (1988).

sure from introduced husbanded animals. The dry, open conditions in the central valleys in both California and Chile feature many annuals. Raven (1971) contends that annual plants reach the zenith of their diversity in mediterranean-climate regions. Worldwide, annuals comprise about 13% of the flora (Raven and Axelrod 1978), but in the Californian floristic province the proportion is 27%. Arroyo et al. (1990, 1993) have closely examined the Californian and central Chilean floras and reported incidences for annuals of 33% and 20% respectively, including introduced species, and 29% and 16% considering only the native flora. By comparison, the southern Australian flora is much poorer in annuals (Hopper 1979), as low as 7% (Beard 1981) for the southwest of Western Australia.

Anthropogenic Influences

An important difference in the recent histories of the three regions has been the differential influence of indigenous and colonial people. The idea that a special relationship connects mediterranean ecosystems and their degradation by human disturbance has been widely canvassed, although most attention has been given to the Mediterranean basin (e.g., le Houerou 1981; Trabaud 1981; Zohary 1983).

Humans beings first reached the southwest in North America and southern South America at least 10,000 years B.P. (Aschmann and Bahre 1973; Bahre 1979); by contrast, the first people were living in southern Australia at least 40,000 years B.P. (Bowler 1990). However, they used the harsh mallee landscape with little water only opportunistically (Harris 1990). But these indigenous people had quite different effects on the environments in each region. North America had some domestication of plants but in California it was not extensive. The principal influence was the use of fire to maintain grazing lands for game. Fire was also important to the Indians in Chile, but was often employed as a clearing tool for agriculture. By 2000 years B.P., three winter-rain crop plants had been developed (Aschmann and Bahre 1973) and just before European settlers arrived, the llama and alpaca had been domesticated (Table 6.4).

In the sixteenth century the Spanish and Portuguese began their conquest of South America. The Spanish began settling Chile late in the 1530s, establishing Santiago in 1541. The early massive invasion by Mediterranean annuals (Raven 1971; Aschmann 1973) is to be expected because the colony and source country match climatically. The invasive species were often in contaminated grain seed.

Whereas California was discovered as early as 1542, the Spanish began settling in isolated missions only around 1770. The first in a series of missions was established in San Diego in 1769, but for fifty years or more settlement was confined to the immediate vicinity of the missions. The Californian Gold Rush in 1848 saw a rapid influx of population and increased demand for

Table 6.4. Principal aspects of human intervention in the natural vegetation of the three Pacific mediterranean regions (based on Fox and Fox 1986a)

	California	Central Chile	Southern Australia
Arrival of indigenes	> 10,000 y B.P.	> 10,000 y B.P.	> 40,000 y B.P.
Indigenous use of fire	Regular	Regular	Locally frequent
Indigenous agriculture and animal husbandry	Limited	3 crops (2000 B.P.) llama (recent)	nil
Other indigenous habitat modification	hunting and gathering	cutting and burning brush	hunting and gathering
European settlement	1769 (Mission Period)	1550	1828 (mediterranean region)
European fire regime	Fewer	Fewer	Fewer
Period of major influence	220 years	430 years	160 years
Principal weeds and pest animals	Old World annuals and grasses, deer and rabbits	Old World annuals and grasses, rabbits	Old World annuals and grasses, South African shrubs and geophytes, rabbits and goats

food crops (Mooney et al. 1972). The invasion of annuals from the Mediterranean Basin to California echoed what had happened in Chile 300 years earlier. Along with that influx of new species, though, began an exchange of species between the two Americas (Raven 1973). Raven reports 130 important herbaceous species common to both California and Chile, and the most abundant ones are the "European" weeds. Analyzing the California weed flora (Raven and Axelrod 1978) shows that many were Old World weeds. Of 674 introductions, 559 are from the Old World (principally the Mediterranean Basin). Of the few from the New World, only 38 species are from South America.

The mediterranean region in southern Australia was settled by Europeans early in the nineteenth century (Fox 1982). For a long time the mallee lands were shunned as waterless and inhospitable (Harris 1990). Then artesian bores provided stock with water and pastoralism slowly spread into the mallee. Cropping on mallee soils awaited further technological changes, such as special ploughs and fertilizer (Fox 1982; Harris 1990). Vegetation change included direct loss from clearing but also modification by changes in the fire regime, newly introduced grazers and browsers, and weed species (Adamson and Fox 1982). Disturbance in the form of alteration to natural cycles enhances the establishment of weed species (Fox and Fox 1986b).

Although many of the settlers themselves were from northern Europe, once again the crop plants and associated weeds were Mediterranean in origin (Fox 1990b). Kloot (1987) analyzes the origins of 903 alien plant species in South Australia (the settled portion of which is mediterranean),

32% of which originated in the Mediterranean Basin. Specht (1972) had analyzed a smaller weed flora (654 species) and also found 32% from the Mediterranean Basin, with only 5% from Chile and 4% from California.

Contact Between Regions

The first intensive contact among the three Pacific mediterranean regions was prompted by the gold rushes in California (1848) and Australia (1867). This coming together promoted trade between southern Australia and California, which still continues. To a lesser extent it also promoted trade between Chile and southern Australia, but neither is a major trading partner with the other and they have separate trade routes to markets in the Northern Hemisphere. Similarly, a short period of trade between Chile and California diminished with completion of the transcontinental railway in the United States, and opening of the Panama Canal.

A number of biological exchanges have occurred, however, among the homologous climatic zones in California, Chile, and Australia and invasions continue. Overwhelmingly, though, the numbers of annual weed species from the Mediterranean Basin outnumber additions from the other mediterranean regions. Fox (1990b) suggests that the invasion occurs as three distinct waves: the primary invasion of the Mediterranean agrestal weeds, a secondary exchange of weedy woody species between mediterranean regions (such as *Pinus radiata* from California to Chile), and a tertiary invasion *within* each region (e.g., east–west invasions in southern Australia). Possibly the first wave has spent its force in the three regions being reviewed here, although parts still being settled such as Baja California are still receiving Old World annual weeds. For the older settled regions in Chile, California, and southern Australia the continuing flux in species invasions is by woody species between them and by native and introduced species within each expanding their ranges. Such expansions generally respond to anthropogenic habitat modification but may also accompany changes in climate.

The three mediterranean regions compared in this chapter comprise an interesting subset of the five world regions. Di Castri (1981) suggests a grouping of the five based on the pivotal role of the Mediterranean Basin, and then two pairs comprising the older Gondwanan pair of southern Australia and South Africa, and the 'Pleistocene' assemblage (Naveh and Whittaker 1979) of Chile and California. Fox (1990b) builds on this scheme to emphasize the common patterns of settlement and trade: northern Europeans in the former and Spanish settlement in the latter. In this chapter I contrast the two Pleistocene regions with the older Australian region, but when we consider the history of European contact, Australia and California have been settled for comparable periods, though Chile has had more than two centuries of European influence. Fuentes and Muñoz (this volume) however, found that evidence of convergence persists in spite of the severe and pervasive landscape modifications introduced by people. Modern people, though, have influenced convergence, by introducing the same suite of plants and animals around the world wherever European people have made permanent settlements, and the same types of habitat modification (clearing, frequent burning) continue to operate.

Conclusions

The two parts of the mediterranean region in southern Australia have shrubland formations dissimilar in characteristics of community structure and also partly in floristic composition. Although α-diversity values are similar (somewhat lower for the eastern sites), β-diversity is far higher in the west, as is the rate of change of β-diversity. Future research could further explore these trends, and particularly look at γ-diversity across the continent. How much of these differences in diversity can be attributed to climate, soil characteristics, or evolutionary history could also be further investigated.

Where data are available for the other two regions, indications are that the Californian communities are less rich than their Australian counterparts. If indeed species richness is inversely correlated with soil nutrient status, this could be the predicted order. The few data available for Chile, however, do not confirm this trend. With the highest soil-nutrient status, they also have the richest communities. How much of this richness is supplemented by introduced species, or how much by opening up of the communities' structure by the grazing done by introduced mammals, remains to be fully explored. Perhaps because many species are not shared between these regions, higher plant taxa could very profitably be employed (Elsol 1985). As a step toward better understanding the similarities and differences between regions, and to validate or refute evidence for convergence, more data are needed on α-, β-, and γ-diversity in all mediterranean regions.

Acknowledgments

I thank Mary Kalin Arroyo and the organizers of the Pacific Science Inter-Congress for the invitation to present the original form of this paper in 1989 at the meeting in Vina del Mar and for financial assistance to visit Chile. Barry Fox helped in analyzing the data and the final preparation of the manuscript, assistance that is sincerely appreciated; Kevin Maynard drafted the maps.

References

Adamson DA, Fox MD (1982) Change in Australasian vegetation since European settlement. In Smith JMB (ed) *A History of Australasian Vegetation*. McGraw-Hill, Sydney, pp 109–146

Aplin TEH (1977) A checklist of *Eucalyptus* of Western Australia. W.A. Department of Agriculture Tech. Bull. No. 33

Arroyo MTK, Marticorena C, Muñoz M (1990) A checklist of the native annual flora of continental Chile. Gayana (Botanica) 46:121–151

Arroyo MTK, Armesto JJ, Squeo F, Gutierrez J (1993) Global change: flora and vegetation of Chile. In Mooney HA, Fuentes E, Kronberg B (eds) *Earth System Response to Global Change: Contrasts Between North and South America.* Academic Press, New York, pp 239–263

Aschmann H (1973) Distribution and peculiarity of mediterranean ecosystems. In di Castri F, Mooney HA (eds) *Mediterranean-Type Ecosystems: Origin and Structure.* Chapman & Hall, London, pp 11–19

Aschmann H, Bahre C (1973) Man's impact on the wild landscape. In Mooney HA (ed) *Convergent Evolution in Chile and California.* Dowden, Hutchinson & Ross, Stroudsburg, PA, pp 73–84

Bahre CJ (1979) *Destruction of the Natural Vegetation of North-Central Chile.* U California, Berkeley

Beadle NCW (1981) *The Vegetation of Australia.* Cambridge U Press, Cambridge

Beard JS (1981) Vegetation Survey of Western Australia, 1 : 1 000 000 series sheet 7, "Swan," U Western Australia Press, Nedlands

Beard JS (1990) The mallee lands of western Australia. In Noble JC, Joss PJ, Jones GK (eds) *The Mallee Lands: A Conservation Perspective.* CSIRO, East Melbourne, pp 29–33

Bell DT, Stephens LJ (1984) Seasonality and phenology of kwongan species. In Pate JS, Beard JS (eds) *Kwongan: Plant Life of the Sandplain.* U Western Australia Press, Nedlands, pp 205–226

Bell DT, Watson LE (1988) Species richness of plant communities in southwestern Western Australia. In Specht RL (ed.) *Mediterranean-Type Ecosystems: A Data Source Book.* Kluwer, Dordrecht

Blackburn G, Wright MJ (1989) Soils. In Noble JC, Bradstock RA (eds) *Mediterranean Landscapes in Australia: Mallee Ecosystems and Their Management,* CSIRO, Melbourne, pp 35–53

Bond W (1983) On alpha diversity and the richness of the Cape Flora: a study in southern Cape fynbos. In Kruger FJ, Mitchell DT, Jarvis JUM (eds) *Mediterranean-Type Ecosystems: The Role of Nutrients.* Springer-Verlag, New York, pp 337–356

Bowler JM (1982) Aridity in the late Tertiary and Quaternary of Australia. In Barker WR, Greenslade PJM (eds) *Evolution of the Flora and Fauna of Arid Australia.* Peacock Publications, Adelaide, pp 35–45

Bowler JM (1990) Human occupation and environmental change: the ancient record from Willandra Lakes. In Noble JC, Joss PJ, Jones GK (eds) *The Mallee Lands: A Conservation Perspective.* CSIRO, East Melbourne, pp 152–161

Brooker MIH, Kleinig DA (1989) *Field Guide to Eucalyptus,* Vol 2, Western Australia (southern part, below 260° latitude), northern South Australia, Eyre Peninsula, and Kangaroo Island, New South Wales (west and north of the Darling river), Inkata Press, Sydney

Burbidge NT (1947) Key to the South Australian species of *Eucalyptus* L'Herit. Trans Royal Soc S Aust 71:137–167

Carnahan JA, Deveson T (1990) Vegetation. In *Atlas of Australian Resources,* Vol 6, 3rd ser. Department of Administrative Services, Canberra

Cody ML (1986) Diversity, rarity, and conservation in mediterranean-climate regions. In Soulé, ME (ed) *Conservation Biology: The Science of Scarcity and Diversity.* Sinauer Associates, Sunderland, pp 123–152

Darwin C (1894) *Journal of Researches into the Geology and Natural History of the Various Countries visited by H.M.S. Beagle,* (Rev. ed.). Ward, Lock & Bowden, London di Castri F (1981) Mediterranean-type shrublands of the world. In di Castri F, Goodall DW, Specht R (eds) *Mediterranean-Type Shrublands*. Elsevier, New York, pp 1–52

Dick RS (1975) A map of the climates of Australia: according to Köppen's principles of definition. Queensland Geog J, Ser. 3,3: 33–69 and map

Doing H (1981) Phytogeography of the Australian floristic kingdom. In Groves RH (ed) *Australian Vegetation*. Cambridge U Press, Cambridge, pp 3–25

Elsol J (1985) Illustrations of the use of higher plant taxa in biogeography. J Biogeog 12:433–444

Fox BJ, Fox MD (1986a) Resilience of animal and plant communities to human disturbance. In Dell B, Hopkins AJM, Lamont BB (eds) *Resilience in Mediterranean-Type Ecosystems*. Dr W Junk Publishers, Dordrecht, pp 39–64

Fox MD (1982) Vegetation changes in mediterranean Australia since European settlement. In Conrad CE, Oechel WC (eds) *Dynamics and Management of Mediterranean-Type Ecosystems*, Gen. Tech. Rep. PSW-58, Pacific Southwest Forest and Range Experiment Station, Forest Service, USDA, Berkeley, CA, pp 112–117

Fox MD [contributor] (1988). Vegetation, nutrition and climate. (3) Species richness. In Specht RL (ed.) *Mediterranean-Type Ecosystems: A Data Source Book*. Kluwer, Dordrecht, pp 81–92

Fox MD (1990a) Composition and richness of New South Wales mallee. In Noble JC, Joss PJ, Jones GK (eds) *The Mallee Lands: A Conservation Perspective*. CSIRO, Melbourne, pp 8–11

Fox MD (1990b) Mediterranean weeds: exchanges of invasive plants between the five mediterranean regions of the world. In di Castri F, Hansen AJ, Debussche M (eds) *Biological Invasions in Europe and the Mediterranean Basin*. Kluwer Academic Publishers, Dordrecht, pp 177–198

Fox MD (1991) The natural vegetation of the Ana Branch–Mildura 1 : 250 000 map sheet. Cunninghamia 2(3), 443–493 (with 1 : 250 000 coloured vegetation map)

Fox MD, Fox BJ (1986b) The susceptibility of natural communities to invasion. In Groves RH, Burdon JJ (eds) *The Ecology of Biological Invasions: An Australian Perspective*. Aust Acad Sci, Canberra, pp 57–66

George AS, Hopkins AJM, Marchant NG (1979) The heathlands of Western Australia. In Specht RL (ed) *Heathlands and Related Shrublands: Descriptive Studies*. Elsevier, Amsterdam, pp 211–230

Green JW (1985) *Census of the Vascular Plants in Western Australia*, 2nd ed. Government Printer, Perth, 57 pp

Greenwood G, Boardman R (1990) Climatic change and some possible effects upon the terrestrial ecology of South Australia. In Noble JC, Joss PJ, Jones GK (eds) *The Mallee Lands: A Conservation Perspective*. CSIRO, East Melbourne, pp 135–139

Hanes TL (1981) California chaparral. In di Castri F, Goodall DW, Specht RL (eds) *Mediterranean-Type Shrublands*. Elsevier, Amsterdam, pp 139–174

Harris CR (1990) The history of mallee land use: Aboriginal and European. In Noble JC, Joss PJ, Jones GK (eds) *The Mallee Lands: A Conservation Perspective*. CSIRO, East Melbourne, pp 147–151

Hill KD (1989) Mallee eucalypt communities: their classification and biogeography. In Noble JC, Bradstock RA (eds) *Mediterranean Landscapes in Australia: Mallee Ecosystems and Their Management*. CSIRO, Melbourne, pp 93–108

Hill KD (1990) Biogeography of the mallee eucalypts. In Noble JC, Joss PJ, Jones GK (eds) *The Mallee Lands: A Conservation Perspective*. CSIRO, East Melbourne, pp 16–20

Hopkins AJM, Griffin EA (1984) Floristic patterns. In Pate JS, Beard JS (eds)

Kwongan: Plant Life of the Sandplain. U Western Australia Press, Nedlands, pp 69–83

Hopper SD (1979) Biogeographical aspects of speciation in the south west Australian flora. Ann Rev Ecol Syst 10:399–422

Hopper SD (1990) Conservation status of mallee eucalypts in southern Western Australia. In Noble JC, Joss PJ, Jones GK (eds) *The Mallee Lands: A Conservation Perspective*. CSIRO, East Melbourne, pp 21–24

Hopper SD (1992) Patterns of plant diversity at the population and species level in south-west Australian mediterranean ecosystems. In Hobbs RJ (ed) *Biodiversity in Mediterranean Ecosystems in Australia*. Surrey Beatty and Sons Pty Ltd, Sydney, pp 27–46

Kloot PM (1987) The naturalised flora of South Australia 3. Its origin, introduction and distribution, growth forms and significance. J Adelaide Bot Gard 10:99–111

Köppen W (1923) *Die Klimate der Erde*. Borntraeger, Berlin

Lamont BB, Hopkins AJM, Hnatiuk RJ (1984) The flora: composition, diversity and origins. In Pate JS, Beard JS (eds) *Kwongan: Plant Life of the Sandplain*. U Western Australia Press, Nedlands, pp 27–50

le Houerou HN (1981) Impact of man and his animals on mediterranean vegetation. In di Castri F, Goodall DW, Specht RL (eds) *Mediterranean-Type Shrublands*. Elsevier, New York, pp 479–521

Mabry TJ, Difeo DR (1973) The role of secondary plant chemistry in the evolution of the mediterranean scrub vegetation. In di Castri F, Mooney HA *Mediterranean-Type Ecosystems: Origin and Structure*. Chapman & Hall, London, pp 121–155

Markham NT, Noble JC (1989) Essential oil. In Noble JC, Bradstock RA (eds) *Mediterranean Landscapes in Australia: Mallee Ecosystems and Their Management*. CSIRO, East Melbourne, pp 353–361

Milewski AV (1979) A climate basis for the study of convergence of vegetation structure in mediterranean Australia and southern Africa. J Biogeog 6:293–299

Miller PC (1981) *Resource Use by Chaparral and Matorral: A Comparison of Vegetation Function in Two Mediterranean-Type Ecosystems*. Springer-Verlag, Berlin, 455 pp

Montenegro G, Teillier S (1988) Species richness of plant communities in the mediterranean-climate region of Chile. In Specht RL (ed) *Mediterranean-Type Ecosystems: A Data Source Book*. Kluwer, Dordrecht, p 87

Mooney HA (ed.) (1977) *Convergent Evolution in Chile and California: Mediterranean Climate Ecosystems*. Dowden, Hutchinson & Ross, Stroudsburg PA

Mooney HA, Dunn EL, Shropshire F, Song L. (1972) Land use history of California and Chile as related to the structure of the sclerophyll scrub vegetation. Madrono 21:305–319

Naveh Z, Whittaker RH (1979) Structural and floristic diversity of shrublands and woodlands in northern Israel and other mediterranean areas. Vegetatio 41:171–190

Newbey KR, Dell J, How RA, Hnatiuk RJ (1984) The Biological Survey of the Eastern Goldfields of Western Australia. Part 2 Widgiemooltha–Zanthus Study Area. rec. West Aust Mus Suppl No. 18, pp 21–157

Parkinson G (ed.) (1986) Climate. Volume 4. *Atlas of Australian Resources* (3rd Ser), Division of National Mapping, Canberra

Parsons DJ, Moldenke AR (1975) Convergence in vegetation structure along analogous climatic gradients in California and Chile. Ecology 56:950–957

Parsons RF (1981) *Eucalyptus* scrubs and shrublands. In Groves RH (ed) *Australian Vegetation*. Cambridge U Press, Cambridge, pp 227–252

Pate JS, Beard JS (eds) (1984) *Kwongan: Plant Life of the Sandplain*. U Western

Australia Press, Nedlands

Pittock AB (1988) Actual and anticipated changes in Australia's climate. In Pearman GI (ed.) *Planning for Climate Change*. CSIRO, East Melbourne, pp 35–51

Raven PH (1971) The relationships between "Mediterranean" floras. In Davis DH, Harper PC, Hedge IC (eds) *Plant Life of South-West Asia*. Bot Soc Edinburgh, pp 119–134

Raven PH (1973) The evolution of mediterranean floras. In di Castri F, Mooney HA (eds) *Mediterranean Type Ecosystems: Origin and Structure*. Springer-Verlag, Berlin, pp 213–224

Raven PH, Axelrod DI (1978) *Origin and Relationships of the Californian Flora*. U California, Publications in Botany 72

Rundel PW (1981) The matorral zone of central Chile. In di Castri F, Goodall DW, Specht RL (eds) *Mediterranean-Type Shrublands*. Elsevier, Amsterdam, pp 175–201

Saunders DA, Hobbs RJ (1992) Impact on biodiversity of changes in land use and climate. In Hobbs RJ (ed) *Biodiversity in Mediterranean Ecosystems in Australia*. Surrey Beatty and Sons Pty Ltd, Sydney, pp 61–75

Schodde R (1981) Bird communities of the Australian mallee: composition, derivation, distribution, structure and seasonal cycles. In di Castri F, Goodall DW, Specht RL (eds) *Mediterranean-Type Shrublands*. Elsevier Scientific Publishing, Amsterdam, pp 387–415

Specht RL (1972) *The Vegetation of South Australia*, 2nd Ed, Government Printer, Adelaide, 328 pp

Specht RL (1973) Structure and functional response of ecosystems in the mediterranean climate of Australia. In di Castri F, Mooney HA (eds) *Mediterranean Type Ecosystems: Origin and Structure*. Chapman & Hall, London, pp 113–120

Specht RL (1981a) Mallee ecosystems in southern Australia. In di Castri F, Goodall DW, Specht RL (eds) *Ecosystems of the World*, Vol 11, *Mediterranean-Type Shrublands*. Elsevier, Amsterdam, pp 203–231

Specht RL (1981b) Heathlands. In Groves RH (ed.) *Australian Vegetation*. Cambridge U Press, Cambridge, pp 253–275

Specht RL, Moll EJ (1983) Mediterranean-type heathlands and sclerophyllous shrublands of the world: an overview. In Kruger FJ, Mitchell DT, Jarvis JUM (eds) *Mediterranean-Type Ecosystems: The Role of Nutrients*. Springer-Verlag, New York, pp 41–65

Specht RL, Rayson P (1957) Dark Island heath (Ninety Mile Plain, South Australia) I. Definition of the ecosystem. Aust J Bot 5, 52–85

Specht RL, Specht A (1989) Species richness of overstorey strata in Australian plant communities—the influence of overstorey cover. Aust J Bot 37:337–350

Thrower NJW, Bradbury DE (1973) The physiography of the mediterranean lands with special emphasis on California and Chile. In di Castri F, Mooney HA (eds) *Mediterranean-Type Ecosystems: Origin and Structure*. Springer-Verlag, Berlin, pp 37–60

Thrower NJW, Bradbury DE (eds) (1977) *Chile–California Mediterranean Scrub Atlas*. Dowden, Hutchinson & Ross, Stroudsburg, PA

Trabaud L (1981) Man and fire: impacts on mediterranean vegetation. In di Castri F, Goodall DW, Specht R (eds) *Mediterranean-Type Shrublands*. Elsevier, New York, pp 523–538

Westman WE (1978) Evidence for the distinct evolutionary histories of canopy and understorey in the *Eucalyptus* forest-heath alliance of Australia. J Biogeog 5:365–376

Westman WE (1981) Diversity relations and succession in Californian coastal sage scrub. Ecology 62:170–184

Westman WE (1983) Plant community structure—spatial partitioning of resources. In Kruger FJ, Mitchell DT, Jarvis JUM (eds) *Mediterranean-Type Ecosystems: The Role of Nutrients*. Springer-Verlag, Berlin, pp 417–445

Whittaker RH (1972) Evolution and measurements of species diversity. Taxon 21: 213–251

Whittaker RH, Niering WA, Crisp MD (1979) Structure, pattern and diversity of a mallee community in New South Wales. Vegetation 39:65–76

Williams CH, Raupach M (1983) Plant nutrients in Australian soils. In *Soils: An Australian Viewpoint*. CSIRO, Melbourne and Academic Press, London, pp 777–794

Zohary M (1983) Man and vegetation in the Middle East. In Holzner W, Werger MJA, Ikusima I (eds) *Man's Impact on Vegetation*. Dr W Junk, The Hague, pp 287–295

7. Ecomorphological Characters as a Resource for Illustrating Growth-Form Convergence in Matorral, Chaparral, and Mallee

Gloria Montenegro and Rosanna Ginocchio

One successful approach to refining the interaction between form and environment in plant communities has been to examine the extent of convergence in structural and functional attributes of phylogenetically distinct biotas in climatically similar areas (Parsons 1976).

Only small portions of the globe can be described as mediterranean climate areas (di Castri 1990). The geographic isolation of these regions and strong dissimilarity in the prevailing plant taxa there is evidence of distinct evolutionary history. For these reasons mediterranean ecosystems around the world have been studied as classic examples of vegetation convergence under similar climatic conditions (Mooney and Dunn 1970; di Castri and Mooney 1973; Parsons 1976; Mooney 1977; Thrower and Bradbury 1977; Specht 1979; di Castri et al. 1981; Miller 1981; Quézel 1982; Schulze 1982; Kruger et al. 1983; Shmida 1984; Tenhunen et al. 1987; di Castri 1990).

The similarity in form between phylogenetically unrelated plant species in disjunct areas of mediterranean climate has been ascribed to the limited adaptive solutions possible in environments with multiple stresses (Mooney and Dunn 1970). Mediterranean scrub regions are semiarid and often nutrient poor, and have vegetative canopies that shade the lower strata and the soil. Resources that can be optimized include water, nitrogen, and light energy (Mooney et al. 1983). These communities have greatly varied growth forms that have evolved to optimally use available resources and withstand temporary water deficits (Rundel 1981; Orshan et al. 1984a; 1984b; Montenegro et al. 1985). These plant growth forms are the end result of several growth processes (Orshan 1983a; 1989) that result in biomass increment, and the architectural arrangement of this biomass in space.

To avoid subjectivity in selecting growth form characteristics for vegetation comparison, Orshan (1983a, 1983b) has developed the monocharacter approach for studying plant growth forms. With this approach plant species are described according to their morphological, anatomical, phenomorphological, and physiological attributes. Attributes of the shoot and assimilating organs, the root system and underground stems, longevity and seasonality, and those related to regeneration and reproduction may be used. With the monocharacter approach several growth form types are defined for each ecomorphological character, forming a monocharacter growth-form system. Ecomorphological characters that reflect structural and functional aspects of major vegetation types in mediterranean climate regions are considered relevant parameters (Catling et al. 1988).

Our aim in this chapter is to assess similarities and differences among static and dynamic features of the dominant woody species in the shrub vegetation in mediterranean areas of central Chile, southern California, and southern Australia by comparing the distribution of single monocharacter types of growth form. Such an approach provides a quantitative test of the hypothesis that under similar climatic regimes comparable structural systems will evolve regardless of the distinct evolutionary histories of the biotas.

Selected Sites and Methods

The growth-form monocharacter system was applied to plant communities representing coastal and inland vegetation in the mediterranean regions in central Chile, southern California, and southwestern Australia. Table 7.1 shows the location, vegetation structure, and age of the stands since the last fire for each selected site. The most abundant woody species examined are listed in Table 7.2.

Table 7.3 shows the ecomorphological characters used in these comparisons (modified after Orshan et al. 1984b). Each character has a series of independent monocharacters, selected to be mutually exclusive by definition. Ecomorphological characters were subdivided according to attributes of the whole plant, shoot, and assimilating organs; attributes of the root system and underground stems; longevity and seasonality; and aspects related to regeneration and reproduction. Lack of information for most species prevented comparisons of the root system.

Qualitative monocharacters belonging to each ecomorphological character were assigned in an orderly fashion according to their presumed evolutionary stages, ancestral monocharacters having lower values than derived ones (cf. Orshan 1983a, 1983b; Orshan et al. 1984a; 1983b). Code numbers for each monocharacter type are shown in brackets in Table 7.3.

Table 7.1. Characteristics of study sites

Communities		Location	Vegetation structure	Date of last fire	References
Central Chile	Inland	Fundo Santa Laura (33°04′S, 71°00′W) 1000 m	Typical matorral Sclerophyll evergreen scrub	25 years	Orshan G, Montenegro G,[a] Avila G, Aljaro ME, Walkowiak A, Mujica AM (1985)
	Coastal	Papudo-Zapallar-Cachagua (32°35′S, 71°28′W) 20–28 m	Coastal scrub on ocean bluffs and slopes	20 years	
Southern California	Inland	San Diego State University Sky Oaks Biological Field Station (33°21′N, 116°34′W) 1500 m	Typical chaparral Sclerophyll evergreen scrub	55–60 years	Zedler PH,[a] Zammit CA, Scheid GA[a] (1986)
	Coastal	San Diego County Rancho San Bernardo Foothills on Woodson Mountain (33°00′N, 117°00′W)	Coastal sage scrub	21 years	
Southwestern Australia	Inland	Dark Island Soak, near Keith (36°02′S, 140°31′E) Brookfield Conservation Park (34°20′S, 139°23′E)	Mallee open scrub	29 years	Specht RL[a] (1966)
	Coastal	Mount Lesueur, Western Australia (33°00′S, 117°00′E)	Kwongan (heathland)	20 years	Griffin EA, Hopkins AJM[a] (1985)

[a] Data compilers in each location. In Specht RL (1988).

Table 7.2. Plant species analyzed

Communities	Coastal vegetation	Inland vegetation
Central Chile	*Peumus boldus* Mol. *Flourensia thurifera* (Mol.) D.C. *Proustia cuneifolia* D. Don *Lobelia tupa* L. *Podanthus mitiqui* Lindl. *Fuchsia lycioides* Andr. *Baccharis concava* D.C. *Noticastrum sericeum* (Less) Less ex Phil. *Bahia ambrosioides* Lag. *Haplopappus foliosus* D.C. *Puya chilensis* Mol.	*Quillaja saponaria* Mol. *Cryptocanya alba* (Mol.) Looser *Lithrea caustica* Hook. et Arn. *Kageneckia oblonga* R. et P. *Trevoa trinervis* Miers *Talguenea quinquenervia* (Gill et Hook.) Johnst. *Trichocereus chiloensis* (Colla) B. et R. *Colliguaja odorifera* Mol. *Satureja gilliesii* (Graham) Briq. *Muehlenbeckia hastulata* M. Johnst.
Southern California	*Artemisia californica* Less. *Salvia apiana* Jepson *Salvia mellifera* Greene *Keckiella antirrhinoides* (Benth) Straw *Rhus laurina* Nutt *Mimulus longiflorus* (Nutt) Grant *Haplopappus squarrosus* H. et A. *Eriogonum fasciculatum* Benth.	*Arctostaphylos pringlei* McBr. *Arctostaphylos glandulosa* Eastw. *Adenostoma sparsifolium* Torr. *Adenostoma fasciculatum* H. et A. *Ceanothus greggii* Gray *Quercus dumosa* Nutt. *Tauschia parishii* (Coult et Rose) McBr. *Gnaphalium californicum* Gray *Lonicera subspicata* H. et A. *Garrya flavescens* Wats.
Southwestern Australia	*Banksia tricuspis* Meissn. *Dryandra sessilis* (Knight) Domin *Calothamnus sanguineus* Labill. *Daviesia aff. striata* Turcz. *Lambertia multiflora* Lindl. *Kingia australis* R. Br. *Hakea megalosperma* Meissn. *Banksia micrantha* A.S. George *Calothamnus torulosus* Schau. *Dryandra nivea* R.Br. *Petrophile striata* R.Br.	*Eucalyptus oleosa* F. Muell. *Eucalyptus incrassata* Labill. *Eucalyptus foecunda* Schau. *Eucalyptus gracilis* F. Muell. *Melaleuca uncinata* F. Muell. *Hakea muelleriana* J.M. Black *Baeckea crassifolia* Lindl. *Calytrix tetragona* Labill. *Atriplex suberecta* Verdoon *Melaleuca acuminata* F. Muell.

Table 7.3. Ecomorphologic characters (modified after Orshan et al. 1984a)

I. Whole plant, shoot, and assimilating organs

Character	(0)	(1)	(2)	(3)	(4)	(5)	(6)	(7)	(8)
1. Renewal buds		Chamephyte (<25 cm)	Nanophaneophyte (25 cm–2 m)	Microphaneophyte (2–8 m)	Mesophaneophyte (8–30 m)				
2. Plant height		<10 cm	10–25 cm	25–50 cm	50–100 cm	1–2 m	2–5 m	5–10 m	10–20 m
3. Crown diameter		<10 cm	10–25 cm	25–50 cm	50–100 cm	1–2 m	2–5 m	5–10 m	10–20 m
4. Photosynthetic organs		Leaves	Leaves and Stem	Phyllodes	Cladodes	Absent			
5. Leaf size		Subleptophyll (<0.3 cm)	Leptophyll (0.1–0.25)	Nanophyll (0.25–2.25)	Nanomicrophyll (2.25–12.25)	Microphyll (12.25–20.25)	Micromesophyll (20.25–56.25)	Mesophyll (56.25–180)	Macrophyll (>180)
6. Leaf surface resins oils and mucilage		Absent	Present						
7. Leaf glands and ducts		Absent	Present						
8. Leaf consistency		Malacophyll	Semisclerophyll	Sclerophyll	Succulent	Water succulent	Resin succulent		
9. Leaf tomentosity		Nonhairy	Lower side hairy	Upper side hairy	Both sides hairy				
10. Stem lignification		Axyles	Semixyles	Holoxyles					
11. Spinescence		Absent	Leaves	Stem	Stem and leaves				

II. Longevity and seasonality

Character	(0)	(1)	(2)	(3)	(4)
1. Leaf seasonality		Evergreen	Summer shedders	Winter shedders	
2. Organs periodically shed		Leaf	Branch	Shoot	Whole plant
3. Organ shed seasonality		Spring	Summer	Autumn	Winter
4. Shoot-growth seasonality		Spring	Summer	Autumn	Winter
5. Flowering seasonality		Spring	Summer	Autumn	Winter

III. Regeneration and reproduction

Character	(0)	(1)	(2)	(3)	(4)	(5)	(6)	(7)
1. Presence of underground stems	None	Rhizomes	Rhizodes	Sobols	Stem tubers	Bulbs	Corms	Lignotubers
2. Vegetative multiplication		None	Root/Shoot splitting	Above-ground organs	Below-ground organs			
3. Vegetative regeneration after fire		Plant killed	Epicormic bud below ground	Epicormic bud above ground	Nonepicormic buds below ground	Nonepicormic buds above ground		
4. Flowering pyrogenic		Absent	Present					

Table 7.4. Inland vegetation: Ecomorphological character nomenclature as in Table 7.3. Numbers in body of table correspond to mono-character type, which are numbers in brackets in Table 7.3

Community	Species	I-1	I-2	I-3	I-4	I-5	I-6	I-7	I-8	I-9	I-10	I-11	II-1	II-2	II-3	II-4	II-5	III-1	III-2	III-3	III-4
Central Chile	Q. saponaria	4	8	7	1	4	1	1	3	1	3	1	1	1	3	1	2	7	1	2	1
	C. alba	4	7	6	1	5	1	1	3	1	3	1	1	1	3	1	2	7	1	2	1
	L. caustica	3	7	6	1	4	1	1	3	2	3	1	1	1	4	2	1	7	1	2	1
	K. oblonga	3	6	5	1	5	1	1	3	1	3	1	1	1	4	2	1	7	1	2	1
	T. trinervis	3	6	5	2	4	1	1	1	1	2	3	3	2	2	4	1	7	1	2	2
	T. quinquenervia	3	5	6	1	5	1	1	1	4	3	3	3	2	3	4	2	7	4	2	1
	T. chiloensis	4	7	4	4	—	1	1	—	1	2	3	—	—	3	4	4	—	3	1	1
	C. odorifera	2	5	5	1	4	1	1	1	1	3	1	3	2	3	4	1	7	2	2	1
	S. gilliesii	2	4	4	1	3	1	2	1	4	3	1	3	2	3	4	1	7	2	3	1
	M. hastulata	1	4	6	1	3	1	1	1	1	3	1	1	2	3	2	1	7	3	3	1
Southern California	A. pringlei	3	6	6	1	4	1	1	3	4	3	1	1	1	—	1	1	—	1	2	1
	A. glandulosa	3	5	6	1	4	1	1	3	4	3	1	1	1	—	1	1	7	1	2	1
	A. sparsifolium	3	6	6	2	1	1	1	3	1	3	1	1	1	1	1	2	7	1	2	1
	A. fasciculatum	2	5	5	1	3	1	1	3	2	3	1	1	1	—	1	2	—	1	—	1
	C. greggii	2	5	5	1	1	1	1	3	1	3	1	1	1	—	4	1	7	4	2	1
	Q. dumosa	3	5	6	1	4	1	1	3	2	3	1	1	1	1	1	1	7	4	2	1
	T. parishii	3	2	3	1	5	1	1	1	4	1	1	2	4	2	1	1	1	3	4	1
	G. californicum	2	2	1	1	4	1	2	1	4	1	1	2	4	2	1	1	—	1	1	1
	L. subspicata	3	5	5	1	3	1	1	1	4	3	1	2	4	3	1	2	—	4	2	1
	G. flavescens	3	6	6	1	4	1	1	3	4	3	1	1	4	—	1	2	7	1	3	1
Southwestern Australia	E. oleosa	3	6	6	1	5	1	2	3	1	3	1	1	1	2	2	4	7	1	2	1
	E. incrassata	3	6	6	1	4	1	2	3	1	3	1	1	1	2	2	1	7	1	2	1
	E. foecunda	2	5	4	1	2	1	2	3	1	3	1	1	1	2	1	1	7	1	2	1
	E. gracilis	2	5	6	1	3	1	1	3	1	3	2	1	1	2	1	1	—	1	1	1
	M. uncinata	1	2	2	1	1	1	2	2	1	3	1	1	1	2	1	1	—	1	1	1
	H. muelleriana	2	2	2	1	1	1	2	2	2	3	1	1	1	2	1	3	7	1	1	1
	B. crassifolia	3	7	7	1	3	1	2	2	3	3	1	1	1	2	2	1	—	1	1	1
	C. tetragona	2	3	2	1	3	1	2	4	1	2	1	1	1	2	1	3	7	1	1	2
	A. suberecta	3	6	6	1	4	1	2	3	3	3	1	1	1	2	2	3	7	2	2	1
	M. acuminata	2	5	6	1	2	1	1	3	2	3	2	1	1	—	1	1	—	2	2	1

Table 7.5. Coastal vegetation: Ecomorphological character nomenclature as in Table 7.3. Numbers in body of table correspond to mono-character type, which are numbers in brackets in Table 7.3

Community	Species										Ecomorphological character											
		I-1	I-2	I-3	I-4	I-5	I-6	I-7	I-8	I-9	I-10	I-11	II-1	II-2	II-3	II-4	II-5	III-1	III-2	III-3	III-4
Central Chile	P. boldus	3	7	6	1	4	1	2	3	4	3	1	1	2	3	1	2	7	1	2	1
	F. thurifera	2	5	5	1	4	2	6	1	1	3	3	3	1	2	1	1	—	1	2	1
	P. cuneifolia	2	5	4	1	4	2	6	1	1	3	3	3	2	2	4	2	3	1	3	1
	L. tupa	2	5	5	1	4	1	6	1	2	3	1	3	2	3	1	2	7	1	1	1
	P. mitiqui	2	5	5	2	4	1	6	1	3	3	1	1	2	3	1	1	3	1	2	1
	F. lycioides	2	5	5	1	4	1	6	1	1	3	1	3	1	3	1	1	3	1	1	1
	B. concava	1	4	4	1	4	1	6	2	1	3	1	1	2	3	1	2	7	1	2	1
	N. sericeum	1	3	3	1	3	1	2	1	4	3	1	3	2	3	1	2	—	1	—	1
	B. ambrosioides	1	4	4	1	4	1	2	1	1	3	1	1	2	3	1	2	—	1	3	1
	H. foliosus	1	3	3	1	4	2	2	2	4	2	1	1	2	3	1	2	—	3	3	1
	P. chilensis	2	5	6	1	7	1	1	3	2	—	2	1	1	3	1	2	1	4	2	1
Southern California	A. californica	2	2	5	1	1	1	1	1	4	3	1	3	1	2	1	1	7	1	2	1
	S. apiana	2	2	5	1	6	1	1	1	4	3	1	3	1	2	1	1	7	1	2	1
	S. mellifera	2	2	5	1	4	—	1	2	2	3	1	3	1	2	1	1	7	1	2	1
	K. antirrhinoides	2	2	5	1	3	1	1	2	1	3	1	1	1	—	1	1	7	1	2	1
	R. laurina	2	3	6	1	6	1	1	3	1	3	1	1	—	—	1	2	7	1	2	1
	M. longiflorus	2	4	3	1	4	2	2	3	2	3	1	3	1	2	1	2	7	1		1
	H. squarrosus	2	4	2	1	4	2	2	3	2	3	1	1		2	1	2	7	1	2	1
	E. fasciculatum	2	4	5	1	2	—		2	2	3	1	1			1	2	7	1		1
Southwestern Australia	B. tricuspis	3	2	6	1	3	1	1	3	1	3	1	1	1	2	1	3		1	3	1
	D. sessilis	3	2	5	1	4	1	1	3	1	3	2	1	1	2	1	1		1	1	1
	C. sanguineus	3	2	5	1	2	1	2	3	4	3	1	1	1	2	1	1	7	1	2	1
	D. aff. striata	3	2	5	1	2	1	1	3	1	2	2	1	2	2	1	4		1	1	1
	L. multiflora	3	2	4	1	3	1	1	3	1	3	1	1	1	2	1	1	7	1	2	1
	K. australis	3	2	3	1	4	1	1	3	1	3	1	1	1	2	1	1	8	1	5	2
	H. megalosperma	3	2	3	1	4	1	1	3	1	3	1	1	1	2	1	3	7	1	4	1
	B. micrantha	3	1	3	1	2	1	1	3	1	3	1	1	1	2	1	1	7	1	4	1
	C. torulosus	3	1	2	1	2	2	2	3	1	3	1	1	1	2	1	1	7	1	4	1
	D. nivea	3	1	3	1	3	1	1	3	1	3	2	1	1	2	1	1	7	1	4	1
	P. striata	3	1	3	1	3	1	1	3	1	3	2	1	1	2	1	1	7	1	4	1

The monocharacters that typify each species of the communities analyzed are listed in Tables 7.4 and 7.5 (see also Specht 1988). These data were analyzed to determine the distribution of the monocharacter growth-form types within each community. Each ecomorphological character was then compared across the three communities using a one-way nonparametric ANOVA, with the species as replicates. The Tukey test was used as an a posteriori test (Sokal and Rohlf 1981).

Jaccard's Similarity Index was employed to compare communities. Each ecomorphological character was considered equivalent to a species. If the ecomorphological character showed no difference between the communities being compared, it was assumed that the species was present in all communities, but if the ecomorphological character differed between the communities being compared, it was assumed that the species was present in only one community. Chi Square tests were carried out to determine significance levels between coastal and inland communities in each geographic area according to Saiz (1980).

Similarities

In general, the mediterranean vegetation in central Chile, southern California, and southwestern Australia appears to be similar in monocharacters contributing to the physiognomy of the natural vegetation and in those related to regeneration and reproduction, irrespective of coastal or inland location. Most of the species photosynthesize only through leaves, although a few species in Chile and California have photosynthetic stems (e.g., *Trevoa trinervis* in Chile and *Adenostoma sparsifolium* in California). The nanomicrophyll leaf area (2 to 12 cm^2) appears to be the most common type shed in summer or autumn. The absence of leaf surface resins, oils, and mucilage is another common feature of the matorral, chaparral, and mallee species in both coastal and inland sites.

Fires are common in California, where lightning is the natural source of ignition (Keeley and Keeley 1977; 1981), whereas in Chile people are responsible for most wildfires (Araya and Avila 1981; Avila et al. 1981). It has been suggested that the fire-recurrence interval for stands in mediterranean communities is a function of fuel load (Keeley and Keeley 1981) and that for some reason the flora tends to be highly flammable (Specht 1981b; Keeley 1989). Resins on leaf surfaces could increase flammability of the vegetation, but our results did not show dominance by resinous species in these communities. As we will see, Australia differs from Chile and California by the significant presence of glands and ducts on the leaf mesophyll. These structures accumulate essential oils that could contribute to the vegetation's flammability.

The presence of plants having all their branches lignified and remaining alive for more than one season (holoxyle stems), is another common feature in most of the species of the three mediterranean zones analyzed. Although axyle type plants such as *Puya chilensis* in Chile and semixyle plants such as *Atriplex suberecta* are present in Australia, they do not contribute significantly to the overall physiognomy. The absence of spinescence in these communities may be interpreted as a means of maximizing photosynthesis. Spines correspond morphologically to a mass of fibers covered by sclerified epidermis produced by shoot apical meristems, produced either by modification of the whole shoot or one leaf primordium. Thus the cost of producing spines is higher than that for normal leaves. In California spines are completely absent, but in Chile a few species develop spines from lateral buds on their stems. Spines could also have evolved as protection against grazing and browsing. It has been shown that thorns and prickles are effective deterrents to herbivory, reducing the rate of browsing by large mammals (Cooper and Owen-Smith 1986). Conversely, experimental removal of thorns and spines increases the rate of feeding by mammalian herbivores (Cooper and Owen-Smith 1986; Young and Smith 1987). The few species in Chile with spines may have evolved these structures as a result of the greater grazing pressure of large mammals in the Pleistocene (Solbrig 1984).

The absence of phanerophyte species capable of vegetative reproduction appears to be another common character in the communities analyzed, although in Chile some species are able to propagate vegetatively. For example, *Satureja gilliesii* develops an interxyllary cork between adjacent growth rings in older roots and stems (Montenegro et al. 1979) determining suberization that causes the whole plant to split longitudinally. *Lithrea caustica* and *Peumus boldus* grow a true underground network of horizontal reserve stems that connect large patches of each species in the matorral.

Many species develop underground lignified stems or lignotubers that appear to function as a common regenerative strategy in regions with mediterranean climate that experience summer droughts and frequent fires. But this type of regeneration from epicormic underground buds in lignotubers does not necessarily imply vegetative reproduction. Lignotubers have been defined as a source of dormant epicormic buds buried in a modified stem (James 1984). Such buds are capable of resprouting after the crown of the plant is killed, consumed by fire, or removed by mechanical means. The plant then can resprout following damage. Ability to resprout was found in most inland and coastal-species, which display remarkable survival and regeneration capacity after fire. Fires also affect growth form; many potential tree species resprout after fire, giving rise to a multistemmed, shrubby habit due to numerous shoots arising from the lignotuber. These large burls may also function as a source of carbohydrates and nutrients, and for water storage. Previous work (Montenegro et al. 1983) shows that burls appear as a swollen section at the root–stem junction at a very early stage in plant growth a few centimeters below the ground surface. At this stage they show a well-developed axial parenchyma arranged in paratracheal bands containing large amounts of starch. Carbohydrate reserves may function as an important buffer in woody plants (Keeley 1989) because their growth is seasonal. Reserves are critical so that deciduous species can respire and grow when leaves are not present (Oechel and Lawrence 1981). It could well be that the selective pressure to develop lignotubers has not come from fire, but rather seasonality in the mediterranean climate and that lignotubers are a trait "pre-adapted" to regeneration after fire.

According to Trabaud (1980), the adaptive traits that allow species to survive fire or regenerate and reproduce easily after fire are: (1) vegetative reproduction of perennials through sprouts that may appear at any time in the life cycle, depending on the location of dormant buds, and their protection by soil or bark, (2) sexual reproduction promoted by increased stimulation of flowering after fire and, (3) stimulation of germination by fire.

According to the present survey, species that flower profusely after being stimulated by fire are scarce in mediterranean communities. But *Kingia australis* in Australia and *Trevoa trinervis* in Chile are outstanding exceptions. With regard to seeds, in the Californian chaparral at least, seed germination of herbs requires specific chemical stimulus or scarification by fire. Nonsprouting Californian shrubs produce more seedlings after fire than sprouting shrubs (Keeley 1989). The understory of Australian heathlands contains sclerophyllous species that retain seeds for long periods in tough woody fruits, releasing their seeds only when heated during a bush fire; rapid regeneration by seedlings then follows (Specht 1981a). Experiments with Chilean matorral seeds of woody species treated at high temperatures suggest that fire-induced germination is absent (Muñoz and Fuentes 1989).

Coastal Vegetation

The physiognomic similarities in coastal mediterranean vegetation at the study sites is conferred by two ecomorphological characters: dominance by nanophanerophytes with renewal buds located at a height between 1 and 2 m, and a crown diameter of between 1 and 2 m. Most of the species in these coastal scrubs do not develop glands and ducts in their foliar mesophyll. The period of maximum shoot growth in the coastal communities is spring (Kummerow et al. 1981; Montenegro et al. 1981; Specht 1981b; Kummerow 1983; Orshan et al. 1984a; 1984b). Most of the species usually cease vegetative growth at the onset of the hot and dry summer conditions.

Inland Vegetation

The typical physiognomy of the inland chaparral, matorral, and mallee is given by four monocharacters: dominance be nanophanerophytes (renewal buds located up to 2 m in height) and microphanerophytes (renewal buds located between 2 and 8 m) with dominance by plants that reach heights of up to 1 m and develop a crown between 2 and 5 m in diameter.

Table 7.6. Ecomorphological characters that differ significantly (one-way no parametric ANOVA, $p > 0.05$) between the coastal (A) or inland (B) communities in the three regions analyzed. Pairs of similar communities and the ones that differ are indicated (statistical analysis according to Tukey test, $p > 0.05$)

(A) Similar communities		Ecomorphological character	Dissimilar community	Ecomorphological character
Central Chile	Southern California	Plant height: 1 to 2 m Sclerophyll and malacophyll leaves Pubescent leaves Evergreen and winter shedder plants	Southwestern Australia	Plant height: 50 to 100 cm Sclerophyll leaves No pubescent leaves Evergreen plants
Southern California	Southwestern Australia	Flowering in spring	Central Chile	Flowering in summer

(B) Similar communities		Ecomorphological character	Dissimilar community	Ecomorphological character
Central Chile	Southern California	Leaf glands and ducts absent	Southwestern Australia	Leaf glands and ducts present
Southern California	Southwestern Australia	Sclerophyll leaves Leaf-shedder plants	Central Chile	Malacophyll and sclerophyll leaves Leaf- and branch-shedder plants
Central Chile	Southwestern Australia	No pubescent leaves	Southern California	Pubescent leaves

Table 7.7. Degree of similarity between coastal communities and among inland communities for the three regions analyzed using the Jaccard's Similarity Index. Statistical analysis according to the Chi-Square Test ([a] $p > 0.05$, [b] $p > 0.001$) for similarity indices (Saiz, 1980). A statistically significant difference indicates strong similarity

	Coastal communities			Inland vegetation		
	Chile	USA	Australia	Chile	USA	Australia
Chile	1.00	0.90[b]	0.75[b]	1.00	0.90[b]	0.90[b]
USA	—	1.00	0.75[a]	—	1.00	0.85[b]
Australia	—	—	1.00	—	—	1.00

Differences

Table 7.6a shows the ecomorphological characters that differ significantly among the coastal communities in the three areas analyzed. Australia differs from central Chile and southern California in that plants are shorter, with sclerophyll leaves lacking in trichomes. These communities also differ in leaf seasonality: southwestern Australia mainly shows evergreen plants, while central Chile and southern California show evergreens and winter-deciduous species.

Central Chile differs from southern California and southwestern Australia in seasonality aspects only. In Chile, flowering occurs mostly in summer; in California and Australia it occurs mostly in spring.

Table 7.6b shows the ecomorphological characters that differ significantly among the inland communities. Southwestern Australia differs from central Chile and southern California in that the leaves have glands and ducts. Central Chile differs from the other communities in that it has species with either malacophyll or sclerophyll leaves that can shed leaves as well as branches. Species having pubescent leaves are present in southern California only.

In spite of these differences in ecomorphological characters, the Jaccard Similarity Indices (Table 7.7) are significant for neither coastal nor inland habitats. That is to say, the similar physiognomy of the matorral, chaparral, and mallee is supported by statistical analysis and may be interpreted as morphological and structural convergence in response to the mediterranean climate.

Conclusions

Although different land-use histories or genetic backgrounds may result in different solutions to the same problem, many of the same morphological and structural features are found in the coastal and inland vegetation of the matorral, chaparral, and mallee. Comparing the floras of several mediterranean ecosystems showed few common species or genera (Mooney et al. 1970; Margaris 1980). Thus, any similarity in structure in the mediterranean type vegetation clearly must be caused by convergence, because their evolutionary history is so distinct. Our results support the idea that the influence that phylogeny has on the evolution of growth form can be overcome by selective forces in the environment.

In mediterranean environments, growth and plant biomass are variously and greatly limited by availability of water (Kozlowski 1972; Mooney et al. 1977; Miller 1981; Montenegro et al. 1989), temperature (Mooney et al. 1977; Aljaro and Montenegro 1981; di Castri 1981; Miller 1981), and soil nutrients (Beadle 1966; Specht 1979; Mooney 1981; Lamont and Kelly 1988). These environmental factors display seasonal variations, limiting biological activity in plants to specific periods in the year (Mooney et al. 1977) generating growth patterns with clear seasonal patterns (Montenegro 1987; Montenegro et al. 1989). Morphological and physiological changes throughout the year, besides the type of ecomorphological characters favored by natural selection in areas with mediterranean climate, are effective mechanisms and provide means by which these limiting resources can be used optimally during the short growth period, thus enabling plants to survive. Species able to optimize photosynthetic processes within the constraints imposed by the mediterranean climate will be able to allocate more energy to reproduction as well as to defense against herbivores—two important aspects for species competition (Solbrig and Orians 1977).

Many types of growth cycles have been described for the growth forms in mediterranean plant communities (Orshan 1989). These cycles are driven by interaction between the location and development of renewal buds and organs that are lost periodically by senescence in relation to organs that remain on the plant throughout the year. It has been shown, however, that development of the renewal buds may generate different types of shoot morphologies, also called modules, such as long shoots, temporal or absolute leafy short shoots, inflorescences, and thorns (Ginocchio and Montenegro 1992; Montenegro and Ginocchio 1993a). This partitioning of functions into specialized basic architectural units gives more plasticity to the plant, and the proportion of modules assigned to different roles depends strongly on interactions with the environment (Lovell and Lovell 1985; Barthelemy 1986; Gottlieb 1986; Hardwick 1986).

Studies on how the various modules are distributed among different growth forms in central Chile (Ginocchio and Montenegro 1992; Montenegro and Ginocchio 1993a) and in other mediterranean areas (Montenegro and Ginocchio 1993b) show interspecific and local differences in module distribution. Each of these differences is related to the type of environment. Correlations between growth habit and environment have long been recognized, suggesting that not only specific ecomorphological characters have been selected for in mediterranean climate areas but that some architectures and branching patterns are advantageous in specific habitats.

Besides the morphological and structural convergence found in shrub vegetation, convergence was also found in various community properties

(Mooney et al. 1970). One-to-one total equivalence was not, however, found between the constituent species. The present monocharacter analysis of shrub growth forms helps pinpoint problems and reveal areas for future research. Givnish (1987) points out that "the most powerful approach to the study of plant adaptation and ecology should be an integrated one, embracing comparisons at the ecological, phylogenetic, and biogeographic scales, as well as studies in physiological ecology, biomechanics, optimality analysis, ecological genetics and demography."

Acknowledgments

We acknowledge Miss Emma Whittingham's help in reading and improving the English in the manuscript. Work was supported by FONDECYT Grant No 1940655 to G. Montenegro, The Andrew W. Mellon Foundation to P.U.C., and by NIH Grant No 1 vol TW/CA 00316-01 awarded to B. Timmermann. R. Ginocchio holds a Fundación Andes doctoral fellowship.

References

Aljaro ME, Montenegro G (1981) Growth of dominant Chilean shrubs in the Andean cordillera. Mountain Res Dev 1:287–291

Araya S, Avila G (1981) Rebrote de arbustos afectados por el fuego en el matorral chileno. An Mus Hist Nat 14:107–113

Avila G, Aljaro ME, Silva B (1981) Observaciones en el estrato herbáceo después del fuego. An Mus Hist Nat 14:99–105

Barthelemy D (1986) Establishment of modular growth in a tropical tree: *Isertia coccinea* Vahl (Rubiaceae). In Harper J, Rosen BR, White J (eds) *Growth and Form of Modular Organisms*. Phil Tran Royal Soc London B313:89–94

Beadle NC (1966) Soil phosphate and its role in molding segments of the Australian flora and vegetation, with special reference to xeromorphy and sclerophylly. Ecology 47:992–1007

Catling SM, Cowan PE, Green B (1988) The effect of fire and season on the water and energy turnover of small mammals in south eastern Australia. In di Castri F, Floret Ch, Rambal S, Roy J (eds) *Time Scales and Water Stress*. Proc 5th Int Conf Medit Ecosys. I.U.B.S., Paris, pp 655–660

Cooper SM, Owen-Smith N (1986) Effects of plant spinescence on large mammalian herbivores. Oecologia (Berl) 68:446–455

di Castri F (1981) Mediterranean-type shrublands of the world. In di Castri F, Goodall WD, Specht RL (eds) *Mediterranean-Type Shrublands*. Elsevier, Amsterdam, pp 11–43

di Castri F (1990) An ecological overview of the five regions of the world with mediterranean climate. In Groves RH, di Castri F (eds) *Biogeography of Mediterranean Invasions*. Cambridge U Press, Cambridge, pp 3–15

di Castri F, Mooney HA (eds) (1973) *Mediterranean Type Ecosystems: Origin and Structure*. Springer-Verlag, Berlin

di Castri F, Goodall DW, Specht RL (eds) (1981) *Ecosystems of the World*. Vol. 11. *Mediterranean Type Shrublands*. Elsevier, Amsterdam

Ginocchio R, Montenegro G (1992) Interpretation of metameric architecture in dominant shrubs of the Chilean matorral. Oecologia 90:451–456

Givnish TJ (1987) Comparative Studies of leaf form: assessing the relative roles of selective pressures and phylogenetic constraints. New Phytol Suppl 106:131–160.

Gottlieb LD (1986) The genetic basis of plant form. In Harper JL, Rosen BR, White J (eds) *Growth and Form of Modular Organisms*. Phil Trans Royal Soc London B313:197–208

Hardwick RC (1986) Physiological consequences of modular growth in plants. In Harper JL, Rosen BR, White J (eds) *Growth and Form of Modular Organisms*. Philos Trans Royal Soc London B313:161–173

James S (1984) Lignotubers and burls: Their structure, function and ecological significance in mediterranean ecosystems. Bot Rev 50:225–266

Keeley JE, Keeley SC (1977) Energy allocation patterns of sprouting and non-sprouting species of *Arctostaphylos* in the California chaparral. Am Mid Nat 98:1–10

Keeley JE, Keeley SC (1981) Postfire regeneration of California chaparral. Am J Bot 68:524–530

Keeley SC (ed) (1989) *The California Chaparral: Paradigms Reexamined*. Science Series No. 34, Nat Hist Mus LA County, Los Angeles, CA

Kozlowski TT (1972) *Water Deficits and Plant Growth*. Academic Press, New York

Kruger FJ, Mitchell DT, Jarvis JUM (eds) (1983) *Mediterranean Type Ecosystems: Ecological Studies*. Vol. 43. Springer-Verlag, Berlin

Kummerow J (1983) Comparative phenology of mediterranean type plant communities. In Kruger FJ, Mitchell DT, Jarvis JUM (eds) *Mediterranean Type Ecosystems: The Role of Nutrients*. Springer-Verlag, New York, pp 300–317

Kummerow J, Montenegro G, Krause D (1981) Biomass, phenology and growth. In Miller PC (ed) *Resource Use by Chaparral and Matorral. A Comparison of Vegetation Function in Two Mediterranean Type Ecosystems*. Springer-Verlag, New York, pp 69–96

Lamont BB, Kelly W (1988) The relationship between sclerophylly, nutrition and water use in two species from contrasting soils. In di Castri F, Floret Ch, Rambal S, Roy J (eds) *Time Scales and Water Stress*. Proc 5th Int Conf Medit Ecosys, I.U.B.S., Paris, pp 617–621

Lovell PH, Lovell PJ (1985) The importance of plant form as a determining factor in competition and habitat exploitation. In White J (ed) *Studies on Plant Demography: A Festschrift for John L Harper*. Academic Press, London, pp 209–221

Margaris NS (1980) Structure and dynamics of mediterranean type vegetation. Port Acta Biol 16(14):45–58

Miller PC (ed) (1981) *Resource Use by Chaparral and Matorral*. Springer-Verlag, Berlin

Montenegro G (1987) Quantification of mediterranean plant phenology and growth. In Tenhunen JD, Catarino O, Lange L, Oechel WC (eds) *Plant Response to Stress: Functional Analysis in Mediterranean Ecosystems*. Vol 15G NATO ASI Series. Springer, Berlin, pp 469–488

Montenegro G, Ginocchio R (1993) Modular interpretation of architecture in shrub species. Anais do Acad Bras Cien 65(2):189–202

Montenegro G, Ginocchio R (1994) Effects of insect herbivory on plant architecture. In Ariatnoutsou M, Specht RL (eds) *Plant-Animal Interactions in Mediterranean Type Ecosystems*. Kluwer Academic Publishers, Dordrecht, pp 115–122

Montenegro G, Hoffmann AJ, Aljaro ME, Hoffmann AE (1979) *Satureja gilliesii*, a poikilohydric shrub from the Chilean mediterranean vegetation. Can J Bot 57:1206–1213

Montenegro G, Aljaro ME, Walkowiak A, Saenger R (1981) Growth, net-productivity and senescence in herbs and shrubs of the native vegetation in Central Chile. In Conrad CE, Oechel WC (eds) *Dynamics and Management of Mediterranean Type Ecosystems*. USDA For Serv. Gen Tech Rep PSW 58, pp 135–141

Montenegro G, Avila G, Schatte P (1983) Presence and development of lignotubers in shrubs of the Chilean matorral. Can J Bot 61:1804–1808

Montenegro G, Serey I, Gómez M (1985) Growth forms of arid and semi-arid bio-climatic zones in Chile through the monocharacter approach. Medio Amb 7(2): 21–30

Montenegro G, Avila G, Aljaro ME, Osorio R, Gómez M (1989) Chile. In Orshan G (ed) *Plant Phenomorphological Studies in Mediterranean Type Ecosystems.* Kluwer Academic Publishers, Dordrecht, pp 347–387

Mooney HA (ed) (1977) *Convergent Evolution in Chile and California Mediterranean Climate Ecosystems.* Dowden, Hutchinson & Ross, Stroudsburg, PA

Mooney HA (1981) Primary production in mediterranean-climatic regions. In di Castri F, Goodall DW, Specht RL (eds) *Ecosystems of the World 11. Mediterranean Shrublands.* Elsevier, New York, pp 249–255

Mooney HA, Dunn EL, Shropshire F, Song L (1970) Vegetation comparisons between the mediterranean climatic areas of California and Chile. Flora 159:480–496

Mooney HA, Dunn EL (1970) Convergent evolution of mediterranean-climate evergreen sclerophyll shrubs. Evolution 24:292–303

Mooney HA, Kummerow J, Johnson AW, Parsons DJ, Keeley S, Hoffmann A, Hays RI, Giliberto J, Chu C (1977) The producers, their resources and adaptive responses. In Mooney HA (ed) *Convergent Evolution in Chile and California.* Dowden, Hutchinson and Ross, Stroudsburg, PA pp 85–153

Mooney HA, Kummerov J, Moll EJ, Orshan G, Rutherford MC, Sommerville JEM (1983) Plant form and function in relation to nutrient gradients. In Day JA (ed) *Mineral Nutrients in Mediterranean Ecosystems.* S African Nat Sci Prog Rep 71: 55–76

Muñoz MR, Fuentes ER (1989) Does fire induce shrub germination in the Chilean matorral? Oikos 56:177–181

Oechel WC, Lawrence W (1981) Carbon allocation and utilization. In Miller PC (ed) *Resource Use by Chaparral and Matorral. A Comparison of Vegetation Function in Two Mediterranean Type Ecosystems.* Ecological Studies 39. Springer-Verlag, New York, pp 185–233

Orshan G (1983a) Approaches to definition of mediterranean growth forms. In Kruger FJ, Mitchell DT, Jarvis JUM (eds) *Mediterranean-Type Ecosystems: The Role of Nutrients.* Ecological Studies 43, Springer-Verlag, New York, pp 86–99

Orshan G (1983b) Monocharacter growth form types as a tool in an analytic synthetic study of growth forms in mediterranean type ecosystems. Ecol Medit 8:159–171

Orshan G (1989) Shrubs as a growth form. In *Biology and Utilization of Shrubs.* Academic Press, New York, pp 249–265

Orshan G, Montenegro G, Avila G, Aljaro ME, Walkowiak A, Mujica AM (1984a) Plant growth forms of Chilean matorral. A monocharacter growth form analysis along an altitudinal transect from sea level to 2000 m.a.s.l. Bull Soc Bot France 131, Actualités Botaniques (2-3-4-):411–445

Orshan G, Le Roux A, Montenegro G (1984b) Distribution of monocharacter growth form types in mediterranean type communities of Chile, South Africa and Israel. Bull Soc Bot France 131, Actualités Botaniques 131:427–439

Parsons DJ (1976) Vegetation structure in mediterranean scrub communities of California and Chile. J Ecol 64:435–447

Quézel P (ed) (1982) *Définition et Localisation des Ecosystèmes Méditerranéens Terrestres. Ecologia Mediterranea,* Tome VIII, Marseille, France

Rundel PW (1981) The matorral zone in central Chile. In di Castri F, Goodall WD, Specht RL (eds) *Mediterranean Type Shrublands. Ecosystems of the World* Vol. II, Elsevier, New York, pp 175–201

Saiz F (1980) Experiencias en el uso de criterios de similitud en el estudio de comunidades. Arch Med Bio Exp 13:387–402

Schulze ED (1982) Plant life-forms and their carbon, water and nutrient relations. In

Lange OL, Nobel PS, Osmond CB, Ziegler H (eds) *Encyclopedia of Plant Physiology*, New Series, Vol. 12B. Springer, Berlin, pp 615–676

Shmida A (1984) Convergence and non-convergence of mediterranean-type communities. Bull Soc Bot France, Actualités Botaniques 131:465–472

Sokal RR, Rohlf FJ (1981) *Biometry* 2nd Ed. W. H. Freeman, New York

Solbrig OT (1984) Evolution of land-use patterns. Mountain Res Dev 4(2):135–149

Solbrig OT, Orians GH (1977) The adaptive characteristics of desert plants. Am Sci 65:412–421

Specht RL (ed) (1979) *Heathlands and Related Shrublands. Ecosystems of the World*, Vol. 9A & B. Elsevier, Amsterdam

Specht RL (1981a) Ecophysiological principles determining the biogeography of major vegetation formations in Australia. In Keast A (ed) *Ecological Biogeography of Australia*. Dr. Junk Publishers, The Hague, pp 301–332

Specht RL (1981b) Mallee ecosystems in southern Australia. In di Castri F, Goodall DW, Specht RL (eds) *Mediterranean Type Shrublands*. Elsevier, Amsterdam, pp 257–267

Specht RL (1988) *Mediterranean Type Ecosystems: A Data Source Book*. Kluwer Academic Publishers, Dordrecht

Tenhunen JD, Catarino FM, Lange OL, Oechel WC (eds) (1987) *Plant Response to Stress: Functional Analysis in Mediterranean Ecosystems*. Springer-Verlag, Berlin

Thrower NJW, Bradbury DE (eds) (1977) *Chile–California Mediterranean Scrub Atlas: A Comparative Analysis*. Dowden, Hutchinson and Ross, Stroudsburg, PA

Trabaud L (1980) Impact Biologique et Écologique des Feux de Végétation sur l'Organisation, la Structure et l'Evolution de la Végétation des Zones de Garrigues de Bas-Languedoc. Thesis Académie de Montpellier. Centre d'Études Phytosoc Écol Louis Emberger, Montpellier

Young TP, Smith AP (1987) Herbivory on alpine Mount Kenya. In Rundel P (ed) *Tropical Alpine Systems: Plant Form and Function*. Springer-Verlag, Berlin, pp 185–205

8. Underground Structures of Woody Plants in Mediterranean Ecosystems of Australia, California, and Chile

Josep Canadell and Paul H. Zedler

Interest in the possible evolutionary convergence has been the impetus for comparative studies of how organisms function in mediterranean ecosystems (Cody and Mooney 1978). For plants, stem architecture, seasonal growth patterns, nutrient relationships, and some physiological aspects have been examined to determine if convergence occurs under similar climatic constraints.

Despite the obvious importance of the below-ground component of plants, several difficulties have discouraged detailed work on this aspect of the problem. First, the soil environment in each mediterranean region is much more variable than the environment to which the above-ground part of the plant is exposed. Soil properties frequently vary significantly over very small distances. A microenvironmental mosaic is common in California and Chilean ecosystems, where rocks, steep slopes, and changes in types of bedrock are very frequent. Second, and probably most important, information is scarce because of the tedious work required to gather any type of below-ground data.

Despite these difficulties, studies are numerous enough that it is possible to assemble a preliminary picture. The objective in this chapter is to gather information on structure of root systems and on the water relations characteristic of different types of root-system morphologies. We focus exclusively on woody plants, both shrub and tree species, growing in the Pacific mediterranean ecosystems. This area includes the woodlands and heathlands in

Southeast and Southwest Australia, the chaparral and oak woodlands in California, and the matorral in Chile. We compare the root structure and rooting patterns among different regions with all available data.

Root-System Structure and Rooting Patterns

Root-System Morphologies

Because of the summer drought in regions of mediterranean climate, soil moisture in the surface horizons of the soil profile will be depleted for up to six months (Mooney et al. 1974). But in these regions precipitation is also great enough that in many situations moisture reserves can be considerable at greater depths. Two general root-system morphologies have evolved to deal with the strong spatial and temporal limitations on availability of moisture. First, some deep-rooted species can utilize moisture from lower horizons. When these species also have a well-developed shallow-root system, they are said to have a dual root system. The shallow roots in the upper soil layers enable plants to benefit from the generally higher nutrient content in the surface soils as well as to take immediate advantage of rains, especially when deep horizons are not yet recharged after the summer drought. The root systems of many sclerophyllous evergreen shrubs in the mediterranean regions have this type of morphology.

The existence of the dual root system was revealed in one of the earliest studies on root-system structure. Cannon (1914) studied three species of oaks (*Quercus douglasii*, *Q. agrifolia*, and *Q. lobata*) in California. He concluded that dual root development may be fairly common among these species. This type of root system has strong development of shallow roots within the first meter of soil, and roots penetrating to several meters, depending on the depth of the water table and the amount of rain. Cooper (1922) pointed out that the dual system was possible only when soils were friable and easily penetrated by roots. Where rock is at the surface, which it is very frequently, the roots penetrate the crevices wherever they can. The result is a rooting pattern controlled in large part by the substrate and without a clear pattern. The dual system was reported in Cooper's study for *Adenostoma fasciculatum*, *Quercus durata*, and *Arctostaphylos tomentosa* in Jasper Ridge, central California. In an area close by, Wright (1928) also found a dual type of root system in *Baccharis pilularis*. The extension of the taproot, however, varied from slight to 3.2 m.

The second basic root morphology is a shallow system. Shallow-rooted plants occur in the driest part of the water-availability gradient in the mediterranean climate regions. Such plants cope with drought by becoming dormant during dry periods. Many of the summer-deciduous species found in California and Chile are in this category (Mooney and Dunn 1970). Plants with life strategies other than summer deciduous may also display a shallow-root system, such as some species of the Southeast Australia heathland (Specht 1981a), and *Yucca whipplei* in the California chaparral (Hellmers et al. 1955).

Of the two types of root system, the deep-rooting morphology has been described as the commonest in the five mediterranean regions of the world. A deep-root system, with either one dominant taproot or many deep-penetrating roots, clearly seems to be advantageous for maximizing use of water throughout the year. Successful establishment in xeric habitats also seems highly correlated with early development of a taproot because ability to reach deep moisture layers is crucial for surviving the first summer drought after germination. Cannon (1914) observes that in various California chaparral species a deeply penetrating taproot starts to develop immediately, and elaboration of the surface network occurs later. Specht and Rayson (1957) found a seedling of *Banksia ornata* growing in the sand plains in Southeast Australia to have a taproot extending 0.5 m deep, but little development of the lateral roots. Observing one-year-old seedlings of *Quillaja saponaria*, a Chile matorral species, showed a taproot about 1 m long (Riveros et al. 1976).

The taproot as an adaptation to a climate with seasonal drought was studied experimentally by Zimmer and Grose (1958). They collected seeds of 14 *Eucalyptus* species growing across a moisture gradient in Victoria, Australia, and planted them in a greenhouse. The seedlings showed a transition from a fibrous root system for species from moist habitats to a single taproot with weak laterals for species from drier areas. Similarly, Matsuda and McBride (1986) studied how root morphology of seedlings of *Quercus douglasii*, *Q. lobata*, and *Q. agrifolia* might affect distribution of these species in the central coast ranges in California. They found that *Q. douglasii* had the longest main root. They used this along with other plant features to explain the occurrence of *Q. douglasii* on xeric sites, a pattern documented earlier by Griffin (1971). The high degree of plasticity shown by some of these *Quercus* species, however, complicates this story. For example, it has been shown that *Q. douglasii* may either tap deep water and produce few surface roots, or utilize mainly shallow soil water via a dense lateral root system. Thus *Q. douglasii* may occupy many habitats partly because of its root system's morphological plasticity (Callaway 1990).

The dual system and early development of a taproot have been also observed for plants growing in nonmediterranean climatic regions that have a seasonal drought. *Quercus turbinella*, a shrub of the Arizona chaparral, has a root system of the generalized type, with a taproot—actually many deeply penetrating roots—and a strong lateral root system (Davis 1970). Under greenhouse conditions, roots of *Q. turbinella* may grow to a length of 30 cm before leaves develop from shoots only 2 to 3 cm tall (Davis and Pase 1977).

Root-system morphologies other than the dual and the shallow types have been described in mediterranean climate ecosystems. Attempts to describe morphologies have often lacked a systematic approach because the forms are highly variable and few individuals are excavated because of the tedious work required. Some species also have highly plastic root systems that vary depending on both soil environment and stage in development. Examples are *Hibbertia montana* and *Phyllanthus calycinus*, which grow in the understory of jarrah forest in Southwest Australia. Both species initially have a taproot but it often decays, leaving a freely branching root system (Crombie et al. 1988).

Several attempts have been made to classify root-system morphologies and, more recently, their architecture as well, because it is thought to be critical to a mechanistic understanding of soil-resource acquisition (Berntson 1992, Fitter and Stickland 1992). Cannon (1949) devised a classification of root systems in which he attempted to recognize invariable characters by comparing root systems assumed to be genetically similar but grown in different habitats. He described six basic root systems differentiated by such characters as degree of development of primary and lateral roots, rooting depth, and primary root branching types. Kummerow (1981), however, considered the data set to be too narrow for a general classification of roots. He felt that this limitation particularly applied to mediterranean ecosystems, where rocky soils, steep slopes, and irregular moisture patterns provide greatly varied soil conditions that can be exploited by many species and root forms.

We have no comprehensive descriptions of the root systems in southeastern and southwestern Australia heathlands, California chaparral, and Chilean matorral, but basic information is available for the most dominant woody species in all three ecosystems. One of the most extensive studies was done by Specht and Rayson (1957), who studied the root systems of 91 woody and grass species in the heath community in Southeast Australia. Crombie et al. (1988) also looked at 22 species of trees, shrubs, and grasses in the jarrah forest in Southwest Australia, and Dodd et al. (1984) studied 551 root systems from 43 woody species in the low woodland and open forest in Southwest Australia. Unlike Kummerow's (1981) findings, Dodd et al. (1984), in their systems, found that species had a rather consistent root-system morphology, allowing them to be classified according to Cannon's schemes. Following Cannon's scheme, they described five basic root-system types to produce the broadest study on root-system morphologies in any mediterranean region. These are the five types:

Type 1. Well-developed primary (taproot) and lateral roots, neither dominant. The length of the taproot usually less than 1 m.
Type 2. Root system shallow but laterally very extensive, several times greater than its depth. Roots less than 0.5 m deep.
Type 3. Dominant, deep fibrous taproot lacking an extensive lateral root system. Rooting depth usually greater than 3 m.
Type 4. Dominant deep taproot with well-developed lateral roots near the soil surface. Lateral spread as extensive as vertical.

Type 5. Nondominant shallow taproot, with pronounced branching and
forking of roots. Root depth usually less than 1 m.

In Dodd's study, 83% overall of the species studied had a dominant taproot
that reached depths from 1 m to deeper than 4 m. Specht and Rayson (1957)
found that a quarter of the heath species in Southeast Australia had deep
taproots, one half had shallow taproots, and the remaining quarter (mostly
herbaceous species) had fibrous roots.

Several root-excavation studies have been done in the California chaparral (Hellmers et al. 1955, Kummerow et al. 1977, Kummerow and
Mangan 1981, Kummerow and Wright 1987, Miller and Ng 1977). The most
extensive study was done by Hellmers et al. (1955), who studied 68 individuals in 18 species of shrubs and subshrubs in 14 locations in southern California chaparral. The different study locations had variations in soil type,
texture, and depth, bedrock type, and steepness of slope. Because conditions
were so variable, only two basic root-system morphologies for shrubs were
recognized: shrubs with coarse major roots growing downward, where the
depth of downward penetration was greater than the radial spread, and
shrubs with coarse major roots growing laterally, where radial spread of
roots was usually much greater than downward penetration. A third type
was distinguished for subshrubs, which had a fibrous root system with great
radial spread but rooting no deeper than 1.5 m.

Unlike the two other ecosystems considered, few data on root-system
structure are available for Chilean matorral species (Giliberto and Estay
1978, Hoffmann and Kummerow 1978). Data are not adequate to attempt a
classification, but a basic dual system seems to be fairly common for evergreen resprouters.

Dodd's classification is by far the most thorough for mediterranean species. Its five categories include most of the root types in the various classifications presented in this review, so that it could be used as a general classification for shrubs and trees in the mediterranean ecosystems. Changes on root
length—depth and lateral extension—will have to be reconsidered, however,
when this classification is used outside of Dodd's study area. Different soil
types and species may be the cause of these possible differences.

Factors Affecting Rooting Patterns

The rooting pattern of a plant, like other aspects of morphology, is the result
of environmentally influenced development of a basic genetic plan (Pearson
1974). Environmental constraints are usually so strong, however, that they
are often the driving force determining root-system morphology. The primary factors controlling root development are the amount and depth of
moisture and nutrients, and soil physical properties (Curl and Truelave 1986,
Feldman 1984, 1988, Richards 1986). Although soil environmental factors
exert strongest control on root-system morphology, morphology is also influenced, less obviously, by the environmental factors to which the aerial

Table 8.1. Above- and below-ground dimensions for woody species in Australia, California, and Chile mediterranean ecosystems

		Above-ground			Below-ground		
Species	n	Average height (m)	Average crown radius (m)	Maximum horizontal extension radius (m)	Maximum vertical extension (m)	Taproot	Ref.
SOUTHEAST AUSTRALIA							
Adenanthos terminalis	—	0.4	0.3	0.4	0.6	yes	1
Astroloma conostephioides	—	0.1	0.1	0.5	0.2	yes	1
Banksia marginata	—	0.7	0.2	4.1	2.4[a]	yes	1
Banksia ornata	—	1.2	0.3	6.2	2.4[a]	yes	1
Boronia coerulescens	—	0.3	0.2	0.3	0.3	yes	1
Calytrix alpestris	—	0.2	0.2	0.3	0.6[a]	yes	1
Casuarina muelleriana	—	1.3	0.3	0.9	2	yes	1
Casuarina pusilla	—	0.4	0.2	2.8	2.4[a]	yes	1
Correa rubra	—	0.3	0.3	0.5	0.6[a]	yes	1
Cryptandra tomentosa	—	0.2	0.2	0.3	0.4	yes	1
Daviesia brevifolia	—	0.4	0.1	0.2	2	yes	1
Dillwynia hispida	—	0.2	0.1	0.3	0.3	yes	1
Euphrasia collina	—	0.3	0.1	0.2	0.3	yes	1
Gompholobium minus	—	0.2	0.2	0.3	0.3	yes	1
Hibbertia sericea	—	0.3	0.1	0.3	0.5	yes	1
Hibbertia stricta	—	0.3	0.2	0.3	0.6[a]	yes	1
Laudonia behrii	—	0.8	0.1	0.1	2	yes	1
Leptospermum myrsinoides	—	0.9	0.2	1.2	2.3	yes	1
Leucopogon costatus	—	0.3	0.1	0.3	0.3	yes	1
Leucopogon woodsii	—	0.3	0.2	0.1	0.3	yes	1
Phyllota pleurandroides	—	0.7	0.3	2.8	2.3	yes	1
Phyllota remota	—	0.3	0.2	0.3	2.4[a]	yes	1
Pimelea octophylla	—	0.3	0.1	0.2	0.2	yes	1
Spyridium subochreatum	—	0.4	0.2	0.6	1.9	yes	1
Xanthorrhoea australis	—	0.8	0.3	1.7	2.4	yes	1
Average		0.5 ± 0.1	0.2 ± 0.1	1.0 ± 0.3	1.2 ± 0.2		
Range		0.1–1.3	0.1–0.3	0.1–6.2	0.2–2.4		

SOUTHWEST AUSTRALIA

Species							
Acacia barbinervis	8	0.4	0.3	0.5	1.5	yes	2
Acacia huegelii	2	0.2	0.2	1	1	yes	2
Acacia pulchella	15	0.4	0.2	0.3	0.4	yes	2
Adenanthos cygnorum	6	1.3	0.5	1.3	1.6	yes	2
Allocasuarina humilis	3	0.8	1	4	1.5	yes	2
Andersonia heterophylla	8	0.3	0.4	0.2	0.6	no	2
Aotus ericoides	2	0.5	0.5	0.2	0.4	yes	2
Astartea fascicularis	17	0.5	0.2	1.1	0.8	yes	2
Astroloma xerophyllum	4	0.6	0.6	1	0.9	no	2
Beaufortia elegans	21	0.8	0.8	1.2	1.4	yes	2
Bossiaea eriocarpa	22	0.4	0.4	1.1	0.9	yes	2
Calytrix empetroides	11	0.5	0.2	0.3	0.6	yes	2
Calytrix flavescens	13	0.3	0.4	1.6	2	yes	2
Calytrix fraseri	1	0.5	0.2	0.5	0.6	yes	2
Conospermum curvum	3	0.6	0.4	0.6	1.2	no	2
Conostephium minus	2	0.3	0.5	0.5	0.5	no	2
Conostephium pendulum	11	0.3	0.4	1.3	1.5	no	2
Daviesia juncea	3	0.5	0.6	0.5	1.8	yes	2
Eremaea pauciflora	12	0.7	0.5	6	2.4	yes	2
Eriostemon spicatus	14	0.7	0.3	0.6	0.8	yes	2
Euchilopsis linearis	12	0.6	0.9	0.4	0.4	yes	2
Gompholobium tomentosum	11	0.3	0.2	0.4	1.1*	yes	2
Hemiandra pungens	6	0.3	0.4	0.4	0.4	no	2
Hibbertia aurea	13	0.3	0.3	0.6	0.9	yes	2
Hibbertia helianthemoides	13	0.3	0.2	0.3	0.7	yes	2
Hibbertia huegelii	3	0.4	0.3	0.4	0.9	yes	2
Hibbertia hypericoides	15	0.5	0.5	1.8	2.1	yes	2
Hibbertia subvaginata	53	0.3	0.3	0.4	0.7	yes	2
Hypocalymma angustifolium	16	0.3	0.5	1.2	0.9	yes	2
Jacksonia floribunda	20	0.9	0.7	1.4	3.1	yes	2
Jacksonia furcellata	6	1.1	0.7	5	2	yes	2
Lechenaultia floribunda	3	0.1	0.2	0.2	0.3	yes	2
Leptospermum ellipticum	36	0.7	0.5	0.6	0.9	yes	2
Leucopogon conostephioides	31	0.3	0.5	0.8	0.9	no	2
Leucopogon sprengelioides	27	0.5	0.4	0.3	0.5	no	2

(continued)

Table 8.1 (*continued*)

Species	n	Above-ground		Below-ground			Ref.
		Average height (m)	Average crown radius (m)	Maximum horizontal extension radius (m)	Maximum vertical extension (m)	Taproot	
Melaleuca scabra	20	0.4	0.6	1.6	2	yes	2
Melaleuca seriata	7	0.5	0.5	1.9	2.1	yes	2
Oxylobium capitatum	7	0.4	0.3	0.4	0.6	yes	2
Petrophile linearis	15	0.3	0.4	0.6	2	yes	2
Pithocarpa corymbulosa	4	0.5	0.4	0.2	0.8	yes	2
Regelia ciliata	12	0.9	0.8	1.2	0.6	yes	2
Scholtzia involucrata	11	0.3	0.5	1.9	1.9	yes	2
Stirlingia latifolia	11	0.5	0.5	1.8	2.6	yes	2
Average ± error		0.5 ± 0.1	0.4 ± 0.1	1.1 ± 0.2	1.2 ± 0.1		
Range		0.1–1.3	0.2–1.0	0.2–6.0	0.3–3.1		
SOUTH AND CENTRAL CALIFORNIA							
Adenostoma fasciculatum	13	1.5	0.9	4	2.4/7.6	yes	3
Adenostoma fasciculatum	2	1.5	1	2	2.4	yes	4
Adenostoma fasciculatum	4	0.6	0.2	1.2	1	—	5
Adenostoma fasciculatum	3	—	1	3	0.6	no	6, 7
Adenostoma sparsifolium	2	3.3	3.2	3.3	2.1	no	4
Arctostaphylos glandulosa	4	0.7	0.5	2.7	2.7/5.2	yes	3
Arctostaphylos glauca	4	2	1.1	6.4	2.6	no	3
Arctostaphylos glauca	1	1	0.5	0.6	0.3	—	5
Arctostaphylos pungens	2	—	1.5	4.5	0.6	no	6, 7
Baccharis pilularis	1	0.9	0.5	2.1	3.2	yes	8
Ceanothus crassifolius	7	2	0.8	3	1.4	yes	3
Ceanothus greggii	2	0.9	0.2	0.3	0.3	—	5
Ceanothus greggii	1	—	1.5	3.5	0.6	no	6, 7
Ceanothus greggii	2	0.8	0.5	3.4	1.4	yes	3
Ceanothus leucodermis	4	1.5	1.1	3	3.7	yes	3
Ceanothus oliganthus	3	2.1	0.8	1.7	1.8	yes	3
Cercocarpus betuloides	2	1.5	1	2.7	1.5	no	3

(continued)

Species						Roots deeper	Reference
Diplacus longiflorus	2	0.4	0.2	0.9	0.8	no	3
Eriodictyon crassifolium	1	0.9	—	1.4	—	no	3
Eriogonum fasciculatum	4	0.4	0.4	1.5	1.2	no	3
Haplopappus pinifolius	1	—	0.4	0.6	0.3	no	6,7
Heteromeles arbutifolia	1	1.8	0.4	0.3	0.3	—	5
Heteromeles arbutifolia	1	1.2	0.8	1.5	2.1	yes	3
Lotus scoparius	2	0.3	0.3	0.9	1.1	yes	3
Quercus chrysolepis	1	—	—	—	7.3	yes	3
Quercus dumosa	3	1.5	0.8	3.3	2.4/8.5	yes	3
Salvia apiana	1	0.6	0.6	1.8	1.5	no	3
Salvia mellifera	2	1.2	0.7	1.8	0.6	no	3
Yucca whipplei	2	0.6	0.8	3.3	0.8	no	3
Average + error		1.2 ± 0.1	0.8 ± 0.1	2.3 ± 0.3	1.7 ± 0.3		
Range		0.3–3.3	0.2–3.2	0.3–6.4	0.3–8.5		
CENTRAL CHILE							
Colliguaja odorifera	—	1	1	3.5	0.6	no	7,9
Colliguaja odorifera	—	—	—	1.5	1	—	10
Colliguaja odorifera	2	0.4	0.1	—	0.6	—	5
Cryptocarya alba	—	3.1	1.5	3	0.6[a]	no	9
Cryptocarya alba	—	—	—	4	1	—	10
Lithrea caustica	—	1.2	1.8	4	0.6[a]	no	9
Lithrea caustica	—	—	—	7	5	yes	10
Mutisia retusa	—	1.2	—	—	—	no	9
Quillaja saponaria	—	—	—	—	8	yes	10
Retanilla ephedra	—	—	—	1	1	—	10
Satureja gilliesii	—	—	—	1	0.8	—	10
Satureja gilliesii	—	0.6	0.2	1	0.2	no	9
Satureja gilliesii	4	0.9	0.4	1	0.9	—	5
Average ± error		1.2 ± 0.1	0.8 ± 0.3	2.7 ± 0.6	1.7 ± 0.7		
Range		0.4–3.1	0.1–1.8	1.0–7.0	0.2–8.0		

[a] Roots deeper than the given depth.

References: 1. Calculated from Specht and Rayson 1957. 2. Dodd et al. 1984. 3. Hellmers et al. 1955. 4. Hanes 1965. 5. Miller and Ng 1977; 6. Kummerow et al. 1977; 7. Kummerow 1980; 8. Calculated from Wright 1928. 9. Hoffmann and Kummerow 1978. 10. Giliberto and Estay 1978.

parts of the plant are subjected (Burns 1972). Competition between plants, both intraspecific and interspecific, can also contribute to the final morphology of the root system. Atkinson et al. (1976) reported that as canopy overlap among apple trees increased as a result of decreasing space between individuals, trees altered their root systems by producing many more roots growing downward, rather than horizontally. More recently, Mahall and Callaway (1992) report that the roots of desert shrubs in California can inhibit the growth of roots of their own and other species by releasing biochemical inhibitors. It is conceivable that similar interactions may be present in mediterranean regions.

Availability of soil resources, both moisture and nutrients, has been shown to strongly limit plant production in ecosystems of mediterranean South Australia, California, and central Chile (Beadle 1954, Giliberto and Estay 1978, McMaster et al. 1982, Miller and Poole 1979, Mooney and Dunn 1970, Poole and Miller 1975, Specht 1979, Specht and Rayson 1957). Thus, availability of resources in time and space is expected to heavily influence the final root form and distribution. Among the benefits of the dual root system, for example, may be that the deeper roots supply moisture that allows the surficial roots to remain active and able to exploit brief pulses of nutrients that are released when the soils are wetted after prolonged dry periods. Crick and Grime (1987) have suggested that such an ability to utilize nutrient pulses may be a characteristic feature of a stress-tolerant life history.

The three primary factors—moisture, nutrients, and physical properties—can change on a very broad spatial scale (over hundreds of kilometers) as well as on a very local scale (over just a few meters) (Beckett and Webster 1971, Kummerow and Wright 1988). At either scale, root form and distribution often follow the patterns of soil environmental variation. Kummerow (1981) and Richards (1986) suggested the idea of habitat-specific rather than species-level determination of the rooting patterns of plants. This notion led to the habitat-specific root-classification scheme mentioned in the preceding section. An example of the need for this type of root-morphology classification is the California chaparral shrub, *Adenostoma fasciculatum*. This species has been shown to have a highly plastic rooting pattern that can respond to different soil environmental conditions (Table 8.1). Cooper (1922) found roots of this species at a depth of 1 m, but nothing resembling a taproot. Later, *A. fasciculatum* was identified as a deep-rooted shrub, with a taproot extending as deep as 7.6 m (Hanes 1965, Hellmers et al. 1955). In more recent excavations, however, *A. fasciculatum* has been found growing successfully in shallow soils where the roots did not penetrate deeper than 0.6 m (Kummerow et al. 1977, Miller and Ng 1977).

Not all species, however, even in the same plant community, have the plasticity to enable them to adapt to a broad range of soil conditions. In fact, many species in the California chaparral show relatively nonplastic development of their root systems, such as *Quercus dumosa* and *Eriogonum fasciculatum*. *E. fasciculatum* is a shallow-rooting species (Kummerow et al. 1977) and remains shallow-rooted even when soils allow deeper penetration (Kummerow and Mangan 1981). Conversely, *Q. dumosa* shows a consistently deep rooting habit, and the species does not occur on shallow soils (Kummerow and Mangan 1981, Kummerow and Wright 1988). Several authors who studied the low woodland and open forest in Southwest Australia (Dodd et al. 1984) and the jarrah forest in Southeast Australia (Abbott et al. 1989) also do not agree with the idea that root-system development is determined largely by site characteristics. Dodd et al. (1984) studied 551 individual plants of 43 woody species and found that root-system morphologies were rather consistent within species. Abbott et al. (1989) showed the root-system morphology of *Eucalyptus marginata* to be under strong genetic control, because the same basic architecture was expressed regardless of soil type; Kimber (1974), however, reported varied forms for the four individuals he studied.

The explanation for the contradictory findings about genetic versus environmental control of root system morphology in different ecosystems seems to be primarily related to physical characteristics of the soil. Root morphology can be strongly influenced by the soil's physical properties, such as hardness, porosity, and bedrock depth and its degree of weathering, which can completely change the soil environment in which roots are growing. The deep sandy soils of the sand plains in South Australia allow much freer expression of root-system morphologies, than is possible in Californian and Chilean ecosystems, where rocky soils and steep slopes are common. It also seems to be the case that the genetic control of plasticity differs among species.

Depth of Root Penetration

Plant-rooting depth is influenced by many factors involving both plant and soil characteristics. Although above-ground biomass is not perfectly correlated with rooting depth, trees do tend to root more deeply than shrubs, and larger individuals send roots deeper into the soil profile than smaller individuals. The genetic component is important in determining root depth, but factors such as depth of water table and bedrock, amount of rain, hardness of substratum and degree of weathering, including presence of cracks and channels, are primary factors in controlling the rooting habit, and therefore the actual depths from which roots will be able to acquire resources.

Both qualitative and quantitative evidence say that some mediterranean trees have very deep root systems. For instance, jarrah trees (*Eucalyptus marginata*) in Southwest Australia generally have sinker roots, ending with a fine root system, extending to depths of 20 m (Carbon et al. 1980), and to more than 40 m into highly weathered granite (Abbott et al. 1989, Dell et al. 1983). The roots of some species of California oak trees (*Quercus wislizenii*, *Q. douglasii*, *Q. lobata*, and *Q. kelloggii*) took up water during the summer months from depths of at least 21 m (Lewis and Burghy 1964).

There are several examples of shrubs rooting to considerable depths for the Pacific mediterranean regions (Table 8.1). In the California chaparral, fine roots of *Adenostoma fasciculatum*, *A. sparsifolium*, *Quercus dumosa*, and *Q. chrysolepis* were found in tiny cracks and through fractured rocks to a depth of 7 to 9 m (Hellmers et al. 1955, Hanes 1965). Roots of *Rhus* (*Malosma*) *laurina* were found as deep as 13.2 m for one individual at the edge of a road cut (DeSouza et al. 1986). In the Arizona chaparral, *Q. turbinella* penetrated up to 6.4 m through cracks in rocky regolith (Davis 1970, 1978, Davis and Pase 1977) and to a depth of at least 9.1 m in an active mica-schist quarry (Saunier and Wagle 1967). In the Chilean matorral, roots of *Quillaja saponaria* were found 8 m deep in a well-developed fertile soil that originated from igneous porphyritic rock (Giliberto and Estay 1978). Similarly, roots 1 cm in diameter of different species were quite often found at 6 m below the surface in the kwongan of Southwest Australia near Eneabba. Only roots of *Eremaea beaufortioides*, however, were clearly identified to species (Hnatiuk and Hopkins 1980). But conversely, Dodd et al. (1984) in an extensive study of the rooting patterns of the kwongan shrubs, found only 7 of 43 woody species rooting to depths of 2 m or more.

If the soil is deep enough to allow full expression of the root system, patterns of soil moisture often are critical in shaping the final root morphology. The hardness of the substratum, however, can be a strong impediment, so that development will be freer in sandy soils because of their high permeability. Kimber (1974) found that the characteristic root depth of *Eucalyptus marginata* was greatly affected by depth of the water table. Though roots were capable of growing deeper, a water table at a depth of 15 to 20 m would limit further root penetration. Depth of *Quercus douglasii* roots seems to be controlled by the depth to which moisture can penetrate. *Q. douglasii* usually grows in dry areas where the water table is beyond its reach (Cannon 1914, Griffin 1973). Conversely, Lewis and Burghy (1964) found that several *Quercus* species, including *Q. douglasii*, reach water tables somewhat deeper than 21 m. Callaway (1990) emphasized soil-water distribution in shaping the overall root morphology of *Quercus* species in California.

If the soil is shallower than the usual depth of root penetration, and the underlying rock is not sufficiently weathered to permit passage of roots, a layer of roots usually grows at the soil–bedrock interface. If a taproot reaches the impenetrable bedrock, it will grow horizontally. But most commonly the bedrock does allow, to varying degrees, further deep root penetration through highly weathered material or through a network of cracks and channels. Examples of such penetration into the bedrock have been cited in this section. Another example is *Eucalyptus marginata*, for which roots were found to penetrate along vertical channels in lateritic bedrock in the jarrah forest ecosystem in Southwest Australia (Dell et al. 1983, Johnston et al. 1983). The channels, or preferred pathways, are permanent features of the profile, and it has been suggested that they result from dissolution of laterite by humic acid produced by the root itself (Plumb and Gosting 1973). In other cases, the bedrock structure allows passage of roots. Davis (1972) found roots of *Arctostaphylos pallida* following the planes of shale at a depth of 4 m. A contrary example is *Adenostoma fasciculatum*. It has generally been considered a deep-rooted plant, but Kummerow et al. (1977) reported that roots rarely penetrated into the cracks in granite. In fact, the benefit of penetration into granite has been questioned because of the small amount of water held in the rocks of the southern California watersheds. Rowe and Colman (1951) reported that fractured metamorphic rock below the soil profile could hold no more than 1.3 mm of water per 30 cm of rock depth.

Root penetration into unfissured bedrock has been reported for highly porous rocks. Such rocks absorb water in the wet season, releasing it to invading roots in the dry summer (Oppenheimer 1956, 1960). Some limestone rocks have surprisingly high water-storage capacity, and dwarf shrubs of the Mediterranean phrygana were observed to grow on the bare, unfissured rock in the Mediterranean Basin (Orshansky 1951). These roots behaved as if they were growing in soil. Oppenheimer (1960) also reported several cases for *Quercus* spp. and *Pistacia* spp. growing in localities where little or no soil remained. Penetration into porous rocks, however, has not been reported for the Pacific ecosystems of mediterranean type.

The distinction between maximum rooting depths and distribution of root biomass with depth must also be kept in mind. Miller and Ng (1977) observed that even deep-rooted shrubs can have the bulk of the biomass concentrated near the soil surface. Roots are often most abundant in the first 40 cm of the soil profile (Kummerow 1981, Low 1983, Miller and Ng 1977), and sometimes roots are confined mostly between 10 cm and 40 cm, avoiding the driest upper layer and also allowing better anchorage. At other times the whole root system is much shallower, most of the roots occurring in the first 20 cm, as found for six *Arctostaphylos* species in central California (Davis 1972). In species of the Southeast Australia heathland, the largest fraction of the root system is concentrated in the upper 30 cm of soil (Specht and Rayson 1957).

The highly variable data on rooting depth make it difficult to generalize. It is clear that some shrubs in all three regions are capable, at least under some circumstances, of rooting to great depth, and that mean rooting depths extend well beyond the zones of maximum nutrient concentration (2 to 20 m). Table 8.1 shows that for all the plant communities studied, mean values of root depth are between 1.2 and 1.7 m (individual values range from 0.2 to 8.5 m). Shrub species growing in the sand plains in Southwest and Southeast Australia (Table 8.1) show a somewhat shallower rooting pattern (rooting depth averages 1.2 m) than shrubs in California and Chile (where average rooting depth is 1.7 m). Relative to the above-ground development, however, shrubs in southern Australia have a more extensive root system (a smaller ratio between canopy height and root depth). The significance of the deep roots probably lies more in the access they provide to deep moisture during the times of most severe drought than in their contribution to nutrient absorption or to water absorption during times of maximum plant growth.

Horizontal Root Extension

The functions of shallow roots are to provide support for the aerial stems and to acquire water and nutrients, which are more abundant in the upper soil layers. Thus it is expected that the lateral extension of roots will be greater at or near the surface than deeper in the soil, a pattern reported for trees (Cannon 1914, Abbot et al. 1989, Incoll 1969). Roots of *Eucalyptus marginata* 1 m deep were found extending horizontally up to 20 m from the lignotuber (Abbot et al. 1989), although this has to be considered a rather exceptional distance. For trees on steep slopes the lateral root system is highly asymmetrical, with most of the roots occurring on the uphill side of the tree (Kimber 1974, Canadell and Rodà 1991). This asymmetry presumably has more to do with optimal anchoring of the plant than with improved success at obtaining water or nutrients.

In shrubs the shallow root system extends also over areas many times larger than the above-ground projected canopy area. Specht and Rayson (1957) found that species of the heathland explore between 10 and 20 times the area of their canopies. Ratios of root area to shoot area (cross-sectional projected area) for *Adenostoma fasciculatum* were 7 in a shallow soil (Kummerow et al. 1977) and 2 to 3 in deeper soils (Hellmers et al. 1955, Hanes 1965). According to the data gathered in Table 8.1, the area occupied by the roots in mediterranean species is between 2 and 6.5 times greater than that occupied by the canopy. Southeast and Southwest Australia heathland shrubs show the largest horizontal root extension among the three mediterranean regions studied.

Competition has an important part in the extension and depth to which the shallow roots grow. In the dense California chaparral and Chile matorral in shallow soils evidence is strong for species-specific rooting depths (Hoffmann and Kummerow 1978, Kummerow and Wright 1988, Wright 1987). Kummerow et al. (1977) and Hoffmann and Kummerow (1978) found a complicated network of roots, in which different species were exploiting different levels in the upper soil. Such differential rooting patterns may have arisen or at least been maintained to lessen competition.

Fine roots, commonly considered those less than 5 mm in diameter, have specific spatial distributions. Fine roots are found in variable quantities in the uppermost part of the soil profile, with the highest densities under the shrub canopy. This characteristic distribution has been related to the higher availability of nutrients under the shrub canopy (Lamont 1983). In particular cases when the top soil layer is gravel or a loose type of material, the densest layer of fine roots is at greater depth avoiding easily desiccated horizons. There is little overlap of fine roots from different individuals within the restricted area around the root crown (Kummerow et al. 1977).

Plant–Soil Water Relations

The patterns of change in soil moisture content with depth and time, and the corresponding drying and rewetting cycles, are decisive in determining species composition in an environment. Plants with different rooting habits show different seasonal courses of water potential, and the length of water stress and the distribution of soil moisture with depth will determine whether or not a species can succeed in a particular environment.

Species show differential rooting patterns along moisture gradients in the mediterranean-climate regions. Deep- and shallow-rooted plants are adapted to different degrees of aridity, along with other morphological and physiological characteristics. Generally, deep-rooted evergreen species are more abundant in wetter parts of the mediterranean regions, whereas shallow-rooted drought-deciduous species dominate in the driest areas.

Water Relations in Evergreen Species

Evergreen species growing within an area have been shown to differ in rooting habit, implying differences in exploitation of resources with depth. Evidence of such stratification has been shown directly by excavations and indirectly by measurements of drying and rewetting patterns for soil and plant tissue during drought cycles (Poole and Miller 1975, 1978, Riveros et al. 1976, Dodd et al. 1984).

Dodd et al. (1984) studied xylem pressure potential curves of 30 species during the summer drought in the sand plains in Southwest Australia. The close relationships between water stress and root morphologies were found. Davis and Mooney (1986) studied four co-occurring chaparral shrubs, and found strong evidence that different species share the soil-moisture resource by using different portions of it. At the height of the drought, the water potentials of the deep-rooted *Quercus durata* corresponded to soil-moisture potentials at a depth of 2 m. *Adenostoma fasciculatum* and *Heteromeles arbutifolia* had roots intermediate in depth, and water potentials corresponding to soil moisture at 0.75 m, and water-potential values for the shallow-rooted *Rhamnus californica* were equivalent to soil potentials at 0.5 m depth. For the Chilean matorral, seven of the most abundant shrub species showed a seasonal course of water stress closely related to their rooting habit (Giliberto and Estay 1978). *Lithrea caustica* and *Quillaja saponaria,* with deeper and more extensive roots, had higher and less variable xylem potentials throughout the year, but *Satureja gilliesii, Colliguaja odorifera, Retanilla ephedra,* and *Cryptocarya alba,* with shallower root systems, underwent greater and more variable water stress.

Competition for the scarce soil resources also causes partitioning among different species in the semiarid woodlands in East Australia (Hodgkinson 1992). Trees and shrubs compete for soil moisture, and partial stratification of roots is found between life forms. The trees, with their deeper roots, take up water from shallow and deep soil layers, the latter enabling them to tap additional soil water not available to the shrubs. The shrubs thus had the lowest xylem pressure potentials during the dry periods. A similar situation was reported for trees and understory species in the jarrah forest in Southwest Australia (Crombie et al. 1988). Water relations in both examples were closely related to root morphology.

Using data compiled from several studies, the drought response of the different evergreen species from the mediterranean-climate regions can be grouped in two main categories: species with shallow-rooting characteristics (less than 1 m), and those with deep-rooting characteristics (about 2 m and more). An intermediate category combining characteristics of these two groups might also be considered.

Shallow-rooted species become highly stressed, reaching water potentials between -40 and -60 bars by the end of the summer drought (Dodd et al. 1984, Giliberto and Estay 1978, Griffin 1973, Poole and Miller 1975). Summer rains bring relief, and plants resume growth immediately after rains come in early winter. Generally, shallow-rooted species have higher leaf conductances, greater transpiration and photosynthetic rates, and lower osmotic potentials when water is available, and annual fluctuations in plant water potentials and growth are greater than in deep-rooted species (Poole and Miller 1975, 1978). Medium-rooted species behave essentially like shallow-rooted species, though with reduced water stress.

Deep-rooted species show smaller seasonal variation in water potentials. They either undergo no water stress at all, or are stressed for a much shorter time at the end of the summer drought. They often are able to maintain high rates of transpiration during the summer by drawing groundwater from deep soil horizons or directly from the water table. For instance, *Heteromeles arbutifolia* was found to maintain a positive carbon balance all year (Mooney and Chu 1974). Pairs of closely related species have been studied for the influence of some morphological traits, especially the rooting patterns, on the length of the water stress period. *Arbutus menziesii* occurs under conditions of a shorter water-stress period than *H. arbutifolia* is able to endure (Morrow and Mooney 1974). Similarly, *Adenostoma fasciculatum* has fairly restricted activity during the summer, but *A. sparsifolium* remains physiologically active during the summer drought (Hanes 1965). For both, differences in the rooting pattern were partially responsible for such results.

Species dependent upon the water table for moisture supply often show a quite sudden start and end to water stress that does not correspond directly to variations in climate or soil moisture content. Rather, the descent of the water table beyond a critical point seems to cause the abrupt changes (Dodd et al. 1984).

For deep-rooted species, sporadic rains do not always relieve water stress during the summer drought, because the soil profile is not recharged at depth. But even species that also have well-developed shallow roots are not always able to take full advantage of the surface moisture. Hart and Radosevich (1987) showed that surface roots of commonly deep-rooted species, such as *Adenostoma fasciculatum*, seem to lose their ability to respond to increased soil moisture after rainfall. But roots of *Arctostaphylos stanfordiana*, a shallow-rooted shrub, maintain their capacity for quick absorption (Hart and Radosevich 1987).

Water Relations in Drought-Deciduous Species

In the most xeric conditions of the mediterranean-climate regions in California and Chile, summer-deciduous species replace evergreen ones in the chaparral and matorral communities. Sometimes, however, patches of summer-deciduous species are found in the middle of chaparral and matorral, in response to dry microclimatic conditions, highly disturbed areas, barren rocky slopes, road cuts, or peculiar soil types such as heavy clays (Harrison et al. 1971). Overall, the drought-deciduous species occupy areas where the low available moisture during the dry season would not allow evergreen species to succeed.

The defining trait of the drought-deciduous species is reduced leaf area during summer. These species also, compared with evergreens, have higher photosynthetic and transpiration rates, and they are highly responsive to the long summer–fall drought. As a result of their habitat and root system characteristics, drought-deciduous shrubs undergo stronger and earlier water stress than evergreen species (Miller and Poole 1979). In fact, one of the main structural differences between life forms is their root system (Mooney and Dunn 1970). Generally, summer-deciduous shrubs have a shallow root system but evergreen species are deeper rooted. The primary evidence for this difference comes from the study by Hellmers et al. (1955). Overall, summer-deciduous shrubs, considered to be drought evaders, are considered better adapted to withstand prolonged summer drought than evergreen species.

Contrary to the generalities above, Gill and Mahall (1986) found almost identical seasonal courses of water potential in the co-occurring chaparral evergreen *Ceanothus megacarpus* and the coastal sage summer-deciduous *Salvia mellifera*. The latter did not initiate or terminate growth significantly earlier in the season than the evergreen species. These findings suggest that the soil volumes occupied by roots of evergreen and deciduous shrubs may not necessarily be different, as previously thought. But the validity of the generalization about evergreen–deciduous differences can be defended: (1) *C. megacarpus* is a nonresprouting obligate seeder and among the most shallowly rooted evergreen species (Barnes 1979); and (2) *C. megacarpus* belongs to a section of its genus known to have a number of highly evolved drought-resisting traits (Nobs 1963). It is to be expected that the most drought-resistant evergreens are also the most shallowly rooted, and that these species co-occur with summer-deciduous species at specific points along the moisture-availability gradient. More research is needed, particularly in determining root growth and activity patterns of both groups of plants (Mooney 1989).

Drought-deciduous species have not been described in the Southeast and Southwest Australia heathland, although they have a counterpart in evergreen shallow-rooted species. Under water stress, these evergreen species apparently enter a state of anabiosis, but unlike the drought-deciduous species, they retain much of their leaf canopy (Specht 1981b). It is possible that retention of the evergreen habit across the moisture gradient may be explained at least in part by the soils' nutrient deficiency and the resulting high cost of deciduousness (Specht 1963, Stock et al. 1992).

Lignotubers

Occurrence and Terminology

We define the lignotuber as a woody swollen structure at the stem base, from which roots or rhizomes or both grow. Its size can vary among species from a few cm to almost 2 m in diameter (Jacobs 1955). This structure is genetically determined (Mullette and Bamber 1978) and it appears early in seedling development. Seedlings of most *Eucalyptus* species develop a lignotuber within the first one or two growing seasons after germination (Noble 1984), and between 3 and 5 months were needed for the lignotuber to become apparent in 11 shrub species studied in the Chile matorral (Montenegro et al. 1983).

The phenotypic expression of such a structure may be determined by environmental factors, however, because some species have been found both with and without a lignotuber depending on the environmental conditions. Species may have a lignotuber even though closely related species living in the same region do not. Nonlignotuberous and lignotuberous species of the *Lehmannianae* (a group of *Eucalyptus* species) occur side by side in habitats that are prone to fire in Southwest Australia (Carr et al. 1983). In the California chaparral the lignotuber-forming *Arctostaphylos glandulosa* grows in the same habitats as *A. glauca*, which does not form a lignotuber. On the other hand, specific environmental factors that are not yet clear may be important selective forces for such a feature. Orshan et al. (1984) studied a transect through different mediterranean ecosystems in Chile, from the coast to the high mountains. They found that the proportion of species with lignotubers changed from 29% at the immediate coast to 73% in the coastal range, and from there the percentage gradually decreased with increasing altitude until only 5% of the species had lignotubers at 2000 m.

All mediterranean-climate regions of the world have trees and shrubs that develop lignotubers, although it is not a character exclusive to these regions (James 1984). A survey of 429 kwongan species in the open woodland and heathlands in Southwest Australia reported that 44 species were lignotuberous (Dodd et al. 1984). This number accounted for more than a quarter of the 163 kwongan woody shrub and tree species shown to have resprouting capacity after fire. In the heathland of Dark Island in Southeast Australia, however, no species were found with either epicormic buds or lignotubers, even though a large percentage of the plants were able to survive a fire (Specht and Rayson 1957). In the California chaparral, lignotubers are described for many of the commonest species: *Adenostoma fasciculatum, A. sparsifolium, Arctostaphylos glandulosa, A. vestita, A. patula, Ceanothus* sect. *Ceanothus, Quercus dumosa,* and *Rhus (Malosma) laurina* (DeSouza et al. 1986, Hanes 1965, Hellmers et al. 1955, Jepson 1916, Kummerow 1981, Kummerow and Mangan 1981). For the Chile matorral, lignotubers are reported to form in dominant species that include evergreen, summer semideciduous, and summer-deciduous shrubs: *Colliguaja odorifera, Quillaja saponaria, Lithrea caustica, Cryptocarya alba,* and many others (Hoffmann and Kummerow 1978, Montenegro et al. 1983).

Lignotubers are also called burls, root crowns, or rootstocks by different authors, even when referring to the same species (see *Adenostoma fasciculatum* in Hellmers et al. 1955 and Kummerow 1981). The name burl has often been used to describe the woody swellings when applied to the Ericaceae (Jepson 1916, Garland and Marion 1960). Root crowns and rootstocks have been used as nonspecific, general labels for any type of massive woody structure at the junction of stem base and roots, regardless of whether or not it is genetically determined. At present, anatomy and possible functions of the so-called lignotubers and burls are not known well enough for them to be considered different structures. In fact, Kummerow and Ellis (1989) say that the terms lignotuber and burl, at least in California, are commonly used interchangeably. We believe that the distinction between a structure that is genetically determined and one that is induced by environmental conditions is important. Therefore we suggest using lignotuber in the narrow sense for genetically determined swollen woody structures at the stem base. We propose that burl, or basal burl if more precision is desired, be used as the broader name for woody swellings at the stem base applicable to all such structures whether or not they are genetically determined. We justify our suggestion by origin and usage. Burl is a traditional label applied to any pronounced woody swelling on a tree or shrub. Lignotuber is a scientific term and thus may be applied in a restrictive sense without contradicting popular usage.

Anatomy, Morphology, and Development

Initiation, development, and anatomy of lignotubers have been studied for some species of *Eucalyptus* in great detail (Kerr 1925, Carr et al. 1983, 1984a,

1984b). Recently, thorough work has also been done on the lignotuber ontogeny of *Quercus suber* in the Mediterranean Region (Molinas and Verdaguer 1993a,b). But our knowledge is limited for the rest of the lignotuberous species.

Typically the lignotuber arises in the axils of the cotyledons as a pair of lateral outgrowths, which gradually increase in size until they meet, forming a swelling that encircles the stem (Kerr 1925). In lignotuberous species of the *Lehmannianae*, lignotuber formation is preceded by the appearance of a crescent-shaped array of accessory buds, adaxial to the axillary bud. Lignotubers may be formed at successive nodes, but those with only one upper accessory bud (which may be present even in nonlignotuberous species) do not form lignotubers (Carr et al. 1983). Carr et al. (1984a) found four modes of lignotuber initiation after studying the morphological development of the lignotubers of 13 species that represented five taxonomic groups in *Lehmannianae*.

For the California chaparral species, lignotubers have been defined as clumps of secondary wood that develop from the transition zone between the hypocotyl and the main root of seedlings (Kummerow and Mangan 1981). Usually, to this genetically determined woody swollen structure, bases of stems and roots become incorporated into the lignotuber tissue, presenting a complex and contorted structure.

The lignotubers are quite distinct in their structure, and should not be confused with enlarged portions of stems whose tissues retain their characteristic organization. Lignotuber tissues differ from stem tissues in quantitative aspects, but the types of tissues are the same: cambium, phloem, rhytidome, sapwood, heartwood, and kino veins (Bamber and Mullette 1978). For example, the lignotuber of *Eucalyptus gummifera* has shorter wood fibers, a greater proportion of axial parenchyma of the wood, and less expansion of the axial parenchyma of the outer phloem than in the stem tissues. Carr et al. (1984a) found that in eucalypt seedlings the lignotuber cambium differs from the stem cambium in a number of ways, including size and frequency of ray initials, lengths of the fusiform initials, and composition of the xylem produced.

Lignotuber formation is an inherited characteristic, but like other genetically determined characteristics, it may not be expressed under some circumstances (Kerr 1925, Beadle 1968, Mullette and Bamber 1978). Full penetrance of the genes involved in lignotuber formation for *Eucalyptus* spp. is reported to be affected by both climatic and edaphic factors. Beadle (1954, 1968) proposed that the lignotuber was an adaptation to poor soils, because high levels of P and N restricted its development in seedlings of *Eucalyptus oleosa*, *E. saligna*, and *E. gummifera*. Mullette and Bamber (1978), however, found that, on the contrary, an increase in soil P levels stimulated lignotuber development in *E. gummifera*. Jahnke et al. (1983) studied two geographic provenances of *E. camaldulensis*, one normally lignotuberous and the other nonlignotuberous; plants of both provenances were raised under combinations of soil N and P levels. Lignotubers were not formed in plants of the southern provenance regardless of the nutritional treatment. For the northern provenance the percentage of seedlings that developed a lignotuber was highest at high levels of P associated with low to intermediate levels of N. Therefore, although we know that different soil N and P availability affects lignotuber expression and development, a consistent trend in nutritional effects is not apparent, perhaps because of interaction by factors not considered.

Differential lignotuber development is found within species, resulting in varied sizes and morphologies. For many of the *Eucalyptus* species, lignotubers ranging between 15 mm and 150 mm in diameter have been described, but other species like the mallee forms of eucalypts have lignotubers 1 to 2 m in diameter (Jacobs 1955). A multistemmed *E. gummifera* mallee had one massive lignotuber 75 m^2 in area (Mullette 1978). The Chilean matorral shrub and small tree *Lithrea caustica* was reported to have a lignotuber as heavy as 67 kg (dry mass) (Hoffmann and Kummerow 1978). A *Quercus dumosa* lignotuber of 15 kg was found in the California chaparral, and a lignotuber as heavy as 52 kg was reported for *Q. turbinella* in the Arizona chaparral (Davis and Pase 1977). In the Mediterranean Basin, Canadell and Rodà (1991) reported a lignotuber mass of 317 kg for a *Q. ilex* (tree).

The lignotubers of some eucalypts may be evident only in the juvenile stages of plant development, because as plants grow the lignotubers may merge with the main stem (Pryor 1976). If lignotuberous plants are subjected to above-ground removal (e.g., by fire) and subsequent resprouting, though, the lignotuber rapidly enlarges, both because the tissues grow and because the stem bases merge within the lignotuber structure. Repeated stem harvesting for charcoal manufacturing was the cause of the massive lignotubers of *Lithrea caustica* and *Cryptocarya alba* found in the Chilean matorral (Hoffmann and Kummerow 1978).

Frazier's (1993) study of hybridization between a sprouting, lignotuberous species of *Ceanothus* and nonsprouting species provides the only data on inheritance and genetic control of the lignotuber. In a study of naturally occurring hybrids he found that most morphological traits had character states very close to the midpoint of the two parental types. In contrast, he found that degree of lignotuber development, though intermediate, was much closer to that of the nonsprouting species. The presumption is that the hybrids would therefore have little or no capacity to resprout after fire. This finding is surprising since lignotuber structure would be expected to be a quantitative character affected by many genes. Therefore, either the burl is determined by few genes and one or more recessive alleles or some developmental constraint or ecological factor selects against the "half-burl" condition.

A lignotuber should always be at least as old as the aerial part of the plant, and because it has the potential to persist even when the above-ground parts of the plant are lost, it could be much older. Although attempts have been made to determine the age of lignotubers, only a few studies have provided reliable data. Traditional techniques for dating wood are not applicable to lignotubers because their highly contorted wood makes it impossible to read the annual growth rings. In fact, only carbon dating can be considered reliable enough for lignotuber aging. The first radiocarbon dating for a lignotuber was published by Grant Taylor and Rafter (1963). They determined an age of 200 years for a sample taken from the central part of a lignotuber 1.2 m in diameter of the mallee *Eucalyptus oleosa*. Later, Wellington et al. (1979) confirmed the age of large lignotubers of *E. oleosa*. They analyzed samples from different parts of a lignotuber almost 2 m in diameter found in one of the oldest eucalypt populations in Southwest New South Wales. The oldest radiocarbon age was 330 ± 70 years. For the California chaparral shrub *Adenostoma sparsifolium*, an age of 200 ± 65 years was established for a lignotuber 0.3 m in diameter (Hanes 1965).

Other complexities in lignotuber growth and survival may make it difficult or impossible to determine the age of a genet from the lignotuber. In the California chaparral, we have observed that *Arctostaphylos glandulosa* and *Adenostoma fasciculatum* shrubs frequently consist of small, roughly circular clusters of stems that suggest a gradual outward expansion from a central point in a "fairy-ring" pattern. Examining excavated burls often reveals a more axial portion consisting of decayed wood, and an outer portion of new healthy burl. We believe this condition indicates that after fire, central portions of the burl atrophy, and regenerated tissues tend to form on the outer margins of the burl. Repetitions of this cycle could lead to fragmentation of the genets into a more-or-less circular arrangement of ramets. This tendency is particularly strong in *Arctostaphylos glandulosa*, which has extensive populations consisting of thousands of stems in which it is nearly impossible to recognize subclusters corresponding to individual genets. S. Davis (pers. comm.) has discovered almost perfectly symmetrical annular clones of *Malosma laurina* 10-15 meters in diameter in the Malibu area of southern California. If our suspicions are correct, aging living lignotubers in such species may give only a minimal estimate of the age of the genet.

Functions and Adaptive Meaning

Little is known about the actual function of the lignotuber in mediterranean ecosystems. In fact, little or no direct experimental evidence is available on the functions of the lignotuber for any ecosystem in which lignotuberous plants occur. Functions have often been inferred from studies on structure, anatomy, and nutrient contents of the lignotuber tissues, but direct quantitative observations are few.

The lignotuber has been considered an adaptation to uncertainty in the habitat, especially to stresses such as defoliation and prolonged drought or fire, which cause total or partial loss of the plant's above-ground parts (Carr et al. 1983). Frost can cause widespread above-ground mortality and even death in mediterranean-climate regions (O'Brien 1989). It is not, however, the only possible type of adaptation to these sorts of stresses, because ligno-tuberous and nonlignotuberous species often grow with equal success side by side.

The lignotuber has a decisive role in plant survival after disturbances and stresses. Successful regeneration after logging in the jarrah forest in SE Australia depends upon the presence of well-developed lignotuberous growth at the time the disturbance occurs (van Noort 1960). The presence of large subterranean lignotubers in the *Eucalyptus gummifera* mallee, ensuring regeneration after damage, is a clear advantage in the harsh environments in which it grows (Mullette 1978, Gill 1981). Having a well-developed lignotuber is also advantageous in surviving wildfires in the California chaparral, where some lignotuberous species show no mortality at all after fire (Rundel et al. 1987). Even at the seedling stage, a lignotuber can strongly influence survival. Noble (1984) stated that the lignotuber of eucalypts enhances the seedlings' ability to survive perturbations such as drought, grazing, and fire.

The occurrence of lignotubers in the Chilean matorral, a vegetation in which fire, is thought to have been less important than other mediterranean shrub types demonstrates the diversity of environmental pressures that can drive selection for the lignotuber (Montenegro et al. 1983). In Chile, drought and grazing are suggested as primarily selective forces favoring lignotuber development.

We postulate a dual function for the lignotuber. First, it stores concealed buds (a source of meristematic tissue), and second, it stores resources such as carbohydrates and nutrients. Both functions may allow rapid regrowth after above-ground parts are removed (Chattaway 1958, Specht 1981a), and therefore make the lignotuber a structure especially adapted to environments subjected to recurrent disturbance.

It is well known that the rapid resprouting capacity of some plants is related to the large numbers of concealed buds held in the lignotuber. Zammit (1988) studied the capacity of *Banksia oblongifolia* lignotubers to resprout within intervals in which it was assumed that new buds were not produced. About 30% of the buds remained dormant after the first clipping, and about 10% after the second and third clippings. No lignotubers survived four clippings over the 15-month experimental period. Kummerow and Ellis (1989) reported a figure of 120 sprouts per lignotuber on ten 54-year-old *Adenostoma fasciculatum* individuals two years after a fire. And in the Mediterranean Basin a lignotuber 1304 cm^2 in area of a tree heath, *Erica arborea*, was found to have 1268 sprouts nine months after clipping (J. Canadell unpub.).

Without disturbance, the buds held in the lignotuber are inhibited by apical dominance, and only when the stem is injured or cut off is the inhibition removed so that the buds can sprout (Blake and Carrodus 1970). Some authors, however, have found that a few species have the capacity to continuously sprout from the lignotuber, without fire or any other identifiable disturbance (Mesléard and Lepart 1989, Malanson and Westman 1985, Lacey 1983). Clipping studies indicate that it is not the plant's nutritional status that prevents development of the dormant buds (Blake 1972).

A close relationship binds lignotuber characteristics such as its size and the depth at which it is located, to its relative resistance to fire. Plants with a large lignotuber have a better chance to survive a fire than do plants with smaller ones (Rundel et al. 1987). Similarly, plants with a lignotuber deeper in the soil profile have a higher chance of surviving, for the soil's insulating properties prevent the dormant buds from being killed (Auld 1990, Flinn and Wein 1977). Bradstock and Myerscough (1988) showed that in *Banksia serrata* and *Isopogon anemonifolius*, young juveniles are the stage in the life cycle most sensitive to frequent fires. Insufficient lignotuber development was concluded to be the cause of their mortality. Experiments with *Angophora hispida,* a species common in the heathlands, woodlands, and dry sclerophyll forests in Southeast Australia, revealed that size and depth of the developing lignotuber interacted to influence the degree of plant survival. Only plants with lignotuber volumes greater than 5000 mm^3 showed no mortality after clipping and burning treatments (Auld 1990); a similar pattern was found by Noble (1984) for three *Eucalyptus* species. And the size of lignotubers in five-year-old *Banksia serrata* juveniles governed survival under simulated low-intensity fires (Bradstock and Myerscough 1988).

The lignotuber's function as a storage organ for carbohydrates and mineral nutrients is studied much less often. Some of the strongest evidence that the lignotuber acts as a carbon-storage structure comes from studies of anatomical and chemical content. Montenegro et al. (1983) studied the anatomy of one-year-old shrub seedlings in the Chilean matorral. Starch was found not only in the rays and cortex cells, but also in the axial parenchyma of the secondary xylem, the layers of periderm and parenchymatous cells being more numerous in the lignotuber than in the main stem. Important quantities of starch were contained in the parenchymatous cells. In the lignotuber the axial parenchyma is made up of paratracheal bands one to three cells wide, but the normal stem is made up of one cell layer. From these findings it is concluded that the lignotuber can be considered a storage organ for starch. Carrodus and Blake (1970), however, found inconsistent anatomical support for the carbohydrate-storage function of the lignotuber of *Eucalyptus obliqua*. Starch was restricted to the parenchymatous cells of the rays and cortex. No starch was observed in the tracheid or fiber-tracheids, as had been reported by Chattaway (1958), and their starch content was similar to that found in the adjacent taproots and stems. Lopez (1983) reported starch grains in axial and ray parenchyma of *Quercus dumosa* lignotubers in the California chaparral. But tissues concentrations of both nitrogen and nonstructural carbohydrates showed no differences between lignotubers and stems.

Some studies used the tissue chemical composition of the lignotuber as support for the lignotuber function as a store of nutrients. Dell et al. (1985)

found that phosphorus concentrations were three times higher in the lignotubers than in the stem wood of *Eucalyptus marginata*. This excess resulted
from the more abundant storage parenchyma in the lignotuber. It was suggested that the *E. marginata* lignotuber may be a phosphorus storage organ
that enables these trees to survive on nutrient-deficient soils, especially soils
low in phosphorus. On the other hand, Mullette and Bamber (1978) did not
find statistical differences in mineral nutrient concentrations between lignotuber and stem tissues for *E. gummifera*. But considering the great biomass
of the mallee lignotubers, they concluded that the lignotuber is an important
storage organ for inorganic nutrients. In the chaparral, Lopez (1983) also
states that although concentrations in lignotubers may be low, the vast
amount of biomass means that large quantities of both carbohydrates and
nutrients could be present. He suggests that the lignotuber serves as a source
of food reserves, which is especially important for resprouting after fire.

   In conclusion, although it is tempting to accept the dual function of the
lignotuber—protection for concealed buds and storage of resources—the
only direct evidence available is about the lignotuber's function as a bud
store. The concealed buds held in the lignotuber are essential in regeneration
after above-ground destruction. The lignotuber as an organ for carbohydrate and nutrient reserves has been inferred from indirect evidence, which
does not necessarily point out that the lignotuber has storage functions distinct from that of stems and roots. Furthermore, physiology of plant processes cannot be inferred from anatomical and chemical tissue studies. Carbohydrates stored in the lignotuber nevertheless do have a critical role
during the period before plants start to resprout. Regrowth has been observed to be delayed for up to half a year in some species. The carbon stored
in the lignotuber may then be critical in maintaining respiration, although
carbon stored in the roots may be equally important.

A Hypothesis to Explain the Evolutionary Origin of Lignotubers

Lignotubers are widely distributed geographically, and are not unique to
vegetation that experiences severe summer drought (Axelrod 1975). They
also occur in many families, suggesting that one or a very few fundamental
selective pressures mostly independent of the morphology and physiology of
the individual species account for their origin. We suggest that a universal
selective pressure arises from the constraints of plant architecture. Specifically, we believe that a multiple-stemmed woody plant (shrub) has a limited
range of possible responses to damage to its canopy.

   Our model postulates the original condition to be a long-lived, usually
multiple-stemmed woody shrub. A plant of this type must forage for water
or nutrients over a wide area and therefore has a root system that is deep, or
extensive, or both. Maximum size and therefore maximum fitness is achieved
only after a relatively long period of growth and a large investment in roots.
The strategy for such species is based on producing multiple stems from a

central root crown and returning water and nutrient resources to the cluster of stems at the center. This conservative, spatially fixed soil foraging contrasts to a more active strategy such as lateral expansion by producing stems from root sprouts or rhizomes. Such mobile foraging strategies are not compatible with reliance on deep, extensive, and energetically costly root systems.

Assuming that longevity is an element in the strategy, the canopy needs to recover after fire, drought, and herbivory. Several factors would favor having the replacement stems arise from buds lower on the stems. First, buds placed lower would have a higher probability of survival, especially when burned. Second, regeneration from more highly placed buds would tend to produce weaker and more disease-prone stems because deadwood remaining from the old stem would often be incorporated into the new stem (Fig 8.1).

At least two factors would also favor an increase in the number of buds. The simple numerical advantage is probably of primary importance. The more buds, the more likely it is that at least one will survive to produce a new

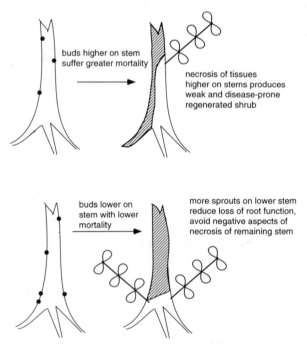

Figure 8.1. Diagrammatic representation of the hypothesized primary causes for the origin of the lignotuber. Buds placed lower have a higher probability of surviving disturbance and regenerating sounder structure. The final stage in the evolution would increase number and storage capacity by enlarging basal diameter.

stem. Returning to the importance of the root system, the proliferation of stems may be useful to maximize the area of photosynthetic canopy and thus maximize the survival of root biomass. Multiple buds and sprouts are also a hedge against herbivory, to which resprouting shrubs are particularly susceptible.

Because bud number is limited by the circumference of the stem, selective pressure to increase the number requires an increase in stem diameter at the base of the plant. Concomitant with the increase in number is the need to ensure that resprouting will be rapid and vigorous. This requirement favors storage of energy reserves in the lignotuber, as suggested earlier.

All these factors acting together will move species toward a lignotuber. Species that lack a burl until after they have suffered severe canopy removal do not contradict our argument. Rather they seem to illustrate an intermediate condition in which buds are concentrated low on the plant, but in which strong selection to increase the number has been lacking.

The absence of a lignotuber in the so-called obligate-seeder species—that is, species that fail to resprout after fire and must reestablish by seed—is consistent with our hypothesis. The regenerative function of the lignotuber matters primarily because of its utility in plant response to relatively rare massive disturbance. Species benefit from a lignotuber only if the interval between massively disturbing events is significantly shorter than the life span.

Our model stresses regeneration of multiple-stemmed woody plants, but tree forms also have a lignotuber. If the probability is reasonably high that the above-ground portion of the tree will be killed by fire, the lignotuber could serve the same function in maximizing bud survival as it does for a multiple-stemmed shrub, in a manner consistent with our hypothesis. This appears to be the sequence in Australian eucalypts, where trees often exist for an extended period in the understory as multiple-stemmed shrub-form individuals with lignotubers (Florence 1981). Such individuals may be repeatedly burned and grazed before an opportunity to recruit into the canopy is presented. Florence (1981) hypothesizes that forming an extensive root system may be a critical aspect of this juvenile stage. If, however, the death of the main stem were very rare, our model would not be valid, and the presence of the lignotuber would have to be explained by some other benefit conferred on the plant.

Conclusions

Data on root structure and morphology are still somewhat scarce for woody species in mediterranean ecosystems, yet basic conclusions can be arrived at from this review.

Root morphology has been shown to have high intra- and interspecific variability. When soils allow free development of the root system, root structure is primarily determined by the species's genetic makeup. Soil features such as rockiness, steep slopes, and changes in bedrock at small spatial scales, however, cause habitat-dependent variability, which may completely alter the root system's structure. This alteration is common in the California chaparral and Chile matorral, where species-level classification has often been unsuccessful. On the other hand, the sand plains in Southeast and Southwest Australia present less constriction to root-system development, enabling the structure of the root system to develop more freely. In that case, classifications characterizing root-system structures at the species level have been more successful.

Of the different root-system morphologies, the dual type is quite common among evergreen species in the Pacific mediterranean ecosystems. It is distinguished by a taproot, which may be either a single dominant root, or multiple nondominant roots that may easily reach between 1.5 m and 2.5 m deep. Examples of roots reaching soil horizons deeper than 4 m have been reported. About 70% of the species in Table 8.1 have shown taproots. The dual type of root system also shows a well-developed lateral root system that may extend over areas several times the size of the canopy.

The dual system seems particularly well suited to mediterranean climate conditions because it enables plants to utilize moisture from deep horizons during part of the summer drought. For that reason, plants are able to photosynthesize all year during wet years. In fact, this type of root system seems to be an adaptation to a seasonal drought regardless of when the drought occurs. Thus, climates other than mediterranean with a pronounced seasonal drought also show a dual type of root system.

Unlike the relatively deep root system of the evergreen species, drought-deciduous species, which occur in the driest extreme of the mediterranean climate, commonly show a shallower root system. They are therefore much more sensitive to drought and usually have a shorter growing season.

Possession of a lignotuber is common among resprouting shrubs in the mediterranean ecosystems studied. *Eucalyptus* species, which dominate the overstory in the mediterranean forest and woodlands in Southeast and Southwest Australia, show one of the most extensive developments of the lignotuber. Although lignotubers are characteristic of mediterranean ecosystems, they are also found in regions with other climatic regimes and therefore probably are a very ancient and very general solution to problems faced by woody plants. Environmental disturbances and stresses such as drought, defoliation, and fire are suggested as possible evolutionary forces.

A dual function is postulated for the lignotuber. With little doubt, a major function is to store concealed buds that resprout after a disturbance. A second function is to store carbohydrates and nutrients, as inferred from anatomical and nutrient tissue-analysis studies. Both functions may allow rapid recovery of the plant after the above-ground parts are removed. It has yet to be experimentally proven, though, if plants use these resources during resprouting, and if so, the extent to which they are used. Also still unclear is the possible role of the lignotuber during periods between disturbances.

Acknowledgments

We thank Wende Rehlaender for discussion and for detailed and careful editing at all stages in the development of this manuscript. Josep Canadell was supported by an M.E.C.–P.D.T.E. Scholarship from Spain. Chaparral studies of P. Zedler were supported by the National Science Foundation (BSR-850769), the California State Department of Parks and Recreation, and the California Department of Forestry. We also owe a debt to our colleague Jochen Kummerow, who has studied root systems of mediterranean species on three continents.

References

Abbott I, Dell B, Loneragan O (1989) The jarrah plant. In Dell B, Havel JJ, Malajczuk N (eds) *The Jarrah Forest: A Complex Mediterranean Ecosystem.* Kluwer Academic Publishers, Dordrecht, pp 41–51

Atkinson D, Naylor D, Coldrick GA (1976) The effect of tree spacing on the apple root system. Hortic Res 16:89–105

Auld TD (1990) The survival of juvenile plants of the resprouting shrub *Angophora hispida* (Myrtaceae) after a simulated low-intensity fire. Aust J Bot 38:255–260

Axelrod DI (1975) Evolution and biogeography of Madrean-Tethyan sclerophyll vegetation. Ann Missouri Bot Garden 62:280–334

Bamber RK, Mullette KJ (1978) Studies of the lignotubers of *Eucalyptus gummifera* (Gaertn. and Hochr.). II. Anatomy. Aust J Bot 26:15–22

Barnes, FJ (1979) Water relations in four species of *Ceanothus.* M.A. Thesis, California State University, San Jose. San Jose, CA

Beadle NCW (1954) Soil phosphate and the delimitation of plant communities in eastern Australia. Ecology 35:370–375

Beadle NCW (1968) Some aspects of the ecology and physiology of Australian xeromorphic plants. Aust J Sci 30:348–355

Beckett PHT, Webster R (1971) Soil variability: a review. Soils Fertil 34:1–15

Berntson GM (1992) A computer program for characterizing root system branching patterns. Plant Soil 140:145–149

Blake TJ (1972) Studies on the lignotubers of *Eucalyptus obliqua* L'Herit. III. The effects of seasonal and nutritional factors on dormant bud development. New Phytol 71:327–334

Blake TJ, Carrodus BB (1970) Studies on the lignotubers of *Eucalyptus obliqua.* L'Herit. New Phytol 69:1073–1079

Bradstock RA, Myerscough PJ (1988) The survival and population response to frequent fires of two woody resprouters *Banksia serrata* and *Isopogon amenonifolius.* Aust J Bot 36:415–431

Burns RE (1972) Environmental factors affecting root development and reserve carbohydrates of Bermuda grass cuttings. Agron J 64:44–45

Callaway RM (1990) Effects of soil water distribution on the lateral root development of three species of California oaks. Am J Bot 77:1469–1475

Canadell J, Rodà F (1991) Root biomass of *Quercus ilex* in a montane Mediterranean forest. Can J For Res 21:1771–1778

Cannon WA (1914) Specialization in vegetation and in environment in California. Plant World 17:223–237

Cannon WA (1949) A tentative classification of root systems. Ecology 30:542–548

Carbon BA, Bartle GA, Murray AM, McPherson DK (1980) The distribution of root length, and the limits to flow of soil water to roots in a dry sclerophyll forest. For Sci 26:656–664

Carr DJ, Jahnke R, Carr SGM (1983) Development of the lignotuber and plant form in *Lehmannianae*. Aust J Bot 31:629–643

Carr DJ, Jahnke R, Carr SGM (1984a) Initiation, development and anatomy of lignotubers in some species of *Eucalyptus*. Aust J Bot 32:415–437

Carr DJ, Jahnke R, Carr SGM (1984b) Surgical experiments on eucalypt lignotubers. Aust J Bot 32:439–447

Carrodus BB, Blake TJ (1970) Studies on the lignotubers of *Eucalyptus obliqua* L'Herit New Phytol 69:1069–1072

Chattaway MM (1958) Bud development and lignotuber formation in eucalypts. Aust J Bot 6:103–115

Cody ML, Mooney HA (1978) Convergence versus nonconvergence in mediterranean ecosystems. Ann Rev Ecol Syst 9:265–321

Cooper WS (1922) The broad-sclerophyll vegetation of California. An ecological study of the chaparral and its related communities. Carnegie Inst Wash Pub 319: 122 pp

Crick JC, Grime JP (1987) Morphological plasticity and mineral nutrient capture in two herbaceous species of contrasted ecology. New Phytol 107:403–414

Crombie DS, Tippett JT, Hill TC (1988) Dawn water potential and root depth of trees and understorey species in south-western Australia. Aust J Bot 36:621–631

Curl EA, Truelave B (1986) The structure and function of roots. In Curl EA, Truelave (eds) *The Rhizosphere*. Adv Agr Sci 15:19–54

Davis CB (1972) Comparative ecology of six members of the *Arctostaphylos andersonii* complex. Ph.D. dissertation, U California, Davis

Davis EA (1970) Root system of shrub live oak in relation to water yield by chaparral. J Ariz Acad Sci 12:62

Davis EA (1978) Root system of shrub live oak in relation to water yield by chaparral. Hydrol Water Resour Ariz Southwest 7:241–248

Davis EA, Pase CP (1977) Root system of shrub live oak: implications for water yield in Arizona chaparral. J Soil Water Cons 32:174–180

Davis SD, Mooney HA (1986) Water use patterns of four co-occurring chaparral shrubs. Oecologia 70:172–177

Dell B, Bartle JR, Tacey WH (1983) Root occupation and root channels of jarrah forest subsoils. Aust J Bot 31:615–627

Dell B, Jones S, Wallace IM (1985) Phosphorus accumulation by lignotubers of jarrah (*Eucalyptus marginata* Donn ex Sm.) seedlings grown in a range of soils. Plant Soil 86:225–232

DeSouza J, Silka PA, Davis SD (1986) Comparative physiology of burned and unburned *Rhus laurina* after chaparral wildfire. Oecologia 71:63–68

Dodd J, Heddle EM, Pate JS, Dixon KW (1984) Rooting patterns of sandplain plants and their functional significance. In Pate JS, Beard JS (eds) *Kwongan: Plant Life of the Sandplain*. U Western Australia Press, Nedlands, pp 146–177

Feldman LJ (1984) Regulation of root development. Ann Rev Plant Physiol 35:223–242

Feldman LJ (1988) The habits of roots. BioScience 38:612–618

Fitter AH, Stickland TR (1992) Architectural analysis of plant root systems. III. Studies on plants under field conditions. New Phytol 121:243–248

Flinn MA, Wein RW (1977) Depth of underground plant organs and theoretical survival during fire. Can J Bot 55:2550–2554

Florence RG (1981) The biology of the eucalypt forest. In Pate JS, McComb AJ (eds) *The Biology of Australian plants*. U Western Australia Press, Nedlands, Western Australia, pp 147–180

Frazier C. (1993) An ecological study of hybridization between chaparral shrubs of contrasting life histories. M.S. Thesis, San Diego State U, San Diego, CA

Garland H, Marion L (1960) California manzanita for smoking pipes. USDA For Serv PSW Misc. Paper 53. Pacific Southwest For and Range Expt Sta, Berkeley, CA

Giliberto J, Estay H (1978) Seasonal water stress in some Chilean matorral shrubs. Bot Gaz 139:236–260

Gill AM (1981) Fire adaptive traits of vascular plants. In Mooney HA, Bonnicksen TM, Christensen NL, Lotan JE, Reiners WA (eds) *Fire Regimes and Ecosystem Properties*. Gen Tech Rep WO-26, pp 208–230

Gill DS, Mahall BE (1986) Quantitative phenology and water relations of an evergreen and a deciduous chaparral shrub. Ecol Mono 56:127–143

Grant Taylor TL, Rafter TA (1963) N.Z. natural radiocarbon measurements I–V. Radiocarbon 5:118–161

Griffin JR (1971) Oak regeneration in the upper Carmel Valley, California. Ecology 52:862–868

Griffin JR (1973) Xylem sap tension in three woodland oaks of central California. Ecology 54:152–159

Hanes TL (1965) Ecological studies on two closely related chaparral shrubs in southern California. Ecol Mono 35:213–235

Harrison AT, Small E, Mooney HA (1971) Drought relationships and distribution of two mediterranean-climate California plant communities. Ecology 52:869–875

Hart JJ, Radosevich SR (1987) Water relations of two California chaparral shrubs. Am J Bot 74:371–384

Hellmers H, Horton JS, Juhren G, O'Keefe J (1955) Root systems of some chaparral plants in southern California. Ecology 36:667–678

Hnatiuk RJ, Hopkins AJM (1980) Western Australian species-rich kwongan (sclerophyllous shrubland) affected by drought. Aust J Bot 28:573–585

Hodgkinson KC (1992) Water relations and growth of shrubs before and after fire in a semi-arid woodland. Oecologia 90:467–473

Hoffmann A, Kummerow J (1978) Root studies in the Chilean matorral. Oecologia 32:57–69

Incoll WD (1969) Root excavation of *Eucalyptus regnans*. Res Act For Com Vict 69:15–16

Jacobs MR (1955) Growth habits of the eucalypts. Commonw For Tim Bur: Canberra

Jahnke R, Carr DJ, Carr SGM (1983) Lignotuber development and growth parameters in *Eucalyptus camaldulensis* (Dehnh.): effects of phosphorus and nitrogen levels. Aust J Bot 31:283–292

James S (1984) Lignotubers and burls—their structure, functions and ecological significance in Mediterranean ecosystems. Bot Rev 50:225–266

Jepson WL (1916) Regeneration in manzanita. Madroño 1:3–11

Johnston CD, Hurle DH, Hudson DR, Height MI (1983) *Water Movement Through Preferred Paths in Lateritic Profiles of the Darling Plateau, Western Australia*. CSIRO, Aust Div Groundwater Res Tech Pap No 1, 4 pp

Kerr LR (1925) The lignotubers of eucalypt seedlings. Proc Roy Soc Victoria 37:79–97

Kimber PC (1974) The root system of jarrah (*Eucalyptus marginata*). For Dept, W.A. Res Paper 10, Perth

Kummerow J (1980) Adaptation of roots in water-stressed native vegetation. In Turner NC, Kramer PJ (eds) *Adaptation of Plants to Water and High Temperature Stress*. John Wiley & Sons, pp 57–73

Kummerow J (1981) Structure of roots and root systems. In di Castri F, Goodall

DW, Specht RL (eds) *Mediterranean-Type Shrublands. Ecosystems of the World*, Vol 11. Elsevier Scientific Publishing, Amsterdam, pp 269–288

Kummerow J, Ellis BA (1989) Structure and function in chaparral shrubs. In Keeley SC (ed) *The California chaparral: Paradigms reexamined*. Sciences series No. 34. Nat Hist Mus LA, pp 141–150

Kummerow J, Mangan R (1981) Root systems in *Quercus dumosa* Nutt. dominated chaparral in southern California. Acta Oecol/Oecol Plant 2:177–188

Kummerow J, Wright CD (1988) Root distribution and resource availability in mixed chaparral of southern California. In Proc 5th Int Conf Medi Ecosys, pp 225–259 (Int U Bio Sci: Paris)

Kummerow J, Krause D, Jow W (1977) Root systems of chaparral shrubs. Oecologia 29:163–177

Lacey CJ (1983) Development of large plate-like lignotubers in *Eucalyptus botryoides* Sm. in relation to environmental factors. Aust J Bot 31:105–118

Lamont BB (1983) Strategies for maximizing nutrient uptake in two Mediterranean ecosystems of low nutrient status. In Kruger F, Mitchell DT, Jarvis JUM (eds) *Mediterranean-Type Ecosystems. The Role of Nutrients*. Springer-Verlag, Berlin, pp 246–273

Lewis DC, Burgy RH (1964) The relationship between oak tree roots and groundwater in fractured rock as determined by tritium tracing. J Geophys Res 69:2579–2588

Lopez EN (1983) Contribution of stored nutrients to post-fire regeneration of *Quercus dumosa*. M.S. thesis, California State U, Los Angeles, CA

Low AB (1983) Phytomass and major nutrient pools in an 11-year post-fire coastal fynbos community. S Afr J Bot 2:98–104

Mahall BE, Callaway RM (1992) Root communication mechanisms and intracommunity distributions of two Mojave desert shrubs. Ecology 73:2145–2151

Malanson, GP, Westman WE (1985) Postfire succession in California coastal sage scrub: the role of continual basal sprouting. Am Mid Nat 113:309–317

Matsuda K, McBride JR (1986) Difference in seedling growth morphology as a factor in the distribution of three oaks in central California. Madroño 33:207–216

McMaster GS, Jow WM, Kummerow J (1982) Response of *Adenostoma fasciculatum* and *Ceanothus greggii* chaparral to nutrient additions. J Ecol 70:745–756

Mesléard F, Lepart J (1989) Continuous basal sprouting from a lignotuber: *Arbutus unedo* L. and *Erica arborea* L., as woody Mediterranean examples. Oecologia 80:127–131

Miller PC, Poole DK (1979) Patterns of water use by shrubs in southern California. For Sci 25:84–98

Miller PC, Ng E (1977) Root:shoot biomass ratios in shrubs in southern California and central Chile. Madroño 24:215–223

Molinas ML, Verdaguer D (1993a) Lignotuber ontogeny in the cork-oak (*Quercus suber*; Fagaceae). I. Late embryo. Am J Bot 80:172–181

Molinas ML, Verdaguer D (1993b) Lignotuber ontogeny in the cork-oak (*Quercus suber*; Fagaceae). II. Germination and young seedling. Am J Bot 80:182–191

Montenegro G, Avila G, Schatte P (1983) Presence and development of lignotubers in shrubs of the Chilean matorral. Can J Bot 61:1804–1808

Mooney HA (1989) Chaparral physiological ecology—paradigms revisited. In Keeley SC (ed) *The California Chaparral. Paradigms Reexamined*. Sci Ser No. 34. Nat Hist Mus LA County, Los Angeles, CA pp 85–90

Mooney HA, Chu C (1974) Seasonal carbon allocation in *Heteromeles arbutifolia*, a California evergreen shrub. Oecologia 14:295–306

Mooney HA, Dunn EL (1970) Convergent evolution of Mediterranean-climate evergreen sclerophyll shrubs. Evolution 24:292–303

Mooney HA, Parsons DV, Kummerow J (1974) Plant development in mediterranean climates. In Lieth H (ed) *Phenology and Seasonality Modeling*. Springer-Verlag, Berlin, pp 255–267

Morrow PA, Mooney HA (1974) Drought adaptations in two California evergreen sclerophylls. Oecologia 15:205–222

Mullette KJ (1978) Studies of the lignotubers of *Eucalyptus gummifera* (Gaertn. & Hochr.). I. The nature of the lignotuber. Aust J Bot 26:9–13

Mullette KJ, Bamber RK (1978) Studies of the lignotubers of *Eucalyptus gummifera* (Gaertn. & Hochr.). III. Inheritance and chemical composition. Aust J Bot 26: 23–28

Noble IR (1984) Mortality of lignotuberous seedlings of *Eucalyptus* species after an intense fire in montane forest. Aust J Ecol 9:47–50

Nobs ME (1963) Experimental studies on species relationships in *Ceanothus*. Carnegie Inst Wash Pub No 623

O'Brien TP (1989) The impact of severe frost. In Noble JC, Bradstock RA (eds) *Mediterranean Landscapes in Australia*. CSIRO, Australia, pp 181–188

Oppenheimer HR (1956) Pénétration active des racines de buissons méditerranéens dans les roches calcaires. Bull Res Counc Israel. D Bot. 5:219–222

Oppenheimer HR (1960) Adaptation to drought: xerophytism. Arid Zone Res XV: 105–138

Orshan G, Montenegro G, Avila G, Aljaro ME, Walckowiak A, Mujica AM (1984) Plant growth forms of Chilean matorral. A monocharacter growth form analysis along an altitudinal transect from sea level to 2000 M.A.S.L. Bull Soc Bot Fr 131, Actual Bot, (2/3/4): 411–425

Orshansky G (1951) Ecological studies on lithophytes. Palest J Bot Jerusalem 5:119–128

Pearson RW (1974) Significance of rooting pattern to crop production and some problems of root research. In Carson WE (ed) *The Plant Root and Its Environment*. U Press, Charlottesville, pp 247–270

Plumb KA, Gosting VA (1973) Origin of Australian bauxite deposits. Bureau of Mineral Resources, Geology and Geophysics. Aust Dept Minerals Energy. Record 1973/156.

Poole DK, Miller PC (1975) Water relations of selected species of chaparral and coastal sage communities. Ecology 56:1118–1128

Poole DK, Miller PC (1978) Water related characteristics of some evergreen sclerophyll shrubs in central Chile. Oecol Plant 13:289–299

Pryor LD (1976) *The Biology of Eucalypts*. Edward Arnold, London

Richards JH (1986) Root form and depth distribution in several biomes. In Carlisle D, Berry WL, Kaplan IR, Watterson JR (eds) *Mineral Exploration: Biological Systems and Organic Matter*. Prentice-Hall, Englewood Cliffs, NJ, pp 83–97

Riveros F, Hoffmann A, Avila G, Aljaro ME, Araya S, Hoffmann AE, Montenegro G (1976) Comparative morphological and ecophysiological aspects of two sclerophyllous Chilean shrubs. Flora 165:223–234

Rowe PB, Colman EA (1951) Disposition of rainfall in two mountain areas of California. USDA Tech Bull 1048, 84 pp

Rundel PW, Baker GA, Parsons DJ, Stohlgren TJ (1987) Postfire demography of resprouting and seedling establishment by *Adenostoma fasciculatum* in the California chaparral. In Tenhunen JD, Catarino FM, Lange OL, Oechel WC (eds) *Plant Response to Stress*. Springer-Verlag, Berlin, pp 575–596

Saunier RE, Wagle RF (1967) Factors affecting the distribution of shrub live oak (*Quercus turbinella* Greene). Ecology 48:35–41

Specht RL (1963) Dark Island heath (Ninety-Mile Plain, South Australia). VII. The effect of fertilizers on composition and growth. 1950–1960. Aust J Bot 11:67–94

Specht RL (1979) The sclerophyllous (heath) vegetation of Australia: the eastern and central states. In: Specht RL (ed) *Ecosystems of the World. Heathlands and Related Shrublands*. Elsevier, Amsterdam, pp 125–210

Specht RL (1981a) Mallee ecosystems in southern Australia. In di Castri F, Goodall DW, Specht RL (eds) *Mediterranean-Type Shrublands*. Elsevier, Amsterdam, pp 203–231

Specht RL (1981b) The water relations of heathlands: morphological adaptations to drought. In Specht RL (ed) *Heathlands and Related Shrublands: Analytical Studies*. Elsevier Scientific, Amsterdam, pp 123–129

Specht RL, Rayson P (1957) Dark Island Heath (Ninety-Mile Plain, South Australia). III. The root systems. Aust J Bot 5:103–114

Stock, WD, van der Heyden F, Lewis OAM (1992) Plant structure and function. In Cowling R (ed) *The Ecology of Fynbos*. Oxford U Press, Cape Town. pp 226–240

van Noort AC (1960) The development of jarrah regeneration. For Dep W.A. Bull. 65:3–12

Wellington AB, Polach HA, Noble IR (1979) Radiocarbon dating of lignotubers from mallee forms of *Eucalyptus*. Search 10:282–283

Wright AD (1928) An ecological study of *Baccharis pilularis*. M.S. thesis, U California, Berkeley

Wright CD (1987) The relief of the soil-rock interface and its effect on plant cover in southern California chaparral. M.S. thesis. San Diego State U, San Diego, CA

Zammit C (1988) Dynamics of resprouting in the lignotuberous shrub *Banksia oblongifolia*. Aust J Ecol 13:311–320

Zimmer WJ, Grose RJ (1958) Root systems and root/shoot ratios of seedlings of some Victorian eucalypts. Aust For 22:13–18

9. Mineral Nutrient Relations in Mediterranean Regions of California, Chile, and Australia

Byron B. Lamont

Given similar climates, levels of soil nutrients may be expected to control the structure of vegetation. Lack of convergence in the structure and function of mediterranean ecosystems in various continents has been attributed to differences in their soil fertility status in previous studies (Miller et al. 1977, Naveh and Whittaker 1979, Cowling and Campbell 1980, Milewski 1982, Lamont et al. 1985). The general impression is that the order of nutrient availability runs Chile = Mediterranean Basin > California > South Africa > Australia. The scientific "folklore" also states that North American ecosystems are N-limited and Australian ecosystems are P-limited (e.g., Specht 1963, Gray and Schlesinger 1983). My purpose in this chapter is to test these hypotheses in the context of the mediterranean regions of California, Chile, and Australia by examining recently collected nutrient data on soils and plants. This analysis is supplemented by ecomorphological attributes of plants considered to reflect the nutrient status of the soil.

California Versus Chile

Few studies have compared these two regions directly. An important exception is the work summarized in Table 9.1. Climatically matched sites with a similar history of disturbance showed significant differences in structure and function of vegetation. Although the parent materials should have varied

Table 9.1. Site descriptions and concentration of mineral nutrients in the upper 0.3 m of soil (μg g^{-1}) and shoots (% dry mass) of four species at Echo Valley, California and seven species at Fundo Santa Laura, Chile

		Echo Valley California	Fundo Santa Laura Chile
Annual rainfall (mm)		475	550
Parent material		quartz diorite	andesite
Soil pH		6.8	6.3
Available NO$_3$		4.7	2.6
Available P		1.6	5.2
Available K		0.8	0.8
Available Ca		29.7	30.5
Available Mg		3.6	5.6
20-year recovery from		fire	woodcutting
Shrub mass (g m^{-2})		3450	1696
Annuals		rare	abundant
Vegetative growth		spring	all year
N (foliage)	x^-	0.98	1.46
	range	0.81–1.32	0.84–2.73
P	x^-	0.13	0.19
	range	0.11–0.14	0.15–0.26
K	x^-	0.55	0.89
	range	0.44–0.75	0.60–1.72
Ca	x^-	0.94	1.92
	range	0.73–1.13	1.06–3.65
Mg	x^-	0.20	0.36
	range	0.16–0.23	0.26–0.64
ΔN (total plant)	x^-	0.02	−0.10
	range	(−0.01–0.06)	(−0.43–0.08)
ΔP	x^-	0.05	−0.01
	range	(0.04–0.07)	(−0.02–0.03)

Δ refers to change in total plant concentration from late autumn to spring. Collated from Miller et al. (1977) and Shaver (1983).

little in P content (Norrish and Rosser 1983), available P was much greater at Fundo Santa Laura, Chile than at Echo Valley, California. The major cations were slightly higher at Santa Laura, but NO$_3$ (levels of which are very season- and treatment-dependent) was 80% higher at Echo Valley. The mean levels of N, P, K, Ca, and Mg in foliage of the dominant perennials were 46% to 100% higher at Santa Laura, with no overlap in levels per species for P or Mg between the two sites. The abundance of annuals at Santa Laura is further evidence of the much higher surface levels of limiting nutrients there, while I suggest that the halving in specific mass of perennials at Santa Laura may be caused by the much shallower soils there (Hoffmann and Kummerow 1978). The change in total plant levels of nutrients through the year is a crude index of the timing of nutrient uptake. The data gathered by Shaver (1983) showed that, contrary to expectations, little or no uptake of

N or P occurred over winter and spring, although it was at least consistently positive for P at Echo Valley. The rise in levels of P in the Californian plants over summer (not studied in Chile) indicates that nutrient uptake from the subsoil may be more important as a nutrient supply than it is usually thought to be.

California Versus Southwestern Australia

In 1987, I took three soils from southwestern Australia to California and compared their attributes with six soils from Jasper Ridge Preserve, 40 km south of San Francisco. Despite many inadequacies in matching of sites and number of soil types examined, this appears to be the first time that intercontinental soil–plant data have been gathered under the same experimental conditions. The parent material included sandstone (three sites), greenstone (two), and serpentine (one) in California and sandstone (one), laterite (one), and gneissic granite (one) in Australia. Bulk density ranged from 1.03 to 1.25 g mL^{-1} in California and 1.30 to 1.47 g mL^{-1} in Australia; pH ranged from 5.50 to 6.38, and 4.35 to 5.78 respectively; humus content ranged from 22 to 53 g L^{-1}, and 9 to 36 g L^{-1} respectively. Total N in the Californian soils was 2.1 to 11.3 times that in the Australian soils (Table 9.2a); total P was 2.7 to 33.6 times; total K was 0.6 to 47.5 times; total Ca was <3.8 to 238.7 times; and total Mg was 10.0 to 1611.3 times. The data support earlier propositions that mediterranean Australian soils are much less fertile than equivalent soils in California (Cowling and Campbell 1980; Lindsay 1985). As expected, P is relatively scarcer in the Australian soils than N. Surprisingly, the difference is even greater for the cations, especially Ca and Mg, for soils that are clearly nonalkaline.

The analysis was developed further by relativizing each nutrient in respect to the others in pairwise comparisons with all other soils. For example, the relative nutrient content index for N (X_N) between two soils i and j was given by: $X_N = N_i N_j^{-1} (P_j P_i^{-1} + K_j K_i^{-1} + Ca_j Ca_i^{-1} + Mg_j Mg_i^{-1})$. The mean of soil i with all other soils was calculated and compared with the means for all other soils, after normalizing by ln x, with the Student-Newman-Keuls test (Table 9.2a). This time N was highest in two Australian soils and lowest in three Californian soils when relativized against the other four nutrients. P was relatively highest in Walyunga clay-loam, and it was lowest in five of the six Jasper Ridge soils. Relative K was also highest in Walyunga soil and lowest in Obispo clay (serpentine). Relative Ca was only significantly lower in Walyunga soil than in the other eight soils. Mg was relatively 2 to 3 orders of magnitude greater in Obispo clay than the others, with two Australian soils clearly lowest. Within soils, Mg was relatively highest and K lowest in Obispo clay, supporting previous trends in serpentine soils (Proctor and Whitten 1975). Ca and Mg were relatively higher than the other major nutrients in the grassland on Gilroy soil. Mg was relatively higher than the other

Table 9.2. Levels of (a) total and (b) available major nutrients ($\mu g\ mL^{-1}$) in the California (A–F) and southwest Australian (G–I) soils used in the glasshouse experiments. Available N is for NH_4 only. Results are means for three subsamples. Values with different letters are different at $P < 0.05$ for the SNK multiple t-test. Letters under values (r/c) are in order of significant differences (i.e., a largest, b next) for the *relative* values, where r is for comparisons across rows (different nutrients for that soil) and c is for columns (different soils for that nutrient)

a. *Total nutrient concentration* ($\mu g\ mL^{-1}$ soil)

Soil	Local name	N	P	K	Ca	Mg
Californian Soils						
A	Francisquito loam	1963^b	653^c	2864^c	12482^b	$4628^{c(d)}$
	(conifer forest)	a/c	a/c	a/bc	a/a	a/bc(b)
B	Stonyford complex	1634^d	626^d	2678^d	11635^c	$6185^{c(c)}$
	(chaparral)	a/c	a/c	a/bc	a/a	a/bc(b)
C	Obispo clay	1579^d	528^e	578^g	2635^f	$74120^{a(a)}$
	(serpentine)	b/bc	b/c	c/c	b/a	a/a(−)
D	Gilroy complex	1626^d	518^e	1734^e	18377^a	$11038^{b(b)}$
	(grassland)	b/c	b/c	b/bc	a/a	a/b(a)
E	Los Gatos loam	3566^a	1211^a	3401^b	6556^d	$2532^{d(e)}$
	(mixed forest)	a/bc	a/bc	a/bc	a/a	a/bc(b)
F	Supan loam	1823^c	691^b	3609^a	3060^e	$4188^{cd(d)}$
	(oak woodland)	a/bc	a/c	a/b	a/a	a/bc(b)
Australian Soils						
G	Eneabba sand	315^g	36^g	76^i	$>414^h$	$46^{e(h)}$
	(kwongan)	a/a	a/bc	a/bc	a/a	b/c(c)
H	Watheroo loam	425^f	181^f	215^h	$>685^g$	$89^{e(g)}$
	(eucalypt woodland)	a/ab	a/ab	a/bc	a/a	b/c(c)
I	Walyunga clay	742^e	193^f	932^f	$>77^i$	$254^{e(f)}$
	(eucalypt woodland)	a/a	a/a	a/a	c/b	b/bc(b)

b. *Available nutrient concentration* ($\mu g\ mL^{-1}$ soil)

Soil	Local name	N	P	K	Ca	Mg
Californian Soils						
A	Francisquito loam	7.6^b	5.9^c	164^b	2372^e	286^d
		b/bc	a/bc	a/ab	a/bc	ab/bcd
B	Stonyford complex	7.4^b	0.1^f	119^d	4162^b	684^b
		b/ab	c/d	ab/a	a/a	a/a
C	Obispo clay	8.7^b	1.4^e	51^f	492^g	1577^a
		b/bc	b/bc	b/b	b/cd	a/a
D	Gilroy complex	7.7^b	1.3^e	95^e	2922^c	653^{bc}
		b/bc	ab/c	a/b	a/b	a/ab
E	Los Gatos loam	20.6^a	26.6^a	299^a	4464^a	249^e
		bc/bc	a/ab	ab/ab	ab/bc	c/d
F	Supan loam	5.7^c	3.0^d	141^c	2621^d	638^c
		b/c	a/bc	a/ab	a/b	a/abc
Australian Soils						
G	Eneabba sand	3.6^d	1.0^e	15^g	414^h	44^f
		a/a	b/bc	b/bc	b/bc	b/cd
H	Watheroo loam	3.3^d	16.6^b	14^g	685^f	49^f
		b/a	a/a	d/c	c/bc	d/d
I	Walyunga clay	4.9^c	0.3^f	16^g	77^i	53^f
		a/a	bc/bc	b/b	c/d	bc/bcd

Upper letters in parentheses based on ln x, lower letters when soil C omitted from analysis.

nutrients in the two sandy Australian soils and, in the third, relative Ca was lowest and N, P, and K were highest.

The disparity between regions was less marked for absolute levels of available nutrients (Table 9.2b). Available N is given as NH_4 because NO_3 values were erroneously high in the Australian soils because they were collected in winter and allowed to incubate before analysis, while the Californian soils were collected at the same time (in summer) and remained dry before analysis. Available NH_4 was 1.2 to 6.2 times higher in the Californian soils. Extractable P (with NH_4F-HCl) was 166 times greater in Watheroo loam than the chaparral Stonyford soil to 88.7 greater in Los Gatos alluvium than the Walyunga soil. Available K was 3.2 to 21.4 times higher in the Californian soils; available Ca was 0.7 to 58.0 times higher; and available Mg was 4.7 to 35.8 times higher. Except for the anomaly of the reversal in the two soils above for P, available nutrients thus followed the trends for total nutrients, except that the values were less extreme.

When relativized, available NH_4 was highest in the three Australian soils and lowest in the Californian Supan loam, which supported deciduous oak woodland (Table 9.2b). Relative exchangeable P was highest, by far, in Watheroo loam, which supported eucalypt woodland, and lowest, by far, in the chaparral soil. In contrast, relative K, Ca, and Mg were highest in the chaparral soil and they were lowest in one of the Australian soils. Within soils, highlights were the low levels of relative N in Francisquito loam, Gilroy clay, and Supan loam (all Californian soils), and high levels of relative N in Eneabba sand and Walyunga clay-loam (both Australian soils). Relative P was lowest in chaparral soil and highest in the Los Gatos and Watheroo loams. K was relatively low only in Watheroo loam, and Ca in Walyunga clay-loam. Relative Mg was low in the Los Gatos and Watheroo loams and highest in the serpentine clay.

To obtain direct information on whether N, P, or K were most limiting in these soils, I conducted a N × P × K interaction pot trial. Five 250-mL cone-shaped plastic pots were filled with each soil. Four germinants of wild oats (*Avena fatua* L.), a widespread weed in California and southwestern Australia (Bartolome et al. 1986, Hobbs and Atkins 1988), were sown in each container. Seeds were collected from near the Gilroy (D) soil site. Nutrient solution was applied twice a week in a complete three-way interaction experiment for each of the nine soils with two levels (treatment plus control) of each nutrient. N was applied as $NaNO_3$ at 2.87 μg N mL^{-1} soil per week, P as $NaH_2PO_4 \cdot H_2O$ at 2.72 μg P mL^{-1} and K as KCl at 36.4 μg K mL^{-1}. Levels were set at the middle-to-upper concentrations recommended for hydroponic culture (Handreck and Black 1984) and proved conservative compared with endogenous available N, and high for P and K (Table 9.2b). Calculations showed that only K would have reached levels exceeding total amounts likely to be absorbed from the unamended soils by the end of the trial. An additional treatment used an equivalent level of NaCl (to confirm that any effect was not due to Na or Cl), but because the results were not different from the controls they are not reported here.

Table 9.3. Shoot mass (mg) per pot of 10-week-old *Avena fatua* grown in California (A–F) and southwest Australian (G–I) soils in a three-way interaction experiment with $\pm N$, $\pm P$, and $\pm K$: Results are means for 5 replicates of 4 bulked plants per pot. Analysis of variance based on $\sqrt{x}$ (squares given here)

Californian Soils

A Francisquito loam	$-N$	$+N$		
$-P - K$	270	367	N***	$(+)$
$-P + K$	257	303	P*	$(+)$
$+P - K$	290	406	K***	$(-)$
$+P + K$	265	337	NK*	$(N+, K-)$
B Stonyford complex				
$-P - K$	126	150	P***	$(+)$
$-P + K$	164	122	PK*	$(P+, K^{0}_{-})$
$+P - K$	204	176	NPK**	$(N\pm, P+, K\pm)$
$+P + K$	163	154		
C Obispo clay				
$-P - K$	183	220	NK*	$(N^{0}_{-}, K\pm)$
$-P + K$	228	192		
$+P - K$	223	204		
$+P + K$	230	195		
D Gilroy complex				
$-P - K$	89	168	N***	$(+)$
$-P + K$	104	134	P**	$(+)$
$+P - K$	90	199	NP***	$(N+, P^{0})$
$+P + K$	99	180	NK***	$(N+, K^{0}_{-})$
E Los Gatos loam				
$-P - K$	146	266	N***	$(+)$
$-P + K$	118	229	P*	$(-)$
$+P - K$	113	256	K**	$(-)$
$+P + K$	110	213		
F Supan loam				
$-P - K$	173	154	P***	$(+)$
$-P + K$	136	190	NPK***	$(N\pm, P+, K\pm)$
$+P - K$	232	266		
$+P + K$	250	220		

Australian Soils

G Eneabba sand			N***	$(+)$
			P***	$(+)$
$-P - K$	105	104	K**	$(+)$
$-P + K$	123	111	NP***	$(N+, P+)$
$+P - K$	104	149	NK*	$(N+, K+)$
$+P + K$	92	213	NPK**	$(N+, P+, K+)$
H Watheroo loam				
$-P - K$	163	220	N***	$(+)$
$-P + K$	188	242	K*	$(+)$
$+P - K$	194	202	NP**	$(N+, P\pm)$
$+P + K$	198	215		
I Walyunga clay				
$-P - K$	94	76	N***	$(+)$
$-P + K$	84	95	P***	$(+)$
$+P - K$	163	294	NP***	$(N^{0}, P+)$
$+P + K$	135	295	NK*	$(N+, K\pm)$

*** $P < .001$, ** $P < .01$, * $P < .05$, $(+)$ positive response to the treatment, $(-)$ negative response, (0) no response to some treatments in interaction

The pots were also watered with deionized water to prevent water stress at other times. Although the cones has drainage holes they were never watered to excess. The plants were grown in an unshaded glasshouse over autumn (October to mid-December) 1987 at Stanford University, Palo Alto. Shoots were harvested and bulked per container, dried at 70°C for 2 days, and weighed. Shoot masses were normalized by taking ln x and the results submitted to three-way analysis of variance. Absolute values between soils could not be compared because the soils were sown over five days using the most recent germinants (i.e., only treatments within soils were in phase) and their position was randomized (twice) on the bench within but not between soil types.

Table 9.3 shows that oats in half the Californian and all the Australian soils responded strongly to the addition of $NaNO_3$ (N) alone, including those in Los Gatos alluvium with by far the highest total and available N. Four Californian soils responded positively to NaH_2PO_4 (P), especially the chaparral soil (with the lowest available P) and Supan loam, and two Australian soils, especially the Walyunga clay. The response in Los Gatos loam, with by far the highest total and available P, was slightly negative to P amendment, but plants in Watheroo loam, the next highest available P and relatively highest of any soil, showed no response. Adding KCl (K) suppressed growth in the two Californian soils with highest available K but it increased growth somewhat in two of the three Australian soils with lowest available K. There was a strong positive N × P interaction for the Gilroy and, especially, Eneabba and Walyunga soils. The other Australian soil showed a slight drop with N + P in contrast to their individual effects. N × K and P × K interaction effects in three Californian soils involved positive responses to N or P and nil or negative responses to K.

The Obispo clay showed almost no response to the treatments, except for increased growth with added K in the absence of added N. These responses are consistent with the low relative total soil K (Table 9.2a) and probable overriding toxic levels of Mg and heavy metals in serpentine soils (Proctor and Whitten 1975). Eneabba sand showed a positive N × K interaction, and Walyunga clay gave a slight decrease in growth in the presence of extra K but not N, consistent with its apparently higher relative total K content. The N × P × K interactions in the two Californian soils involved positive responses to P and variable responses to N and K. The extreme infertility of the Eneabba sand is indicated by maximum growth by far occurring only in the presence of additional N, P, *and* K.

Finally, a small-seeded species from each region was grown in all nine soils. *Eucalyptus loxophleba* Benth. (Myrtaceae), the dominant tree on Watheroo loam, was selected from southwestern Australia because earlier studies had shown it to be very sensitive to nutrient conditions (Lamont and Kelly 1988). *Baccharis pilularis* ssp. *consanguinea* (DC) C.B. Wolf (Asteraceae), native to four of the Jasper Ridge soils used, was selected because it is a fast-growing shrub very sensitive to growing conditions (Williams et al.

Table 9.4. Shoot mass (mg) and total mineral nutrients (mg) in 10-week-old *Eucalyptus loxophleba* (from Watheroo loam) and *Baccharis pilularis* (from Francisquito loam) grown in Californian (A–F) and southwest Australian (G–I) soils: Results are means for 10 plants from five pots. Values with different letters are significantly different (SNK multiple *t*-test) based on $\sqrt{x}$ (with squares given here) for mass and ln x (antilogs here) for nutrients. Letters below the values (r/c) are in order of significant differences (i.e., *a* largest, *b* next) for the relative values, where *r* is for comparisons across rows (different nutrients for the soil) and *c* is for columns (different soils for that nutrient)

Soil	Local name	Shoot mass	N	P	K	Ca	Mg
a.	***Eucalyptus loxophleba***						
A	Francisquito loam	64.1^b	1.58^a	0.68^a	1.35^{ab}	0.57^b	0.11^a
			b/b	a/a	a/ab	a/bc	b/cd
B	Stonyford complex	8.6^d	0.42^c	0.04^f	0.08^e	0.07^d	0.02^d
			a/a	c/c	c/cd	ab/b	b/bc
C	Obispo clay	16.3^d	0.65^b	0.08^{de}	0.23^d	0.08^d	0.06^{bc}
			a/a	b/c	b/bc	b/d	a/a
D	Gilroy complex	12.8^d	0.50^{bc}	0.14^{cd}	0.54^c	0.16^d	0.04^c
			b/b	b/bc	a/a	b/bc	b/bc
E	Los Gatos loam	78.1^a	1.58^a	0.85^a	1.89^a	1.23^a	0.12^a
			b/b	a/a	a/ab	a/a	b/d
F	Supan loam	42.2^c	1.60^a	0.22^{bc}	1.06^b	0.27^c	0.11^a
			ab/a	c/c	a/ab	bc/cd	ab/bc
G	Eneabba sand	5.0^d	0.23^c	0.04^f	0.05^e	0.04^e	0.01^d
			a/a	a/a	b/d	a/bc	a/bc
H	Watheroo loam	59.5^b	1.02^b	0.38^b	0.92^b	0.51^b	0.08^{ab}
			b/b	a/ab	a/ab	a/bc	ab/cd
I	Walyunga clay	5.3^d	0.30^c	0.05^{ef}	0.06^e	0.03^e	0.02^d
			a/a	a/ab	b/cd	b/d	a/ab
b.	***Baccharis pilularis***						
A	Francisquito loam	48.7^b	1.79^a	0.37^a	2.39^b	0.68^a	0.10^a
			a/ab	a/a	a/bc	a/b	a/b
B	Stonyford complex	8.8^c	0.31^c	0.13^b	0.60^d	0.09^b	0.02^c
			ab/abc	a/a	ab/cd	b/b	ab/b
C	Obispo clay	4.6^c	0.18^c	0.01^c	0.23^e	0.02^c	0.03^b
			b/a	d/c	b/abc	c/b	a/a
D	Gilroy complex	14.2^c	0.32^c	0.07^b	0.87^d	0.11^b	0.03^b
			a/abc	a/ab	a/bc	a/b	a/b
E	Los Gatos loam	81.7^a	1.55^{ab}	0.65^a	7.50^a	0.58^a	0.04^b
			ab/ab	a/a	a/a	bc/b	c/b
F	Supan loam	47.5^b	1.44^{ab}	0.39^a	5.30^{ab}	0.54^a	0.04^b
			a/ab	a/ab	a/ab	ab/b	b/b
G	Eneabba sand	0.5^d	$<0.06^d$	$<0.01^d$	0.03^f	0.02^c	$<0.01^d$
			bc/bc	ab/ab	c/d	a/ab	bc/b
H	Watheroo loam	44.5^b	0.91^b	0.50^a	1.72^c	0.76^a	0.03^b
			bc/abc	a/a	bc/cd	ab/a	c/b
I	Walyunga clay	0.8^d	$<0.02^d$	$<0.01^d$	0.05^f	0.03^c	$<0.01^d$
			bc/c	c/bc	bc/cd	a/a	b/b

Most plants died before termination of experiment.

1987). The trial was modeled on the wild oat trial and conducted concurrently. One-week-old seedings were planted two per cone with five replicates. After ten weeks, the plants were harvested and the dry shoots analyzed for major nutrients. Inspecting the root systems indicated time had been insufficient for mycorrhizas to develop.

Both species proved much more sensitive to the soil types than *Avenua fatua*, especially *Baccharis pilularis*, which showed a threefold range in shoot mass (Table 9.4). Taken together, shoot growth was greatest by far in the Los Gatos loam, and least by far in Eneabba sand and Walyunga clay. The other Australian soil, Watheroo loam, however, yielded as much as several Californian soils (Francisquito loam, Supan loam). Total shoot N in these three Californian soils tended to be higher than in the Watheroo loam, though those in the other Californian soils tended to be lower. N uptake tended to be lowest in the other two Australian soils. Total P uptake was as high in the Watheroo as the three Californian soils above for *Baccharis* but not *Eucalyptus*. Plant P content was particularly low in the two Australian soils and Obispo clay, and also the chaparral Stonyford soil for the eucalypt.

As a semisucculent species, *Baccharis* had higher levels of K than the eucalypt in all soils (Table 9.4). Apart from very low uptake in the Stonyford soil for the eucalypt, levels of K were much lower for two of the Australian soils than for the Californian soils. The Watheroo loam provided as much K as three or four Californian soils. Apart from plants in Stonyford and Obispo soils, Ca content of the shoots was much lower in the Eneabba and Walyunga soils. The other Australian soil provided as much Ca as all Californian soils, except for the eucalypt in Los Gatos loam. Mg content tended to be higher in the eucalypt than in *Baccharis*. Except for the eucalypt in Stonyford soil, plants in two Australian soils had lower Mg content, especially *Baccharis*. Only *Baccharis* in Francisquito loam had more Mg than did plants in Watheroo soil.

When shoot N concentrations were relativized, the Obispo clay plants consistently had the highest values. For relative P content, the Francisco and Los Gatos loam plants were consistently highest, and the Obispo clay plants were lowest. For relative K, Eneabba sand plants were consistently lowest, closely followed by those in the Stonyford and Walyunga soils. Relative K in Gilroy soil was highest for the eucalypt and Los Gatos for *Baccharis*. Relative Ca for the eucalypt was lowest in Obispo and Walyunga clays and highest in Los Gatos loam. For *Baccharis*, by contrast, relative Ca was higher in Walyunga and Watheroo soils than the others. As expected, relative Mg was highest by far in the Obispo (serpentine) soil, with the eucalypt having relatively lowest uptake in the Los Gatos loam.

Between nutrients for the same soil, relative Mg was again highest for plants in the serpentine soil, with relative P lowest. Mg was lowest in the Los Gatos loam plants. Relative P was highest in Watheroo loam; K was lowest in Eneabba sand. No other consistent trends appeared among both species. Interesting trends for the eucalypt only were the relatively high N and low P

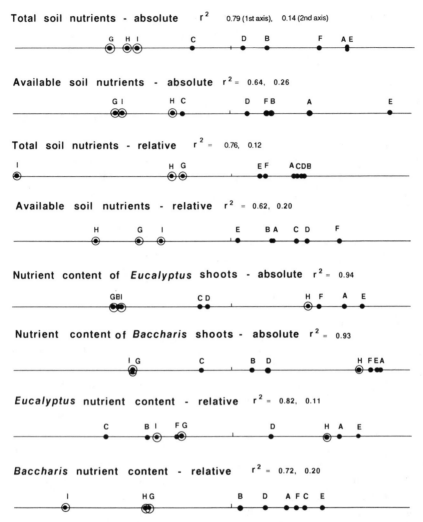

Figure 9.1. One-dimensional (nutrient gradient) principal coordinates analysis of the significance of differences in data in Tables 9.2 and 9.4. Attribute values with successive letters of difference by SNK test were given successive numbers, except where values had two letters when they were given intermediate numbers. The dissimilarity index was a range-standardized, Manhattan type (Lamont 1985). Where the first dimension initially accounted for <80% of the variance, the second dimension was added by dropping coordinates orthogonally onto an axis drawn at an angle corresponding to the relative contribution of the two major dimensions to total variance. The Australian soils are circled.

and K in the chaparral soils; high N (and Mg) in Obispo clay; high K in Gilroy soil; low N and Mg in the Los Gatos loam; and high K and low P in Supan loam. For *Baccharis*, Ca was relatively low in Stonyford soil; P and K were relatively high in Los Gatos loam; Mg was lowest in Supan loam and Watheroo loam; Ca was highest and P lowest in Walyunga clay.

Figure 9.1 summarizes the significant results in Tables 9.2 and 9.4. These show that the three Australian soils were much less fertile than the six Californian soils in total levels of the five major nutrients. A similar pattern is evident for available levels, except the most fertile, Los Gatos loam, is more clearly separated from the rest (Fig. 9.1). Watheroo loam has shifted to the right because of its high extractable P. When relativized, total nutrients in the Australian soils remain separate from the rest, though the Walyunga clay-loam is further from the other two Australian soils than they are from the Californian soils, because of its markedly higher relative total N, P, and K and lower Ca. The relatively high N and P and low Mg separate the Eneabba and Watheroo soils from the rest. The separation again is clear for relative available nutrients, with Walyunga soil now closest to the Californian soils. This condition appears to be caused by the relatively high available N and low K and Mg (especially in Watheroo) for the Australian soils.

Ordination of total shoot content for the five nutrients showed a somewhat different pattern. For both *Eucalyptus loxophleba* and *Baccharis pilularis*, two Australian soils were at one end, and Watheroo loam was close to the most fertile Californian soils. In addition, the chaparral soil was placed among the two Australian soils for the eucalypt. These trends were mostly caused by differences in shoot growth in the various soils (Table 9.4) rather than by major differences in concentrations of nutrients in the shoots. Shoot growth did reflect the general fertility status, except that growth in Watheroo loam was much greater than expected, and growth in the poorest Californian soils (chaparral, serpentine, and Gilroy) was less than expected.

Because the overall relative levels of nutrients also distinguished the Australian and Californian soils, it is interesting to compare relative levels in the shoots (Fig. 9.1). The Australian soils were clearly separated from the Californian for *Baccharis*, in a pattern not unlike that for relative total nutrients in the soils. This result mostly reflected the relatively high Ca and low N and K for shoots in the Australian soils. The pattern was much less clear for the eucalypt. Two Australian soils were positioned near the poorest Californian soils, but Watheroo loam was near the most fertile. This pattern had affinities with that for the eucalypt shoot content, except for the rearrangement of the serpentine and Supan soils. Eneabba and Walyunga shared relatively high shoot levels of N and P and low K; Watheroo shared with the two highest positions relatively low N and Mg, and high P.

Comparing all absolute and relative, soil and plant values, some clear trends appear. The low K content of shoots in the Eneabba and Walyunga soils reflected low absolute levels in these soils, and relatively low K availability in Eneabba soil. Shoot K was also relatively low compared with that

in other soils and for other nutrients, especially in the Eneabba sand. Total
N was low in both soils and species, but relative soil N was high. Total Ca
content of shoots was low in the Walyunga soil only, reflecting low absolute
and relative soil levels of Ca. Relative shoot Ca was also low in the eucalypt
but exceptionally high in *Baccharis*: this reading highlights the differences
in absorptive capacities between species. Shoot P content was also low in
Walyunga clay, reflecting low soil P levels. In the third Australian soil, nutrient content of the shoots was comparatively high, especially P and Ca. This
content was associated with high absolute and relative availability of P in the
soil, but not of Ca. Although total shoot Mg was not low, its concentration
was, as reflected in the low relative level compared with other nutrients in the
shoots. This status was consistent with both relatively and absolutely low soil
Mg levels. Soil and plant N and K were low on an absolute basis, but only
available K was low on a relative basis.

Despite the high N content of shoots in Francisquito loam, relative N was
lowest among the five nutrients in the eucalypt: this level reflected relative N
availability in the soil. Relative N was also low in the shoots of the eucalypt
for Los Gatos loam, but did not reflect absolute levels in the shoots or
absolute or relative levels in the soil. Absolute and relative P in the eucalypt
was low for the chaparral soil, reflecting low absolute and relative availability of soil P. The K content of shoots in the chaparral soil was also low,
especially on a relative basis. This reading, however, was unexpected from
the soil levels. The Ca shoot content was low in the serpentine soil, especially
on a relative basis, reflecting low available soil Ca. High absolute and relative levels in the serpentine soil resulted in high relative Mg in the shoots.

All the data sets for each nutrient were combined and subjected to 2-D
principal coordinate analyses (Fig. 9.2) to determine if there were overall
differences between the Californian and Australian soils. For N, the two
poorest Australian soils were isolated from the rest, with the Watheroo loam
in an intermediate group with three Californian soils (serpentine, chaparral,
and grassland). The other three Californian soils were relatively N-rich. The
three Australian soils were at one end of the P ordination, but the anoma-
lously high P in the Watheroo loam isolated it from all other soils. The low
P levels in the chaparral, serpentine, grassland, and oak woodland soils en-
sured they were closest to the other two Australian soils. The consistently
low K in the two Australian soils explains why they were at one end of the
ordination, with the chaparral, serpentine, and Watheroo soils showing similar low K levels for different attributes. The N-rich Californian soils were
also K-rich.

Ca is clearly lowest for the Walyunga clay, followed by Eneabba sand and
serpentine clay (Fig. 9.2). Watheroo loam is in an intermediate position with
Supan loam. The other Californian soils tend to be high for plant (Los Gatos
forest) or soil (chaparral, grassland) attributes. The extremely high (and
probably toxic) levels of Mg in the serpentine clay isolate it from all other

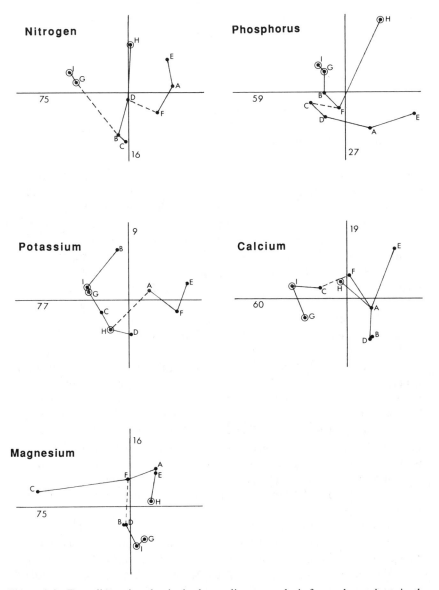

Figure 9.2. Two-dimensional principal coordinates analysis for each nutrient, in the nine soils. Analyses based on total and concentration of that nutrient in shoots of *Eucalyptus loxophleba* and *Baccharis pilularis*, relative shoot contents, total and available soil levels, and relative soil levels. The southwest Australian soils (G, H, I) are circled. Numbers refer to the % variance accounted for by each axis. The minimum spanning tree connects the closest pairs in the dissimilarity matrix, with broken lines representing the second or third closest pair required to complete the tree.

soils. The Australian soils are at the other end, the Watheroo loam showing more affinities with the two most fertile Californian soils, which themselves have some low Mg values.

California Versus Chile Versus Australia

Unlike the pairwise studies described above, no comparable studies have been done for the soils in all three regions. Comprehensive analyses, however, have been undertaken recently on the nutrient contents of leaves for reasonably representative species in mediterranean regions (Rundel 1988). Because mediterranean climates cover two distinct parts of Australia, separated by the Nullarbor Plain, it is appropriate to treat them separately, for it is believed that the southwest (Western Australia) is less fertile than the southeast (South Australia, New South Wales, Victoria) (Westman 1983, Lindsay 1985). To the extent that foliage levels reflect relative soil nutrient status, Table 9.5 shows clear differences among the four regions.

Concentration of N is almost 50% higher in the Chilean species on average than the Californian and southeastern Australian species and more than 100% higher than the southwest Australian species (Table 9.5). P concentration in the Chilean species is more than 50% higher than the California species, 2.83 times that in southeastern Australia and 5.67 times in southwestern Australia. P levels in all Chilean or Californian species exceeded those in the southwestern Australian species. The pattern for K is similar to that for N but less extreme, Chile exceeding southwestern Australia by only 40%. Although Ca concentration is also highest in Chile, concentrations vary greatly in all regions. Surprisingly, it is not significantly higher than in southwestern Australia, but 40% higher than in California and 65% than in southeastern Australia. Mg concentration in Chile is about 80% higher than in the other regions. Unexpectedly, mean Mg shows no difference between plants in the other regions.

I consider that the mass of structural tissue (lignin plus cellulose) divided by living tissue (protein = 6.25 N) is a useful index of the metabolic efficiency of a species. It is, however a poor, if not misleading, index of sclerophylly as used in Rundel (1988): it confounds tissue density (one of two components of sclerophylly, Witkowski and Lamont 1991) with nutrient content (which may be causally related to sclerophylly but is not an index of it). In other words, a hard, thick leaf is highly sclerophyllous (a textural concept) whether its nutrient concentration is low *or* high. The selected southwestern Australian species (all from nonsodic or noncalcareous soils) have a mean metabolic efficiency 43% greater than those from southeastern Australia (which include species from sodic soils), 2.46 times the Californian plants, and 3.83 times the Chilean plants (Table 9.5).

Table 9.5. Concentration of mineral nutrients in leaves of selected species from four mediterranean regions surrounding the Pacific Basin: Results are mean ± SD (minimum–maximum) for all species given in Rundel (1988), except that the Queensland location has been omitted. Significant differences between regions per nutrient are indicated by raised letters based on multiple paired t-tests performed by B. Lamont. Metabolic efficiency is 100 (lignin + cellulose)/6.25 N

	N	P	K	Ca	Mg	Metabolic efficiency
California 13–29 spp.[o]	1.04 ± 0.37^b (0.64–2.18)	0.11 ± 0.06^b (0.05–0.23)	0.69 ± 0.54^{ab} (0.35–2.90)	0.98 ± 0.54^{bc} (0.44–2.64)	0.20 ± 0.06^b (0.10–0.29)	424 ± 134^c (235–688)
Chile 27 spp.	1.49 ± 0.39^a (1.00–2.73)	0.17 ± 0.05^a (0.10–0.26)	0.86 ± 0.31^a (0.36–1.46)	1.38 ± 0.70^a (0.46–3.65)	0.36 ± 0.21^a (0.14–1.17)	274 ± 131^d (41–691)
SE Aust. 49–57 spp.[o]	1.03 ± 0.50^b (0.38–3.28)	$0.06 \pm 0.07^{c,c*}$ (0.01–0.21)	0.65 ± 0.51^{ab} (0.17–2.70)	0.83 ± 0.38^c (0.20–1.68)	0.19 ± 0.11^b (0.04–0.48)	729 ± 428^b (60–1900)
SW Aust. 16–17 spp.[o]	0.67 ± 0.01^c (0.51–0.85)	$0.03 \pm 0.00^{c,d*}$ (0.02–0.04)	0.62 ± 0.15^b (0.35–0.97)	1.07 ± 0.45^{ab} (0.20–1.88)	0.22 ± 0.12^b (0.08–0.58)	1042 ± 303^a (558–1716)

[o] Not all attributes analyzed for the larger number of species.
* Second letter based on ln x.

Ecomorphological Characters

A comprehensive set of morphological and life cycle attributes has been
compiled recently for the dominant plants in mediterranean regions (Specht
1988). I have selected 7 of the 48 characters used, grouped the various site
data by number of species in each category per character for the three regions, separating or combining the southwestern and southeastern Australian results, and compared them for independence by contingency table analysis (Table 9.6). In equivalent climates, these characters should represent
responses to the soil's nutrient-yielding properties. Plant height confounds
the effects of various nutrient attributes, such as availability of the most
limiting nutrients at various positions in the soil profile, soil texture, and
depth, and levels of niche differentiation between the species. Most species in
Australia are woody shrubs < 1 m high, while most species in California are
shrubs and small trees, 1 to 5 m tall, perhaps reflecting poorer soils in Australia. In contrast to California, a significant number exceeded 5 m in Australia. The deep sands, which allow considerable water storage in summer
(Lamont and Bergl 1991)—quite apart from increased access to nutrients—
could sustain taller trees. Alternatively, the fast-growing eucalypts, with their
P-solubilizing ectomycorrhizas, may provide a superior genetic stock in Australia (Lamont 1982). Most plants in Chile were smaller than those in California. Despite their apparently greater nutrient concentrations, Chilean
soils may be shallower than those in California, providing less nutrients
overall (Hoffmann and Kummerow 1978). Alternatively, woodcutting of the
many resprouting species may be more effective than fires in restricting plant
height (Table 9.1) though this tactic has received no direct study.

Comparing the categories for plant height and root depth (Table 9.6)
shows interesting differences among the regions. Most Chilean plants have
similar heights and depths, but roots in Australia and California have shifted
down a category. This variation indicates that equivalent growth forms are
being supported by deeper root systems, especially in southwestern Australia and to a lesser extent in California. These more extensive root systems
may help compensate for lower nutrient availability by exploring a greater
volume of soil. This goal is better achieved, though, by extensive root development near the soil surface, where most available nutrients are located
(Lamont 1973, Low and Lamont 1990). They may also use this strategy
(Lamont and Bergl 1991) but no comparative figures are available. Again,
deep roots may be drought-driven, rather than nutrient-seeking. The deep,
coarse sands with extremely low water-holding capacity not only provide
opportunities to reach the water table (Farrington et al. 1989) but make it
essential for survival of evergreens over summer (Poole and Miller 1978,
Hart and Radosevich 1987, Lamont and Bergl 1991).

Drought avoidance as implicated above is an option only where access to
permanent soil water is possible. Alternatives rely on the presence of sufficient soil nutrients to enable frequent cycling of plant parts (deciduous

Table 9.6. Ecomorphological characters of dominant species in selected sites from mediterranean regions surrounding the Pacific Basin: Results are number of species in each category collated and modified by B. Lamont from Specht (1988). "Sig." refers to the contingency table (χ^2) analysis on the hypothesis that the distribution of character states in the three continental regions is the same

	California	Chile	SE Aust.	SW Aust.	Australia	Sig.
Plant height						
<0.25 m	3	4	26	12	38	$P < 0.001$
0.25–1	10	13	26	23	49	
1–2	13	12	17	7	24	
2–5	20	7	9	5	14	
5–20	1	5	9	3	12	
>20	0	0	5	0	5	
Total species	47	41	92	50	142	
Root depth						
<0.25 m	2	6	4	4	8	$P < 0.005$
0.25–1	3	19	26	2	28	(omitting unknowns)
1–2	2	9	21	3	24	
2–5	11	6	24	14	38	
>5	6	1	17	12	29	
Unknown	23	0	0	15	15	
Leaf-size category						
leptophyll/leafless	10	4	29	22	51	$P < 0.01$
nanophyll	14	16	22	11	33	
nanomicrophyll	15	15	21	11	32	
microphyll	3	4	8	5	13	
micromesophyll	5	0	11	1	12	
mesophyll	0	2	1	0	1	
Leaf texture						
succulent	0	1	2	0	2	$P \ll 0.001$
semisucculent	0	1	4	1	5	
malacophyll	12	22	14	2	16	
semisclerophyll	6	9	16	3	19	
sclerophyll	29	8	56	44	100	

(continued)

Canopy seasonality						
everygreen	27	22	88	48	136	$P \ll 0.001$
deciduous (summer)	14	19	4	2	6	(excluding unknowns)
unknown	6	0	0	0	0	
Plant longevity						
<1 year	0	5	3	2	5	$P < 0.001$
1–5	4	1	2	0	2	
5–50	14	25	31	26	57	
>50	29	10	56	22	78	
Specialized roots						
root clusters	0	0	13	23	36	$P \ll 0.001$
mycorrhizas	7	0	30	5	35	(including unknowns)
nodules (N_2 fixing)	7	4	12	5	17	
none/unknown	33	37	37	17	54	

leaves, stems) or whole plants (annuals, fire ephemerals) as a viable option (Gray and Schlesinger 1983). We find, among the dominants, that species with (summer) deciduous leaves are rare in Australia (4%) but common in California (34%) and Chile (46%) (Table 9.6). This finding supports earlier statements that mediterranean Australia is least fertile, but less certainly that Chile is more fertile than California. Data for plant longevity are less clear. Most dominants by far in these regions are perennial, with annuals (including exotic species in Chile) accounting for 12% of the flora in Chile, <4% in Australia and 0% in California. All regions have substantial annual floras that are most prominent on the denser, cation-rich, thin soils exposed to direct sunlight. Where woody shrubs and trees are present, the consensus appears to be that annuals usually contribute more to biomass in Chile than in California (Table 9.1) or Australia. Short-lived perennials (≤ 5 years) show some tendency to be better represented in California (Table 9.6). The proportion of long-lived (> 50 years) trees and shrubs is greater in California (62%) and Australia (55%) than in Chile (24%), which is consistent with the differences in height and root depth patterns. These results clearly are greatly influenced by the study sites selected, especially size of the introduced flora.

If we can accept that water and light available to the dominants are similar between these regions, then leaf morphology becomes a useful index of nutrient concentration, hence soil nutrient availability (Lamont and Kelly 1988, Specht and Rundel 1990). More than 80% of leaf areas are <1125 mm^2 in all regions, with microphylls-mesophylls accounting for the rest. The most notable differences are in the leafless and minute-leaved (< 25 mm^2) class, with 44% in southwestern Australia, 32% in southeastern Australia, 21% in California, and 10% in Chile. These are well correlated with the leaf P levels mentioned earlier. Most of the dominant species in Chile were soft-leaved (59%), but only 26% were in California and 16% in Australia. The reverse was true for hard leaves, with the greatest proportion of sclerophylls in southwestern Australia (94%). This figure again correlates well with the nutrient data and supports the results of Mooney (1983).

I have argued elsewhere (Lamont 1982, 1984) that "specialized" roots are a response to nutritionally poor soils and that their abundance will be greatest in the impoverished sands and laterites of southwestern Australia. Root clusters called proteoid roots enhance nutrient uptake, especially insoluble P, Fe, and Mn (Grierson and Attiwell 1989). They are particularly common in Australia, especially in the southwest, where 46% of species had proteoid roots because of the Proteaceae-dominated study sites. Root nodules enhance N uptake by fixation of atmospheric N. They contribute 10% to 15% of dominants in all four regions. Most N_2-fixing species in Australia are legumes, and they can make a major contribution to enhancing plant and soil N (Lawrie 1981, Monk et al. 1981, Hingston et al. 1982). Nonlegumes (Casuarinaceae, Cycadaceae) are locally important in southwestern Australia (Halliday and Pate 1976, Grove et al. 1980, Lamont 1984). The much less efficient nonlegumes are the only dominant N_2-fixers in California (Rhamna-

ceae, Rosaceae) and Chile (Rhamnaceae) (Vlamis et al. 1964, Lepper and
Fleschner 1977, Kummerow et al. 1978a, Rundel and Neel 1978). Mycorrhizas supplement general nutrient uptake, especially P and to a lesser extent
N, and their presence is usually a prerequisite for nodule formation. They
appear to be especially abundant in Australia (25% of dominants), a finding
supported by more detailed analyses (Lamont 1982, 1984). The figures of
15% for California and 0% for Chile are clearly underestimates reflecting the
poor state of knowledge on this subject. For example, although they give no
data, Kummerow et al. (1978b) claim mycorrhizas are present in all chaparral
shrubs.

Conclusions

Collating all the evidence presented here, it is probably reasonable to conclude that, among mediterranean regions, the soils in Chile are more fertile
than those in California, which in turn are more fertile than soils in southeastern Australia, which in turn are more fertile than those of southwestern
Australia. This tentative statement is subject to three limitations: (1) the soils
and plants studied so far are not necessarily representative of their regions,
(2) no direct comparisons have been done of environmentally "matched"
sites between these regions, and (3) relative fertility has little meaning except
in the context of particular mineral nutrients. In some ways, the levels of
total and available nutrients in the soil are irrelevant, without correlating
them with plant growth and nutrient uptake. The question then becomes one
of species demand and ability to deal with physical constraints of the soil,
including penetrability and water availability. Some nutrients are relatively
less available than others, and some may be at supraoptimal levels, such as
Mg in the Obispo clay. Some species are able to compensate for perceived
deficiency relative to demand by relatively greater uptake of those nutrients,
such as K by *Baccharis* in all soils, or solubilization of limiting nutrients by
mycorrhizas and root clusters.

In an attempt to identify the growth-limiting nutrients, three approaches
were used in the Californian–Australian soil comparison reported here.
First, total and available nutrients in the soils were relativized by comparing
their levels among the five major nutrients and nine soils. The approach is
more successful as more nutrients and range of sites are used, and so this
must be considered a pilot study. Despite greater absolute levels of N, P, K,
Ca, and Mg by up to two orders of magnitude in the six Californian soils
than in the three southwestern Australian soils, the only clear trend was that
N was relatively higher in Australia. Relative available P was in fact lowest
in the chaparral soil and highest in the Australian Watheroo loam. On the
other hand, relative available Ca was lowest in the Australian Walyunga clay
and highest in the chaparral soil. Relative K was lowest in the Watheroo
loam, and shared lowest Mg with the Los Gatos loam. Levels were also

compared between nutrients in the same soil. The only trends were that relative total K was lowest in the Obispo (serpentine) soil, and Mg was lowest in two of the Australian soils and Ca in Walyunga clay. For relative available soil nutrients, N was lowest of the five in three Californian soils, P in chaparral, and Mg in Los Gatos. Relative available N was in fact highest in the Australian Eneabba sand, Ca was lowest in Walyunga, and K and Mg in Watheroo. Thus all five nutrients may be in relatively short supply, depending on soil type, with only N showing regional trends.

Identifying nutrients in relatively short supply does not take into account plant demand—nutrients at disproportionately low soil levels may still exceed plant requirements, and those at apparently high levels may soon be exhausted because of strong plant demand or inability of the soil to maintain supply. This disparity is recognized by cereal growers, who apply fertilizer in a N:P:K ratio of 10:2.7:1.6 in the U.S.A. and 10:23.3:4.2 in Australia (Williams and Raupach 1983). This imbalance indicates that relative demand for P followed by K is much more likely to exceed supply in Australia. In the absence of fertilizer, relative levels of P should be lower than K in Australian plants, which in turn should be lower than relative N levels.

In the second approach, total uptake of N, P, K, Ca, and Mg was generally lowest in two of the Australian soils, and the third was intermediate. Relative nutrient levels in the shoots showed no regional trends, with relative P lowest on Obispo clay, K on Eneabba sand, and N lowest in *Baccharis* on Walyunga soil. Each nutrient was at lowest shoot concentration relative to the others in at least one Californian and one Australian soil. In only four cases did lowest relative nutrient content of shoots reflect lowest relative supply: N in Francisquito loam, P in chaparral soil, Mg in Los Gatos and Watheroo, and Ca in Walyunga. In addition to the disparity between demand and supply, part of the problem was that rarely was one nutrient significantly lower (more limiting) than another. Furthermore, relative concentrations cannot vary radically within a species; otherwise protoplasmic integrity might fail and the plant would die. Thus total growth is proportional to the nutrient(s) in least supply relative to demand but they are difficult to identify by these methods.

The third approach was a nutrient-addition trial that should pinpoint limiting nutrients, provided that: (1) they are the ones included in the trial, (2) interaction effects are allowed for (to detect higher-order limiting nutrients), and (3) stimulatory rather than toxic levels are used in all treatments. Because N and P, and to a lesser extent K, are usually considered most likely to be limiting in most soils, these were chosen (even with one treatment plus control, including Ca and Mg would have increased the treatments from 8 to 64 per soil). One Californian soil responded positively to N only, two to P only, and two to N and P, and the sixth (Obispo clay) showed little response to any treatment (other nutrients were no doubt limiting). In contrast, one Australian soil responded positively to N and P, one to N and K, and one to N, P, and K (with greatest growth by far at $N \times P \times K$).

It is therefore misleading to treat North American soils as essentially N-limited. For a start, N accumulation depends on the P level of the parent rocks, which varies greatly (Norrish and Rosser 1983). The strong response to P in the chaparral soil is consistent with the relative soil and plant values reported above. Earlier work showed marked responses by chaparral to added N (on the assumption that it was the limiting nutrient) or N × P in particular and sometimes P as well (Hellmers et al. 1955, Christensen and Muller 1975, McMaster et al. 1982, Gray and Schlesinger 1983). I obtained some growth enhancement with K in the absence of other added nutrients to the chaparral soil, whereas Hellmers et al. (1955) observed almost no response in the field. The strong response to P in Supan loam could not be anticipated from the previous analyses, but dominance by *Quercus*, a genus with ectomycorrhizas that specifically enhance P uptake, is consistent. I was not able to duplicate the marked NPK response in Obispo clay by Turitzen (1982), but K did enhance growth in the absence of N—the toxic properties of this serpentine soil may have masked the benefits of additional nutrients (Proctor and Whitten 1975).

It is equally misleading to brand Australian soils as essentially P-limited. The point is more that these ancient, leached soils with minimal humus are generally nutrient-impoverished, so that low P implies low N, K, S, and trace elements as well (Lindsay 1985, Nix 1981, Williams and Raupach 1983). This would explain the strong interaction effects, especially for Eneabba sand (G, Fig. 9.1). Mg and Ca are rarely deficient because they are less soluble (Siddiqi et al. 1976, Williams and Raupach 1983, Lindsay 1985) but both were extremely low in southwestern Australian soils I studied, at both absolute and relative levels. Amendment with Ca and Mg might have proved instructive especially because it would have permitted complete comparison with the relative results. Others have observed a predominant P effect (Teakle and Cariss 1943, Specht 1963) as obtained here for Walyunga clay, although sometimes it was for a legume with decreased dependence on soil N (Grove 1990). Watheroo loam may be exceptional in showing a predominant N effect, no doubt because of its unusually high available P. More usual for the surface horizons of these podzolic soils may be the colimiting levels of K in contrast to the Californian soils, which usually responded negatively to the (high) level of K applied (Colwell and Grove 1976, Siddiqi et al. 1976, Toms and Fitzpatrick 1961, Milewski 1982).

The data provided here demonstrate large differences in mineral content of soils and plants between mediterranean regions on either side of the Pacific Basin. These disparities have produced suites of ecomorphological features characteristic of each region. Thus, soil nutrient supply has the major control over ecosystem structure and functioning when rainfall patterns are matched. Identifying the most growth-limiting nutrient proved more difficult, partly because of the limitations associated with the various approaches used. Results vary greatly between soils, even in the same region, as well as between species. Soils are structurally and chemically very complex, but ecologists ignore their effects on the biotic components at their peril, especially for continents around the Pacific Basin with their very different geological histories.

Acknowledgments

Thanks are due to Hal Mooney for his support while I held a Senior Fulbright Award to undertake the comparative soils study, Curtin University for granting me leave, Heather Lamont, Celia Chu, Douglas Turner, Alex Gilmore, Craig Walton, Sue Radford, and Ed Witkowski for essential technical support, and Bev Jones, Melody Best, and Cherie Taylor for processing the manuscript.

References

Bartolome JW, Klukkert SE, Barry WJ (1986) Opal phytoliths as evidence for displacement of California grassland. Madroño 33:217–222.

Christensen NL, Muller CH (1975) Relative importance of factors controlling germination and seedling survival in *Adenostoma* chaparral. Am Mid Nat 93:71–78

Colwell JD, Grove TS (1976) Assessments of potassium and sulphur fertilizer requirements of wheat in Western Australia. Aust J Exp Anim Husb 16:748–754

Cowling RM, Campbell BM (1980) Convergence in vegetation structure in the mediterranean communities of California, Chile and South Africa. Vegetatio 43:191–198

Farrington P, Greenwood EA, Bartle GA, Beresford JD, Watson GD (1989) Evaporation from *Banksia* woodland on a groundwater mound. J Hydrol 105:173–186

Gray JT, Schlesinger WH (1983) Nutrient use by evergreen and deciduous shrubs in southern California. II. Experimental investigations of the relationship between growth, nitrogen content and nitrogen availability. J Ecol 71:43–56

Grierson PF, Attiwell PM (1989) Chemical characteristics of the proteoid root mat of *Banksia integrifolia*. Aust J Bot 37:137–143

Grove TS (1990) Twig and foliar nutrient concentrations in relation to nitrogen and phosphorus supply in a eucalypt (*Eucalyptus diversicolor* F. Muell.) and an understorey legume (*Bossiaea laidlawiana* Tovey and Morris). Plant Soil 162:265–275

Grove TS, O'Connell AM, Malajczuk N (1980) Effect of fire on the growth, nutrient content and rate of nitrogen fixation of the cycad *Macrozamia riedlei*. Aust J Bot 28:271–281

Halliday J, Pate JS (1976) Symbiotic nitrogen fixation by coralloid roots of the cycad *Macrozamia riedlei*: physiological characteristics and ecological significance. Aust J Plant Physiol 3:349–358

Handreck K, Black N (1984) *Growing Media for Ornamental Plants and Turf*. New South Wales U Press, Sydney. Table 27.2

Hart JJ, Radosevich SP (1987) Water relations of two Californian chaparral shrubs. Am J Bot 74:371–384

Hellmers H, Bonner J, Kellecher JM (1955) Soil fertility: a watershed management problem in the San Gabriel Mountains of southern California. Soil Sci 80:189–197

Hingston FJ, Malajczuk N, Grove TS (1982) Acetylene reduction (N_2-fixation) by jarrah forest legumes following fire and phosphate application. J Appl Ecol 19:631–645

Hobbs RJ, Atkins L (1988) Effect of disturbance and nutrient addition on native and introduced annuals in plant communities in the Western Australian wheatbelt. Aust J Ecol 13:171–179

Hoffman A, Kummerow J (1978) Root studies in the Chilean matorral. Oecologia 32:57–69

Kummerow J, Alexander JV, Neel JW, Fishbeck K (1978a) Symbiotic nitrogen fixation in *Ceanothus* roots. Am J Bot 65:63–69

Kummerow J, Krause D, Jow W (1978b) Seasonal changes of fine root density in the southern Californian chaparral. Oecologia 37:201–212

Lamont B (1973) Factors affecting the distribution of proteoid roots within the root systems of two *Hakea* species. Aust J Bot 21:165–187

Lamont B (1982) Mechanisms for enhancing nutrient uptake in plants, with particular reference to mediterranean South Africa and Western Australia. Bot Rev 48: 597–689

Lamont B (1985) Gradient and zonal analysis of understorey suppression by *Eucalyptus wandoo*. Vegetatio 63:49–66

Lamont BB (1984) Specialized modes of nutrition. In Pate JS, Beard JS (eds) *Kwongan: Plant Life of the Sandplain*. U West Aust Press, Nedlands. pp 126–145

Lamont BB, Bergl SM (1991) Water relations, shoot and root architecture, and phenology of three co-occurring *Banksia* species: no evidence for niche differentiation in the pattern of water use. Oikos 60:291–298

Lamont BB, Collins BG, Cowling RW (1985) Reproductive biology of the Proteaceae in Australia and South Africa. Proc Ecol Soc Aust 14:213–224

Lamont BB, Kelly W (1988) The relationship between sclerophylly, nutrition and water use in two species from contrasting soils. In *Time Scales and Water Stress*. Proc 5th Int Conf Medi Ecosys. Int U Bio Sci, Paris, pp 617–621

Lawrie AC (1981) Nitrogen fixation by native Australian legumes. Aust J Bot 29: 143–157

Lepper MG, Fleschner M (1977) Nitrogen fixation by *Cercocarpus ledifolius* (Rosaceae) in pioneer habitats. Oecologia 27:333–338

Lindsay AM (1985) Are Australian soils different? Proc Ecol Soc Aust 14:83–97

Low AB, Lamont BB (1990) Aerial and below-ground phytomass of *Banksia* scrub-heath at Eneabba, south-western Australia. Aust J Bot 38:351–359

McMaster GS, Low WM, Kummerow J (1982) Response of *Adenostoma fasciculatum* and *Ceanothus greggii* chaparral to nutrient additions. J Ecol 70:745–756

Milewski AV (1982) The occurrence of seeds and fruits taken by ants versus birds in mediterranean Australia and southern Africa, in relation to the availability of soil potassium. J Biogeog 9:505–516

Miller PC, Bradbury DE, Hajek E, LaMarche V, Thrower NJ (1977) Past and present environment. In Mooney HA (ed) *Convergent Evolution in Chile and California: Mediterranean Climate Ecosystems*. Dowden, Hutchinson & Ross, Stroudsburg, PA, pp 27–72

Monk D, Pate JS, Loneragan WA (1981) Biology of *Acacia pulchella* R. Br. with special reference to nitrogen fixation. Aust J Bot 29:579–592

Mooney HA (1983) Carbon-gaining capacity and allocation patterns of mediterranean-climate plants. In Kruger FJ, Mitchell DT, and Jarvis JU (eds) *Mediterranean-Type Ecosystems: The Role of Nutrients*. Springer-Verlag, Berlin, pp 103–119

Naveh Z, Whittaker RH (1979) Structural and floristic diversity of shrublands and woodlands in northern Israel and other mediterranean areas. Vegetatio 41:171–190

Nix HA (1981) The environment of *Terra Australis*. In Keast A (ed) *Ecological Biogeography of Australia*. Junk, The Hague, pp 104–133

Norrish K, Rosser H (1983) Mineral phosphate. In *Soils: An Australian Viewpoint.* CSIRO, Melbourne/Academic Press, London, pp 335–361

Poole DK, Miller PC (1978) Water-related characteristics of some evergreen sclerophyll shrubs in central Chile. Oecol Plant 13:289–300

Proctor J, Whitten K (1975) A population of the valley pocket gopher (*Thomomys bottae*) on a serpentine soil. Am Mid Nat 85:517–521

Rundel PW (1988) Foliar analyses. In Specht RL (ed) *Mediterranean-Type Ecosystems: A Data Source Book.* Kluwer Academic, Dordrecht, pp 63–75

Rundel PW, Neel JW (1978) Nitrogen fixation by *Trevoa trinervis* (Rhamnaceae) in the Chilean matorral. Flora 167:127–132

Shaver GR (1983) Mineral nutrient and nonstructural carbon pools in shrubs from mediterranean-type ecosystems of California and Chile. In Kruger FJ, Mitchell DT, Jarvis JU (eds) *Mediterranean-Type Ecosystems: The Role of Nutrients.* Springer-Verlag, Berlin, pp 286–299

Siddiqi MY, Myerscough PJ, Carolin RC (1976) Studies in the ecology of coastal heath in New South Wales. IV. Seed survival, germination, seedling establishment and early growth in *Banksia serratifolia* Salisb., *B. aspleniifolia* Salisb. and *B. ericifolia* L.f. in relation to fire: temperature and nutritional effects. Aust J Ecol 1:175–183

Specht RL (1988) Natural vegetation-ecomorphological characters. In Specht RL (ed) *Mediterranean-Type Ecosystems: A Data Source Book.* Kluwer Academic, Dordrecht, pp 13–45

Specht RL (1963) Dark Island heath (Ninety-Mile Plain, South Australia). VII. The effect of fertilizers on composition and growth, 1950–60. Aust J Bot 11:67–94

Specht RL, Rundel PW (1990) Sclerophylly and foliar nutrient status of mediterranean-climate plant communities in southern Australia. Aust J Bot 38:459–474

Teakle LJ, Cariss HG (1943) Superphosphate requirements for growing wheat in Western Australia. J Dept Agric West Aust 20:1–28

Toms WJ, Fitzpatrick EN (1961) Potassium deficiency in medium rainfall areas. Dept Agric West Aust Bull No 2929

Turitzen SN (1982) Nutrient limitations to plant growth in a California serpentine grassland. Am Mid Nat 107:95–99

Vlamis J, Stone EC, Young CL (1964) Nitrogen fixation by root nodules of western mountain mahogany. J Range Man 17:73–74

Westman WE (1983) Plant community structure: spatial partitioning of resources. In Kruger FJ, Mitchell DT, Jarvis JU (eds) *Mediterranean-Type Ecosystems: The Role of Nutrients.* Springer-Verlag, Berlin, pp 419–445

Williams CH, Raupach M (1983) Plant nutrients in Australian soils. In *Soils: An Australian Viewpoint.* CSIRO, Melbourne/Academic Press, London, pp 777–794

Williams K, Hobbs RJ, Hamburg SP (1987) Invasion of an annual grassland in Northern California by *Baccharis pilularis* spp. *consanguinea.* Oecologia 72:461–465

Witkowski ET, Lamont BB (1991) Leaf specific mass confounds leaf density and thickness. Oecologia 88:486–493

IV. Seed and Fruits

10. Seed-Germination Patterns in Fire-Prone Mediterranean-Climate Regions

Jon E. Keeley

Wildfires are a known feature of mediterranean ecosystems and most plant communities are resilient to such perturbations; many species have evolved adaptations to periodic fires (Keeley 1986). Resilience to a perturbation such as wildfire does not, however, imply that resilient species have evolved traits adapted to such disturbance. In all mediterranean ecosystems many woody and herbaceous species recover after wildfires by resprouting from below-ground vegetative structures, but resprouting is a widespread trait in plants and is probably a preadaptation to surviving fires (Wells 1969).

In mediterranean ecosystems three modes of plant life history are recognizable:

1. Species that resprout after fire but fail to recruit seedlings in the postfire environment. Such taxa are resilient to fires and may require extended fire-free periods for seedling recruitment (Keeley 1992a). With respect to fire, such species are described in the literature as obligate resprouters or fire persisters. For seedling recruitment, these taxa are best described as having "disturbance-free recruitment," because opportunities for recruitment and population expansion occur in the absence of disturbance.
2. Species that restrict seedling recruitment to the immediate postfire environment. Depending on whether or not these species resprout after fire, they have been described as obligate seeders or facultative seeders. Unlike the first group of plants, these taxa have "disturbance-dependent re-

cruitment." For these species, opportunities for recruitment and population expansion are restricted to a short time immediately after fire.
3. Species that do not recruit seedlings immediately after fire but potentially exploit subsequent postfire years for seedling recruitment. Some of these species resprout and flower in the first year after fire. Others may disperse into burned sites. I describe these species as having "disturbance-dependent recruitment" but in later postfire years.

My focus in this chapter is to compare the seed germination behavior of "disturbance-dependent" and "disturbance-free" modes of life history. As has been demonstrated for California chaparral (Keeley 1991), many other characteristics are correlated with these two life history syndromes.

Seed Dormancy and Germination

If a viable seed does not germinate when incubated under appropriate moisture and temperature ranges, it is considered dormant. Different forms of dormancy are possible (Harper 1977). "Induced" dormancy is acquired inability to germinate, due to an environmental condition experienced by the seed after dispersal from the parent, and may persist even after that condition disappears. Many authors call this state secondary dormancy. "Enforced" dormancy is imposed by an environmental restraint such as lack of light or presence of an ambient inhibitor to germination. "Innate" dormancy is what most other authors call primary dormancy.

Little is known about induced seed dormancy in seeds of mediterranean-species. One observation that suggests its presence is the report that for some California chaparral species, fresh seeds germinate readily whereas older seeds require treatment to break dormancy (e.g., Emery 1988).

Enforced seed dormancy imposed by allelopathic compounds has been suggested as an important factor controlling germination in many mediterranean plants. The role of this factor in most ecosystems is a matter of much controversy.

Innate dormancy describes embryos that require an after-ripening period for development, or embryos that require some biochemical transformation that is cued by: (1) an environmental stimulus, (2) an inhibitor in the seed coat that needs to be leached from the seed, or (3) the seed coat, which acts as a barrier to water or oxygen uptake and requires scarification.

Most studies of seed germination of mediterranean species have focused on delineating not the mechanisms responsible for dormancy, but rather the extent to which dormancy could be broken by fire. Refractory seeds are those in which germination depends upon the stimulus of either intense heat shock or chemicals leached from charred wood (Keeley 1991). Thus, my primary focus in this review will be to evaluate the distribution of refractory and nonrefractory seeds in mediterranean ecosystems and the life-history characteristics associated with these two patterns of germination.

California Chaparral

Table 10.1 gives examples of chaparral taxa that illustrate the patterns of germination response evident in the California flora. These germination responses, as we would expect, correlate well with whether the species has disturbance-free or disturbance-dependent seedling recruitment.

Disturbance-Free Recruitment

Species present after fire strictly from resprouts fail to recruit seedlings because they lack a persistent seed bank (Parker and Kelly 1989). The transiency of the seed banks of such species derives from the nonrefractory nature of these seeds and thus do not require fire-related stimuli, e.g., *Quercus dumosa* (Fagaceae) (Table 10.2). Seeds germinate readily at maturity and are

Table 10.1. Representative examples of species exhibiting different germination characteristics from the chaparral. Apparently all herbaceous perennials, including geophytes, have nonrefractory seeds

Nonrefractory seeds	Heat-stimulated seeds	Charred-wood–stimulated seeds
Shrubs		
Fagaceae	Anacardiaceae	Ericaceae
Quercus dumosa	*Rhus* spp.	*Arctostaphylos* spp.
Rhamnaceae	Rhamnaceae	Garryaceae
Rhamnus spp.	*Ceanothus* spp.	*Garrya flavescens*
Rosaceae	Sterculiaceae	Rosaceae
Prunus ilicifolia	*Fremontodendron* spp.	*Adenostoma fasciculatum*
Subshrubs		
Agavaceae	Cistaceae	Papaveraceae
Yucca whipplei	*Helianthemum scoparium*	*Romneya* spp.
Polygonaceae	Fabaceae	Lamiaceae
Eriogonum fasciculatum	*Lotus scoparius*	*Salvia mellifera*
Scrophulariaceae	Lamiaceae	Hydropyllaceae
Keckiella spp.	*Salvia apiana*	*Eriodictyon crassifolium*
Herbaceous perennials		
Amaryllidaceae		
Allium spp.		
Liliaceae		
Zigadenus fremontii		
Paeoniaceae		
Paeonia californica		
Annuals		
Asteraceae	Apiaceae	Papaveraceae
Microseris heterocarpa	*Apiastrum angustifolium*	*Papaver californica*
Polygonaceae	Fabaceae	Caryophyllaceae
Chorizanthe spp.	*Lotus salsuginosus*	*Silene multinervia*
Scrophulariaceae	Onagraceae	Hydropyllaceae
Cordylanthus spp.	*Camissonia hirtella*	*Emmenanthe penduliflora*

Table 10.2. Germination patterns of species with nonrefractory seeds, heat-shock–stimulated germination and charred-wood–stimulated germination (data from Keeley 1987 and Keeley et al. 1985)

	Percentage germination									
	Light						Dark			
Temperature (C)		70	80	100	120	150		70	100	120
Time (min)	Control	120	60	5	5	5	Control	120	5	5
Quercus dumosa (shrub)										
Control	77	0	—	57	55	—	66	0	60	47
Charred wood	68	0	—	53	27	—	70	0	40	50
Artemisia californica (subshrub)										
Control	73	56	—	47	56	—	0	10	3	0
Charred wood	78	80	—	81	87	—	62	49	64	50
Ceanothus megacarpus (shrub)										
Control	11	41	—	48	80	—	6	54	53	88
Charred wood	2	40	—	40	61	—	3	40	56	67
Camissonia hirtella (annual)										
Control	30	—	49	—	66	69	—	—	—	—
Charred wood	26	—	33	—	22	20	—	—	—	—
Romneya trichocalyx (suffrutescent)										
Control	0	0	—	0	0	—	0	0	0	1
Charred wood	24	33	—	34	24	—	17	11	13	10

short-lived; probably none survive beyond nine months in the soil. Examples of other such species, all long-lived evergreen shrubs, with this syndrome include *Heteromeles arbutifolia* (Rosaceae), *Rhamnus* spp. (Rhamnaceae), and *Prunus* spp. (Rosaceae).

Shrub species with nonrefractory seeds produce fleshy fruits or acorns that mature in autumn and winter and are animal dispersed. Germination typically occurs within weeks of the first autumn or winter rains, although seedling establishment is generally rare. The only reports of seedling recruitment come from stands free of fire for more than fifty years (Patric and Hanes 1964, Zedler 1981, Keeley et al. 1986, Keeley 1992a, 1992b).

Physiologically, these "disturbance-free recruiters" are the least drought tolerant of the shrub species in chaparral (Keeley 1992b, 1992d). Adults survive because the established root crowns maintain a root system that allows the plant to avoid extreme stress from summer drought. Seedling recruitment, however, is vulnerable and their recruitment pattern is viewed as a mechanism for escaping the extreme aridity of open disturbed sites. It is argued (Keeley 1992d) that animal dispersal is a necessary correlate of this recruitment syndrome because safe sites for seedlings are a relatively small portion of this arid landscape. Autumn and winter dispersal would take advantage of migrating birds as well as provide for dispersal of these short-lived seeds during the appropriate season for successful seedling establishment.

Disturbance-Dependent Recruitment: Immediate Postfire

For many chaparral species, seedling establishment is restricted to the first spring after fire. In all but a few species, seeds are refractory and seedlings arise from a persistent seed bank. Correlated with this syndrome must be great seed longevity and a germination mechanism for sensing fire. Two mechanisms have evolved in chaparral species for detecting when a fire has passed through a site. One involves germination stimulated by intense heat shock, and the other involves germination stimulated by chemicals produced by combustion of wood. Examples of species with dormant seeds that germinate in response to one or the other of these two stimulii are shown in Table 10.2. Under natural conditions refractory seeds "require" a fire-related stimulus, either alone or coupled with other conditions such as a cold stratification period.

Shrubs in the genus *Ceanothus* (Rhamnaceae) have been known for many years to be highly refractory and to germinate only after exposure to intense heat (Table 10.2). *Ceanothus* typically have a very small percentage of nonrefractory seeds, whereas in other taxa such as *Camissonia hirtella* (Onagraceae), a significant fraction of the seed pool is nonrefractory and germinates readily upon wetting (Table 10.2). In species with heat-stimulated germination, dormancy is imposed by the seed coat, which prevents uptake of water and gases (Stone and Juhren 1951). These seeds are commonly referred to as "hard" and typically are covered with an unbroken cuticle that prevents imbibition (Fig. 10.1A) Heat shock melts or cracks the cuticle or integuments and allows imbibition (Fig. 10.1B). Such "hard-seeded" species with disturbance-dependent seedling recruitment are not unique to mediterranean ecosystems. For example, Martin and Cushwa (1966) describe species from the southeastern United States with "hard" seeds that lie dormant in soil until germination is stimulated by heat shock from fire.

Many chaparral species have dormant seeds that are stimulated to germinate after fire by the presence of charred wood (Table 10.1). *Romneya trichocalyx* (Papaveraceae) is one example of a species that exhibits nearly obligatory dependence upon charred wood to overcome dormancy (Table 10.2). More than 40 chaparral species have been reported to have refractory seeds that germinate in response to incubation in the presence of charred wood (Wicklow 1977, Jones and Schlesinger 1980, Keeley et al. 1985, Keeley 1987, 1991, Keeley and Keeley 1987). Heat shock is ineffective in stimulating germination in most of these species. As with heat shock-stimulated germination, some species have polymorphic seed pools in which some seeds are nonrefractory and germinate readily and others are refractory and germinate only in the presence of charred wood.

The mechanism behind germination stimulated by charred wood is unknown. Experiments thus far have revealed that the chemical is a water-soluble organic compound present in soils of recently burned chaparral (Keeley and Nitzberg 1984, Keeley et al. 1985). Most experiments have been

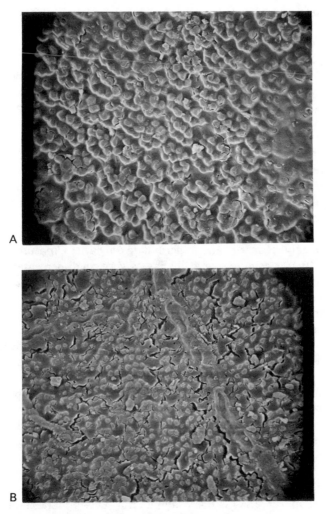

Figure 10.1. SEM (2000x) of *Lotus scoparius* seed coats (A) before, and (B) after heat treatment (120°C for 5 minutes) (from Keeley unpub. data). This species has heat shock–stimulated germination (Keeley 1991).

unsuccessful at pinpointing the precise chemical involved, despite tests of more than twenty potential breakdown products of lignin and cellulose (S. Keeley and Pizzorno 1986, Keeley unpub. data). Recent experiments, however, reveal that for one species, *Romneya coulteri*, the charred wood–stimulated germination response can be duplicated by using either smoke or gibberellic acid (Keeley unpub. data).

For species with heat-stimulated seeds and species with charred-wood–stimulated seeds, sometimes an additional environmental cue other than fire is required for germination. For example, some of these species require a cold stratification period of several months to overcome innate dormancy of the embryo. In the genus *Ceanothus*, high elevation species require this cold treatment, but low elevation species do not (Quick 1935). Some montane chaparral species of *Arctostaphylos* (Ericaceae) require a cold stratification period (Mirov 1936, Carlson and Sharp 1975), in addition to charred wood (Keeley 1991), to overcome dormancy. Populations of *Salvia mellifera* (Lamiaceae) from the coast, mountains, and desert, however, have no stratification requirement (Keeley 1986).

Shrub species with refractory seeds have similar demographic patterns, regardless of whether germination is stimulated by heat or charred wood. All such species establish seedlings in the first growing season after fire from a persistent soil-stored seed bank. Successful seedling establishment after the first postfire year is rare (Horton and Kraebel 1955, Keeley and Zedler 1978, Mills and Kummerow 1989). Many of these species are incapable of resprouting and thus populations are even aged (Keeley 1992a).

Many annual species with refractory seeds are highly specialized "fire followers," and seedling recruitment is mostly if not entirely restricted to the first year after fire (Sweeney 1956, S. Keeley et al. 1981).

Disturbance-dependent recruiters typically disperse their seeds in spring and summer, but often decades pass before germination occurs. Most of these species do not have propagules specialized for dispersal and illustrate the "temporal disperser syndrome" described by Angevine and Chabot (1979). In such species, dispersal is localized and persistent seed banks lie dormant until a disturbance occurs.

Once the chaparral canopy is regrown, seedling establishment by species with refractory seeds (Table 10.1) is seldom successful, except occasionally in gaps (Keeley 1992a, 1992b). Species that produce refractory seeds have polymorphic seed pools of both refractory and nonrefractory seeds. The ratio of refractory to nonrefractory seeds varies with the species. Even within a species there may be ecotypic differences in the ratio, and following dispersal, induced or enforced dormancy may cause changes in this ratio with time.

The role of polymorphic seed pools may be to take advantage of gaps in the chaparral canopy as well as store seeds until fire. Another strategy is illustrated by the germination behavior of *Artemisia californica* (Asteraceae) (Table 10.2). Seeds on the soil surface in gaps would germinate readily following the first autumn rains. Buried seeds, however, remain dormant and persist in the seed bank until charred wood overcomes the dark-enforced dormancy.

Great seed longevity is necessary for refractory seeds that make up the persistent seed banks. *Ceanothus* seeds are known to survive twenty years in laboratory storage (Quick 1935). Longevity is inferred for many other chaparral taxa that may die out on a site, but many years later reestablish seedlings following fire (Keeley 1991).

Enforced seed dormancy has been proposed as a major factor in California chaparral. McPherson and Muller (1969) concluded that nearly all seeds in the soil beneath *Adenostoma fasciculatum* (Rosaceae) are prevented from germinating by a toxin leached from the overstory shrubs; i.e., these allelopathic compounds enforced dormancy on the seeds. Other authors however, have not been able to demonstrate such an effect on the seeds of the most common chaparral shrubs or herbs (Christensen and Muller 1975a, 1975b, Kaminsky 1981, Keeley et al. 1985, Keeley and Keeley 1989). For refractory seeds, allelopathy is not a relevant factor because the seeds remain dormant in the absence of such chemicals. For species with polymorphic seed pools, allelopathic chemicals may suppress germination of the nonrefractory portion of the seed bank. Theoretically such a mechanism would be effective in preventing germination of seeds in such unfavorable sites as under the shrub canopy (Keeley 1991).

A few coniferous taxa, with widely disjunct populations in chaparral, have serotinous cones. These taxa, in genera such as *Cupressus* (Cupressaceae) and *Pinus* (Pinaceae) form even-aged stands with recruitment restricted to the first postfire year (Vogl et al. 1977, Zedler 1977). In these species seeds are not refractory; instead a persistent seed bank accumulates in the canopy because of dormancy imposed by retention within the fruit.

Disturbance-Dependent Recruitment: Later Postfire

All herbaceous perennials in chaparral are abundant on recently burned sites because they resprout from deeply buried rhizomes, corms, or bulbs, but seedlings are not produced in the first postfire year (Keeley and Keeley 1984, unpub. data). All these species flower in the first postfire season and produce nonrefractory seeds (Keeley 1991). Therefore, it is likely that seedling establishment occurs in the second or later postfire years. As the chaparral canopy closes, most of the geophytes are dormant and rarely flower and the few leaves that are produced are consumed by rodents (Stone 1951, Sweeney 1956, Christensen and Muller 1975a). Other herbaceous perennials may persist and flower in gaps in the canopy, and these also have nonrefractory seeds.

Certain resprouting subshrubs, typical of the coastal sage vegetation, do not maintain persistent seed banks and, because they are subligneous, persistence on a site is a function of fire intensity. Examples include *Encelia californica* (Asteraceae) *Eriogonum* spp. (Polygonaceae), and *Haplopappus* spp. (Asteraceae). On coastal sites, where fires are commonly of moderate intensity, these subshrubs resprout and flower profusely in the first year. Because the seeds are nonrefractory, seedling establishment is abundant in the second year (Keeley and Keeley 1984). These same species often do not resprout on the more arid interior sites, possibly because of greater fire intensity, and thus fires may result in localized extinction (Westman and O'Leary 1986).

Many nonnative and a few native annual species disperse into and colonize burned sites in years subsequent to fire. These species have nonrefractory seeds (Keeley and Keeley 1987), although other factors such as light quality and soil-nutrient status may affect germination.

In summary, refractory seeds, which require a fire-related stimulus, are not typical of species that recruit seedlings in the second or subsequent postfire years.

Mediterranean Basin

Wildfires are a natural feature of the garrigue, maquis, and phrygana vegetation in Europe (Naveh 1974). As in the California chaparral, many taxa are obligate resprouters after fire and seedling establishment occurs if not disturbed. Other taxa depend upon fires for establishing seedlings from a soil-stored seed bank.

Disturbance-Free Recruitment

Postfire obligate resprouting shrubs and lianas include all *Quercus* spp. (Fagaceae), *Pistacia* spp. (Anacardiaceae), *Rhamnus* spp. (Rhamnaceae), *Phillyrea latifolia* (Oleaceae), *Sorbus* spp. (Rosaceae), *Olea europaea* (Oleaceae), *Osyris alba* (Santalaceae), *Crataegus aronia* (Rosaceae), *Erica arborea* (Ericaceae), *Myrtus communis* (Myrtaceae), *Nerium oleander* (Apocynaceae), *Juniperus phoenicea*, *J. oxycedrus* (Cupressaceae), *Smilax* spp. (Liliaceae), and *Rubia* spp. (Rubiaceae) (Le Houerou 1974, Naveh 1975, Gratani and Amadori 1991). None of these species is known to produce refractory seeds and many have been shown to germinate readily under appropriate moisture and temperatures (Mitrakos 1981, Mesleard and Lepart 1991, Izhaki and Safriel 1990, J. Herrera 1991, Keeley unpub. data). Lack of refractory seeds would result in a transient seed bank (Harif 1978), which would account for the lack of seedling establishment after fire.

Results of detailed studies of *Arbutus unedo* are probably typical of others with this syndrome. Fleshy fruits are dispersed by birds in winter but no evidence shows that passage through the gut enhances germination beyond removal of the pericarp (Debussche 1985, Keeley 1987). Seeds are nonrefractory and germinate readily on moist substrate at temperatures below 15°C (Ricardo and Veloso 1987, Mazzoleni 1989, Mesleard and Lepart 1991). Seeds retained within fruits may survive in the soil until the following rainy season, although other seeds that fail to germinate may not survive long in the soil. Seedlings do not tolerate summer drought stress on exposed sites and thus seedling establishment is restricted to dense vegetation (Mesleard and Lepart 1991).

The life history characteristics of these postfire obligate resprouters is remarkably similar to characteristics of obligate resprouters in the California chaparral, many of which are in the same genera—e.g., *Quercus*, *Rhamnus*, and *Prunus*. As with the California taxa, all these Mediterranean species produce fruits that are animal dispersed. Dispersal is in the autumn and winter, taking advantage of migrating bird populations (C. Herrera 1984a, 1984b, 1987, Izhaki and Safriel 1985, Debussche et al. 1980, Izhaki et al. 1991, J. Herrera 1991). Winter dispersal is a further advantage because seeds are nonrefrectory and germination occurs soon after they encounter moist substrates. Successful seedling recruitment appears to require disturbance-free conditions under the vegetation canopy.

This "disturbance-free recruitment" syndrome is apparently quite old and seems to predate the present mediterranean climate. C. Herrera (1992a, 1992b) points out that many species with this character syndrome represent old phylogenetic lines. This observation is supported by the remarkable similarity in seed germination, dispersal, and seedling recruitment patterns observed in Californian and Mediterranean species of *Quercus*, *Rhamnus*, and *Prunus*. Herrera (1992b) suggests there is little reason to accept an "adaptationist" interpretation of this syndrome when it is clearly a result of historical effects.

Much about the seedling recruitment patterns of these taxa suggests that they are best adapted to moist microsites and consequently are not able to fully exploit much of the mediterranean-climate landscape. Under a former summer-rain climate, these taxa would undoubtedly have flourished. Today they persist because (1) they are long lived, (2) they resist destruction by wildfires by resprouting, and (3) older undisturbed vegetation generates sufficient moist microsites to allow occasional seedling recruitment. I suggest that the coevolved life history characteristics of seed germination, seed dispersal, and seedling physiology are just as adaptive in such microsites as they were under former climates, albeit on a far more limited scale.

Disturbance-Dependent Recruitment: Immediate Postfire

Although subshrubs such as *Cistus* spp. (Cistaceae) are capable of resprouting under moderately intense burning, they also establish seedlings in the first year after fire from soil-stored seeds (Martin and Juhren 1954, Naveh 1975, Troumbis and Trabaud 1986, Alonso et al. 1992). It is interesting that these *Cistus* spp. germinate in the autumn following fire, and poor survival is predicted for spring-germinating seeds (Troumbis and Trabaud 1986). This pattern is apparently true for other mediterranean species and contrasts markedly to the chaparral, where postfire seeding species establish in spring following fire.

Seeds of Cistaceae are known to be "hard-seeded"—i.e., dormancy is imposed by an impervious seed coat that prevents imbibition, and heat shock from fire is sufficient to overcome this seed coat imposed dormancy. Traubaud and Oustric (1989) report that germination of the refractory seeds of *Cistus albidus*, *C. monspeliensis*, *C. salvifolius* was increased from a few percent for controls to $> 70\%$ with certain heat treatments. They mention too that species differed in the optimal temperature treatment.

Thanos and Georghiou (1988) also report heat-shock–stimulated germination in the eastern Mediterranean *Cistus incanus* and *C. salvifolius*. They demonstrated that light and incubation temperature were unimportant factors in germination. These authors noted that the seed pools of these *Cistus* spp. were polymorphic. Some germinate readily and are capable of establishing on open sites. Others have water-impermeable seed coats and are stimulated to germinate by intense heat shock. They suggest that seed coat hardness was variable in the seed pool and would allow for germination under many conditions. Similar seed-pool polymorphism has also been seen in *C. ladanifer* from the western Mediterranean, where it was demonstrated that nearly 60% of the seeds were nonrefractory (Valbuena et al. 1992). Germination of this species, as well as *C. laurifolius*, was significantly increased, however, with heat shock.

Aronne and Mazzoleni (1989) used microscopic examination of seeds to study the effect of heat shock on seed coat integuments of *Cistus incanus* and *C. monspeliensis*. They demonstrated exposure to 120°C for 90 seconds was sufficient to form cracks on the inner layer of the seed coat, and this was necessary for imbibition and subsequent germination.

Other shrubs that recruit seedlings immediately after fire from persistent soil-stored seed banks include taxa in the Fabaceae (*Ulex* spp., *Cytisus* spp., *Calicotome* spp., and *Genista* spp.), Lamiaceae (*Rosmarinus officinalis*, *Satureja thymbra*, *Teucrium* spp., *Thymus* spp., and *Salvia* spp.), Rosaceae (*Sarcopoterium spinosum*), and *Anacardiaceae* (*Rhus coriaria*) (Naveh 1974, 1975, Papanastasis and Romanas 1977, Belhassen et al. 1987, Trabaud 1987, Izhaki et al. 1992).

Based on the patterns observed with California taxa, I predict that the "hard-seeded" Fabaceae species will have heat shock-stimulated germination, and some evidence supports this belief; germination is stimulated by high temperatures in *Calicotome villosa* (Keeley, unpublished data), *Cytisus scoparius* (Bossard 1993), and *Genista florida* (Tarrega et al. 1992). The latter two authors also show temperatures above 150°C are lethal. Other species that exhibit heat shock–stimulated germination include *Phlomis lanata*, *Lavandula stoechas*, and *Sarcopoterium spinosum* (Table 10.3).

Of more than fifty species from the eastern Mediterranean, only one, *Stachys tymihaus*, has tested positive for enhanced germination in the presence of charred wood (Table 10.3).

Sarcopoterium spinosum is a species that germinates prolifically after fire, and yet results from Table 10.3 and other studies (Papanastasis and Romanas 1977) indicate that a very high proportion of seeds germinate without heat shock. Other factors may be involved in controlling germination, such as an increase in the ratio of red:far-red radiation, because of canopy removal by fire, could cue germination (Margaris 1981). That this effect might

Table 10.3. Seed germination response of selected Greek maquis species (from Keeley unpub. data): Seeds were incubated in the light on filter paper in petri dishes at 5°C for 2 months, followed by 20°C/12°C (12 hrs/12 hrs) for 1 month. Values represent the mean of 3 dishes of 50 seeds each; numbers within the same row with the same superscript are not significantly different at $P > 0.05$

Species	Family	Growth form	Percentage germination			
			Control	Heated 80°C/30 min	Heated 115°C/5 min	Charred wood
Nonrefractory seeds						
Allium sp.	(Amaryllidaceae)	Geo-	90[a]	19	3	72[a]
Asphodelus aestivus	(Liliaceae)	Geo-	57[a]	55[a]	0	32[a]
Heat-shock–stimulated seeds						
Cistus salvifolius	(Cistaceae)	Phano-	3[a]	39	80	0[a]
Phlomis lanata	(Lamiaceae)	Chamae-	8[a]	72[b]	62[b]	8[a]
Lavandula stoechas	(Lamiaceae)	Chamae-	19[a]	59	17	21[a]
Sarcopoterium spinosum	(Lamiaceae)	Chamae-	29[a]	73	50	39[a]
Charred-wood–stimulated seeds						
Stachys tymihaus	(Lamiaceae)	Chamae-	0[a]	0[a]	0[a]	13

be a factor controlling the nonrefractory portion in the seed pool of *S. spinosum* is supported by the results of Roy and Arianoutsou-Faraggtitaki (1985), who showed that germination was stimulated by an increase in the red:far-red radiation ratio from 0.3 to 1.1. Similar patterns have been observed for *Cistus* species (Roy and Sonie 1992).

The factors controlling the postfire germination of *Rhus coriaria* are unknown, but if comparative analysis is of any value I predict that it may be heat, as it is with its Californian congener, *R. ovata* (Stone and Juhren 1951). An important observation is that these two taxa have fleshy fruits and are exceptions (Izhaki et al. 1992) to the generalization that disturbance-dependent recruiters are not animal-dispersed. Bird dispersal may also be one of the factors accounting for distribution of seedlings after fires. Ne'eman et al. (1992) noted a concentration of seedlings within the shadow of the former canopy of burned pines, but other factors have not been ruled out.

As in California, the Mediterranean has relatively few serotinous species with canopy-stored seed banks. The few that occur are gymnosperms, such as *Pinus brutia* (Thanos et al. 1989). Seed dormancy is imposed by retention in the cone and, once released, the seeds are nondormant and germinate optimally at 20°C in darkness (Thanos and Skordilis 1987).

It is noteworthy that many of the postfire seeder species are drought deciduous malacophyllous subshrubs, which are relatively short lived. Conversely, obligate resprouters that establish seedlings under disturbance-free conditions are long lived, broad-leaf, evergreen sclerophyllous shrubs. The subligneous nature of the disturbance-dependent recruiters probably leads to greater susceptibility to destruction by fire and hence strong selective advantage in evolving a persistent seed bank.

Disturbance-Dependent Recruitment: Later Postfire

Many geophytes are endemic to the Mediterranean Basin, although none have been reported to recruit seedlings after fire from a persistent seed bank. Although relatively little is published on germination, the patterns observed for *Asphodelus aestivus* and a species of *Allium* (Table 10.3) indicate non-refractory seeds in these geophytes from the eastern Mediterranean. This finding, coupled with the knowledge that geophytes resprout and flower after fire, make it likely that seedling recruitment occurs in later postfire years.

Two phenological types have been described for Mediterranean geophytes: those which flower in autumn and those flowering in spring. Although both have nonrefractory seeds, the former seeds germinate immediately after dispersal whereas for the latter, germination is delayed a year by lack of moisture for germination (Dafni et al. 1981).

Many annual species are mainly opportunistic colonizers that disperse into burned areas (Papio 1988). Examples of such species shown to have nonrefractory seeds include *Daucus carota* (Apiaceae), *Crepis* spp. (Asteraceae), *Linaria* sp. (Scrophulariaceae), *Anagallis arvensis* (Primulaceae), *Filago* sp. (Asteraceae), and *Hedypnois cretica* (Asteraceae) (Keeley unpub. data).

Central Chile

Wildfires are not presently widespread in Chilean matorral vegetation. Because convective thunderstorms and associated lightning are rare, a frequent natural ignition source is lacking (Aschmann and Bahre 1977, Rundel 1981), although vulcanism has been suggested as a natural source of wildfire ignitions (Fuentes and Espinosa 1986). Overall it does not appear that fire has had a major part in shaping the vegetation characteristics of Chile.

Disturbance-Free Recruitment

All Chilean woody species resprout from root crowns or lignotubers (Mooney et al. 1977, Montenegro et al. 1983) and thus are resilient to fires ignited by local residents. One study of a burned matorral site revealed very little cover by native annual species and no seedling regeneration by the shrub flora, with the one exception of *Trevoa trinervis* (Rhamnaceae) (S. Keeley and Johnson 1977). Herbaceous perennials in Liliaceae, Amaryllidaceae, Alstroemeriaceae, and Iridaceae are common components of unburned matorral (Rundel 1981) and are present as resprouts after fire.

Most of the common matorral shrubs such as *Lithrea caustica* (Anacardiaceae), *Quillaja saponaria* (Rosaceae), *Colliguaja odorifera* (Euphorbiaceae), *Kageneckia oblonga* (Rosaceae), and *Muehlenbeckia hastulata* (Polygonaceae) establish seedlings in undisturbed vegetation (Fuentes et al. 1984, 1986, Del Pozo et al. 1989). Seedling recruitment tends to occur under the canopy of shrubs, although survival is precarious and those not preyed upon typically succumb to summer drought (Jaksic and Fuentes 1980).

The demographic pattern of seedling establishment without fire suggests that these species have transient soil-stored seed banks of nonrefractory seeds. *Colliguaja* spp. and *Quillaja saponaria* are known to germinate readily without treatment, but *Lithrea caustica* seeds are reportedly refractory and may require scarification in concentrated sulfuric acid (Montenegro et al. 1983, Muñoz and Fuentes 1989, but cf. Jimenez and Armesto 1992). In nature, *Lithrea* seeds seem to germinate readily after passage through the gut of bird dispersers (Fuentes et al. 1984). As in California, many of these species are fleshy fruited and most tend to be more important on mesic sites (Hoffmann et al. 1989, Jimenez and Armesto 1992).

Enforced dormancy imposed by allelopathic chemicals is not present in evergreen matorral (Montenegro et al. 1978) but may be a factor in certain drought-deciduous vegetation (Fuentes et al. 1987).

Disturbance-Dependent Recruitment

Other than the report mentioned above of postfire seedling recruitment by *Trevoa trinervia*, this mode seems to be poorly represented in Chile. *Baccharis* spp. and other Asteraceae taxa are capable of dispersing in and establishing after fires and other disturbances, but nothing is known of their germination biology.

South Africa

Wildfires are at least as frequent in South African fynbos as in the Californian chaparral, and the vegetation responses show many similarities (Keeley 1992c, Versfeld et al. 1992).

Disturbance-Free Recruitment

Species that are present immediately after fire from resprouts, but have no seedling recruitment, are found in all life forms: shrubs, suffrutescents, and herbaceous perennials (van Wilgen and Forsyth 1992). Such shrubs include members of the Anacardiaceae (*Rhus* spp. and *Heeria*) and *Olea europaea* (Oleaceae) and a number of evergreen tree species restricted to moist ravines, e.g., *Cunonia capensis* (Cunoniaceae), *Kiggelaria africana* (Flacourtiaceae).

The few data available on seed germination indicate nonrefractory seeds, and some of these species are known to establish seedlings in unburned fynbos (Manders et al. 1992). As with the California obligate resprouters, these species have fleshy fruits dispersed in the autumn and winter by birds, strongly affecting distribution of seedlings. This strategy may explain why, as in the California chaparral, these obligate resprouting species tend to be distributed patchily within a matrix of seeding species (Kruger 1979).

Other evergreen shrubs, which persist on burned sites strictly as obligate resprouters, include species of *Protea*, *Leucospermum* and *Leucadendron* in the Proteaceae, *Widdringtonia* spp. in the Cupressaceae, species in the Rhamnaceae, Asteraceae, Anacardiaceae, Fabaceae, and Restionaceae (Wicht 1948, Scriba 1976, Bond 1980, 1985, Hoffmann et al. 1987, Higgins et al. 1987). Some, such as *Protea grandiflora*, *Widdringtonia cedarbergensis*, and *Gymnosporia laurina*, survive fires in protected locations and because of their thick bark and self-pruning (Wicht 1948, Manders 1987). Many of these taxa have nonrefractory seeds, but little is known about patterns of seedling recruitment. One exception is the observation by Taylor (1978b) that in an area protected from fire for thirty-five years, *Protea neriifolia* seedlings established in gaps created by the death of shrubs, along with some re-cruitment of woody shrubs and small tree species derived from the forest flora. Other *Protea* species routinely establish seedlings without fire. These taxa occur on more arid sites and probably experience a longer fire-return interval than most fynbos sites. *Widdringtonia cedarbergensis* is not serotinous, and the nonrefractory seeds germinate within a few weeks of the first autumn rains (Manders 1987). Germination, however, is inhibited by litter from the parent plant and thus little is known about the exact patterns of seedling recruitment for this species.

Disturbance-Dependent Recruitment: Immediate Postfire

Many shrubs establish seedlings after fire from persistent seed banks in the soil or canopy (Le Maitre and Midgley 1992). Seed-germination behaviors are quite different, depending upon the mode of seed storage.

Soil-Stored Seeds

The proteaceous genus *Leucadendron* includes some species that have soil-stored seed, along with the serotinous taxa discussed below (Bond 1985). Other taxa with soil-stored seed include *Leucospermum* spp., *Mimetes stokoei* and *Orothamnus zeyheri* (Proteaceae), *Cliffortia* spp. (Rosaceae), *Phylica* spp. (Rhamnaceae), *Euryops abrotanifolius*, *Metalasia muricata* and *Elytropappus rhinocerotis* (Asteraceae), *Selago corymbosa* (Selaginaceae), *Passerina* spp. (Thymelaceae), *Aspalathus* spp. (Febaceae), and *Anthospermum aethiopicum* (Rubiaceae) (Adamson 1935, Levyns 1935, Martin 1966, Kruger 1979, Bond 1985, Higgens et al. 1987, Kilian and Cowling 1992).

Refractory seeds of *Leucadendron* may be stimulated to germinate by temperatures of 100°C (Williams 1972). Heat-stimulated germination has also been shown for *Elytropappus rhinocerotis* (Levyns 1927) and *Agathosma* spp (Rutaceae) (Blommaert 1972). *Erica hebecalyx* (Ericaceae) germination is greatly stimulated by a 3-min treatment at 96.5°C, and this treatment is capable of overcoming the almost total inhibition imposed by darkness (van de Venter and Esterhuizen 1988). *Podalyria calyptrata* (Fabaceae) germination is increased from 9% (control) to 72% when seeds are exposed for just 1 min at 60°C (Jeffery et al. 1988). Other species with heat shock-stimulated germination include *Phylica ericoides* (Rhamnaceae) and *Hermania* spp. (Sterculiaceae) (Table 10.4).

It is often assumed that the mechanism for heat-stimulated germination lies in breaking the seed coat (or pericarp), which in most hard seeds imposes dormancy by inhibiting entry of water or movement of gases. This description is very likely to apply to the refractory seeds of *Protea*, *Leucodendron*, and *Leucospermum*, which have been shown to lack leachable chemical inhibitors (van Staden and Brown 1977). In some cases when the hard outer pericarp inhibits gaseous exchange, scarification greatly improves germination, and incubation of unscarified seeds under pure oxygen may overcome the pericarp-imposed dormancy (van Staden and Brown 1977, Deall and Brown 1981). Brits (1986b) and Brits and van Niekerk (1986) show that refractory seeds of *Leucadendron* species and *Leucospermum* species can be stimulated to germinate by soaking them 24 hr in 1% H_2O_2, and they sug-

Table 10.4. Seed germination response of selected South African fynbos species (from Keeley and Bond, unpub. data): Seeds were incubated in the light on filter paper in petri dishes at 5°C for 2 months, followed by 20°C/12°C (12 hrs/12 hrs) for 1 month. Values represent the mean of 3 dishes of 50 seeds each; numbers within the same row with the same superscript are not significantly different at $P > 0.05$

Species	Family	Growth form	Percentage germination			
			Control	Heated 80°C/30 min	Heated 115°C/5 min	Charred wood
Nonrefractory seeds						
Wachendorfia paniculata	(Hyacinthaceae)	Geo-	80[a]	72[a]	0	80[a]
Moraea sp.	(Iridaceae)	Geo-	74[a]	77[a]	0	60[a]
Heat-shock–stimulated seeds						
Phylica ericoides	(Rhamnaceae)	Phano-	9[a]	44	84	11[a]
Hermannia alnifolia	(Sterculiaceae)	Chamae-	0[a]	17	67	1[a]
Charred-wood–stimulated seeds						
Pharnaceum elongatum	(Aizoaceae)	Chamae-	0[a]	4[a]	1[a]	47
Nemesia cf. *lucida*	(Scrophulariaceae)	Thero-	7[a]	9[a]	2[a]	84

gest that dormancy is imposed by barriers to oxygen uptake. In other species the pericarp physically restricts growth of the embryo and elevated oxygen does not improve germination (Brown and Dix 1985).

Other temperature regimes that stimulate germination include fluctuating soil temperatures, as on recently burned sites (Brits 1986a, 1987). Most seeds sown in summer remain dormant until fall (Brits and van Niekerk 1986). Whether they do so because of dormancy induced by high temperatures or a germination requirement for cool temperatures is unknown. Tests with *Elytropappus rhinocerotis* indicate that an afterripening may be required following dispersal of seeds (Levyns 1927).

The seeds of some fynbos species are stimulated to germinate by the presence of charred wood (Table 10.4). Along with species reported in Table 10.4, at least five other fynbos taxa have charred-wood–stimulated germination (Keeley and Bond unpub. data). Other studies, however, have not found a stimulatory effect for charred wood. Pierce (1990) reported no such effect on germination of various coastal fynbos species, but these species had nonrefractory seeds that germinated readily under alternating temperatures. Van de Venter and Esterhuizen (1988) demonstrated that gases such as ethylene and ammonia, which are products of combustion, can stimulate germination of *Erica hebecalyx*. This species, however, seems also to be stimulated by heat shock, though it showed no response to charred wood (Keeley and Bond unpub.). Germination of *Audouinia capitata* (Bruniaceae) is stimulated by exposure to smoke (de Lange and Boucher 1990) although it is unknown whether this reaction is caused by gases produced by the smoke or particulate matter suspended in the smoke (Le Maitre and Midgley 1992).

Soil-stored seeds are often dispersed by ants, whereas serotinous species of both *Protea* and *Leucodendron* are more often wind dispersed (Bond 1985). For both genera, dispersal is relatively localized (Kruger 1983, Slingsby and Bond 1985, Manders 1986). Unlike canopy-stored seeds, which are generally short-lived once they are dispersed, soil-stored seeds are thought to have great longevity.

In many species the controls on seed germination are not clearly worked out. For example, in coastal dune fynbos, species such as *Muraltia squarrosa* (Polygalaceae) and *Passerina vulgaris* have few or no seedlings on unburned sites but abundant recruitment immediately after fire (Pierce and Cowling 1991). Seeds, however, appear to germinate readily without fire-related stimulus. *Passerina paleaceae* is another species with strict postfire seedling recruitment. Seeds are highly refractory, but various heat treatments and charred wood have not been successful in stimulating germination (Kilian and Cowling 1992, Keeley and Bond unpub. data).

Canopy-Stored Seeds

Serotinous fruits are produced by species of *Protea*, *Leucadendron*, *Aulax* (Proteaceae), *Widdringtonia* (Cupressaceae), *Nebelia* (Bruniaceae), *Phaeno-*

coma, Helipterum (Asteraceae), *Brunia, Berzelia* (Bruniaceae), *Cliffortia* (Rosaceae), and *Erica sessiliflora* (Ericaceae) (Wicht 1948, Gill 1975, Frost 1984, Bond 1985, E. Oliver pers. comm.). Seeds are retained for variable lengths of time, and all these species disperse some seeds before as well as after fire. Thus, these species have both a transient and persistent seed bank. In the absence of fire, fruits often open after several years and release seed that is nonrefractory and germinates in the first rainy season after being shed (Bond 1984, Le Maitre 1990).

Although seeds may be dispersed both before and after fire, in mature fynbos, seedling establishment is rare (Bond 1980). Strictly speaking, *Protea arborea* does not produce serotinous cones and most germination occurs before fire. But only the seeds retained in the fruits until fire result in successful seedlings (Kruger 1977). In older stands, predation of seedlings appears to be a major factor limiting establishment (Bond 1983, Breytenbach 1984). *Protea laurifolia* (Manders 1986) has seedling regeneration mostly restricted to the first year after fire, but some seedlings may establish in the second or third year (Bond 1984, Bond et al. 1984). Seeds released after spring burns seem less successful because of predation than seeds released after fall burns (van Wilgen and Viviers 1985).

Seeds in serotinous cones are nonrefractory, but are maintained in a quiescent state by enclosure in the fruit. If seeds are dispersed in summer, most do not germinate until fall (Bond 1984). It is unclear if this delay is caused by dormancy induced by high temperature or a germination requirement for low temperature. It appears that some *Leucadendron* species require cold stratification (Deall and Brown 1981). This characteristic prevents germination during the summer drought and it is noteworthy that *Protea rouppelliae* from the summer-rain eastern Cape region does germinate in summer (Bond 1984). Although seeds do not appear to survive in the soil for more than one season, *Leucadendron xanthoconus*, a weakly serotinous nonsprouter, is perhaps an exception to this generalization (Davis 1992).

Disturbance-Dependent Recruitment: Later Postfire

Apparently geophytes in the South African flora share with their chaparral counterparts the characteristic of being obligate resprouters after fire. Included are taxa of Geraniaceae, Oxalidaceae, Poaceae, Orchidaceae, Liliaceae, Iridaceae, Amaryllidaceae and related families (Adamson 1935, Hoffmann et al. 1987). These species flower profusely in the first year after fire, and some in fact flower "only" after fire and have gained the name "fire-lilies." One of these, *Cyrtanthus angustifolius* (Amaryllidaceae) produces flower heads almost immediately after burning, prior to vegetative growth (Bond 1980, Le Maitre and Brown 1992). Other geophytes observed to have flowering mainly restricted to the first postfire season include *Brunsvigia orientalis* (Amaryllidaceae) (Taylor 1978a), *Androsymbirium leucanthum* (Liliaceae) (Wicht 1948), *Haemanthus canaliculatus* (Amaryllidaceae) (Levyns

1966), *Bobartia spathacea* (Iridaceae) (Adamson 1935) and *Watsonia pyramidata* (Iridaceae) (Kruger 1977). It has been suggested that the stimulus for postfire flowering comes from the changed rhythm in diurnal fluctuations in soil temperature (Martin 1966) or higher soil temperatures (Frost 1984). One experiment with the strict fire-lily *Cyrtanthus ventricosus*, however, induced flowering by treating planted bulbs with smoke from burning wood (Keeley 1993).

Although geophytes and other herbaceous perennials apparently do not store persistent seed banks until fire, the precise timing of seedling establishment is not known. Kruger (1977) did report that for *Watsonia pyramidata*, seedling establishment occurred in the second growing season after fire. It is quite likely that this delay is typical of other geophytes because flowering is abundant in the first year and most species apparently have nonrefractory seeds; *Wachendorfia paniculata* (Haemodoraceae) and *Geissorhiza Moraea* sp. (Iridaceae) germinate readily and this readiness has been observed for more than a dozen other geophytes from the fynbos (Keeley and Bond unpub. data). Some species, however, are inhibited in the dark, as illustrated by the germination response of *Wachendorfia paniculata* (Fig. 10.2). Others are apparently inhibited by high temperatures. *Watsonia fourcadei* germinates readily at 10°C (in light and dark), but is mostly inhibited at temperatures above 25°C (Esterhuizen et al. 1986).

Other species may be absent on recent burns but disperse on to sites in particular annuals subligneous species of *Erica* (Ericaceae) (Adamson 1935). This species may remain in gaps in broad-leaved Proteaceae dominated fynbos (Wicht 1948, Hoffmann et al. 1987) or form a more heathlike fynbos,

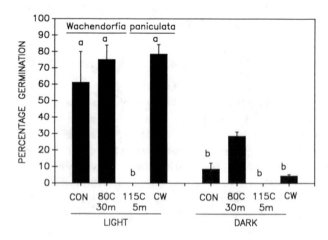

Figure 10.2. Germination of a South African fynbos geophyte in light and darkness (methods as described in Table 10.4) (from Keeley and Bond unpub. data).

as on coastal dunes (Pierce 1990). For some, seed banks are relatively transient because seeds germinate readily under a diurnal temperature range of 10°C to 20°C (Pierce 1990). Seedling recruitment in mature fynbos seems to be rare, although seedlings of *Agathosma* spp. (Rutaceae) have been reported from mature fynbos (Pierce and Cowling 1991).

Australia

After wildfires in heath, mallee, and other sclerophyllous shrublands in Australia, some species reestablish by seed only, some by resprouts and seed, and rarely, strictly by resprouts.

Disturbance-Free Recruitment

It is not apparent from the literature that clear examples of this mode appear in the mediterranean ecosystems in Australia. Most Australian woody plants, whether they resprout or not, seem to establish seedlings after fire.

Disturbance-Dependent Recruitment: Immediate Postfire

Shrubs, suffrutescents, and annual taxa are seeders, many of which establish seedlings only in the first postfire year. Many of the woody taxa both resprout and establish seedlings. As in South Africa, seed banks contributing to this flush of seedlings are either soil-stored or canopy-stored.

Soil-Stored Seeds

Woody taxa with soil-stored seed include many members of the Fabaceae, a family known for its hard-seeded character, such as *Acacia, Bossiaea, Dillwynia, Daviesia, Kennedia,* and *Pultenaea* (Floyd 1966, Warcup 1980, Pieterse and Cairns 1986, Bell et al. 1993). *Leptospermum* (Myrtaceae) includes species that are serotinous and others that are nonserotinous; the latter species are hard-seeded and form a persistent seed bank. Circumstantial evidence of seed longevity for soil-stored seed suggests fifty years or more (Mott and Groves 1981). Although annuals are not a highly diverse group in Australia, some species are present after fires from a persistent seed bank (Specht 1981, Bell et al. 1984).

   Germination of hard-seeded legumes, such as species of *Acacia, Albizzia, Chorizema, Gompholobium, Hardenbergia, Kennedia, Labichea, Mirbelia,* and *Oxylobium* is stimulated after brief heating treatments of 80°C and 100°C or boiling water (Shea et al. 1979; Jeffery et al. 1988; Bell et al. 1993). *Acacia suaveolens* germination is nil without exposure to temperatures above 60°C and is stimulated by temperatures of up to 100°C for durations of a hour (Auld 1986b). Brief exposure to temperatures of 120°C were also stimulatory, but temperatures above 150°C were lethal. Apparently the heat breaks the seedcoat, allowing imbibition (Cavanagh 1980). There is some evidence that *Acacia aneura* seeds have an afterripening period; with scarification, germination is low immediately after being shed but increases over several months (Preece 1971). Hodgkinson and Oxley (1990) show that the optimal temperature for germination of this species is a function of numerous factors, including depth of seed burial, fuel levels, and soil moisture.

Other families with taxa having persistent seed banks that show heat-stimulated germination included Convolvulaceae, Epacridaceae, Euphorbiaceae, Geraniaceae, Haemodoraceae, Rhamnaceae, Rubiaceae, and Sapindaceae (Warcup 1980) and a fuller description of these species has recently been published by Bell et al. (1993).

Bell et al. (1987) have tested the stimulatory effect of charred wood on germination of forty nonleguminous understory species of the jarrah (*Eucalyptus marginata*) forest. Only one of the species, *Burchardia umbellata* (Liliaceae) increased germination; 9% for controls and 35% with charred wood (heating 2 min at 85°C had no stimulatory effect). This species normally does not establish seedlings in the first season after fire, but, like other geophytes, it flowers from resprouts in the first postfire season. It is unclear what role charred wood–stimulated germination has in the life history of this herb. Tests of species in other vegetation are needed to evaluate the importance of charred wood–stimulated germination in Australia.

Enforced seed dormancy imposed by allelopathic compounds was suggested as an important phenomenon in Californian populations of *Eucalyptus* (del Moral and Muller 1969). Some have questioned whether allelopathy is important in Australia (Mott and Groves 1981), although experimental studies suggest it may be a factor in suppressing understory species in *Eucalyptus* forests (May and Ash 1990) and other vegetation types (Hobbs and Atkins 1991). Many species in the understory of Australian forests, though, are hard-seeded and thus dormancy is imposed by the seed coat.

Although passive dispersal is common, many of the Australian *Acacia* species have seeds with appendages that are attractive to either birds or ants, and thus the seeds are relatively specialized for dispersal (Davidson and Morton 1984, Auld 1986a). Dispersal, however, is fairly localized. Potential advantages of such myrmechochorous seed dispersal include transport of the seed to nutrient-rich sites and burial of the seed, which reduces loss of seed to predation or to intense temperatures during fire. One test did not support the nutrient-enrichment hypothesis (Rice and Westoby 1986). One additional advantage may be to reduce predation by using the elaiosome as a form of bribe so that ants are selected to leave the seed unscathed (Keeley 1992c).

Canopy-Stored Seeds

Taxa with serotinous woody fruits that open after fire include *Eucalyptus* (Myrtaceae), *Casuarina* (Casuarinaceae), *Banksia* (Proteaceae), *Hakea* (Proteaceae), and *Leptospermum* (Myrtaceae) (Specht 1981).

Most species of the shrubby genus *Banksia* produce woody "cones" that retain seeds for several to many years after cones mature, forming a canopy seed bank that persists until fire (Bradstock and Myerscough 1981, Bell et al. 1984, Cowling et al. 1987, Zammit and Westoby 1987a, 1988, Bradstock and O'Connell 1988, Lamont and Barker 1988). Serotinous cones open after exposure to temperatures of 145°C to 390°C (Gill 1976, Enright and Lamont 1989), and disperse seeds over several months (Cowling and Lamont 1985a). Dispersal is relatively localized and subject to intense predation. Consequently, fires in spring provide more time for predation than fires in fall, and thus seasonal effects on seedling regeneration are reported (Cowling and Lamont 1987). It is suggested that successful seedling recruitment may be tied to particular weather patterns following fire (Lamont et al. 1991).

Seeds of *Banksia*, and of the closely related genus *Hakea*, are nonrefractory and germinate readily (Siddiqi et al. 1976, Sonia and Heslehurst 1978, Abbott 1985a, Zammit and Westoby 1987b, Richardson et al. 1987, Lamont and Barker 1988, Lamont et al. 1993), soon after autumn rains. Temperatures above 150°C are lethal and thus the thick woody follicles are important in protecting seeds during fires. Germination is light-insensitive and, as a rule, temperatures below 12°C and above 25°C suppress germination. Seeds in cones 1 to 3 years of age show the highest germination, but substantial germination is still possible from seeds as old as 9 years of age (Cowling and Lamont 1985b). Seeds exposed on the soil surface for 5 months over spring and summer germinated poorly when tested at 15°C. It is unclear whether this failure was caused by an induced dormancy, loss of seed viability, or change in the optimal temperature for germination caused by these storage conditions.

In mature stands, very old cones, or cones on dead branches, open and release their seeds. In most unburned stands, successful seedling establishment is rare (Brown and Hopkins 1984, Abbott 1985b, Cowling and Lamont 1985b, Lamont and Barker 1988), although it has been reported (Specht 1981, Zammit and Westoby 1987a).

Several genera of Myrtaceae are known for congregation of serotinous cones in clusters, and this has been demonstrated to be an important factor in shielding seeds from excessive heating during fires (Judd and Ashton 1991).

Eucalyptus spp. have woody capsules that release seed following maturation, but also retain seed within fire-resistant capsules so that seeds will disperse after fire. As in other serotinous species, seeds of *Eucalyptus* spp. are nonrefractory and germinate readily over a wide range of temperatures (Prescott 1941, Ladiges 1974, Zohar et al. 1975). Some *Eucalyptus* species require light and others require darkness for germination (Clifford 1953, Grose 1963), and an afterripening period has been recorded for others. In nature, germination occurs soon after the first autumn rains following release (Abbott 1984, Wellington and Noble 1985b). Consequently, these weakly serotinous *Eucalyptus* species do not accumulate a persistent soil seed bank (Vlahos and Bell 1986).

In unburned *Eucalyptus* stands, some seedling recruitment occurs, but successful seedling establishment is more likely in burned areas. Wellington and Noble (1985a) reported 7000 *E. incrassata* seedlings/ha in a one-year-old burn, but only 30 seedlings/ha in unburned mallee. Ants are important seed predators in unburned stands, which may be an important factor limiting recruitment (O'Dowd and Gill 1984, Andersen 1987).

Disturbance-Dependent Recruitment: Later Postfire

Geophytes apparently survive fire by resprouting and do not establish seedlings from a persistent seed bank (Purdie and Slatyer 1976, Purdie 1977, Bell et al. 1984). These include taxa in the Orchidaceae (*Caladenia* and *Prasophyllum*), Liliaceae (*Blandfordia, Dasypogon, Lomandra,* and *Thysanotus*), and Droseraceae (*Drosera*) (Pate and Dixon 1981). These resprouts flower in the first year after fire. From what is known about geophytes in other mediterranean type ecosystems, it is expected that Australian geophytes would have nonrefractory seeds and establish seedlings in subsequent postfire years. Some dependence upon fire is suggested by the observation that some geophyte species may disappear from infrequently burned sites (Nieuwenhuis 1987).

Some shrubs lack abundant seedling establishment after fire, but resprouts (or survivors) flower in the first and second year after fire. Seeds of these species are nonrefractory and seedlings establish in the first rainy season after dispersal, not unlike the coastal sage scrub of California. Examples include *Xanthorrhoea* spp. (Xanthorrhoeaceae), *Telopea speciosissima* (Proteaceae), *Lambertia formosa* (Proteaceae) and *Angophora hispida* (Myrtaceae) (Gill and Ingwersen 1976, Pyke 1983, Auld pers. comm.).

Conclusions

A remarkable degree of convergence appears in seed-germination syndromes in the five mediterranean-climate ecosystems. The main areas of convergence are summarized here, and those statements I believe to be much in need of further study are indicated (*).

1. *Disturbance-free recruitment.* All but the Australian region have shrub species with transient seed banks, and persistence of such species after fire depends entirely on their resprouting capacity. Most of these have fleshy fruits that are dispersed by birds in the autumn and winter. Seeds are nonrefractory and germinate soon after dispersal. In California, successful seedling recruitment seems to depend on long fire-free conditions. *In other regions the conditions required for successful seedling establishment by these taxa needs further study.

2. *Disturbance-dependent recruitment: soil seed banks.* All regions except Chile have diverse flora of species that cue their seedling recruitment to

the immediate postfire environment. Species that establish seedlings after fire must have a persistent seed bank, either in the soil or on the canopy. Soil-stored seeds are refractory. Heat-shock–stimulated germination is found in species of all regions, except possibly Chile. Seeds of such species are usually hard-seeded, meaning that the seed coat or pericarp imposes dormancy by preventing imbibition or gas exchange.

Some taxa with persistent soil seed banks are not hard-seeded, and many of these have charred wood–stimulated germination. This strategy is important in California and South Africa.

3. *Disturbance-dependent recruitment: canopy seed banks.* In South Africa and Australia, many species have persistent seed banks in the canopy (i.e., serotinous fruits) and seed germination is cued by timing of dispersal. Heat stimulates the opening of fruits, and the seeds are usually nonrefractory. For some species, seeds may have innate or induced dormancy mechanisms that prevent germination during summer. Serotinous species are unknown in Chile and uncommon in California and Europe. Why canopy-stored seed banks are more important in South Africa and Australia than elsewhere requires some thought. Seeds of serotinous species are typically much larger than those of soil-stored taxa, possibly because they have the luxury of retaining seeds within the protection of woody fruits.* Perhaps the low nutrient content of the South African and Western Australian soils has selected for large seeds that can store critical elements such as nitrogen and phosphorous.

4. *Polymorphic seed pools.* Postfire seeders, regardless of where the persistent seed bank is stored, have both a transient and a persistent seed bank. For soil-stored seeds this condition arises because of polymorphic seed-germination behavior. For canopy-stored seeds this arises because of incomplete retention of seeds within fruits until fire. Both have the potential for some seed germination without fire, although the most successful recruitment occurs after fire.

5. *Geophytes.* Apparently, in all five regions, geophytes do not recruit seedlings in the immediate postfire environment, and their presence is due to resprouts from bulbs, corms, or rhizomes. Seeds are nonrefractory and * are likely to germinate soon after the first autumn or winter rains following dispersal. For many species, flowering is restricted to, or more profuse in, the first year after fire. *The second year after fire is probably an important time for seedling establishment. One exception may be *Cyrtanthus angustifolius*, a fynbos geophyte that flowers within a few days after fires, and thus may recruit seedlings in the first postfire season.

The uniformity among regions in the response of geophytes to fire suggests that intense selective pressure is acting on this life form. Geophytes are unable to compete with the dominant shrub vegetation, and thus growing conditions are optimal after fire. Corms and bulbs are a reliable mode of establishing after fire; they are deeply buried and thus not subject to damage by fire or predation between fires. Flowering from stored carbohydrates produces an abundant seed pool in the first postfire year. *The subsequent post-fire years are the optimal time for seedling recruitment and therefore this life form has not evolved refractory seeds. A weakness of the geophyte strategy is that bulbs and corms need to persist in a relative dormant state until fire. If bulbs and corms were not long-lived then infrequent fires would select against this life form. *The very high fire frequency in South Africa and Australia may account for the abundance of geophytes in those regions, relative to other ecosystems of mediterranean type. *Other factors may be involved, for example, the low phosphorus and nitrogen content of soils in these Southern Hemisphere regions may also select for life forms capable of retaining these nutrients in underground corms or bulbs.

Acknowledgments

Support was provided by NSF grants SPI-7926576, TFI-8100529, and RII-8304946, and by a fellowship from the John Simon Guggenheim Foundation. I thank David Bell, Byron Lamont, Cheryl Swift, and Costas Thanos for very useful comments on an earlier version.

References

Abbott I (1984) Emergence and early survival of seedlings of six tree species in mediterranean forest of Western Australia. For Ecol and Man 9:51–66

Abbott I (1985a) Reproductive ecology of *Banksia grandis* (Proteaceae). New Phytol 99:129–148

Abbott I (1985b) Recruitment and mortality in populations of *Banksia grandis*. Austra J Bot 33:261–270

Adamson RS (1935) The plant communities of Table Mountain. III. A six years' study of regeneration after burning. J Ecol 23:43–55

Alonso I, Luis E, Tarrega R (1992) First phases of regeneration of *Cistus laurifolius* and *Cistus ladanifer* after burning and cutting in experimental plots. Int J Wildl Fire 2:7–14

Andersen AN (1987) Effects of seed predation by ants on seedling densities at a woodland site in SE Australia. Oikos 48:171–174

Angevine MW, Chabot BF (1979) Seed germination syndromes in higher plants. In Solbrig OT, Jain S, Johnson GB, Raven PH (eds) *Topics in Plant Population Biology*. Columbia U Press, New York, pp 188–206

Aronne G, Mazzoleni S (1989) The effects of heat exposure on seeds of *Cistus incanus* L. and *Cistus monspeliensis* L. Gior Bot Ital 123:283–289

Aschmann H, Bahre C (1977) Man's impact on the wild landscape. In Mooney HA (ed) *Convergent Evolution of Chile and California Mediterranean Climate Ecosystems*. Dowden, Hutchinson and Ross, Stroudsburg, PA, pp 73–84

Auld TD (1986a) Population dynamics of the shrub *Acacia suaveolens* (Sm.) Willd.: dispersal and the dynamics of the soil seed-bank. Austr J Ecol 11:235–254

Auld TD (1986b) Population dynamics of the shrub *Acacia suaveolens* (Sm.) Willd.: fire and the transition to seedlings. Aust J Ecol 11:373–385

Belhassen E, Pomente D, Trabaud L, Gouyon PH (1987) Récolonisation après incendie chez *Thymus vulgaris* (L.): résistance des graines aux températures élevées. Oecol/Oecol Plant 22 (new series vol 8):135–141

Bell DT, Hopkins AJM, Pate JS (1984) Fire in the kwongan. In Pate JS, Beard JS (eds) *Kwongan: Plant Life of the Sandplain*. U Western Australia Press, Nedlands, W.A., pp 178–204

Bell DT, Plummer JA, Taylor SK (1993) Seed germination ecology in southwestern Western Australia. Bot Rev 59:24–73

Bell DT, Vlahos S, Watson LE (1987) Stimulation of seed germination of understorey species of the northern Jarrah forest of Western Australia. Aust J Bot 35:593–599

Blommaert KLJ (1972) Buchu seed germination. J S African Bot 38:237–239

Bond WJ (1980) Fire and senescent fynbos in Swartberg, southern Cape. S African For J 114:68–71

Bond WJ (1983) Fire survival of Cape Proteaeceae: influence of fire season and seed predators. Vegetatio 54:65–74

Bond WJ (1984) Seed predators, fire and Cape Proteaceae: Limits to the population approach to succession. Vegetatio 56:65–74

Bond WJ (1985) Canopy-stored seed reserves (serotiny) in Cape Proteaceae. S African J Bot 51:181–186

Bond WJ, Vlok J, Viviers M (1984) Variation in seedling recruitment in fire-adapted Cape Proteaceae after fire. J Ecol 72:209–221

Bossard CC (1993) Seed germination in the exotic shrub *Cytisus scoparius* (Scotch broom) in California. Madroño 40:47–61

Bradstock RA, Myerscough PJ (1981) Fire effects on seed release and the emergence and establishment of seedlings in *Banksia ericifloria* L.f. Aust J Bot 29:521–531

Bradstock RA, O'Connell MA (1988) Demography of woody plants in relation to fire: *Banksia ericifolia* L.f. and *Petrophile pulchella* (Schrad) R.Br. Aust J Ecol 13:505–518

Breytenbach GJ (1984) Single agedness in fynbos: a predation hypothesis. In Dell B (ed), Medecos IV. Proc 4th Int Conf Medit Ecosys. Bot Dept, U Western Australia, Nedlands, W.A., pp 14–15

Brits GJ (1986a) Influence of fluctuating temperatures and H_2O_2 treatment on germination of *Leucospermum cordifolium* and *Serruria florida* (Proteaceae) seeds. S African J Bot 52:286–290

Brits GJ (1986b) The effect of hydrogen peroxide treatment on germination in Proteaceae species with serotinous and nut-like achenes. S African J Bot 52:291–293

Brits GJ (1987) Germination depth vs. temperature requirements in naturally dispersed seeds of *Leucospermum cordifolium* and *L. cuneiforme* (Proteaceae). S African J Bot 53:119–124

Brits GJ, van Niekerk MN (1986) Effects of air temperature, oxygenating treatments and low storage temperature on seasonal germination response of *Leucospermum cordifolium* (Proteaceae) seeds. S African J Bot 52:207–211

Brown JM, Hopkins AJM (1984) Regeneration after fire in 170-year-old vegetation on Middle Island, south-western Australia. In Dell B (ed), Medecos IV. Proc 4th Int Conf Medit Ecosys. Bot Dept, U Western Australia, Nedlands, W.A., pp 18–19

Brown NAC, Dix L (1985) Germination of the fruits of *Leucadendron*. S African J Bot 51:448–452

Carlson JR, Sharp SC (1975) Germination of high elevation manzanitas. USDA For Serv, Tree Planters' Notes 26(3):10–11, 25–26

Cavanagh AK (1980) A review of some aspects of the germination of acacias. Proc Royal Soc Victoria 91:161–180

Christensen NL, Muller CH (1975a) Effects of fire on factors controlling plant growth in *Adenostoma* chaparral. Ecol Mono 45:29–55

Christensen NL, Muller CH (1975b) Relative importance of factors controlling germination and seedling survival in *Adenostoma* chaparral. Am Mid Nat 93:71–78

Clifford HT (1953) A note on the germination of *Eucalyptus* seed. Aust For 17:17–20

Cowling RM, Lamont BB (1985a) Seed release in *Banksia*: the role of wet-dry cycles. Aust J Ecol 10:169–171

Cowling RM, Lamont BB (1985b) Variation in serotiny of three *Banksia* species along a climatic gradient. Aust J Ecol 10:345–350

Cowling RM, Lamont BB (1987) Post-fire recruitment of four co-occurring *Banksia* species. J Appl Ecol 24:645–658

Cowling RM, Lamont BB, Pierce SM (1987) Seed bank dynamics of four co-occurring *Banksia* species. J Ecol 75:289–302

Dafni A, Cohen D, Noy-Meir I (1981) Life-cycle variation in geophytes. Ann Missouri Bot Gar 68:652–660

Davidson DW, Morton SR (1984) Dispersal adaptations of some *Acacia* species in the Australian arid zones. Ecology 65:1038–1051

Davis G (1992) Regeneration traits in the weakly serotinous obligate seeder *Leucadendron xanthoconus* (Proteaceae). S African J Bot 58:125–128

Deall GB, Brown NAC (1981) Seed germination in *Protea magnifica* Link. S African J Sci 77:175–176

Debusshe M (1985) Rôle des oiseaux desseminateurs dans la germination des graines de plantes à fruits charnus en région Méditerranéenne. Acta Oecol/Oecol Plant 20 (new series vol 6):365–374

Debusshe M, Escarre J, Lepart J (1980) Changes in Mediterranean shrub communities with *Cytisus purgans* and *Genista scorpius*. Vegetatio 43:73–82

de Lange JH, Boucher C (1990) Autecological studies on *Audouinia capitata* (Bruniaceae). I. Plant-derived smoke as a seed germination cue. S African J Bot 56:700–703

del Moral R, Muller CH (1969) Fog drip: a mechanism of toxin transport from *Eucalyptus globulus*. Bull Torrey Bot Club 96:467–475

Del Pozo AH, Fuentes ER, Hajek ER, Molina JD (1989) Zonación microclimática por efecto de los manchones de arbustos en el matorral de Chile central. Rev Chil Hist Nat 62:85–94

Emery DE (1988) *Seed Propagation of Native California Plants*. Santa Barbara Bot Gard. Santa Barbara, CA

Enright NJ, Lamont BB (1989) Fire temperatures and follicle opening requirements of ten *Banksia* species. Aust J Ecol 14:107–113

Esterhuizen AD, van de Venter HA, Robbertse PJ (1986) A preliminary study on seed germination of *Watsonia fourcadei*. S African J Bot 52:221–225

Floyd AG (1966) Effect of fire on weed seeds in wet sclerophyll forests of northern New South Wales. Aust J Bot 14:243–256

Frost PGH (1984) The responses and survival of organisms in fire-prone environments. In Booysen PdV, Tainton NM (eds) *Ecological Effects of Fire in South African Ecosystems*. Springer, Heidelberg, pp 273–309

Fuentes ER, Espinosa G (1986) Resilience of central Chile shrublands: a vulcanism-related hypothesis. Interciencia 11:164–165

Fuentes ER, Espinosa G, Gajardo G (1987) Allelopathic effects of the Chilean matorral shrub *Flourensia thurifera*. Rev Chil de Histor Nat 60:57–62

Fuentes ER, Hoffman AJ, Poiani A, Alliende MC (1986) Vegetation change in large clearings: patterns in the Chilean matorral. Oecologia 68:358–366

Fuentes ER, Otaiza RD, Alliende MC, Hoffmann A, Poiani A (1984) Shrub clumps of the Chilean matorral vegetation: structure and possible maintenance mechanisms. Oecologia 42:405–411

Gill AM (1975) Fire and the Australian flora: a review. Aust For 38:4–25

Gill AM (1976) Fire and the opening of *Banksia ornata* F. Muell. follicles. Aust J Bot 24:329–335

Gill AM, Ingwersen F (1976) Growth of *Xanthorrhoea australis* R. Br. in relation to its fire tolerance. J Appl Ecol 13:195–203

Gratani L, Amadori M (1991) Post-fire resprouting of shrubby species in Mediterranean maquis. Vegetatio 96:137–143

Grose RJ (1963) Effective seed supply for the natural regeneration of *Eucalyptus delegatensis*. I. Germination and seed dormancy. School For, U Melbourne, Bull No. 2

Harif I (1978) The effect of burning on the germination and development of the maquis plants. Israel J Bot 27:44

Harper JL (1977) *Population Biology of Plants*. Academic Press, New York

Herrera CM (1984a) A study of avian frugivores, bird-dispersed plants, and their interaction in Mediterranean scrublands. Ecol Mono 54:1–23

Herrera CM (1984b) Adaptation to frugivory of Mediterranean avian seed dispersers. Ecology 65: 609–617

Herrera CM (1987) Vertebrate-dispersed plants of the Iberian Peninsula: a study of fruit characteristics. Ecol Mono 57:305–331

Herrera CM (1992a) Historical effects and sorting processes as explanations for contemporary ecological patterns: character syndromes in Mediterranean woody plants. Am Nat 140:421–446

Herrera CM (1992b) Mediterranean plant–bird seed dispersal systems: The roles of history and adaptation. In Thanos CA (ed) Medecos VI Proc 6th Int Conf Medit Clim Ecosys *Plant-Animal Interactions in Mediterranean-type Ecosystems*, held at Maleme, Crete (Greece) September 23–27, 1991. U Athens, Greece, pp 241–250

Herrera J (1991) The reproductive biology of a riparian mediterranean shrub, *Nerium oleander* L. (Apocynaceae). Bot J Linnean Soc 106:147–172

Higgins KB, Lamb AJ, van Wilgen BW (1987) Root systems of selected plant species in mesic mountain fynbos in the Jonkershoek Valley, south-western Cape Province. S African J Bot 53:249–257

Hobbs RJ, Atkins L (1991) Interactions between annuals and woody perennials in a Western Australian nature reserve. J Veg Sci 2:643–654

Hodgkinson KC, Oxley RE (1990) Influence of fire and edaphic factors on germination of the arid zone shrubs *Acacia anerua*, *Cassia nemophila* and *Dodonaea viscosa*. Aust J Bot 38:269–279

Hoffmann AJ, Teillier S, Fuentes ER (1989) Fruit and seed characteristics of woody species in mediterranean-type regions of Chile and California. Rev Chil Hist Nat 62:43–60

Hoffmann MT, Moll EJ, Boucher C (1987) Post-fire succession at Pella, a South African lowland fynbos site. S African J Bot 53:370–374

Horton JS, Kraebel CJ (1955) Development of vegetation after fire in the chamise chaparral of southern California. Ecology 36:244–262

Izhaki I, Safriel UN (1985) Why do fleshy-fruit plants of the Mediterranean scrub intercept fall—but not spring—passage of seed-dispersing migratory birds? Oecologia 67:40–43

Izhaki I, Safriel UN (1990) The effect of some Mediterranean scrubland frugivores upon germination patterns. J Ecol 78:56–65

Izhaki I, Lahav H, Ne'eman G (1992) Spatial distribution patterns of *Rhus coriaria* seedlings after fire in a Mediterranean pine forest. Acta Oecol 13:279–289

Izhaki I, Walton PB, Safriel UN (1991) Seed shadows generated by frugivorous birds in an eastern Mediterranean scrub. J Ecol 79:575–590

Jaksic FM, Fuentes ER (1980) Why are native herbs in the Chilean matorral more abundant beneath bushes: microclimate or grazing? J Ecol 68:665–669

Jeffery DJ, Holmes PM, Rebelo AG (1988) Effects of dry heat on seed germination in selected indigenous and alien legume species in South Africa. S African J Bot 54:28–34

Jimenez H, Armesto JJ (1992) Importance of the soil seed bank of disturbed sites in Chilean matorral in early secondary succession. J Veg Sci 3:579–586

Jones CS, Schlesinger WH (1980) *Emmenanthe penduliflora* (Hydrophyllaceae): further consideration of germination response. Madroño 27:122–125

Judd TS, Ashton DH (1991) Fruit clustering in the Myrtaceae: seed survival in capsules subjected to experimental heating. Aust J Bot 39:241–245

Kaminsky R (1981) The microbial origin of the allelopathic potential of *Adenostoma fasciculatum* H. and A. Ecol Mono 51:365–382

Keeley JE (1986) Seed germination patterns of *Salvia mellifera* in fire-prone environments. Oecologia 71:1–5

Keeley JE (1987) Role of fire in seed germination of woody taxa in California chaparral. Ecology 68:434–443

Keeley JE (1991) Seed germination and life history syndromes in the California chaparral. Bot Rev 57:81–116

Keeley JE (1992a) Demographic structure of California chaparral in the long-term absence of fire. J Veg Sci 3:79–90

Keeley JE (1992b) Recruitment of seedlings and vegetative sprouts in unburned chaparral. Ecology 73:1194–1208

Keeley JE (1992c) A Californian's view of fynbos. In Cowling RM (ed) *The Ecology of Fynbos: Nutrients, Fire and Diversity*. Oxford U Press, Cape Town, RSA, pp 372–388

Keeley JE (1993) Smoke-induced flowering in the fire-lily *Cyrtanthus ventricosos*. S Afr J Bot 59:638.

Keeley JE (1992d) Temporal and spatial dispersal syndromes. In Thanos CA (ed) Medecos VI Proc 6th Int Conf Medit Clim Ecosys *Plant-Animal Interactions in Mediterranean-Type Ecosystems* held at Maleme, Crete (Greece) September 23–27, 1991. U Athens, Greece, pp 251–256

Keeley JE, Brooks AJ, Bird, T, Cory S, Parker H, Usinger E (1986) Demographic structure of chaparral under extended fire-free conditions. In DeVries JJ (ed) Proc Chaparral Ecosys Res Conf. California Water Resources Center, U California, Davis, Report No. 62, pp 133–137

Keeley SC, Johnson AW (1977) A comparison of the pattern of herb and shrub growth in comparable sites in Chile and California. Am Mid Nat 97:120–132

Keeley JE, Keeley SC (1984) Postfire recovery of California coastal sage scrub. Am Mid Nat 111:105–117

Keeley JE, Keeley SC (1987) The role of fire in the germination of chaparral herbs and suffrutescents. Madroño 34:240–249

Keeley JE, Keeley SC (1989) Allelopathy and the fire induced herb cycle. In Keeley SC (ed) *The California Chaparral: Paradigms Re-examined*. Nat Hist Mus LA County, CA, Sci Ser, No. 34, pp 65–72

Keeley JE, Morton BA, Pedrosa A, Trotter P (1985) Role of allelopathy, heat, and charred wood in the germination of chaparral herbs and suffrutescents. J Ecol 73:445–458

Keeley JE, Nitzberg ME (1984) The role of charred wood in the germination of the chaparral herbs *Emmenanthe penduliflora* (Hydrophyllaceae) and *Eriophyllum confertiflorum* (Asteraceae). Madroño 31:208–218

Keeley JE, Zedler PH (1978) Reproduction of chaparral shrubs after fire: a comparison of sprouting and seeding strategies. Am Mid Nat 99:142–161

Keeley SC, Keeley JE, Hutchinson SE, Johnson AW (1981) Postfire succession of the herbaceous flora in southern California chaparral. Ecology 62:1608–1621

Keeley SC, Pizzorno M (1986) Charred wood stimulated germination of two fire-following herbs of the Califonia chaparral and the role of hemicellulose. Am J Bot 73:1289–1297

Kilian D, Cowling RM (1992) Comparative seed biology and co-existence of two fynbos shrub species. J Veg Sci 3:637–646

Kruger FJ (1977) Ecology of Cape fynbos in relation to fire. In Mooney HA, Conrad

CE (eds) Proc Sym Environ Cons Fire Fuel Man Medit Ecosys. USDA For Serv, Gen Tech Rep WO-3, pp 230–244

Kruger FJ (1979) Plant ecology. In Day J, Siegfried WR, Louw GN, Jarman ML (eds) *Fynbos Ecology: A Preliminary Synthesis.* S African Nat Sci Prog Rep No. 40, pp 88–126

Kruger FJ (1983) Plant community diversity and dynamics in relation to fire. In Kruger FJ, Mitchell DT, Jarvis JUM (eds) *Mediterranean-type Ecosystems. The Role of Nutrients.* Springer-Verlag, New York, pp 446–472

Ladiges PY (1974) Differentiation in some populations of *Eucalyptus viminalis* Labill. in relation to factors affecting seedling establishment. Aust J Bot 22:471–487

Lamont BB, Barker MJ (1988) Seed bank dynamics of a serotinous, fire-sensitive *Banksia* species. Aust J Bot 36:193–203

Lamont BB, Connell SW, Bergl SM (1991) Seed bank and population dynamics of *Banksia cuneata*: the role of time, fire, and moisture. Bot Gazette 152:114–122

Lamont BB, Witkowski ETF, Enright NJ (1993) Post-fire litter microsites: safe for seeds, unsafe for seedlings. Ecology 74:501–512

Le Houerou HN (1974) Fire and vegetation in the Mediterranean basin. Proc Tall Timbers Fire Ecol Conf 13:237–277

Le Maitre DC (1990) The influence of seed ageing on the plant on seed germination in *Protea nerifolia* (Proteaceae). S African J Bot 56:49–53

Le Maitre DC, Brown PJ (1992) Life cycles and fire-stimulated flowering in geophytes. In van Wilgen BW, Richardson DM, Kruger FJ, van Hensbergen JH (eds) Fire in South African Mountain Fynbos. Springer-Verlag, New York, pp 145–160

Le Maitre DC, Midgley JJ (1992) Plant reproductive ecology. In Cowling RM (ed) *The Ecology of fynbos: Nutrients, Fire and Diversity.* Oxford U Press, Cape Town, RSA, pp 135–174

Levyns MR (1927) A preliminary note on the rhenoster bush *Elytropappus rhinocerotis* and the germination of its seed. Trans Royal Soc S Africa 14:383–388

Levyns MR (1935) Veld burning experiments at Oakdale, Riversdale. Trans Royal Soc S Africa 23:231–244

Levyns MR (1966) *Haemanthus canaliculatus*, a new fire-lily from the western Cape Province. J S African Bot 32:73–75

Manders PT (1986) Seed dispersal and seedling recruitment in *Protea laurifolia*. S African J Bot 52:421–424

Manders PT (1987) Is there allelopathic self-inhibition of generative regeneration within *Widdringtonia cedarbergensis* stands? S African J Bot 53:408–410

Manders PT, Richardson DM, Masson PH (1992) Is fynbos a stage in succession to forest? Analysis of the perceived ecological distinction between two communities. In van Wilgen BW, Richardson DM, Kruger FJ, van Hensbergen HJ (eds) *Fire in South African Mountain Fynbos: Ecosystem, Community and Species Response at Swartboskloof.* Springer-Verlag, New York, pp 81–107

Margaris NS (1981) Structure and dynamics of mediterranean type vegetation. Portugaliae Acta Biolog 16(4):45–58

Martin ARH (1966) The plant ecology of the Grahamstown Nature Reserve. II. Some effects of burning. J S African Bot 32:1–39

Martin LB, Juhren M (1954) *Cistus* and its response to fire. Lasca Leaves 4:65–69

Martin RE, Cushwa C (1966) Effects of heat and moisture on leguminous seed. Proc Tall Timb Fire Ecol Conf 5:159–176

May FE, Ash JE (1990) An assessment of the allelopathic potential of *Eucalyptus*. Aust J Bot 38:245–254

Mazzoleni S (1989) Fire and mediterranean plants: germination responses to heat exposure. Anna Bot 47:227–233

McPherson JK, Muller CH (1969) Allelopathic effects of *Adenostoma fasciculatum*, "chamise," in the California chaparral. Ecol Mono 39:177–198

Mesleard F, Lepart J (1991) Germination in seedling dynamics of *Arbutus unedo* and *Erica arborea* on Corsica. J Veg Sci 2:155–164

Mills JN, Kummerow J (1989) Herbivores, seed predators and chaparral succession. In Keeley SC (ed) *The California Chaparral: Paradigms Re-examined*. Nat Hist Mus LA County, CA, Sci Ser, No. 34, pp 49–55

Mirov NT (1936) Germination behaviour of some California plants. Ecology 17:667–672

Mitrakos K (1981) Temperature germination responses in three Mediterranean evergreen sclerophylls. In Margaris NS, Mooney HA (eds) *Components of Productivity of Mediterranean-climate Regions: Basic and Applied Aspects*. Dr. W. Junk, The Hague, pp 277–279

Montenegro G, Avila G, Schatte P (1983) Presence and development of lignotubers in shrubs of the Chilean matorral. Can J Bot 61:1804–1808

Montenegro G, Rivera O, Bas F (1978) Herbaceous vegetation in the Chilean matorral. Dynamics of growth and evaluation of allelopathic effects of some dominate shrubs. Oecologia 36:237–244

Mooney HA, Kummerow J, Johnson AW, Parsons DJ, Keeley S, Hoffmann A, Hays RI, Giliberto J, Chu C (1977) The producers: their resources and adaptive responses. In Mooney HA (ed), *Convergent Evolution of Chile and California Mediterranean Climate Ecosystems*. Dowden, Hutchinson and Ross, Stroudsburg, PA, pp 85–143

Mott JJ, Groves RH (1981) Germination strategies. In Pate JS, McComb AJ (eds) *The Biology of Australian Plants*. U Western Australia Press, Nedlands, W.A., pp 307–341

Muñoz MR, Fuentes ER (1989) Does fire induce shrub germination in the Chilean matorral? Oikos 56:177–181

Naveh Z (1974) Effect of fire in the Mediterranean region. In Kozlowski TT, Ahlgren CE (eds) *Fire and Ecosystems*. Academic Press, New York, pp 401–434

Naveh Z (1975) The evolutionary significance of fire in the Mediterranean region. Vegetatio 29:199–208

Ne'eman G, Lahav H, Izhaki I (1992) Spatial pattern of seedlings 1 year after fire in a Mediterranean pine forest. Oecologia 91:365–370

Nieuwenhuis A (1987) The effect of fire frequency on the sclerophyll vegetation of the West Head, New South Wales. Aust J Ecol 12:373–385

O'Dowd DJ, Gill AM (1984) Predator satiation and site alteration following fire: mass reproduction of alpine ash (*Eucalyptus delegatensis*) in southeastern Australia. Ecology 65:1052–1066

Papanastasis VP, Romanas LC (1977) Effect of high temperatures on seed germination of certain Mediterranean half-shrubs. Thessaloniki For Res Inst Bull 86 (in Greek)

Papio C (1988) Respuesta al fuego de las principales especies de la vegetación de Garraf (Barcelona). Orsis 3:87–103

Parker VT, Kelly VR (1989) Seed banks in California chaparral and other mediterranean climate shrublands. In Leck MA, Parker VT, Simpson RL (eds) *Ecology of Soil Seed Banks*. Academic Press, San Diego, pp 231–255

Pate JS, Dixon KW (1981) Plants with fleshy underground storage organs—a Western Australian survey. In Pate JS, McComb AJ (eds) *The Biology of Australian Plants*. U Western Australia Press, Nedlands, W.A., pp 181–215

Patric JH, Hanes TL (1964) Chaparral succession in a San Gabriel Mountain area of California. Ecology 45:353–360

Pierce SM (1990) Pattern and process in south coast dune fynbos: population, community and landscape level studies. Ph.D. Thesis, U Cape Town, Rondebosch, South Africa

Pierce SM, Cowling RM (1991) Dynamics of soil-stored seed banks of six shrubs in fire-prone dune fynbos. J Ecol 79:731–747

Pieterse PJ, Cairns ALP (1986) The effect of fire on an *Acacia longifolia* seed bank in the south-western Cape. S African J Bot 52:233–236

Preece PB (1971) Contributions to the biology of mulga. II. Germination. Aust J Bot 19:39–49

Prescott EE (1941) Germination of the seed of mallee eucalypts. Vict Nat 58:8

Purdie RW (1977) Early stages of regeneration after burning in dry sclerophyll vegetation. II. Regeneration by seed germination. Aust J Bot 25:35–46

Purdie RW, Slatyer RO (1976) Vegetation succession after fire in sclerophyll woodland communities in south-eastern Australia. Aust J Ecol 1:223–236

Pyke GH (1983) Relationship between time since the last fire and flowering in *Telopea speciosissima* R. Br. and *Lambertia formosa* Sm. Aust J Bot 31:293–396

Quick CR (1935) Notes on the germination of *Ceanothus* seeds. Madroño 3:135–140

Ricardo CPP, Veloso MM (1987) Features of seed germination in *Arbutus unedo* L. In Tenhunen JD, Catarino FM, Lange O, Oechel WC (eds) (1987) *Plant Response to Stress: Functional Analysis in Mediterranean Ecosystems*. Springer, Berlin, pp 565–572

Rice B, Westoby M (1986) Evidence against the hypothesis that ant-dispersed seeds reach nutrient-enriched microsites. Ecology 67:1270–1274

Richardson DM, van Wilgen BW, Mitchell DT (1987) Aspects of the reproductive ecology of four Australian *Hakea* species (Proteaceae) in South Africa. Oecologia 71:345–354

Roy J, Arianoutsou-Faraggtaki M (1985) Light quality as the environmental trigger for the germination of the fire-promoted species *Sarcopoterium spinosum* L. Flora 177:345–349

Roy J, Sonie L (1992) Germination and population dynamics of *Cistus* species in relation to fire. J Appl Ecol 29:647–655

Rundel PW (1981) Fire as an ecological factor. In Lange OL, Nobel PS, Osmond CB, Ziegler H (eds) *Physiological Plant Ecology*. I. Springer-Verlag, New York, pp 501–538

Scriba JH (1976) The effects of fire on *Widdringtonia modiflora* (L.) Powrie on Mariepskop. S African For J 97:12–17

Shea SR, McCormick J, Portlock CC (1979) The effect of fires on regeneration of leguminous species in the northern jarrah (*Eucalyptus marginata* Sm.) forest of Western Australia. Aust J Ecol 4:195–205

Siddiqi MY, Myerscough PJ, Carolin RC (1976) Studies in the ecology of coastal heath in New South Wales. IV. Seed survival, germination, seedling establishment and early growth in *Banksia serratifolia* Salisb., *B asplenifolia* Salisb. and *B. ericifolia* L.F. in relation to fire: temperature and nutritional effects. Aust J Ecol 1:175–183

Slingsby P, Bond WJ (1985) The influence of ants on the dispersal distance and seedling recruitment of *Leucospermum concocarpodendon* (L.) Buek (Proteaceae). S African J Bot 51:30–34

Sonia L, Heslehurst MR (1978) Germination characteristics of some *Banksia* species. Aust J Ecol 3:179–186

Specht RL (1981) Responses to fires in heathlands and related shrublands. In Gill AM, Groves RH, Noble IR (eds) *Fire and the Australian Biota*. Aust Acad Sci, Canberra, Australian Capital Territory, pp 394–415

Stone EC (1951) The stimulative effect of fire on the flowering of the golden brodiaea (*Brodiaea ixiodes* Wats. var. *lugens* Jeps.). Ecology 32:534–537

Stone EC, Juhren G (1951) The effect of fire on the germination of the seed of *Rhus ovata* Wats. Am J Bot 38:368–372

Sweeney JR (1956) Responses of vegetation to fire. A study of the herbaceous vegetation following chaparral fires. U California Pub Bot 28:143–216

Tarrega R, Calvo L, Trabaud L (1992) Effect of high temperatures on seed germination of two woody Leguminosae. Vegetatio 102:139–147

Taylor HC (1978a) Notes on the vegetation of the Cape Flats. Bothalia 10:637–646

Taylor HC (1978b) Capensis. In Werger MJA (ed) *The Biogeography and Ecology of Southern Africa*. Dr. W. Junk, The Hague, pp 171–229

Thanos CA, Georghiou K (1988) Ecophysiology of fire-stimulated seed germination in *Cistus incanus* ssp. *creticus* (L.) Heywood and *C. salvifolius*. Plant, Cell Env 11:841–849

Thanos CA, Marcou S, Christodoulakis D, Yannitsaros A (1989) Early post-fire regeneration in *Pinus brutia* forest ecosystems of Samos Island (Greece). Acta Oecol Ecol Plant 19:79–94

Thanos CA, Skordilis A (1987) The effects of light, temperature, and osmotic stress on the germination of *Pinus halepensis* and *P. brutia* seeds. Seed Sci Tech 15:163–174

Trabaud L (1987) Natural and prescribed fire: survival strategies of plants and equilibrium in mediterranean ecosystems. In Tenhunen JD, Caterino FM, Lange OL, Oechel WC (eds) *Plant Response to Stress. Functional Analysis in Mediterranean Ecosystems*. Springer-Verlag, Berlin, pp 607–621

Trabaud L, Oustric J (1989) Heat requirements for seed germination of three *Cistus* species in the garrigue of southern France. Flora 183:321–325

Troumbis A, Trabaud L (1986) Comparison of reproductive biological attributes of two *Cistus* species. Acta Oecol/Oecol Plant 21 (new ser vol 7):235–250

Valbuena L, Tarrega R, Luis E (1992) Influence of temperature on germination of *Cistus laurifolius* and *Cistus ladanifer*. Int J Wildland Fire 2:15–20

van de Venter HAJ, Esterhuizen AD (1988) The effect of factors associated with fire on seed germination of *Erica sessiliflora* and *E. hebecalyx* (Ericaceae). S African J Bot 54:301–304

van Staden J, Brown NAC (1977) Studies on the germinating of South African Proteacae—a review. Seed Sci Tech 5:633–643

van Wilgen BW, Forsyth GG (1992) Regeneration strategies in fynbos plants and their influence on the stability of community boundaries after fire. In van Wilgen BW, Richardson DM, Kruger FJ, van Hensbergen HJ (eds) *Fire in South African Mountain Fynbos*. Springer-Verlag, New York, pp 54–80

van Wilgen BW, Viviers M (1985) The effect of fire on serotinous Proteaceae in the western Cape and the implications for fynbos management. S African For J 133:49–53

Versfeld DB, Richardson DM, van Wilgen BW, Chapman RA, Forsyth GG (1992) The climate of Swartboskloof. In van Wilgen BW, Richardson DM, Kruger FJ, van Hensbergen HJ (eds) *Fire in South African Mountain Fynbos*. Springer-Verlag, New York, pp 21–36

Vlahos S, Bell DT (1986) Soil seed-bank components of the northern jarrah forest of Western Australia. Aust J Ecol 11:171–179

Vogl RJ, Armstrong WP, White LL, Cole KL (1977) The closed-cone pines and cypresses. In Barbour MJ, Major J (eds) *Terrestrial Vegetation of California*. John Wiley, New York, pp 295–358

Warcup JH (1980) Effect of heat treatment of forest soil on germination of buried seed. Aust J Bot 28:567–571

Wellington AB, Noble IR (1985a) Post-fire recruitment and mortality in a population of the mallee *Eucalyptus incrassata* Labill. J Ecol 73:645–656

Wellington AB, Noble IR (1985b) Seed dynamics and factors limiting recruitment of the mallee *Eucalyptus incrassata* in semi-arid, south-eastern Australia. J Ecol 73:657–666

Wells PV (1969) The relation between mode of reproduction and extent of speciation in woody genera of the California chaparral. Evolution 23:264–267

Westman WE, O'Leary JF (1986) Measures of resilience: the response of coastal scrub to fire. Vegetatio 65:179–189

Wicht CL (1948) A statistically designed experiment to test the effects of burning on a sclerophyll shrub community. Trans Royal Soc S Africa 31:479–501

Wicklow DT (1977) Germination response in *Emmenanthe penduliflora* (Hydrophyllaceae). Ecology 58:201–205

Williams IJM (1972) A revision of the genus *Leucadendron* (Proteaceae). Cont Bolus Herbr 3:1–425

Zammit C, Westoby M (1987a) Population structure and reproductive status of two *Banksia* shrubs at various times after fire. Vegetatio 70:11–20

Zammit C, Westoby M (1987b) Seedling recruitment strategies in obligate-seeding and resprouting *Banksia* shrubs. Ecology 68:1984–1992

Zammit C, Westoby M (1988) Pre-dispersal seed losses, and the survival of seeds and seedlings of two serotinous *Banksia* shrubs in burnt and unburnt heath. J Ecol 76:200–214

Zedler PH (1977) Life history attributes of plants and the fire cycle: a case study in chaparral dominated by *Cupressus forbesii*. In Mooney HA, Conrad CE (eds) Proc Symp Environ Cons Fire Fuel Man Mediter Ecosys. USDA For Serv, Gen Tech Rep WO-3, pp 451–458

Zedler PH (1981) Vegetation change in chaparral and desert communities in San Diego county, California. In West DC, Shugart HH, Botkin D (eds) *Forest Succession. Concepts and Applications.* Springer-Verlag, New York, pp 406–430

Zohar Y, Wisel Y, Karschon R (1975) Effects of light, temperature and osmotic stress on seed germination of *Eucalyptus occidentalis* Endl. Aust J Bot 23:391–397

11. Distribution and Ecological Significance of Seed-Embryo Types in Mediterranean Climates in California, Chile, and Australia

Nancy J. Vivrette

The ability of a seed to germinate and grow quickly under favorable conditions provides a competitive advantage. In semiarid climates, rapid germination improves the chances that a seedling will be able to utilize the brief period when moisture is available. The ability of an embryo within a seed to develop rapidly is partly related to the type of food storage in that seed. Seeds differ in internal morphology, ranging from those with small embryos and storage materials in adjacent tissues such as endosperm, to those with large embryos and storage materials within the cotyledons of the embryo. The possession of embryos and copious endosperm has been interpreted as a primitive condition in seeds (Stebbins 1974). In his discussion of the evolutionary trend from food storage in endosperm to food storage in the cotyledons, Stebbins states: "Most families and genera having [food storage in the cotyledons] consist of woody plants that possess vigorous seedlings, suggesting that the substitution of food material in the cotyledons for that in the endosperm somehow increases the speed and efficiency of germination and the vigor of the seedlings." These observations have stimulated questions about the effects of embryo size and development on germination ecology and plant distribution, and suggest that placement of food storage is a character that has been acted upon by selective forces.

A number of published investigations are useful in understanding the relationship between seed morphology and germination behavior. Martin (1946) described the internal morphology of seeds and classified their em-

CLASSIFICATION OF EMBRYO TYPES

I. EMBRYO SMALL, DEPENDENT UPON ENDOSPERM. Embryo location:

A. Basal, rudimentary

B. Axile: large seeded

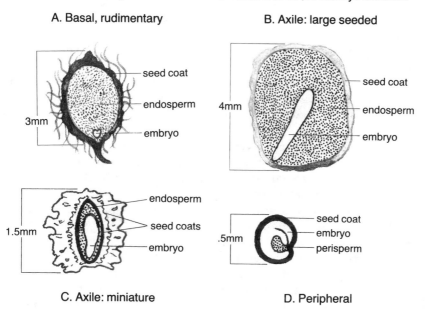

C. Axile: miniature

D. Peripheral

II. EMBRYO LARGE, INDEPENDENT OF ENDOSPERM. Seed cover protection:

A. Hard

B. Mucilaginous

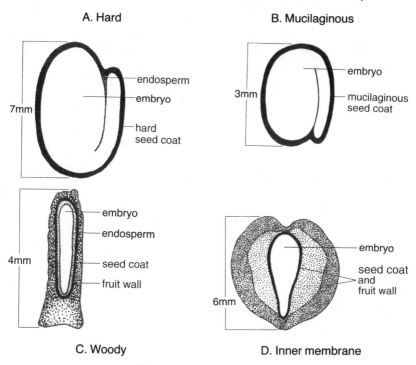

C. Woody

D. Inner membrane

Figure 11.1.

bryos based on their shape, position, and orientation. Corner (1976) described the similarity of the internal morphology of seeds within plant families and between closely related families. Nikolaeva (1969) described the similar physiological responses of seeds with similar morphologies. Atwater (1980) described the close association between plant family, internal seed morphology, and germination response under laboratory conditions; and modified Martin's classification for herbaceous dicotyledons to a system of eight embryo types. Atwater and Vivrette (1987) grouped these eight types in two major classes (Fig. 11.1). The first class includes seeds with small embryos that depend upon copious endosperm. The second major class includes seeds with fully developed embryos and small amounts of endosperm. In this group, the storage materials are found primarily in the cotyledons, making the embryo independent of the endosperm.

This accumulated knowledge of the nature and importance of embryo types leads to these questions: Can the embryo type of the seed be used to predict components of seed germination and seed ecology? What are the selective forces that lead to greater storage materials in the embryo? Do regions with different climates have plants with different embryo types? Using mediterranean climates as a test case, do regions with similar environmental conditions have similar proportions of embryo types? How do these proportions compare to those in more extreme climatic regimes such as deserts? The first step in answering these questions was to examine the rates of germination and emergence of seeds classified by embryo types. The second step was to describe the geographic and ecological distribution of species classified by these embryo types. The third was to compare distribution of embryo types in similar climatic regions. The fourth was to compare distribution of embryo types in adjacent climatic regions with different environmental constraints to seedling establishment.

Classification of Embryo Types

The data in this investigation were obtained by surveying the literature. For this study, only the two major classes "dependent" and "independent" embryo type were used. Dependent embryo types were defined as those that have small embryos relative to the copious endosperm or when the embryo was completely surrounded by a distinct layer of endosperm. Independent embryo types have embryos that are large relative to the size of the seed with little or no endosperm.

Descriptions of plant families were obtained from Heywood (1978). These were compared with the embryo classification of Atwater and Vivrette (1987) and the descriptions by Corner (1976). When it was not possible to make a clear classification using these references, live material was examined.

In the analysis, all genera and species within a family were assumed to have the same embryo and endosperm characteristics. Based on Corner's

(1976) descriptions, this is a reasonable assumption for most plant families. The classification into only two embryo types is a simplification imposed for ease of analysis. In reality, there is a gradient among plant families from seeds with copious endosperm to seeds with little or none. Some of the classification decisions were therefore somewhat arbitrary.

Classification into embryo type is most difficult in families with peripheral embryo types (Fig. 11.1) which grade from seeds with large amounts of endosperm to seeds with almost no endosperm or perisperm. Peripheral embryos are characteristic of the entire order Caryophyllales, all of which is here included within the dependent-embryo category. A similar problem is found within the hard-seed type, in the independent-embryo category, because some hard seeds still retain large amounts of endosperm (Fig. 11.1).

Incomplete information necessitated omitting some families and groups. Some of the references on which this study is based are restricted to herbaceous dicotyledons, so that species in woody dicotyledonous families that have only shrubs and trees were omitted from some of the analyses. All monocotyledons and gymnosperms were omitted. Herbaceous dicotyledons were omitted from the analysis if the descriptions of the families did not include the internal morphology of the seed, and if seeds were not available for determination. Because of these limitations, 7 percent of the species were omitted from the worldwide analysis, leaving a remainder of 114,316 species in this survey.

Ecology and Biogeography of Embryo Types

Germination and Rate of Emergence by Embryo Type

To test if the type of food storage in seeds affects the speed of germination, the species of herbaceous dicotyledons listed in the Rules of the Association of Official Seed Analysts (AOSA) (1981) were classified by embryo type (defined by the amount of endosperm) and the time needed for germination. Two periods are listed in the AOSA rules for testing seed: the first count (the number of days on which a first count of seedlings is to be made in a germination test) and the final count (the number of days when a final count of seedlings is made) (Table 11.1). The number of days for the first count indicates early germination (the emergence of intact seedlings) and the number of days for the final count indicates the completion of germination. Species belonging to families with dependent-embryo types require longer both for early emergence and complete emergence compared to the average for all species. Species in families with independent-embryo types require a shorter time for both early emergence and complete emergence. Species with food storage in the embryo (independent-embryo types) on average germinate faster and therefore require a shorter time to seedling establishment. Food storage in the embryo could be an advantage in regions with lower or less-predictable rainfall.

Table 11.1. Mean number of days to first and final counts for seedling emergence in germination tests of herbaceous dicotyledons based on rules of the Association of Official Seed Analysts. First count indicates days to emergence; final count indicates completion of emergence

	First count		Final count	
Embryo type	No. species	Days (x)	No. species	Days (x)
Dependent	92	7.5	120	16.9
Independent	205	5.1	243	12.9
Total and mean	297	5.8	363	14.3

General Distribution of Embryo Types

If there were no selective advantage for either the dependent or independent embryo type, the proportion of species with each type would be expected to be similar to the proportion of plant families found in each group. However, families with independent-embryo types are more species-rich than families with dependent-embryo types (Table 11.2). This imbalance results in a distinct worldwide bias toward independent-embryo types on a species basis. Worldwide, 69% of families of herbaceous dicotyledons have dependent embryos, but only 45% of the species have dependent embryos and 55% have independent embryos (Table 11.2). The higher percentage of species with independent-embryo types worldwide and the larger number of species in families with independent embryos suggests an advantage to food storage in the cotyledons compared to storage in the endosperm.

Temperate, Tropical, and Cosmopolitan Affinity of Embryo Types

Using Heywood's (1978) descriptions of the distributions of plant families, each family was classified as temperate, tropical, or cosmopolitan in geographic affinity. If Heywood's description did not allow separation into these categories, a decision was based on the map of the range. Families with representatives in both temperate and tropical regions were scored as cosmopolitan for this analysis. Of the herbaceous dicotyledonous species worldwide, 63% are in plant families with a cosmopolitan distribution (Table 11.3A). Of these species, 65% have independent embryos, higher than expected from the worldwide values. Of the species of herbaceous dicotyledons

Table 11.2. Worldwide distribution of embryo types of herbaceous dicotyledons.

	Embryo type			
	Family		Species	
	Dependent	Independent	Dependent	Independent
Number	75	36	51,679	62,637
Percentage	69	31	45	55

Table 11.3A. Distribution of embryo types of herbaceous dicotyledons by geographic affinity of plant family

Region	No. species	Embryo type (%)	
		Dependent	Independent
Temperate	20,275	65	35
Tropical	21,555	80	20
Cosmopolitan	72,486	35	65
Total	114,316	45	55

B. Distribution of embryo types of herbaceous dicotyledons in temperate North America

Region	No. species	Embryo type (%)	
		Dependent	Independent
Boreal	674	41	59
Eastern Deciduous	249	37	63
Mediterranean	1,400	38	62
Desert	980	36	64
Worldwide	114,316	45	55

C. Distribution of embryo types of herbaceous dicotyledons of the Santa Barbara, California flora by geographic affinity of the plant families

Region	No. species	Embryo type (%)	
		Dependent	Independent
Temperate	218	40	60
Tropical	95	52	48
Cosmopolitan	1087	36	64
Total flora	1400	38	62

worldwide, 18% are in plant families with temperate affinities. Thirty-five percent of these species have independent-embryo types, lower than expected. Nineteen percent of the species of herbaceous dicotyledons worldwide are in families with tropical affinities. Of these species, 80% have embryos dependent on endosperm. This unusually high proportion of species with dependent-embryo types may be due in part to the concentration of phylogenetically primitive plant families in the tropics and may be evidence for the relative primitiveness of the dependent-embryo types. It could also reflect some advantage, or at least no disadvantage under tropical conditions, for the slower emergence seen in dependent-embryo types. Regions with continuous moisture and favorable temperatures for growth are well suited to the slower germination and emergence of seeds with food storage in the endosperm.

Embryo Types in Different Climatic Regions

Comparing the different climatic regions in North America it was found that each region had similar proportions of species with the two embryo types

(Table 11.3B). These distributions, however, had a higher percentage of independent-embryo types compared to the worldwide distribution (Table 11.2). Analyzing the flora in the mediterranean climatic region near Santa Barbara, California, U.S.A., which includes 1400 species (Smith 1976) showed a distribution of 38% dependent and 62% independent embryo types. Polunin and Huxley's (1966) flora of the Mediterranean region in Europe lists 1255 species, 33% with dependent and 67% with independent embryo types. Both of these regions have a pronounced summer drought, and both showed a higher proportion of independent-embryo types than the worldwide average. Analyzing other very different climatic regions in temperate North America revealed a distribution of the two embryo types similar to that found in Mediterranean climates (Table 11.3B). The floras in boreal (Marie-Victorin 1964), eastern deciduous (Dwelley 1973), and desert (DeDecker 1984) regions in North America all had a higher percentage of independent-embryo types than the worldwide average. Storing food in the cotyledons appears to be an advantage in regions with strong seasonality and periods unfavorable to germination and establishment. Although the percentage of embryo types is similar between climatic regions in North America, differences could appear in the proportion of species in families with different geographic affinities (Table 11.3A).

Geographic Affinity of Embryo Types in Mediterranean Areas

Distribution of Embryo Types in California

The distribution of embryo types by geographic affinity was analyzed for Santa Barbara, California flora (Smith 1976). Seventy-eight percent of the native herbaceous dicotyledonous species in the flora are in plant families with cosmopolitan affinities (Table 11.3C). The proportion of independent embryos in this group (64%) is almost the same as for the flora as a whole. Species in families with temperate affinities make up 16% of the Santa Barbara flora. Of these species, 60% have independent-embryo types. This figure is higher than expected for species in families with temperate affinities worldwide, where 35% of species have independent-embryo types, but is similar to the percentage in the Santa Barbara flora as a whole (62%). Species in families with tropical affinities make up only 7% of the Santa Barbara flora. Forty-eight percent of these have independent embryo types, which is higher than the 20% expected from the worldwide proportion for tropical embryo types, but lower than the 62% average overall of the Santa Barbara flora. There is a shift toward independent embryo types in families with temperate and tropical affinities. These patterns suggest that in this region of relatively low and unpredictable rainfall, there is a selective advantage to the independent embryo types, with their capacity for more rapid emergence.

Embryo Types in Mediterranean Climates in California, Chile, and Australia

The distribution of embryo types was compared among three regional floras in the mediterranean climatic regions in the Pacific Basin: Santa Barbara, California, U.S.A. (Smith 1976); Santiago, Chile (Navas Bustamante 1973 and 1979); and Perth, Western Australia (Marchant and Perry 1981). The Santa Barbara flora with 62%, and the Santiago flora with 61%, show a higher percentage of species in families with independent-embryo types than the 55% expected from the worldwide distribution (Table 11.4). The Perth flora, with 47%, shows a lower percentage of independent-embryo types than the worldwide average. To investigate the low frequency of independent-embryo types in Australia, the analysis was extended to include the percentage of independent-embryo types at the family, genus, and species level for all three regions (Table 11.4). All three regions show an increase in the percentage of independent-embryo types at the genus and species level compared to the family level. The percentage of independent-embryo types at the genus level is similar to the percentage of independent-embryo types at the species level within each region (Table 11.4). In other words, families with independent-embryo types have more genera and species than families with dependent-embryo types in all three areas, even though the average number of species per family ranges from 9 to 23 among the three regions. The lower percentage of independent-embryo types at the species level in Australia than in California and Chile seems to be due to a lower percentage of families with independent-embryo types.

Embryo Types at the Genus Level Worldwide, Regionally, and Locally

As shown by the flora of the Perth region, the presence or absence of families with a given embryo type will necessarily affect the percentage of genera and species with that embryo type. To what degree is the representation of genera with dependent or independent embryo types in a region consistent within a family? One way to answer this question is to compare the number of genera

Table 11.4. Distribution of independent embryo types at the family, genus, and species levels in native herbaceous dicotyledons from three Mediterranean climatic regions in the Pacific Basin

| | Independent embryo type | | | | | | |
| | Family | | Genus | | Species | | Average N |
Region	N	$\%$	N	$\%$	N	$\%$	sp/family
Santa Barbara, USA	28	44	199	62	649	62	23.18
Santiago, Chile	29	47	122	61	278	61	9.59
Perth, Australia	17	33	89	48	331	47	19.47

Table 11.5. Distribution of embryo types in families represented near Santiago, Chile

A. Herbaceous genera

Area	Number of genera	Embryo type (%)	
		Dependent	Independent
Worldwide	5,343	40	60
Chile	612	38	62
Santiago: all genera	296	38	62
Santiago: natives	200	41	59

B. Woody genera

Area	Number of genera	Embryo type (%)	
		Dependent	Independent
Worldwide	443	37	63
Chile	29	69	31
Santiago: all genera	16	63	37
Santiago: natives	15	67	33

with each embryo type in a given set of families on a worldwide, regional, and local basis. This was done by tabulating the number of genera with independent or dependent embryo types for the families represented in the flora of the Santiago region in Chile (Navas Bustamante 1973, 1979). The number of genera with dependent or independent embryo types found worldwide, within Chile as a whole, and in the Santiago region were tabulated for the families found in the flora in the Santiago region (Table 11.5A and 11.5B). Genera were used because the distributions of embryo types for genera and species were similar (Table 11.4), and the number of species for Chile as a whole is not known.

Elsewhere in the world 60% of the herbaceous genera in families occurring near Santiago have independent-embryo types (Table 11.5A), higher than the percentage for all species in all families (Table 11.2). Similar percentages of herbaceous genera within the Chilean flora as a whole and from the Santiago region had independent-embryo types (Table 11.5A). Restricting consideration to the native genera near Santiago does not significantly change the percentages of embryo types.

If the percentage of woody genera with independent-embryo types in families occurring near Santiago is compared to that of the herbaceous genera in these families worldwide, the percentages are similar (Table 11.5B). In the Chilean flora, however, the percentage of woody genera with independent-embryo types is markedly lower than that of herbaceous genera: 63% worldwide versus 31% in Chile (Table 11.5B). The percentage of woody genera in the vicinity of Santiago with independent-embryo types is similar to that of woody genera in Chile as a whole. When only native woody genera near Santiago are considered, the percentages of embryo types change little.

Embryo types of woody plants seem to be under different selective pres-

Table 11.6. Distribution of embryo types of herbaceous dicotyledons by habitat in the mediterranean climatic region of Santa Barbara, California, USA

Habitat	Number of species	Embryo type (%)	
		Dependent	Independent
Vernal pools	23	48	52
Freshwater marsh	47	45	55
Oak woodland	81	44	56
Chaparral	204	44	56
Conifer forest	67	43	57
Coastal sage	23	39	61
Creeks and seeps	49	37	63
Sand dunes	146	34	66
Grassland	111	23	77
Total flora	1400	38	62

sures than those of herbaceous plants, as shown by consistent differences at the generic level. Because the woody genera are few and have different proportions of embryo types than the herbaceous genera, all other comparisons in this study are restricted to native taxa in herbaceous families.

Distribution of Embryo Types by Habitat

The proportion of independent embryos in herbaceous dicotyledonous families apparently responds to the similar selective pressures imposed by the mediterranean climate in Chile, California and Australia. Because very low summer rainfall is probably an important selective pressure a difference may also be apparent in the distribution of embryo types in wet and dry habitats within each region. Are wetter habitats more favorable to species with dependent embryos?

The percentages of dependent and independent embryo types were determined for a variety of habitats within the Santa Barbara region (Table 11.6) which were then arranged by increasing percentages of independent-embryo type. The wetter habitats, vernal pools and freshwater marsh, have the highest percentages of species in families with dependent-embryo types. The habitats with least moisture, grassland and sand dunes, have the highest percentages of species in families with independent-embryo types.

Embryo Types in Desert Regions Adjacent to Mediterranean Areas

Regional Affinity of Embryo Types in the Mojave Desert

The correlation between higher percentages of independent-embryo types and drier habitats was further examined using the species listed in the DeDecker (1984) flora of the Mojave Desert, California, U.S.A. (Table 11.7). Overall, 64% of the herbaceous dicotyledonous species in the Mojave flora

Table 11.7. Distribution of embryo types of herbaceous dicotyledons in the Mojave Desert, USA, by regional affinity of the family

Region	Number of species	Embryo type (%)	
		Dependent	Independent
Temperate	221	54	46
Tropical	84	23	77
Cosmopolitan	675	36	64
Total	980	36	64

have independent embryo types (Table 11.7), more than the worldwide over-all proportion, 55% (Table 11.3A). In the Mojave flora, species in families with cosmopolitan affinities had a percentage of independent-embryo types equivalent to the overall Mojave percentage and to the worldwide cosmopolitan percentage. In contrast, 23% of the Mojave species are in families with temperate affinities. These species show 46% independent-embryo types, lower than the 64% independent-embryo types in the Mojave flora as a whole, but higher than the 35% independent-embryo type worldwide for species in families with temperate affinities. Species in families with tropical affinities are only 9% of the herbaceous dicotyledons in the Mojave flora. Seventy-seven percent of these species have independent-embryo types compared to the world wide figures of 64% and 20% for the flora as a whole and for species in families with tropical affinities respectively. Selection against dependent-embryo types is seen both in the smaller number of species in families with temperate and tropical affinities, and in the increase in the percentage of species with independent-embryo types within these groups in the desert.

Distribution of Embryo Types by Season of Germination and Growth in the Mojave Desert

Went (1948) recognized two groups of annual species in the desert: winter annuals and summer annuals. Beatley (1976) added a third category, spring annuals. Germination and establishment precedes the season of vegetative growth and flowering that Went and Beatley used to categorize these annuals. Thus, spring annuals germinate in the preceding winter, a period with cooler temperatures and more moisture. The winter annuals germinate in the preceding late fall and early winter, seasons with less predictable rain punctuated by periods of higher temperature.

The winter-germinating spring annuals have fewer than expected species in families with independent-embryo types (52%), compared to 62% for all annuals (Table 11.8). Conversely, annuals that germinate in late fall and early winter (winter annuals), have more species than expected in families with independent-embryo types (73%). Summer annuals germinate after summer showers and grow during the period of least moisture availability

Table 11.8. Distribution of embryo types of herbaceous dicotyledons by season of growth in the Mojave Desert, USA

Season of germination	Season of growth	Number of species	% Embryo type	
			Dependent	Independent
Fall and early winter	Winter	154	27	73
Winter	Spring	162	48	52
All annuals		316	38	62

and highest temperatures. Only four of these species are listed in Beatley's (1976) analysis, and all these are in the Asteraceae with independent embryos. Thirty-eight percent of the annual flora in the Mojave Desert are classified as having dependent-embryo types, with 65% of these species germinating in the winter, the time of maximum water availability. This pattern is consistent with the requirement of dependent-embryo types for longer periods of moisture for germination.

Distribution of Embryo Types by Habitat in the Mojave Desert

There are striking differences among habitats with respect to soil moisture in the Mojave Desert. Habitats range from those in which water is available for long periods, to sporadically wet habitats, to areas in which fresh water is available for germination during a very brief period (Beatley 1976). The percentage of embryo types was determined for each habitat, and the results are presented in increasing order of percentage of independent-embryo types (Table 11.9). The wettest habitats (springs and seeps, arroyos-washes, and mountains) have the highest percentages of species in families with dependent-embryo types. The driest habitats (*Larrea* bajadas, Great Basin sagebrush scrub, and alkali flats) have the highest percentages of species in families with independent-embryo types. The habitat with the highest percentage of species in families with independent-embryo types, alkali flats, are domi-

Table 11.9. Distribution of embryo types of herbaceous dicotyledons in the Mojave Desert, by habitat

Habitat	Number of species	Embryo type (%)	
		Independent	Dependent
Springs and seeps	42	45	55
Arroyos-washes	28	43	57
Mountain	30	40	60
Larrea bajadas	39	30	70
Great Basin sagebrush scrub	64	27	73
Alkali flats	26	19	81
Total flora	671	34	66

nated by *Atriplex*, and are characterized by high alkalinity, which prevents germination except during brief periods when fresh water input is high enough to leach the salts. A similar phenomenon occurs in saline maritime habitats (Vivrette and Muller 1977). The ability to germinate and grow rapidly when conditions are favorable is critical for establishment in this habitat.

Conclusions

The rapid germination and emergence that characterizes the independent-embryo types can be an advantage in environments with low or sporadic moisture availability. Because the food is stored in the cotyledons, the seed's imbibition is sufficient to allow radicle emergence and elongation. Such seeds may exhibit root emergence within hours of contact with moisture, if the seed is not dormant. But if the food is stored in the endosperm, it must first be absorbed by the embryo before it can be utilized for growth. This transfer of stored food from the endosperm to the embryo slows the rate of radicle emergence as well as subsequent growth. As a result, longer periods of external moisture are required for the seedlings to establish.

The advantage of rapid emergence is reflected by both the relatively high percentages of species with independent-embryo types in dry habitats and by their germination during drier periods of the year. Conversely, the higher percentages of dependent-embryo types in moist habitats, and their germination during more predictably moist periods of the year reflects their slower germination and early growth rate. The requirement for longer periods of moisture may also in part explain the high percentage of species with dependent-embryo types in regions of the tropics with high levels of moisture and long periods favorable to germination, growth, and establishment.

The relative proportions of embryo types in a region is partly a function of the phylogenetic status of the plant families found in that region (many primitive families have dependent-embryo types), and partly by the geographic affinity of the families. Thus the lower percentage of herbaceous dicotyledonous families that have independent-embryo types near Perth, Australia results in lower percentages of genera and species than the similar climatic regions in California and Chile. All three mediterranean climatic regions show a higher percentage of independent-embryo types in the native genera and species than expected from the percentages found at the family level. This bias suggests a consistent pattern of selection for independent-embryo types in all three regions of mediterranean climate.

The number of species per family is similar in herbaceous dicotyledonous families with independent-embryo types at Perth, Australia and Santa Barbara, California. Santiago, Chile has markedly fewer species per family compared to Perth and Santa Barbara. The percentage of independent-embryo types found in a region is partly a function of the percentage of families represented that have independent embryos and partly a function of the degree of speciation in these families within the region. The percentage of independent-embryo types in the genera found in the Santiago area is similar to the percentage of independent-embryo types found in all the genera of these families in the whole flora of Chile. The percentage of independent-embryos for all genera worldwide in the families found within the Santiago area are again similar to the percentage of independent-embryo types in the Chilean flora. The similarity of the percentage of independent-embryo types in families at the worldwide, regional, and local level suggests that selection for independent-embryo types in herbaceous dicotyledonous families within the Santiago, Chile area has not resulted from increased speciation.

Woody genera in families found near Santiago have a higher than expected percentage of dependent-embryo types than the worldwide percentage of dependent-embryo types in these woody families. This surplus suggests that different selective forces are acting on embryo type in woody species than on herbaceous species. The advantage of rapid emergence by independent-embryo types may be especially important for survival in annual or herbaceous species that reproduce primarily by seed (*r*-selection). Selection for longevity of the parent plant or vegetative propagation may be favored in the woody species (*K*-selection).

The proportion of embryo types of herbaceous dicotyledons in a region is a function of the phylogenetic status of the families represented, the geographic affinity of the families, and the moisture availability of the habitat. Wet habitats show higher percentages of species in families with dependent-embryo types, and dry habitats have higher percentages of species in families with independent-embryo types. In dry regions, species in families with dependent-embryo types germinate more frequently in the moister periods of the year. These distributions are correlated with slower germination and growth of dependent embryo types. It seems surprising that a classification as crude as that provided by this analysis can reveal such large differences in the distributions of species both in time and in space. The role of embryo types in the distribution and germination of species warrants further investigation.

Acknowledgments

Special thanks go to Betty Ransom Atwater for years of discussions about embryo types. Helpful editorial comments came from J. Robert Haller, Paul H. Zedler, and Susan Mazer.

References

Association of Official Seed Analysts (1981) Rules for testing seeds. J Seed Tech 6:1–126
Atwater BR (1980) Germination, dormancy and morphology of the seeds of herbaceous ornamental plants. Seed Sci Tech 8:523–573

Atwater BR, Vivrette, NJ (1987) Natural protective blocks in the germination of seeds. In Germination of Ornamental Plant Seeds. Acta Hort 202:57–68

Beatley JC (1976) *Vascular Plants of the Nevada Test Site and Central-Southern Nevada: Ecologic and Geographic Distributions.* U.S. Energy Res and Dev Admin. Div Biomed Environ Res. TID-26881

Corner EJH (1976) *The Seeds of Dicotyledons.* Vols I, II. Cambridge U Press, London

DeDecker M (1984) *Flora of the Northern Mojave Desert, California.* California Native Plant Society Special Publication No. 7

Dwelley M (1973) *Spring Wildflowers of New England.* Down East Enterprise, Camden, ME

Heywood VH (1978) *Flowering Plants of the World.* Oxford U Press, Oxford

Marchant NG, Perry G (1981) A checklist of the vascular plants of the Perth region, Western Australia. West Aust Herb Res Notes. 5:111–134

Marie-Victorin F (1964) *Flore Laurentienne.* Les Presses de L'Université de Montréal, Canada

Martin AC (1946) The comparative internal morphology of seeds. Am Mid Nat 36:513–660

Navas Bustamante LE (1973) *Flora de la Cuenca de Santiago de Chile.* Tomo II. Ediciones U Chile

Navas Bustamante LE (1979) *Flora de la Cuenca de Santiago de Chile.* Tomo III. Ediciones U Chile

Nikolaeva AG (1969) *Physiology of Deep Dormancy in Seeds.* Israel Prog Sci Transl. Jerusalem, Israel

Polunin O, Huxley A (1966) *Flowers of the Mediterranean.* Houghton Mifflin, Boston, USA

Smith CF (1976) *A Flora of the Santa Barbara Region, California.* Santa Barbara Mus Nat Hist, Santa Barbara, CA

Stebbins GL (1974) *Flowering Plants: Evolution above the Species Level.* Harvard U Press. Cambridge, MA

Vivrette NJ, Muller CH (1977) Mechanism of invasion and dominance of coastal grassland by *Mesembryanthemum crystallinum.* Ecol Mono. 47:301–318

Went FW (1948) Ecology of desert plants I: Observations on germination in the Joshua Tree National Monument, California. Ecology 29:242–253

12. Modes of Seed Dispersal in the Mediterranean Regions in Chile, California, and Australia

Alicia J. Hoffmann and Juan J. Armesto

Evolutionary convergence results from similarities developing between distantly related taxa that occur in similar environments (Cody and Mooney 1978). In mediterranean regions, functionally significant characters of vegetation, such as growth form spectra, leaf duration, leaf size, and presence of spines, are mostly convergent (Mooney et al. 1970; Mooney and Parsons 1973; Parsons and Moldenke 1975). Dissimilarities also occur, however, when comparing plants, animals, and their interactions in mediterranean regions (Cody and Mooney 1978). Some authors have proposed that historic constraints, and the interactions among organisms and their abiotic environment, give rise to complex communities in which evolutionary convergence is unlikely (Whittaker 1977; Wiens 1983).

The degree of convergence in the modes of dissemination of plant propagules in mediterranean-region floras has seldom been investigated. To our knowledge, the only attempt to assess convergence in plant-dispersal modes between continents is the study by Milewsky and Bond (1982), who found similar adaptations to myrmecochory in plants from mediterranean habitats in Australia and South Africa. Seeds and fruits respond to diverse and sometimes opposite selective forces. Fruit traits related to seed dispersal (e.g., type, size, and color) probably result from compromises between phylogeny, abiotic environment, and selective pressure from dispersal agents (Baker 1972; Gentry 1982; Howe and Smallwood 1982; Willson et al. 1990).

Here we compare the dispersal syndromes of plants from the mediterranean vegetation in Chile, California, and Australia, at two spatial scales: biogeographic and the local community. In Chile and California the principal type of lowland mediterranean vegetation is dense evergreen scrub, of mainly broad-leaf shrubs. Evergreen scrub occurs in both regions at intermediate positions in a soil-moisture gradient (Mooney 1977). Toward the drier parts of the gradient the vegetation becomes more open, and broad-leaved, evergreen sclerophylls are replaced by microphyllous and drought-deciduous shrubs and members of the Cactaceae. More humid sites are dominated by broad-leaved trees in both Chile and California. Of all Australian vegetation types, the mallee is considered to be more similar to the sclerophyllous vegetation in other mediterranean regions (Specht 1981). Mallee is an open scrub dominated by several *Eucalyptus* species (5 to 8 m tall) found on more fertile soils. An open heath, which includes several shrubs from the mallee understory, is found toward the top of sand hills on poorer soils.

The following aspects of the Chilean matorral, Californian chaparral, and Australian mallee are analyzed: (1) the overall frequency of different propagule types in each region, (2) the abundances of propagule types in local communities along parallel environmental gradients within each region, (3) the size characteristics and color of propagules, and (4) the available information on dispersal agents. If dispersal modes are convergent, the overall representation of propagule types is expected to be similar in the three regions (and also similar to other regions with a mediterranean climate), regardless of phylogenetic relatedness. Similar trends in distribution of propagule types in plant communities along analogous environmental gradients should also be expected. Finally, we discuss the probable cause of convergent and nonconvergent patterns.

Floristics

Floristic lists for mediterranean vegetation were gathered from various published studies (see Tables 12.3, 12.4, and 12.5) in each of the three regions. The sites selected ranged from semiarid desert to the mediterranean scrub transition, cool-montane sites, and subhumid sites in Chile and California.

Table 12.1. Numbers of woody species and genera in the mediterranean vegetation of Chile, California, and Australia. The numbers of genera common to each pair of regions are also shown. See Tables 12.3 to 12.6 for references

	Number of species	Number of genera	Number of genera in common with		
			Chile	California	Australia
Chile	169	113	—	13	3
California	272	109	—	—	2
Australia	140	66	—	—	—

No analogous environmental gradient was evident to us for the sites compared in Australia. Examining these floristic lists indicates little taxonomic overlap among regions (Table 12.1). No combination of the three regions has species in common. At the generic level, 13 genera are common to Chile and California, only three to Chile and Australia, and two are shared by California and Australia (Table 12.1). This low taxonomic affinity has been attributed to the largely independent origins of mediterranean floras (Mooney 1977). Consequently, similar dispersal spectra in these areas are likely to represent convergent evolution rather than a product of floristic relatedness.

Propagule Types

Modes of dispersal can often be inferred with fair reliability from propagule morphology (Howe and Smallwood 1982; van der Pijl 1982; Murray 1986; Willson et al. 1990). Dispersal syndromes of matorral plants are based on Hoffmann et al. (1989), supplemented by studies of fruit morphology in field-collected material; dispersal syndromes of chaparral plants are based on descriptions in Munz and Keck (1965); for Australian plants we follow descriptions given by Black (1960; 1963; 1965). Propagules are assigned to one of seven broad categories (Table 12.2), similar to those defined by other authors (Armesto and Pickett 1985; Hoffmann et al. 1989; Willson et al. 1990). Major criteria for assignments are propagule size and presence of associated dispersal structures.

Small, Fleshy Propagules

Fleshy propagules measuring 3 to 15 mm in diameter, including seeds having large (more than 2 to 3 mm), coiled, colored arils. Because of their size these propagules were considered to be dispersed mainly by avian frugivores that consume the fruit pulp and disperse the seeds. Reports of birds eating fruit were not available for all species in this category, but this assumption is supported by data on sizes of propagules consumed by birds in other temperate and mediterranean regions (Herrera 1984; Johnson et al. 1985; Armesto

Table 12.2. Categories of diaspore morphology and their potential dispersal agents (Adapted from Armesto and Pickett 1985; Hoffmann et al. 1989; Willson et al. 1990)

Diaspore type	Size range	Potential disperser
Small, fleshy	<15 mm	Birds, mammals (internal)
Small, dry	<15 mm	None or multiple
Large, fleshy or dry	>15 mm	Large vertebrates (internal)
Winged or plumed	—	Wind
Arillated or with elaeosome	2–3 mm	Ants
Explosive dehiscent	—	Ballistic
Hooked or spiny	—	Animals (external)

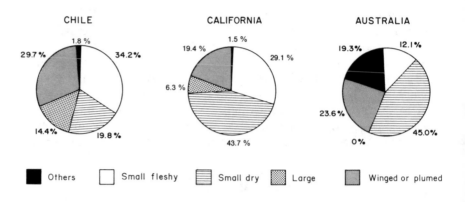

Figure 12.1. Relative importance of different propagule types in the mediterranean floras in Chile, California, and Australia. See Table 12.1 for definition of propagule types. The category "others" includes arillate, ballistic, and spiny or hooked propagules.

1987; Armesto et al. 1987). The overall frequency of small, fleshy fruits was slightly higher in Chile than in California, but much lower in Australia (Fig. 12.1).

Small, Dry Propagules

This group includes propagules generally lacking specialized means of dispersal, such as nuts and nutlets, achenes without a pappus, dry berries and drupes, indehiscent pods, loments, and seeds without arils released from capsules or follicles. Dispersal mechanisms of these propagules are not well understood. Many simply drop to the ground when ripe. They apparently fit the category of "nothing special" of Willson et al. (1990). A number of these small, dry propagules may benefit from chance transport by various biotic and abiotic vectors, thus fitting the "multiple-dispersal" category of Villagrán et al. (1986), defined for very small (1 mm) propagules. The frequency of small, dry propagules is lower in the woody flora of Chile than in California and Australia (Fig. 12.1), where they are clearly the most important propagule type.

Large Propagules

This category includes a number of dry or fleshy propagules with a diameter larger than 15 mm, corresponding with the maximum gape width of potential avian frugivores. Janzen and Martin (1982), Herrera (1985), and Janzen (1986) have all argued that extinct megafauna might have greatly influenced the evolution of these usually thick-husked, leathery fruits. Many such fruits are presently consumed and dispersed by domestic livestock (Gutiérrez and Armesto 1981; van der Pijl 1982; Janzen 1986). Large propagules are more frequent in Chile than in California, and they are absent in Australia (Fig. 12.1).

Winged or Plumed Propagules

These are dry propagules having wings, plumes, or a many-barbed pappus that slow the rate of fall. This kind of propagule is represented in the three regions, but with higher frequency in Chile than in Australia and California, respectively (Fig. 12.1).

Other Propagule Types

This heterogeneous group includes arillate, ballistic, and spiny or hooked propagules. Arillate propagules measure 2 to 3 mm in diameter and have food bodies, mostly white arils or elaiosomes that are associated with consumption and dispersal by ants (O'Dowd and Gill 1986; Willson et al. 1990). Ballistic types are known to be propelled by explosive dehiscence of the fruit, usually a legume or a capsule. Propagules with hooks or spines are defined by the presence of devices that adhere the propagule to the fur or feathers of animals that carry them externally (epizoochory). The frequency of "other" types of propagules, including small arillate, ballistic, and hooked or spiny propagules is less than 2% in Chile and California, but they are represented in 20% of the species in Australia (Fig. 12.1). This higher proportion is due exclusively to the abundance of elaiosome-bearing, ant-dispersed propagules in Australia. Ballistic and epizoochorous types are rare in all three regions.

Dispersal Spectra Within Regions

Chile

Species with small, fleshy propagules were present in all Chilean plant communities (Table 12.3), increasing in importance toward subhumid sites and coastal woodlands. Fleshy-fruited species account for 20% to 53% of the local flora, except in one site (dry desert transition). Small, dry propagules varied in importance between a maximum of about a third of the flora in a xeric coastal community and a minimum in a coastal community. The frequency of large propagules is variable, with a maximum of nearly 16% in a desert transition site, from several species of Cactaceae. In subhumid sites, the frequency of large propagules varied between 5% and 11%. Such propagules commonly occur in woody vines (e.g., *Lardizabala*, *Lapageria*) and trees (*Beilschmiedia*). Large propagules were absent from 2 communities. Winged or plumed propagules were present with similar frequencies in all sites. The frequency of propagules grouped under "other" types was generally low (up to 5.6% in a coastal woodland site).

California

Frequencies of small, fleshy propagules varied between 12.5% and 43% depending on location (Table 12.4). Small, dry propagule types were a high

Table 12.3. Importance of propagule types in mediterranean communities in Chile

Vegetation type	Latitude (°S)	Propagule type										Sources
		Small, fleshy		Small, dry		Large		Winged		Others[a]		
		N	%	N	%	N	%	N	%	N	%	
Desert transition	30.15	2	10.5	6	31.6	3	15.8	7	36.8	1	5.3	(1)
Montane matorral	31.40	10	31.3	6	18.8	4	12.5	11	34.4	1	3.1	(2)
Coastal scrub	32.06	8	30.8	5	19.3	3	11.5	9	34.6	1	3.8	(3)
Coastal woodland	32.30	31	41.9	13	17.6	8	10.8	20	27.0	2	2.7	(4)
Coastal woodland	32.38	9	33.3	8	29.6	2	7.4	7	25.9	1	3.7	(5)
Coastal woodland	32.57	30	44.1	11	16.2	4	5.9	21	30.9	1	2.9	(6)
Montane matorral	33.04	7	25.9	6	22.2	3	11.1	10	37.0	1	3.7	(7)
Montane matorral	33.04	11	34.4	8	25.0	2	6.3	10	31.3	1	3.1	(1)
Coastal woodland	33.26	6	33.3	5	27.8	0	0.0	6	33.3	1	5.6	(8)
Matorral	33.26	7	30.4	5	21.7	2	8.7	8	34.8	1	4.3	(9)
Matorral	33.30	5	20.0	9	36.0	0	0.0	10	40.0	1	4.0	(10)
Matorral	33.33	20	35.7	11	19.6	4	7.1	20	35.7	1	1.8	(11)
Subhumid matorral	34.41	16	38.1	10	23.8	2	4.8	12	28.6	2	4.8	(2)
Subhumid matorral	35.02	26	53.1	9	18.4	1	2.0	12	24.5	1	2.0	(12)
Subhumid matorral	35.30	21	43.8	7	14.6	4	8.3	15	31.3	1	2.1	(13)
Coastal woodland	35.40	19	43.3	5	11.4	4	9.2	14	31.5	2	4.6	(14)

[a] Includes ballistic and hooked or spiny propagules.

Sources: (1) Thrower and Bradbury (1977), (2) Hoffmann and Hoffmann (1982), (3) Mooney and Schlegel (1967), (4) Johow (1945). (5) Armesto and Martínez (1978), (6) Villaseñor (1980), (7) Parsons (1976), (8) Villagrán et al. (1980), (9) Muñoz (1959), (10) Steward and Webber (1981), (11) Schlegel (1966), (12) Aravena (1974), (13) Castro (1985), (14) San Martín et al. (1984).

Table 12.4. Importance of different propagule types in mediterranean communities in California

Vegetation type	Latitude (°N)	Propagule type										Sources
		Small, fleshy		Small, dry		Large		Winged		Others[a]		
		N	%	N	%	N	%	N	%	N	%	
Desert transition	30.45	10	32.3	11	35.5	1	3.2	8	25.8	1	3.2	(1)
Chaparral	30.45	6	20.0	8	26.7	7	23.3	8	26.7	1	3.3	(2)
Chaparral	31.54	1	12.5	3	37.5	2	25.0	2	25.0	0	0.0	(3)
Montane chaparral	32.50	4	22.2	11	61.1	0	0.0	3	16.7	0	0.0	(1)
Chaparral	32.54	8	27.6	15	51.7	0	0.0	4	13.8	2	6.9	(1)
Coastal sage scrub	33.40	3	30.0	4	40.0	1	10.0	2	20.0	0	0.0	(1)
Montane chaparral	33.49	3	16.7	9	50.0	0	0.0	6	33.3	0	0.0	(4)
Interior chaparral	33.54	10	31.3	17	53.1	0	0.0	5	15.6	0	0.0	(5)
Interior chaparral	33.55	15	40.5	12	32.4	1	2.7	9	24.3	0	0.0	(6)
Interior chaparral	34.00	18	37.5	18	37.5	0	0.0	12	25.0	0	0.0	(7)
Interior chaparral	34.06	9	42.9	7	33.3	0	0.0	5	23.8	0	0.0	(8)

[a] Includes arillate, ballistic, and hooked or spiny propagules.

Sources: (1) Thrower and Bradbury (1977), (2) Mooney and Harrison (1972), (3) Mooney et al. (1974), (4) Vogl and Schorr (1972), (5) Parsons (1976), (6) Keeley and Keeley (1988), (7) Hanes (1971), (8) Hanes and Jones (1967).

proportion of the flora, between 27% in a chaparral community and 61% in a montane site. Large propagules were present only in 5 of the species lists, mainly in the more xeric sites including Cactaceae. Between 14% and 33% of the species had wind-dispersed propagules; variability among sites was apparently lower than that found for other types of propagules. No arillate propagules were recognized in the chaparral. Only one species had epizoochorous fruit. Explosive fruits are present in the chaparral (*Ceanothus* spp.) but are not treated as separate category in Table 12.4.

Australia

The frequency of small, fleshy propagules was generally quite low and variable among sites (Table 12.5). They were found in all four communities in the mallee in southern Australia. In Victoria, small, fleshy propagules were found in many sites, varying in frequency between 0% and 38%. It is noteworthy that an important proportion of the flora of the mallee communities had small, dry propagules. In contrast, no large propagules were found in the communities examined. Winged or plumed propagules were found in most sites. Propagules with arils or elaiosomes also were found in most sites, but their frequency was extremely variable. Ballistic and hooked or spiny propagules were rare or absent.

Size and Color of Propagules

Size

The size ranges of fleshy propagules differ among Chile, California, and Australia (Fig. 12.2). Overall, diameters vary between 2 and almost 33 mm, with bimodal distribution of sizes in both Chile and California. This pattern was consistent with our distinction of small and large fleshy propagules in a separate category (Table 12.2). The tendency for bimodality was more pronounced in Chile than in California (Fig. 12.2). In Chile, one peak is in the size range of 4.1 to 8 mm (about a third of the species) and the second is in the range of over 25 mm. In California, one peak is in the range 0.1 to 4 mm and the second is in the same range as in Chile. Consequently, fleshy propagules tend to be larger on the average in Chile than in California (Fig. 12.2). In the Australian mallee, only very small, fleshy propagules (up to 6 mm diameter) associated with ant dispersal were present.

Sizes of dry propagules vary between 0.1 and more than 25 mm in diameter. A first group, measuring between 0.1 and 16 mm in diameter (small, dry, not wind-dispersed), comprises 70% of the dry propagules in the Chilean list and 90% of those in the Californian list. Within this size range, the largest frequency in Chile (almost 30%) is in the 4.1- to 8-mm range, and in Califor-

Table 12.5. Importance of different propagule types in mediterranean communities in Australia

Vegetation type	Propagule type										
	Small, fleshy		Small, dry		Winged		Arillate		Others[a]		Sources
	N	%	N	%	N	%	N	%	N	%	
South Australia (30 to 35° S)											
Eucalyptus diversifolia	6	15.4	21	53.8	8	20.5	3	7.7	1	2.6	(1)
E. dumosa	2	5.4	25	67.6	3	8.1	7	18.9	0	0.0	(1)
E. incrassata	1	5.3	8	42.1	6	31.6	2	10.5	2	10.5	(1)
E. oleosa-gracilis	8	22.9	14	40.0	9	25.7	3	8.5	1	2.9	(1)
Victoria (30 to 36° S)											
Chenopodium mallee	3	17.6	8	47.1	4	23.5	0	0.0	2	11.8	(2)
Shallow-sand mallee	0	0.0	7	87.5	0	0.0	1	12.5	0	0.0	(2)
East/west dune mallee	0	0.0	5	55.5	1	11.1	2	22.2	1	11.1	(2)
Mallee heath	0	0.0	12	52.2	6	26.1	4	17.4	1	4.3	(2)
Shallow-sand mallee heath	2	7.7	12	46.2	6	23.1	5	19.2	1	3.8	(2)
Loamy-sand mallee	1	6.3	11	68.7	2	12.5	2	12.5	0	0.0	(2)
Sandstone-rise mallee	0	0.0	8	80.0	1	10.0	1	10.0	0	0.0	(2)
Brumbush mallee	1	4.5	13	59.2	4	18.2	3	13.6	1	4.5	(2)
Red-swale mallee	1	5.6	12	66.7	3	16.5	1	5.6	1	5.6	(2)
Big mallee	3	37.5	4	50.0	0	0.0	0	0.0	1	12.5	(2)
Gravelly-sediment mallee	0	0.0	0	0.0	3	50.0	3	50.0	0	0.0	(2)

[a] Includes ballistic and hooked or spiny propagules.
Sources: (1) Sparrow (1989). (2) Cheal and Parkes (1989).

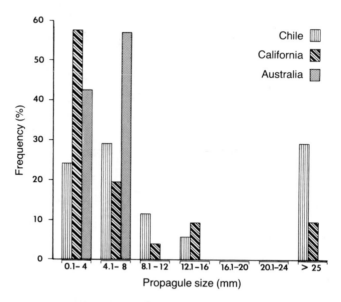

Figure 12.2. Size distributions of fleshy propagules in the mediterranean vegetation in Chile, California, and Australia.

nia 74% are in the 0.1–4 mm range. Large, dry propagules (>25mm) are 30% of the propagules in Chile and 10% in California, ranging from 30 to 120 mm in width. No data on sizes of dry propagules were available for the Australian mallee.

Color

The overall color distributions of ripe, fleshy fruits (including seeds with colored arils) are shown in Fig. 12.3. In Chile, violet/black is the most frequent color (48% of species), and red, green, and pink occur in similar proportions (10% to 20% of species); other colors are less frequent. In California, red is the predominant color (43% of species), violet/black is found in 27% and brown in 15% of this species. An analysis of color could not be made for Australian mallee fruits.

Dispersal Agents

Biotic Dispersal

Very limited information is available on seed dispersal by animals for the mediterranean regions compared. The information in this section is based mainly on anecdotal observations of the activity and feeding behavior of frugivores.

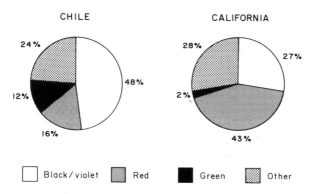

Figure 12.3. Relative frequencies of colors of fleshy fruits in the mediterranean vegetation in Chile and California.

Birds

Although numerous species consume fleshy fruit, no specialized frugivores exist in the Chilean matorral. Some studies (references in Table 12.6) suggest that *Mimus thenca*, *Turdus falklandii*, and *Curaeus curaeus* are the major avian dispersers of seeds. More detailed information is available on the dispersal of propagules of *Tristerix tetrandrus*, a widespread hemiparasite that depends exclusively on birds for its dispersal. Birds feed opportunistically on the fruit of this species and evacuate the seed on rocks or tree branches (Hoffmann et al. 1986). The capacity of the seeds of *Tristerix* to establish on their host increases after the seeds are voided through regurgitation or defecation. The main effect of the passage of the fruit through the bird's gut is mechanical removal of the seed coat and not enhancement of germination (Hoffmann et al. 1986). *Mimus thenca* also consumes fruits of other species of *Tristerix*, including *T. aphyllus*, *quintral del quisco*, which grows endophytically on *Trichocereus chiloensis* and other members of the Cactaceae.

Perching behavior of birds may affect seed dispersal and the structure of vegetation. For instance, the sclerophyllous vegetation in central Chile is often composed of multispecific shrub clumps of various sizes (Fuentes et al. 1984). Of nine species found in such clumps, seven had fleshy propagules, presumably bird-dispersed.

Limited information is available on birds as seed dispersers in California and Australia. Keeley (1987) reported that migrant birds consume the brightly colored berries of some common chaparral shrubs such as *Heteromeles arbutifolia*, *Prunus ilicifolia*, *Rhus integrifolia*, *Rhamnus crocea*, and *R. californica*. Scrub jays commonly feed on the seeds of various shrub species including oaks (Keeley 1987), but their effectiveness as dispersers has not been evaluated.

In the Australian mallee, Willson and O'Dowd (1989) show that fruits of

Table 12.6. Woody species of the Chilean matorral with fleshy fruits reported to be consumed by local avian frugivores

Plant species	Avian frugivores
Aristotelia chilensis	Curaeus curaeus
	Turdus falklandii
Cryptocarya alba	Turdus falklandii
	Mimus thenca
Drimys winteri	Elaenia albiceps
	Pyrope pyrope
	Sicalis luteola
Lithrea caustica	Turdus falklandii
	Mimus thenca
	Phytotoma rara
Maytenus boaria	Diuca diuca
	Pyrope pyrope
Myrceugenia exsucca	Mimus thenca
	Curaeus curaeus
	Turdus falklandii
Persea lingue	Columba araucana
Schinus latifolius	Curaeus curaeus
	Mimus thenca
Trichocereus chiliensis	Mimus thenca
Tristerix spp.	Mimus thenca
	Curaeus curaeus
	Turdus falklandii

Sources: Solar (1975), Armesto et al. (1987), Hoffmann et al. (1986), and unpub. observations of A.H.

Rhagodia parabolica are consumed by several bird species. Individuals of this plant occur under the canopy of trees and shrubs more often than other species that are not bird-dispersed.

Mammals

Evidence shows that Chilean mammals eat fruit and possibly contribute to seed dispersal (Armesto et al. 1987). Foxes, particularly *Pseudalopex spp.*, feed on fruits of *Cryptocarya alba, Aristotelia chilensis*, and *Lithrea caustica* (Jaksic et al. 1980; León and Arroyo 1994); the introduced rabbit (*Oryctolagus cuniculus*) is also known to feed on *Peumus boldus* (E. Fuentes pers. comm.). There are doubts about the effectiveness of mammals as dispersal agents, however (Bustamante et al. 1992). Although foxes can travel long distances with seed in their stomach, they tend to leave their droppings containing seeds mostly in open spaces (Bustamante et al. 1992; León and Arroyo 1994), where seedling survival is low (Fuentes et al. 1986). Rodents may feed on fruit and seeds but we know no evidence of scatter-hoarding behavior that may contribute to dispersal. Native and introduced rodents consume the fruit of the Chilean palm (*Jubaea chilensis*), probably without hoarding.

Today, however, less than 5% of the fruit crop of this species falls to the ground because of harvesting for human consumption (L. Yates, pers. comm.).

In California, acorns are commonly dispersed by squirrels (Keeley 1987), but they are also collected by wood rats (Horton and Wright 1944; fide Mills and Kummerow 1989). Again, the effectiveness of mammals as seed dispersers has not been assessed. Seed dispersal by mammals has not been reported for species in the Australian mallee.

Ants

Species with arillate propagules are absent or rare in the mediterranean vegetation in Chile and California (Berg 1966; Bullock 1978; Westoby et al. 1990). In the sclerophyllous vegetation in California, harvester ants act as dispersal agents for species of *Ceanothus* by storing their seeds in galleries near the soil surface, where they germinate after fire (Mills and Kummerow 1989). On the other hand, Australia has a very high incidence of ant-dispersed plants (Milewski and Bond 1982; Westoby et al. 1982). For instance, in dry sclerophyllous vegetation in New South Wales (33° 55' S, 151° 10' W), just outside the mediterranean zone, 30% of all species and more than 80% of the shrubs are myrmecochorous (Westoby et al. 1982). Most of these ant-dispersed species are endemic, suggesting evolution of this syndrome in situ after isolation of the Australian flora (Berg 1975).

The effectiveness of ants as seed dispersers appears to be related to propagule morphology and size (Westoby et al. 1990). Anderson (1982) reported that seed removal by ants involves many species, as well as seeds with and without elaiosome-like structures. Hughes and Westoby (1990) showed that almost all elaiosome-bearing seeds in dry sclerophyll vegetation are removed within a few days of release. Moreover, fallen *Eucalyptus* seeds that are devoid of elaiosomes are removed by harvester ants that store the seeds underground. In *Acacia*, where arillate species are primarily associated with bird and ant dispersal, aril color is probably the best discriminator. White or brown arils characterize ant-dispersed propagules, and brightly colored arils are related to bird dispersal (O'Dowd and Gill 1986).

Westoby et al. (1990) established that, on infertile soils, species with smaller seed mass or growing to less that 2 m tall are likely to be adapted for dispersal by ants rather than by vertebrates.

Abiotic Dispersal

In California, the life histories of some chaparral species are keyed to wildfire. *Heteromeles arbutifolia*, *Quercus dumosa*, *Prunus ilicifolia*, *Cercocarpus betuloides*, and *Rhamnus* spp. produce seeds borne in fleshy or winged fruit that are fire labile—that is, they rarely survive high temperatures (Bullock 1978; Keeley 1987). Many species, however, show some fire adaptation. The small seeds of *Ceanothus greggii* (ca. 0.007 g) are readily dispersed to sites affected by fire (Keeley and Zedler 1978). *Ceanothus* is collected by harvester ants that store the seeds in galleries where they survive and germinate after fire (Mills and Kummerow 1989). In this species heat has been shown to stimulate germination (Keeley and Keeley 1988; Zedler and Zammit 1989). An even finer adjustment to the fire cycle is seen in serotinous species, in which seeds are held in cones that remain on the plants for several years, to be released only after fire. In California, serotiny has been reported only for conifers such as *Cupressus abramsii*, *Pinus attenuata*, and *P. muricata*. In *C. abramsii*, cones with viable seed are known to survive intact more than forty years (Keeley and Keeley 1988).

Fire is also an important factor for seed dispersal in Australia. On-plant seed storage (serotiny) and fire-stimulated dispersal and seed germination are frequent among Australian mediterranean plants (Gill 1977). Many shrubs in mallee and heath vegetation, mainly Proteaceae, but also Myrtaceae and Ericaceae (Gill 1977), retain viable seed on the plant, constituting "aerial seed banks" (Noble et al. 1980). Most of these are small, dry propagules. These plants release their seed soon after fire. Serotiny seems to have evolved independently in different lineages of the Proteaceae (*Banksia* and *Hakea* in Grevilloideae, *Isopogon*, *Protea* and *Leucadendron* in Persooniodeae), reflecting evolutionary convergence (Kruger 1983). In *Banksia*, Gill (1977) found that only 1% of the follicles retained on intact mature plants had ruptured; in contrast, release was substantially higher after fire. The same occurs in *Hakea*. Local postfire dispersal may be enhanced by wind (Whelan 1986; Fox 1988).

In *Eucalyptus*, seeds are stored in woody capsules that do not fall to the ground when ripe, but remain on the plants for two to three years. Fruits of several annual crops can coexist on the trees (Kruger 1983; Wellington 1989). The seeds are slowly released and dispersed by wind (Cremer 1977). Seed release increases following twig death and subsequent drying of capsules. Fire causes mallee seed to be released in large quantities from the capsules, thus playing an important role in seed dispersal. Hence, the aerial seed bank would release considerable genetic diversity (Noble 1982). The dependence of seed release on fire has led some workers to speculate that fire-promoting characteristics of mallee communities (accumulation of large quantities of litter fuel beneath the canopy) are the result of natural selection (Noble et al. 1980).

There are no reports of seed dispersal in relation to fire in central Chile (see Armesto et al., this volume).

Discussion

Convergent Versus Nonconvergent Patterns

Only a few of the patterns found indicate convergence in dispersal systems at the biogeographic or community scale in central Chile, California, and Australia. The predominant propagule types are different in each region and some types are present in only one region. In the Chilean matorral, bird- and wind-dispersed propagules, adapted for long-range dispersal predominate in the woody flora, whereas small, dry propagules represent the major dispersal category in California and Australia. Ant-dispersed propagules are largely restricted to the Australian flora.

Evidence for convergence is limited to parallel trends in distribution of propagule types in two of the three regions. Small, fleshy propagules appear to have similar overall importance in Chile and California and both regions have bimodal distribution of propagule sizes. These patterns are not, however, restricted to mediterranean regions; similar patterns have been documented for fleshy fruits in other temperate regions (Howe and Smallwood 1982; Herrera 1984; 1985; Willson et al. 1990). Another suggestion of convergence is seen in the increase in fleshy fruited plants from xeric to mesic sites in Chile and California. Again, this tendency is not exclusive to mediterranean regions (e.g., Gentry 1982). On the other hand, species with small, dry propagules are found in similar proportions (45%) in California and Australia, also suggesting convergent trends in two of the three regions. Winged or plumed propagules appear similarly represented in all three regions compared (20% to 30% of the flora). This presence cannot, however, be taken as indicating convergence among mediterranean regions, because similar proportions of wind-dispersed propagules are found in many tropical or temperate floras (Howe and Smallwood 1982; Willson et al. 1990), where they may be related to exposure to wind, openness of the vegetation, and moisture availability.

Other patterns clearly do not support the convergence hypothesis. First, large, leathery fruits are found in 14% of the mediterranean flora in Chile, in a few species in California, and are lacking in Australia. Second, the size distribution of fleshy propagules in Australia is very different from that of Chile and California. Third, arillate propagules are lacking among woody species in Chile and only a few are found in California. In contrast, they occur in nearly 20% of the species in Australia. It is possible that further research will show that some small, dry propagules falling directly beneath the parent plant may also be ant-dispersed. For instance, Mills and Kummerow (1989) reported that ants collect and possibly disperse seeds of some chaparral species regardless of the presence of elaiosomes. Although ant dispersal is normally associated with short-range dispersal, i.e. a narrow seed rain, this syndrome may be effective in reducing mortality caused by seed or seedling predation near the parent (Reichman and Oberstein 1977; O'Dowd and Hay 1980), and burial of seeds by ants may reduce mortality of the seed bank following fire.

Causes of Convergent and Nonconvergent Patterns

Possible explanations for the patterns described above are discussed as several, not mutually exclusive, hypotheses:

1. *The soil-moisture–nutrient-availability hypothesis.* If the relative importance of fleshy propagules in a local flora were determined primarily by soil-moisture availability, we would expect similar overall proportions of fleshy-fruited plants in the three mediterranean-climate regions and at the local scale, and similar correlations between the frequency of fleshy propagules in a community and its position in a soil-moisture gradient. Whereas the abundances of fleshy propagules in Chile and California tend to support this hypothesis, fleshy fruits appear to be underrepresented in Australia. A possible reason for the lower frequency of fleshy propagules here relative to Chile and California could be the lower availability of soil nutrients in the mediterranean region of Australia (Kruger 1983). Producing fleshy fruits requires an adequate supply of several mineral nutrients, particularly potassium, which is known to be deficient in Australian soils (Rice and Westoby 1981; Milewski and Bond 1982).

The higher frequency of small elaiosome-bearing propagules in Australia than in Chile and California may also be a consequence of the lower nutrient supply in Australian mallee soils (Westoby et al. 1982). Nutrient-poor habitats may have selection for propagules with small food bodies, suited to ant dispersal. Ant-dispersed propagules seem to have derived from bird-dispersed propagules (O'Dowd and Gill 1986) and attributes responsible for ant dispersal in *Acacia* would have evolved as a response to various selective forces, including nutrient-deficient soils, fire, and distance-dependent seed predation (O'Dowd and Gill 1986).

2. *The disturbance-regime hypothesis.* Different disturbance regimes may be reflected in plant assemblages that differ in their major seed-dispersal modes in regions with similar climates. For instance, in California and Australia natural fires have been historically frequent and extensive gaps result at predictable intervals. Fire frequency in California has been estimated as one every forty years (Armesto and Pickett 1985). Each site is likely to burn at least once within a lifetime of a dominant shrub species. Convergence in the abundance of small, dry propagules in California and Australia might be associated with recurrence of fire disturbance in these two regions. Strong selection for long-distance dispersal is unlikely in this environment. Small, dry, often fire-resistant propagules that accumulate directly beneath, or near the parent plant, will replace their parents when they are destroyed by fire (Keeley and Zedler 1978). Fire disturbance may be an additional reason for the low abundance of fleshy propagules in the Australian mallee. Fleshy fruits may be less adaptive to fire-prone environments than dry propagules because they are not resistant to high temperature (Keeley and Keeley 1988). In contrast, fleshy propagules, adapted for long-distance dispersal, may be favored in stabler environments where wildfire is not a major disturbance.

Under this last hypothesis fleshy, primarily avian-dispersed propagules should predominate in the mediterranean region of Chile. With no large-scale natural fires, regeneration of the matorral would occur primarily within small gaps unpredictable in time and space (Murray 1986). As a consequence, the production of a broad seed shadow such as that exhibited by bird- and wind-disseminated propagules should be highly adaptive (Howe and Smallwood 1982). Even though dispersal patterns of fleshy fruits are generally leptokurtic, with most fruit falling within 50 m of the parent plant (Smith 1975; Fenner 1985; Murray 1986), the distribution has a long tail. Other individuals in disturbed areas also may act as recruitment foci, producing a patchy seed rain (McDonnell and Stiles 1983; Fuentes et al. 1984).

Another factor contributing to the importance of fleshy propagules in the Chilean matorral might be absence of marked climatic fluctuations caused by the maritime influence (Arroyo et al. 1993), which has favored persistence of relict taxa derived from the Tertiary tropical flora, especially in coastal sites (Troncoso et al. 1980; Villagrán and Armesto 1980; Arroyo et al. 1993; this volume). Strong phylogenetic inertia of propagule morphology, as postulated by Herrera (1986), might contribute to maintain a high proportion of fleshy fruits in the mediterranean flora of Chile.

3. *The megafaunal hypothesis.* Fruit traits often remain unaltered during long evolutionary periods, even though the plants may be exposed to drastic changes in climate and to high faunal turnover (Herrera 1986). This phylogenetic inertia may reduce the probability of evolutionary convergence in the characteristics of propagules found in similar communities.

The category of large propagules found in the flora of some mediterranean regions possibly evolved in association with a rich but now extinct megafauna during the Pleistocene (Janzen and Martin 1982; Herrera 1985; Janzen 1986; but see Howe 1985). Without their original biotic vectors, these fruits are now consumed by a variety of modern vertebrates that have learned that the fruit is edible. The persistence of large propagules in the mediterranean flora in central Chile, after the demise of the large mammalian herbivores that were abundant during the Pleistocene (Simonetti 1984), suggests that there is not a strong selection for changing fruit morphology to suit the present-day fauna. The lack of large propagules in Australia could be related to the absence of an extensive megafauna in the Australian continent.

Conclusions

The examples analyzed above suggest that strong convergence at the regional and community level is unlikely because of the complex interplay of historical and phylogenetic constraints over directional selection. The Myrtaceae in mediterranean regions illustrate the role of these selective forces in shaping dispersal characteristics. Neotropical Myrtaceae differ from the Australasian species in fruit morphology (Stebbins 1974). Neotropical species have fleshy fruits, mostly berries or drupes (except for one species, *Tepualia stipularis*, in temperate Chile). Australian species, in turn, bear predominantly dry fruit, mostly serotinous capsules. Thus, the Myrtaceae in the

Chilean mediterranean region bear bird-dispersed fleshy fruits, whereas in the fire-prone, nutrient-poor mediterranean region in Australia, they bear dry propagules. Hence, this family provides valuable material for comparative analyses of phylogeny vs. abiotic environment in the evolution of propagule morphology. If fleshy fruits are more primitive among flowering plants than small, dry propagules, as often believed (Stebbins 1974), then the fruit traits of Chilean Myrtaceae could be explained by maintenance of ancient characteristics, whereas those of Australian Myrtaceae could represent a more recent evolutionary line responding to a different disturbance history.

In another context, our results indicate that disturbance regimes may be important in the evolution of dispersal modes. Convergent propagule types may have evolved in response to recurrent fire disturbance in California and Australia. We suggest too that nutrient-poor soils may represent important abiotic constraints on fruit-trait evolution in mediterranean regions. Specht (1979) and Cowling and Campbell (1980) have also emphasized that differences in availability of soil nutrients can explain lack of convergence in other characteristics of plant communities in these areas.

Acknowledgments

We appreciate the invitation of the editors to write this chapter. Professors C. Marticorena and M. Westoby kindly provided access to bibliographic material. Manuscript preparation was funded in part by FONDECYT, Chile, Grant 92-1135 to JJA.

References

Anderson A (1982) Seed removal by ants in the mallee of northwestern Victoria. In Buckley RC (ed) *Ant–plant Interactions in Australia*. Dr W Junk Publishers, London, pp 31–44

Aravena P (1974) Presentación en análisis breve de 61 familias dicotiledóneas y monocotiledóneas del bosque de Alto Vilches (Andes de Talca). U Católica de Chile, Sede Regional del Maule

Armesto JJ (1987) Mecanismos de diseminación de semillas en el bosque de Chiloé: una comparación con otros bosques templados y tropicales. An IV Cong Latinoamer Bot, Medellín, Colombia, pp 7–24

Armesto JJ, Martínez JA (1978) Relations between vegetation structure and slope aspect in the mediterranean region of Chile. J Ecol 66:881–889

Armesto JJ, Pickett STA (1985) A mechanistic approach to the study of succession in the Chilean matorral. Rev Chil Hist Nat 58:9–17

Armesto JJ, Rozzi R, Miranda P, Sabag C (1987) Plant/frugivore interactions in South American temperate forests. Rev Chil Hist Nat 60:321–336

Arroyo MTK, Armesto JJ, Squeo F, Gutiérrez J (1993) Global change: Flora and vegetation of Chile. In Mooney HA, Fuentes E, Kronberg B (eds) *Earth System Response to Global Change: Contrasts Between North and South America*. Academic Press, New York, pp 239–263

Baker HG (1972) Seed weight in relation to environmental conditions in California. Ecology 53:997–1010

Berg RY (1966) Seed dispersal in *Dendromecon*: Its ecologic, evolutionary, and taxonomic significance. Am J Bot 53:61–73

Berg RY (1975) Myrmecochorous plants in Australia and their dispersal by ants. Aust J Bot 23:475–508

Black JM (1960) *Flora of South Australia* Part I. Woolman DJ Government Printer, South Australia

Black JM (1963) *Flora of South Australia* Part II. Hawes WL Government Printer, Adelaide

Black JM (1965) *Flora of South Australia* Part IV. Hawes WL Government Printer, Adelaide

Bullock SH (1978) Plant abundance and distribution in relation to types of seed dispersal in chaparral. Madroño 25:104–105

Bustamante RO, Simonetti JA, Mella JE (1992) Are foxes legitimate and efficient seed dispersers? A field test. Acta Oecol 13:203–208

Castro E (1985) *Perfil ecológico del Embalse Colbún*. Empresa Nacional de Electricidad, S.A., Santiago

Cheal DC, Parkes DM (1989) Mallee vegetation in Victoria. In Bradstock RA, Noble JC (eds) *Mediterranean Landscapes in Australia: Mallee Ecosystems and Their Management*. Brown Prior Anderson Pty Ltd, Melbourne, pp 125–140

Cody ML, Mooney HA (1978) Convergence vs. non-convergence in mediterranean-climate ecosystems. Ann Rev Ecol System 9:265–321

Cowling RM, Campbell MB (1980) Convergence in vegetation structure in the mediterranean communities of California, Chile, and South Africa. Vegetatio 43:191–197

Cremer KW (1977) Distance of seed dispersal in eucalypts estimated from seed weights. Aust For Res 7:225–228

Fenner M (1985) *Seed Ecology*. Chapman and Hall, London

Fox MD (1988) Understorey changes following fire at Myall Lakes, New South Wales. Cunninghamia 2:85–95

Fuentes ER, Otaíza RD, Alliende MC, Hoffmann AJ, Poiani A (1984) Shrub clumps in the Chilean matorral vegetation: Structure and possible maintenance mechanisms. Oecologia 64:405–411

Fuentes ER, Hoffmann AJ, Poiani A, Alliende MC (1986) Vegetation change in large clearings: Patterns in the Chilean matorral. Oecologia 68:358–366

Gentry HA (1982) Patterns of neotropical plant species diversity. Evol Bio 15:1–84

Gill AM (1977) Plant traits adaptive to fire. Proc Sym Env Cons Fire Fuel Man Medit Ecosys. USDA For Serv. Gen Tech Rep WO-3, pp 17–26

Gutiérrez JR, Armesto JJ (1981) El rol de ganado en la dispersión de las semillas de *Acacia caven* (Leguminosae). Cien Invest Agr 8:3–8

Hanes TL (1971) Succession after fire in the chaparral of southern California. Ecol Mono 41:27–51

Hanes TL, Jones HW (1967) Postfire chaparral succession in southern California. Ecology 48:260–264

Herrera CM (1984) A study of avian frugivores, bird-dispersed plants and their interactions in mediterranean scrublands. Ecol Mono 54:1–23

Herrera CM (1985) Determinants of co-evolution: The case of mutualistic dispersal of seeds by vertebrates. Oikos 44:132–141

Herrera CM (1986) Vertebrate-dispersed plants: Why they don't behave the way they should. In Estrada A, Fleming TH (eds) *Frugivores and Seed Dispersal*. Dr W Junk Publishers Dordrecht, pp 5–18

Hoffmann AJ, Hoffmann AE (1982) Altitudinal ranges of phanerophytes and chamaephytes in central Chile. Vegetatio 48:151–163

Hoffmann AJ, Fuentes ER, Cortés I, Liberona F, Costa V (1986) *Tristerix tetrandrus* (Loranthaceae) and its host-plants in the Chilean matorral: patterns and mechanisms. Oecologia 69:202–206

Hoffmann AJ, Teiller S, Fuentes ER (1989) Fruit and seed characteristics of woody species in mediterranean-type regions of Chile and California. Rev Chil Hist Nat 62:43–60

Horton JS, Wright JT (1944) The woodrat as an ecological factor in southern California watersheds. Ecology 25:341–351

Howe HF (1985) Gomphotere fruits: A critique. Am Nat 124:853–865

Howe HF, Smallwood J (1982) Ecology of seed dispersal. Ann Rev Ecol System 13:201–228

Hughes L, Westoby M (1990) Removal rates of seeds adapted for dispersal by ants. Ecology 71:138–148

Jaksic FM, Schlatter RP, Yáñez JL (1980) Feeding ecology of central Chilean foxes, *Dusicyon culpaeus* and *D. griseus*. J Mam 61:254–260

Janzen DH (1986) Chihuahuan desert nopaleras: defaunated big mammal vegetation. Ann Rev Ecol System 17:595–636

Janzen DH, Martin PS (1982) Neotropical anachronisms: The fruits that the Gomphoteres ate. Science 215:19–27

Johnson RA, Willson MF, Thompson JN, Bertin RI (1985) Nutritional values of wild fruit and consumption by migrant frugivorous birds. Ecology 66:819–827

Johow F (1945) Flora de Zapallar. Rev Chil Hist Nat 49:1–566

Keeley JE (1987) Role of fire in the seed germination of woody taxa in California chaparral. Ecology 68:434–443

Keeley JE, Keeley SC (1988) Temporal and spatial variation in fruit production by California chaparral shrubs. In Castri FD, Floret C, Ramball S, Roy J (eds) *Time Scales and Water Stress*, Proc 5th Int Conf Medit Ecosys. Int U Bio Sci, Paris, pp 457–463

Keeley JE, Zedler PH (1978) Reproduction of chaparral shrubs after fire: a comparison of sprouting and seeding strategies. Am Mid Nat 99:142–161

Kruger FJ (1983) Plant community diversity and dynamics in relation to fire. In Kruger FJ, Mitchell DT, Jarvis JUM (eds) *Mediterranean-Type Ecosystems: The Role of Nutrients*. Springer Verlag, Berlin, pp 446–472

León PM, Arroyo MTK (1994) Germinación de semillas de *Lithrea caustica* (Mol.) H. et A. (Anacardiaceae) dispersadas por *Pseudalopex* spp. (Canidae) en el bosque esclerófilo de Chile central. Rev Chil Hist Nat: in press

McDonnell MJ, Stiles EW (1983) The structural complexity of old-field vegetation and the recruitment of bird-dispersed plant species. Oecologia 56:109–116

Milewski AU, Bond WJ (1982) Convergence of myrmecochory in mediterranean Australia and South Africa. Geobotany 4:89–98

Mills JN, Kummerow JK (1989) Herbivores, seed predators and chaparral succession. In Keeley SC (ed) *The California Chaparral. Paradigms Reexamined*. Los Angeles Sci Ser No. 34, Nat Hist Mus LA County, Los Angeles, pp 49–56

Mooney HA (ed) (1977) *Convergent Evolution in California and Chile: Mediterranean Climate Ecosystems*. Dowden, Hutchinson and Ross, Stroudsburg, PA

Mooney HA, Harrison AT (1972) The vegetational gradient on the lower slopes of the Sierra San Pedro Martir in northwest Baja California. Madroño 21:139–145

Mooney HA, Parsons DJ (1973) Structure and function of the California chaparral: an example from San Dimas. In Mooney HA, di Castri F (eds) *Mediterranean Type Ecosystems: Origin and Structure*. Springer Verlag, Berlin, pp 88–112

Mooney HA, Schlegel F (1967) La vegetación costera del Cabo Los Molles en la provincia de Aconcagua. Bol U Chile 75:27–32

Mooney HA, Dunn EL, Shropshire F, Song L (1970) Vegetation comparisons between the mediterranean climatic areas of California and Chile. Flora 159:480–496

Mooney HA, Gulmon SL, Parsons DJ, Harrison AT (1974) Morphological changes within the chaparral vegetation as related to elevational gradients. Madroño 22:281–285

Muñoz J (1959) Flora de los cerros de Renca. Tesis, Facultad de Química y Farmacia, U Chile

Munz PA, Keck DD (1965) *A California Flora.* U California Press, Berkeley, CA

Murray KG (1986) Consequences of seed dispersal for gap-dependent plants: relationships between seed shadows, germination requirements, and forest dynamic processes. In Estrada A, Fleming TH (eds) *Frugivores and Seed Dispersal.* Dr W Junk Publishers, Dordrecht, pp 187–198

Noble JC (1982) The significance of fire in the biology and evolutionary ecology of mallee Eucalyptus populations. In Barker WR, Greenslade PSM (eds) *Evolution of the Flora and Fauna of Arid Australia.* Peacock Publications, Adelaide

Noble JC, Smith AW, Leslie HW (1980) Fire in the mallee shrublands of western New South Wales. Aust Range J 2:104–114

O'Dowd DJ, Gill AM (1986) Seed dispersal syndromes in Australian *Acacia.* In Murray DR (ed) *Seed Dispersal.* Academic Press, Sydney, pp 87–121

O'Dowd DJ, Hay ME (1980) Mutualism between harvester ants and a desert ephemeral: Seed escape from rodents. Ecology 61:531–540

Parsons DJ (1976) Vegetation structure in the mediterranean scrub communities of California and Chile. J Ecol 64:435–447

Parsons DJ, Moldenke AR (1975) Convergence in vegetation structure along analogous climatic gradients in California and Chile. Ecology 56:950–957

Reichman OJ, Oberstein O (1977) Selection of seed distribution types by *Dipodomys merriami* and *Perognathus amplus.* Ecology 58:636–643

Rice BL, Westoby M (1981) Myrmecochory in sclerophyll vegetation of the West Head, New South Wales. Aust J Ecol 6:291–298

San Martín J, Figueroa H, Ramírez C (1984) Fitosociología de los bosques de ruil (*Nothofagus alessandri* Espinosa) en Chile central. Rev Chil Hist Nat 57:171–200

Schlegel F (1966) Pflanzensoziologische und floristische Untersuchungen über Hartlaubgehölze in La Plata-Tal bei Santiago de Chile. Berichte Oberhessische Gesellschaft für Natur- und Heilkunde zu Giessen, Neue Folge, Nat Abt 34:183–204

Simonetti JA (1984) Late Pleistocene extinctions in Chile: A Blitzkrieg? Rev Chil Hist Nat 57:107–110

Smith AJ (1975) Bird-dissemination of woody plants in a temperate forest. Ecology 56:19–34

Solar V (1975) *Las Aves de la Ciudad.* Gabriela Mistral, Santiago, Chile

Sparrow A (1989) Mallee vegetation in South Australia. In Noble JC, Bradstock RA (eds) *Mediterranean Landscapes in Australia, Mallee Ecosystems and Their Management.* Brown Prior Anderson Pty Ltd, Melbourne, pp 109–124

Specht RL (1979) The sclerophyllous (heath) vegetation of Australia: the eastern and central states. In Specht RL (ed) *Ecosystems of the World*, Vol 9A. *Heathlands and Related Shrublands. Descriptive Studies.* Elsevier, Amsterdam, pp 125–210

Specht RL (1981) Mallee ecosystems in Southern Australia. In Specht RL (ed) *Ecosystems of the World*, Vol 11. *Mediterranean Type Shrublands.* Elsevier, Amsterdam, pp 203–229

Stebbins GL (1974) *Flowering Plants, Evolution Above the Species Level.* Belknap Press of Harvard U, Cambridge, MA

Stewart D, Webber PJ (1981) The plant communities and their environments. In Miller PC (ed) *Resource Use by Chaparral and Matorral.* Springer-Verlag, New York, pp 43–68

Thrower NJW, Bradbury DE (1977) *Chile–California Mediterranean Scrub Atlas.* Dowden, Hutchinson and Ross, Stroudsburg, PA

Troncoso A, Villagrán C, Muñoz M (1980) Una nueva hipótesis acerca del origen y edad del bosque de Fray Jorge (Coquimbo, Chile). Bol Mus Nac Hist Nat, Santiago, Chile 37:117–152

Van der Pijl L (1982) *Principles of Dispersal in Higher Plants.* Springer-Verlag, New York

Villagrán C, Armesto JJ (1980) Relaciones florísticas entre las comunidades relictuales del Norte Chico y la zona central con el bosque del sur de Chile. Bol Mus Hist Nat, Chile 37:87–101

Villagrán C, Riveros M, Villaseñor RV, Muñoz M (1980) Estructura florística y fisionómica de la vegetación boscosa de la Quebrada de Córdoba (El Tabo) Chile central. An Mus Hist Nat, Valparaíso, Chile, 13:71–89

Villagrán C, Armesto JJ, Leiva R (1986) Recolonización postglacial de Chiloé insular: Evidencias basadas en la distribución geográfica y en los modos de dispersión de la flora. Rev Chil Hist Nat 59:19–39

Villaseñor R (1980) Unidades fisionómicas y florísticas del Parque Nacional La Campana. An Mus Hist Nat, Valparaíso, Chile, 13:65–70

Vogl RJ, Schorr PK (1972) Fire and manzanita chaparral in the San Jacinto Mountains, California. Ecology 53:1179–1188

Vuilleumier F (1985) Forest birds of Patagonia: ecological geography, speciation, endemism and faunal history. Ornith Mono 36:255–302

Wellington AB (1989) A study of the population dynamics of the mallee *Eucalyptus incrassata* Habill. Ph.D. Thesis, Australian National U

Westoby M, Rice BL, Shelley JM, Haig D, Kohen JL (1982) Plants' use of ants for dispersal at West Head, New South Wales. Geobotany 4:75–87

Westoby M, Rice BL, Howell J (1990) Seed size and plant growth form as factors in dispersal spectra. Ecology 71:1307–1315

Whelan RJ (1986) Seed dispersal in relation to fire. In Murray DR (ed) *Seed Dispersal.* Academic Press, Sydney, pp 237–271

Whittaker RH (1977) Evolution of species diversity in land communities. Evol Bio 10:1–67

Wiens JA (1983) Avian community ecology: an iconoclastic view. In Brush AH, Clark GA (eds) *Perspectives in Ornithology.* Cambridge U Press, Cambridge, pp 355–403

Willson MF, O'Dowd DJ (1989) Fruit color polymorphism in a bird-dispersed shrub (*Rhagodia parabolica*) in Australia. Evol Ecol 3:40–50

Willson MF, Rice BL, Westoby M (1990) Seed dispersal spectra: a comparison of temperate plant communities. J Veg Sci 1:547–562

Zedler PH, Zammit CA (1989) A population-based critique of concepts of change in the Chaparral. In Keeley SC (ed) *The California Chaparral. Paradigms Reexamined.* Los Angeles Science Series No. 34, Nat Hist Mus LA County, pp 73–83

13. Effect of Seed Predation on Plant Regeneration: Evidence from Pacific Basin Mediterranean Scrub Communities

Svaťa M. Louda

Plant communities are distinguished by the diversity and dynamics of their component species. A few species generally account for most of the biomass and cover in a community, and the majority of species in any plant community are sparse or rare. As a consequence, the richness, dynamics, and special character of a plant community are determined by its uncommon species, those immersed interstitially within the matrix of dominants. Most ecological studies of plant communities, however, have focused on the dominant species, including those within mediterranean climate systems (e.g., Mooney 1977, di Castri et al. 1981, Kruger et al. 1983, Specht et al. 1988). Clearly, understanding the diversity of a plant community also requires knowledge about the demography, density, and distribution of the less common, interstitial species, those which contribute most to community richness. Furthermore, many contemporary environmental issues, including maintenance of biodiversity, sustainable use of natural systems, and rehabilitation of damaged ecosystems, require understanding of the processes that define and limit the performance of the characteristic, less common species. Herbivory, or feeding on living plants, is one such process which can influence relative plant success and which deserves more attention in this context (Harper 1977, Crawley 1983, Weis and Berenbaum 1989, Louda et al. 1990a, Huntly 1991).

The influence of herbivory on plant reproductive success is potentially very important (Harper 1977, Hairston 1989), but usually underevaluated

(Hendrix 1988, Louda 1989a,b). Yet, experiments show that reproductive herbivory limits seed or the number of seedlings of several characteristic, but sparse, interstitial plants (e.g., Hendrix 1979, Louda 1982a,b, 1983, Kinsman and Platt 1984, Louda et al. 1990b). Such reproductive herbivory includes destruction of developing inflorescences, flowers in various stages of presentation, developing ovules, and matured seed. Both direct and indirect effects of reproductive herbivory have been documented. The direct effects of reproductive herbivory alter key estimators of fitness as well as important parameters of population dynamics. For example, reproductive herbivory often drastically reduces the number of viable seeds (e.g., Janzen 1971, Waloff and Richards 1977, Louda 1982a,b,c, 1983, Scott 1982, Kinsman and Platt 1984, Auld 1986, Zammit and Hood 1986, Lamont and van Leeuwen 1988, Wallace and O'Dowd 1989, Louda et al. 1990b). Interestingly, the quality of the remaining viable seeds can also be altered (e.g., Hendrix and Trapp 1989, Gange et al. 1989). Postdispersal seed predation often modifies the number and spatial array of seeds surviving to germinate (e.g., Janzen 1971, Brown et al. 1979a,b, 1986, Auld 1983, Hughes and Westoby 1992). Feeding by floral- and seed-exploiting insects can also directly limit the number of seedlings that are recruited (e.g., Louda 1982a,c, 1983, Louda et al. 1990b, Louda and Potvin 1994). Such effects occur in addition to other, occasionally obvious, negative effects of foliage herbivory on adult plant survival and flowering (Harper 1977, Crawley 1983, Huntly 1991).

Significant indirect effects of reproductive herbivory change plant interactions (Louda et al. 1990a). Differential consumption within a subset of competing plants is a mechanism for strong indirect effects. Not all plant species are equally palatable or similarly damaged (Janzen 1971, Huntly 1991). Consistent differences in damage occur among individuals and genotypes, among plants in different environments, and among species. Such differences can change expectations of competitive outcome based on other criteria, such as physiological tolerances or competitive abilities (Harper 1969, 1977, Lubchenco 1978, Louda et al. 1990a).

The importance of recruitment for persistence of a species should increase as disturbance increases or plant tolerance of competition decreases. If so, the significance of processes that affect recruitment will also increase. The interstitial species in a plant community are generally shorter-lived and less competitive than the dominant species. And, as expected, many of these species show high dependence on seed regeneration for persistence (Grubb 1977). Because reproductive herbivory often strongly affects seed availability, it can be expected to play a critical part in the population dynamics and relative abundance of many subdominant and rare species in plant communities. Although the effects of physical factors and competition on regeneration of plants are well investigated, however (e.g., Harper 1977, Fenner 1985, 1992, Crawley 1990), those of reproductive herbivory are not (Louda 1982a,c, 1983, Auld and Myerscough 1986, Hendrix 1988, Louda 1989b).

My purpose here is to stimulate discussion and further consideration of the effects of reproductive herbivores, especially inflorescence- and seed-feeding insects, on individual, population and community patterns of seed production and regeneration, particularly in Pacific Basin mediterranean communities. I suggest that seed-consumer interactions are likely to be fundamental in the dynamics of at least one set of interstitial plant species. This set includes the short-lived perennials with inflorescence and seed predators and without permanent seed banks. The hypothesis relies primarily on a limited number of experimental studies on the effects of inflorescence- and seed-feeding consumers on seed number and seedling recruitment. It suggests a way to order the expected population outcome of interactons. I also propose a conceptual model to predict the probability of influence by seed losses on plant demography and abundance. This scheme systematizes factors known to influence plant reliance on current seed production for population persistence. The evidence is suggestive, but far from definitive; the model is preliminary. Thus, the ideas are presented as hypotheses meant to challenge and stimulate further work by ecologists interested in the role of interactions with consumers in modifying plant population dynamics and community diversity in broadly distributed but threatened systems, such as mediterranean shrublands.

The chapter has four main parts. In the first part I briefly review vulnerability to early mortality in the plant life cycle. I then review predictions of the expected demographic outcome of seed loss, discussing the critical, experimental evidence required for definitively assessing the effects of consumers on recruitment plant dynamics. In the third part I review evidence on the frequency and influence of granivory on key demographic transitions in mediterranean-climate vegetation, emphasizing experimental studies in Pacific Basin systems. My own data from the coastal scrub and chaparral in southern California suggest that herbivore limitation of individual reproduction and population regeneration can both limit local recruitment and determine distribution of plants along a regional environmental gradient from coast to inland mountains. I conclude with my working model for predicting the relative importance of reproductive herbivory on plant population dynamics. The model is an attempt to distinguish key life-history characteristics of plants that alter the probability of seed limitation of recruitment and persistence and, thus, the potential effects of consumers on occurrence. In this model, the cumulative effects of the combined plant traits determine demographic vulnerability. The goal is to increase the precision with which we can predict the plants and the situations in which the interaction with consumers, especially those affecting plant reproductive effort, will be fundamental in explaining abundance and distribution of plants.

Early Mortality in Plant Population Dynamics

Plant Vulnerability

Plant development can be divided into two phases: vegetative growth and sexual reproduction. Research on the effect of herbivores in the former phase, that is, effects of foliage consumption on plant survival and growth, is relatively common (Huntly 1991). Catastrophic (outbreak) levels of foliage removal on dominants are often the only circumstances under which large-scale effects of foliage-feeding insects on plant dynamics and community structure are obvious. Many parameters of individual plant performance are often altered by foliage-feeding herbivores, however (Harper 1977, Crawley 1983, Louda 1984, Belsky 1986, Hendrix 1988, Louda et al. 1990a). The relative importance of foliage removal by herbivores, and the conditions under which such removal affects plant performance, are becoming clearer. Such feeding often alters growth and reduces reproductive allocation (Crawley 1983, Whitham and Mopper 1985, Huntly 1991). But plants respond and many compensate, to some degree, for damage to foliage and vegetative meristems (Rockwood 1973, McNaughton 1979, Whitham et al. 1991). Some species may compensate fully for foliage removal (Paige and Whitham 1987), whereas others definitely do not (Louda 1984) or do so only under limited conditions (Maschinski and Whitham 1989, Belsky et al. 1993, Meyer and Root 1993). Thus, there appears to be a continuum of plant response to foliage destruction (Maschinski and Whitham 1989) that is likely to be selected by the combination of factors that can cause serious leaf damage (Belsky et al. 1993).

Research on the effects of herbivores in the second phase, such as effects of inflorescence consumption on sexual regeneration, has lagged behind other research on herbivory (Harper 1977, Fenner 1985, Hendrix 1988, Louda 1989b). Reported levels of seed reduction, however, including those by both pre- and postdispersal consumers, are often large. Thus, the magnitude of loss (Janzen 1971, Naylor 1984), the ubiquity of the seed-feeding strategy (Janzen 1971), and the importance of seed regeneration for persistence of many species (Harper 1977, Grubb 1977, Fenner 1985), argue for the potential importance of seed mortality for some plant populations (Janzen 1971, Hendrix 1988, Louda 1989b). Relevant data at the population level are limited, though. Most analyses of insect herbivory, including studies of floral and predispersal seed predation, focus on individual plant performance rather than examine variation within and among populations. Thus, although reductions in seed production are widespread and often severe (Salisbury 1942, Janzen 1971, Louda 1989b), the connection between the seed-crop losses and subsequent plant population densities remain either unclear or disputed.

Traditionally, it has often been assumed that seed numbers are ample. Therefore, plant recruitment and density are usually expected to be limited by competition or by regeneration ("safe") sites (Grubb 1977, Harper 1977, Crawley 1990, 1992). This assumption is no doubt encouraged by the observation that, for many conspicuous plants, more viable seeds appear to be dispersed than could establish, and more seedlings established than could survive. In fact, this surplus does occur in dominant species. We have, however, no reason to suppose that it holds for all, or even most, species.

Theoretically, seed availability can also limit plant numbers (Ehrlich and Birch 1967, Harper 1977, Louda 1982a,c, 1983, Crawley 1992), and may do so for obligate-seeding species in fire-prone systems (Keeley 1991, Hughes and Westoby 1992). Thus, ovule reduction could impose a quantitative limit on plant recruitment (Janzen 1971, Harper 1977, Fenner 1985). Even proponents of the trophic limitation hypothesis for herbivore productivity acknowledge that seed feeders may be limited by their food resource (seeds) and thus may limit the standing crop of that resource (Slobodkin et al. 1967, Hairston 1989), and thus regeneration. In sum, seed limitation needs to be evaluated as a potential controlling factor in plant recruitment. Because some evidence already suggests that seed limitation can occur, we need a more critical evaluation of the processes that determine seed availability, including inflorescence and seed consumption, and their effects in complex assemblages of plants.

Demographic Effects of Seed Predation

The population consequences of mortality at one stage cannot be assessed independently of the plant's fate in successive stages in the life cycle (Harper 1977, Crawley 1983, Louda 1989b, Louda and Potvin 1994; Fig. 13.1). The life cycle can be partitioned into three major transitions: flower bud to seed, seed to seedling, and seedling to mature adult. Each of the three links is critical to plant regeneration, especially for plants dependent on current seed production. It is cumulative mortality, including that at the earliest stages of seed production, that ultimately limits plant recruitment. Data on the transi-

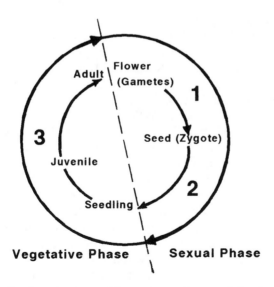

Figure 13.1. Seed production and loss as integral parts of the plant life cycle.

tion from seedling to adult is relatively common, but evidence on the inter-generational effects of the first two transitions is much rarer and often lower in quality. The experimental evidence is reviewed here, by key demographic transitions.

Flower Bud to Seed

Extensive and variable damage to developing flowers and seed is well documented for many species, including those of mediterranean climate ecosystems (Salisbury 1942, Janzen 1971). The population consequences are not well documented, however (Hendrix 1988, Louda 1989b). Patterns in the data include major differences in seed reduction: (1) among species, (2) among individuals within a species, and (3) along spatial and temporal gradients (Janzen 1971, Louda 1982b, 1983, 1989a,b, Louda et al. 1990a). If such differences persist to the adult stage, these seed losses could alter the relative abundance, population structure, and gradient distribution of the most vulnerable plant species.

Seed to Seedling

The next transition in the life cycle entails dissemination, dormancy, escape from predators, and location of germination site. Release from the plant increases exposure to a large group of more generalized, postdispersal seed predators. In Australian mediterranean climate systems, the ants that are the predominant seed consumers are also seed dispersers. There is some question, however, about their effectiveness as dispersal agents (Hughes and Westoby 1992). In Californian chaparral, ants can also be important. But the predominant seed predators and dispersers are vertebrates, such as rodents and birds (e.g., Keeley and Hays 1976, Quinn 1990). After dispersal, consumers significantly and preferentially decrease the persistence of larger seeds (Horton and Wright 1944, Keeley and Hays 1976, Fenner 1985, Foster 1986). But experimental studies on the effect of such predation on plant recruitment and community composition are rare (Westoby et al. 1992). Seed dormancy also influences the transition from seed to seedling. Long-term dormancy leads to development of a permanent, buffering seed bank. Alternatively, transient seed banks, where seed turnover is high and seed persistence is low, increase reliance on current reproductive effort. In fact, current models suggest that the intensity of selection on annuals by seed predators depends on: (1) magnitude of the seed bank, (2) relative contribution of fresh versus buried seed to recruitment, and (3) variation in rates of loss (Brown and Venable 1991). Thus, increased reliance on fresh seed for establishment and variation in loss caused by inflorescence-feeding herbivores should also increase the probability that the reduction in seed will create significant variation in seedling establishment. Annuals and monocarpic perennials are likely to be more susceptible than long-lived plants.

Seedling to Adult

Seedling survival to flowering adult is a third stage in the transition series from flower to flower. Extensive data are available on seedling survivorship and the seedling–adult link in greenhouse experiments and in experimental gardens (see Harper 1977, Fenner 1985). Comparable data from natural vegetation are not as abundant, but are increasing (e.g., Gross and Werner 1982, Goldberg and Werner 1982, Gross 1984, Crawley 1990). These studies show that the probability of success for this transition varies with plant species, plant life history, and competitive context. High survival and great longevity of flowering age classes decrease the influence of seed consumers on population persistence relative to the factors influencing this transition.

Prediction and Evaluation of Seed Predation

Conceptual Approach

To clarify how and when seed consumers affect the probability of regeneration and the occurrence of a plant species within the community matrix, we need to know the determinants of success in seedling recruitment. Such factors could include the relation between seedling establishment and: (1) quantity and quality of viable seed released, (2) seed-bank dynamics, and (3) conditions under which germination and establishment occur. Few direct tests have been made on the influence of insect consumers on each of these steps in seedling establishment (Harper 1977, Fenner 1985, Hendrix 1988, Louda 1989b). Therefore, the general consequences of flower and seed predation on plant dynamics, especially predation by insects, remain controversial (Louda 1989b, Crawley 1990, 1992).

When might seed consumption limit plant numbers? The significance of mortality at any stage in the life cycle depends in large measure upon processes at subsequent stages (Harper 1977). At which stage, though, would the effect of seed loss on the population be most evident? Harper proposed that seed number was crucial (1977: 460): "Predation is relevant in the control of population size if it carries the seed density below that to which the plant population will be reduced by later density-dependent processes" (Fig. 13.2A). Clearly, average seed production over the life span below that required to establish potential adult densities will limit plant numbers. Although this criterion is sufficient to demonstrate a predation effect, is it really the necessary criterion? Our studies imply that it is too severe (Louda 1982a,b, Louda and Potvin 1994). I propose, instead, that seedling number is more important, that predation is critical in controlling population size if it carries the seedling density below that to which the plant population will be reduced by later density-dependent processes (Fig. 13.2B). This hypothesis takes the low and highly stochastic nature of seedling establishment into account and partitions pre- versus postestablishment mortality. Though less

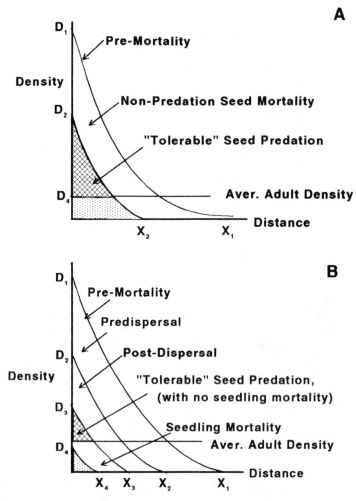

Figure 13.2. Models of the effect of seed predators on seed and seedling distribution
and density. (A) Harper's model, suggesting that tolerable amounts of seed predation
could be identified as the seeds in excess of sustainable adult density (Modified from
Harper 1977). (B) Modified model, distinguishing pre- from postdispersal mortality
and loss in seed and illustrating a less restrictive criterion for significant effect by seed
consumers—that which decreases seedling density required to establish subsequent
adult density (Modified from Louda and Potvin 1994).

restrictive, it is still clear that if seedling density is higher than that leading to the usual adult regeneration rate, then seed losses to insects before seedlings are established will have little quantitative effect on adult density. Alternatively, if seedling densities are lower than the number of adults that could mature to flower, then seed losses to insects are a numerically important source of mortality.

I expect also that the effects of seed reduction on plant population dynamics and community composition will be proportional to the relative importance of seed regeneration for persistence of the plant species. The relative importance of seed loss must be intimately related to the plant's life history and strategy for regeneration. Short generation time with high adult plant mortality and turnover and little seed bank buffering of variation in seed input should increase the relative importance of seed regeneration in plant abundance and persistence. At one extreme, some annual plants have ruderal or ephemeral life histories characterized by rapid population cycling and by a reproductive strategy involving high numbers of small seeds and a permanent seed bank. At the other extreme, over a comparable time long-lived perennial plants have life histories characterized by root and canopy development and vegetative spread, and proportionately less energy invested in yearly seed reproduction. Intermediate life histories, such as those of monocarpic or short-lived perennials, tend to be most successful in disturbed, resource-rich habitats. The dynamics of plants with intermediate life histories are typically fugitive and their distributions are patchy. Characteristic reproductive and seed traits of fugitive species include: large inflorescences, fewer and larger seeds than annuals, relatively synchronous seed production, and small or transient seed banks. Most of these traits also increase the value of the inflorescence resource to predispersal flower- and seed-feeders.

In sum, the effect of floral herbivores on the dynamics of plants with different life histories is expected to be nonlinear. Populations of short-lived perennials with intermediate, fugitive life history strategies, large seeds, and transient seed banks should be most vulnerable to seed reduction by consumers (Louda 1989b, Louda et al. 1990a). Disturbance on various scales and over a range of frequencies is an integral part of the dynamics of these species.

Further, the influence of granivory—seed consumption—on community and ecosystem scales should be higher in systems with: (1) spatially heterogeneous vegetation structure, such as systems in which small-scale, local disturbance or larger-scale community disturbance are relatively predictable; (2) a large complement of short-lived perennials; and (3) a subset of such species that have only transient seed banks, or permanent seed banks that decline steeply with time. Interestingly, most mediterranean climate scrublands fit this scenario. Species richness of these shrublands is directly related to disturbance, such as fire (e.g., Westman and O'Leary 1986). Each shrubland has a characteristic array of relatively short-lived perennials. For example, the herbaceous perennials comprise 25% (range = 14 to 37%) of the coastal scrub in California and 34–39% of matorral in Chile (Westman 1988). Although many of these species seem to have permanent seed banks, a subset of species such as *Haplopappus* species (Asteraceae) do not (Louda 1978; Zedler, this volume). Establishing new plants in disturbances, the main mode of persistence for short-lived perennials in mature vegetation, requires large numbers of seeds.

Evaluating Seed Predator Effects

What types of data are necessary to unambiguously determine the role of inflorescence feeders on plant density? If such consumers have a demographic influence, it must be initiated by the end of seed maturation. Thus, the question really has two important components: (1) To what degree does seed availability limit densities of seedlings and adults? (2) To what degree do insects limit seed availability in cases where seedling density is proportional to seed density?

The first component question can be answered, at least as a first approximation, by determining the rate of seedling establishment and survival from experimentally sown seed (e.g., Gross and Werner 1983, Gross 1984, Crawley 1990, 1992, De Steven 1991). If regeneration probability is limited by current seed production, then carefully designed additions of seed should lead to increased plant recruitment and density. If regeneration is not limited by seed availability under the vegetation density in the experiment, then seed addition should have no effect on seedling density or subsequent adult density. Seed input experiments to determine the role of seed availability in seedling establishment have a long tradition (see Harper 1977, Fenner 1985).

The advantages of seed input experiments include ability to control ambient plant density and to specify soil surface conditions. The interpretation of seed addition experiments for assessing seed predator effects, however, is subject to at least five major problems that make designing an experiment difficult. First, dispersal needs to be within the potential seed shadow of real plants, to mimic the effects of the typical localized dispersion observed. Second, it is also important to mimic the density and spatial array of seed that is sown naturally, because seed removal may be density dependent. Third, it is also important, but extremely difficult, to approximate the gradual timing of dispersal of naturally sown seed, for much the same reason. Fourth, vegetation density must bracket the conditions relevant to the species dynamics, because the density and character of the background vegetation influence interpretation of the experimental results. If plant establishment is usually in small-scale disturbances (e.g., Goldberg and Werner 1982, Rabinovitz et al. 1989), then testing seed rain effects in homogeneous, dense vegetation is irrelevant, and it will lead to erroneous conclusions on the importance of seed in regeneration. Fifth, it may also be important to ensure that secondary dispersal and its influence are just as probable from hand-sown as from naturally dispersed seed. Plant establishment patterns may be more influenced by secondary dispersal phenomena than by the initial dispersal and shadow. Finally, many tests of seed establishment criteria have been done under greenhouse or garden conditions rather than field conditions. These results document potential, rather than actual, germination constraints. Their relevance to the field needs to be evaluated directly. Thus, although hand sowing of seed is an appropriate test for some questions, it is not the definitive one for seed predator influence on seedling establishment and population dynamics.

Excluding inflorescence insects, with documentation of subsequent seedling and adult densities, is the only unequivocal method to quantify the effect of insects on plant reproduction and density (Harper 1969, Louda 1982a,c, 1983, Louda et al. 1990b, 1992). In situ exclusion is required to determine both the effect on seed number and its influence on seedling establishment (Brown et al. 1979a,b, Louda 1982a,c, 1983, Brown et al. 1986, Brown and Heske 1990). Three main methods have been used to exclude insects: bagging (DeBach 1964), hand removal (Breedlove and Ehrlich 1968, 1972, Jordano et al. 1990), and insecticide exclusion (Waloff and Richards 1977, Louda 1982a,b, 1983, 1984, Brown et al. 1986). Bagging of inflorescences excludes herbivorous insects; however, it may also preclude other interactions, such as with mutualists like pollinators. Hand removal of insects also reduces insect density; however, it is effective primarily against diurnal, conspicuous, and large, usually late-instar insects. Given the magnitude of seed reduction often caused by early feeding (below), quantifying the full effect of insect feeding is likely to require insecticide exclusion. This method has the greatest chance of excluding all stages of multiple types of insects, rather than only the late instars of conspicuous insect herbivores.

Advantages of the insecticide exclusion test include its being a direct test of the hypothesis that insects modify the seed-to-seedling linkage. Also, it is done within the context of the existing vegetation and it allows natural sowing of seed. Two possible disadvantages are: (1) potential toxic effects of the insecticide on seed and seedling quality, and (2) interference with mutualists such as pollinators or insect predators. Interestingly, however, the prediction under both of these potentially confounding circumstances would be that insecticided plants should do worse than control plants. Thus, both disadvantages bias the results against the hypothesis that reproductive herbivores critically influence plant recruitment success and subsequent density.

Only a few studies have directly quantified seedling establishment and survival in response to the exclusion of inflorescence-feeding insects. Those I found from Pacific Basin mediterranean climates are reviewed in the next section. Only one study, in grassland, has experimentally quantified the influence of insect-induced seed loss on the number of flowering adult progeny in the next generation (Louda et al. 1990b, 1992, Louda and Potvin 1994).

This study was designed to assess the intergenerational fitness effects of pre-dispersal seed predation on a native species (*Cirsium canescens*), a species that represents the intermediate, fugitive life history. Also, by following recruitment by individuals, we now have the data to examine variation in fitnesses created by differences in seed predator load (Louda and Potvin, in prep.). In this study, inflorescence insects strikingly and significantly affected the number of progeny that survived to flower. We conclude that inflorescence insects had a direct, long-term effect on individual plant fitness (Louda and Potvin 1994). The cumulative effect of altering the magnitude of the earliest linkage in the life history is clear in the age–stage transition diagram (Fig. 13.3). This fate diagram directly quantifies the influence of insect herbivores on the three demographic transitions discussed above. It clearly illustrates perpetuation of insect-induced early mortality throughout the life cycle of this native short-lived perennial. Interestingly, these results contradict the expectations of the current paradigm (Crawley 1989a,b) that insects have little, if any, influence on plant fitness and population dynamics.

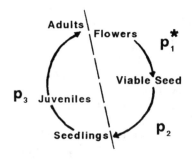

	Insecticide	Control	Transition
p_1^* =	0.155	0.063	Floral bud→seed
p_2 =	0.021	0.020	Seed→seedling
p_3 =	0.170	0.165	Seedling→adult

Figure 13.3. Transition probabilities for key life cycle stages of *Cirsium canescens*, illustrating the highly significant difference in seed production caused by inflorescence-feeding predispersal flower and seed insects, an effect that persisted over the lifetime of the attacked plants (Louda and Potvin 1994).

Granivory in Mediterranean-Climate Vegetation

Analysis of seed predation and its effects on plant regeneration in Pacific Basin mediterranean climate ecosystems has concentrated on a limited though fundamental set of questions. The most prevalent have been: (1) How much seed predation occurs? (2) Does the loss occur differentially among plant species? (3) What effects do the main seed feeders, such as ants or rodents, have on seed disperal and survival seed? Most of the studies I found focused on postdispersal seed predation. Several experimental studies, though, examined predispersal flower and seed consumption. As in most ecosystems, the research has been heavily concentrated on the dominant plant species. Yet the dominant plant species are the least likely to be regulated or limited primarily by seed availability (Briese and Macauley 1981, Wellington and Noble 1985, Andersen 1989a, Louda 1989b). I found no tests of selectivity or relative vulnerability to predation between seeds of common, dominant species and those of sparse species.

Postdispersal Seed Predation

Rates of postdispersal seed losses to consumers, and their effect on seed or seedling regeneration in mediterranean systems of the Pacific Basin, have been examined extensively. Losses have been studied in relation to three main processes: vertebrate foraging, insect foraging, and effects on plant regeneration after fire.

Vertebrate Foraging and Feeding

Studies of this process clearly show that postdispersal predation by vertebrates can be an important source of early plant mortality in mediterranean climate systems, such as chaparral (see Horton and Wright 1944, Parker and Kelly 1989). In addition, because the intensity of postdispersal predation usually varies among sites or habitats (e.g., Price and Jenkins 1987, Whelan et al. 1991), systematic variation in small mammal foraging rates or densities along environmental gradients in mediterranean systems is likely to contribute to variation in the density of preferred plant species along those gradients. But no intensive studies of this possibility appear to have been done. In fact, Ostfeld (1992: 45) in a recent review of the small-mammal/plant interaction literature as a whole suggested that "the need for studies of the impact of small-mammal herbivores on plant distribution should be stressed." Alternatively, increased or more-effective dispersal could counterbalance some of the negative consequences of vertebrate seed feeding (Janzen 1969, Harper 1977). For example, Lloret and Zedler (1991) report that, in long unburned coastal chaparral, animals promoted seedling establishment of *Rhus integrifolia* outside the parental canopy. Zedler and Black (1992) found viable seeds of poorly dispersed vernal-pool plant species in the droppings of rabbits in Californian coastal scrub. Thus, even incidental consumption of seed

by herbivorous species, such as rabbits, may contribute to dispersal of plants
with hard seeds.

Experimental exclosures of vertebrates have been done in Californian
coastal sage scrub and chaparral (Bartholomew 1970, Halligan 1973, 1974,
Louda 1983, 1988, Mills and Kummerow 1989, Kelly and Parker 1990,
Swank and Oechel 1991) and Chilean matorral (Fuentes and Le Boulenge
1977, Jaksic and Fuentes 1980). In general, excluding vertebrate foragers led
to increases in herbaceous plant cover and density, suggesting that vertebrates reduced early survival significantly in both systems. The general experimental design, however, does not allow separation of seed versus seedling consumption as the main mechanism explaining the observation.

In the Chilean system, given the observations above and knowing that
specialized seed-feeding mammals appear absent or scarce (Cody et al. 1981,
Catling 1988), the most plausible hypothesis is that introduced rabbits now
often retard successful seedling establishment in typical matorral (Jaksic and
Fuentes 1980). In the Californian coastal scrub, it is likely that the vertebrates had their effect by reducing seedling survival more than seed survival
because specialized seed-feeding mammals are absent (Catling 1988) or at
least uncommon. In the more inland Californian chaparral, however, the
larger number of mammal species frequently includes two specialized seed-
feeding heteromyid rodents (Catling 1988). Also, at least three studies in
chaparral document high rates of seed removal (Keeley and Hays 1976,
Kelly and Parker 1990). Thus, it is likely that significant levels of postdispersal seed predation contribute to lowering recruitment in the chaparral
(Horton and Wright 1944, Parker and Kelly 1989).

Vertebrate exclosures in California shrublands also suggest that consumers can alter survivorship (Mills 1986). This study also showed that the
first month is the critical period for establishment for *Adenostema fasciculatum* and *Ceanothus greggii*, two of the dominant species. Five years of small
mammal exclosure significantly increased recruitment and density for these
two species in this coastal mediterranean system (Mills and Kummerow
1989). Finally, an elegant factorial experiment in the same system demonstrates that important interaction is possible among nutrient availability,
water availability, and the effect of herbivores on plant recruitment and
persistence (Swank and Oechel 1991).

A study of vertebrate seed predation by Kelly and Parker (1990) verifies
and extends an earlier study by Keeley and Hays (1976). Keeley and Hays
(1976) quantified seed disappearance from artificial caches of two chaparral
species in southern California: *Arctostaphylos glauca*, which reseeds after
fire, and *A. glandulosa*, which resprouts after fire. Their experiments showed
that the rate of seed removal was higher for *A. glauca*. They also found that
seeds of both species that remained in the soil had less dry mass on average
than mature seeds on the shrubs, suggesting faster loss of the larger seeds
from the seed bank, possibly to seed predators. One problem, however, with
extrapolating seed disappearance data into seed destruction estimates is that

the ultimate fate of the seed is unknown. Removal is not necessarily equivalent to destruction. In chaparral in central California, Kelly and Parker (1990) compared seed banks and seedling densities of a reseeding species, *Arctostaphylos canescens*, with those of the same resprouter, *A. glandulosa*, when vertebrates were excluded versus not excluded. They unexpectedly found higher total densities of seed in the *A. glandulosa* seed bank than in the *A. canescens* seed bank, suggesting higher seed reproductive effort in their region by the resprouting species. In parallel with Keeley and Hays's earlier results, though, they found that the *A. canescens* seed bank had a lower proportion of viable seed than did that of *A. glandulosa*, suggesting the seed of the obligatory reseeding species was more vulnerable to depletion and destruction by seed predation.

Insect-Predation Effects on Regeneration

Studies on this second process suggest that invertebrate consumers in mediterranean climate systems can have effects comparable to those of vertebrate seed consumers on regeneration. The most commonly studied insect seed feeders are ants. I found no data, for example, on the effects of ground-dwelling beetles and other insects that can damage, consume, or transport seed. Some ants, the elaiosome-collecting species, are primarily seed dispersers. Other ant species are primarily consumers. Thus, the relative magnitudes of opposing processes, reflecting relative abundances and activity of the ant species present, appear to determine the net effect of such ants on plant dynamics.

In the mediterranean-climate regions in Australia, ants are conspicuous seed feeders, and they are likely to be the most important postdispersal seed predators (O'Dowd and Gill 1984, Abbott and van Heruck 1985, Morton 1985). Andersen (1987) found that seed predation by ants significantly reduced both seed survival and seedling densities. Some ant species also disperse seed, however. Andersen and Ashton (1985) suggest that the rate of ant seed removal depends on: seed species, size of the seed clump, season of exposure, duration of exposure, and ant composition, abundance, and foraging behavior. They also propose that the effect of ant seed predation on seedling recruitment depends on seed size, seed numbers, weather, timing of seed fall, location of seed fall, availability of alternative ant foods, patterns of seedling mortality, and fire. Auld (1986), for example, showed that seed removal did not necessarily lead to increased seedling establishment because subsequent seed predation by *Pheidole* ant species was intense. The outcome of ant–seed interaction depended on the characteristics of the interacting species.

Experiments in Australian communities have been used to assess both the rates and effects of seed removal by ants. These studies substantiate the observation that ants cause major depletion of the soil surface store of seed in heath and mallee vegetation (e.g., O'Dowd and Gill 1984, Auld 1986, Hughes and Westoby 1992). Rates and patterns of depletion may thus influence temporal and spatial patterns of plant regeneration. O'Dowd and Gill (1984) hypothesize that predator satiation, caused by synchronized release of seed by serotinous species in response to fire, is required for successful seed regeneration in these systems. Reported seed disappearance rates during interfire periods are consistent with the predator satiation hypothesis. Although some seed release occurs and some seedlings establish during the between-fire intervals, germination conditions limit most recruitment to the immediate postfire period (e.g., Wellington and Noble 1985).

Experiments by Hughes and Westoby (1992) extend the earlier studies. They found that average dispersal distance by ants, even of elaisome-bearing seeds, was short (1.1 m). Also, the three main ant species had different effects on initial dispersion of seed: *Pheidole* 1 ate most of the seeds, whereas *Rhytidoponera* "*metallica*" and *Aphaenogaster longiceps* deposited most of the seeds intact outside their nest after removing the elaiosome. By using simulated seed and nest excavation, they showed that most stored seeds were generally located too deep to germinate successfully. Furthermore, their transition matrix suggested that intact seeds moved to the surface by ants were likely to be secondarily removed and eaten by *Pheidole*. Their small-scale experiment comparing seedling establishment from different nests after burning suggested that very few seeds ($< 1\%$) germinate after fire, but germination is more likely on non-*Pheidole* nests.

Few studies have been done on ant predation and dispersal in Californian and Chilean scrub vegetation. Perhaps this lack reflects the lower species diversities, or densities, of ants in these two systems. Australian mediterranean climate communities have higher ant diversities than California communities, which, in turn, have higher ant species richness than Chilean ones (Cody et al. 1981). In Chile, the lowest number of species was recorded in succulent coastal scrub and the highest in evergreen sclerophyllous scrub (Cody et al. 1981). In mediterranean shrub communities in California, ant dispersal and predation is characteristic of a few plants, such as *Dendromecon rigida* (Bullock 1974). Dispersal agents, such as ants, determine the level of spatial aggregation of several herbaceous species of plants in the coastal scrub in southern California (Bullock 1981, 1989). As in Australia, when seed collection is by seed-feeding harvester ants, as for *D. rigida*, the cost of dispersal is major seed losses from predation.

Mills and Kummerow (1989) report that their experiments revealed that harvester ants (*Pogonomyrmex subnitidus*) removed large quantitites of seeds of two chaparral dominants from the litter. These ants could differentially affect the seed supply of these competing dominants because they exhibited a distinct preference for seeds of one, *Ceanothus greggii*, over those of the other, *Adenostoma fasciculatum*. Mills and Kummerow (1989) argue that regeneration processes determine relative recruitment rates and abundance of these two dominants, and that consumers are important in determining these rates. Their exclusion experiment lends support to this interpretation. Interestingly, comparable patterns of seed removal and potential influence of ants and of rodents have been reported in fynbos vegetation in the mediterranean-climate region in South Africa (e.g., Pierce and Cowling 1991).

Another interesting aspect of ant–plant interactions that deserves more study in the field is suggested by a recent greenhouse study. Harmon and Stamp (1992) studied the influence of ant foraging on size hierarchies of plants from dense versus sparse seed populations. They concluded that ant foraging patterns, especially at high seed density, significantly altered size variation among plants within the experimental plant population. Because plant size hierarchies influence competitive interactions among plants (e.g., Weiner 1985, 1988), these results suggest that other subtle, potentially important consequences of foraging by small, postdispersal seed predators are presently being overlooked in the field.

Finally, some evidence suggests that insect herbivores that feed on very young seedlings of mediterranean scrub species can reduce recruitment success (Mills 1986). Mills's study showed that damage by sap-feeding insects within the first month of establishment significantly increased mortality, slowed growth, and decreased the probability of seedling establishment. Little else has been published about insect consumers in early seedling survival of native mediterranean climate plants.

Reestablishment, Regeneration, and Succession After Fire

Many studies of seed predation have been done about the fire cycle in Australia (see above, and Cowling and Lamont 1987, Zammit and Westoby 1988, Hughes and Westoby 1992) and in California (e.g., Keeley and Hays 1976, Kelly and Parker 1990). The evidence suggests that most of the common, characteristic species of these systems show substantial recruitment only after fire (Ashton 1979, Wellington and Noble 1985, Auld 1986, Kelly and Parker 1990, Hughes and Westoby 1992). Although several explanations are possible for these patterns, reduced predation after fire is one of the hypotheses that requires more evaluation.

Multiple factors that affect plant establishment change after a fire, including predation pressure on seeds. In California, small mammal densities can be reduced in burned areas for up to eight months; return to preburn densities was related to redevelopment of vegetation structure (Catling 1988). Thus, reduced small mammal seed predation could contribute to seedling establishment and growth after fire. I found no estimate on the effect of fire on densities or activities of vertebrate seed feeders in Australia. In Chile, exclosing vertebrates increased regeneration (Jaksic and Fuentes 1980), but I found no data on response of vertebrate densities to fires there. The next step should be to do more multifactor experiments (e.g., Wallace and O'Dowd 1989). These experiments should use seed bank assessments, seed additions, soil moisture and fire manipulations, as well as predator exclusions of various types, to evaluate the relative contribution of each of these potentially important factors to regeneration after fire in Mediterranean shrublands.

Predispersal Flower and Seed Predation

Rates of consumption of flowers and seeds prior to seed release and their influence on sexual reproductive success have been evaluated experimentally for several characteristic species of the Pacific Basin mediterranean climate shrublands. In Australia, these experiments assessed the effects of inflorescence feeders on the reproductive output of species of *Acacia* (Auld 1986, Auld and Myerscough 1986), *Eucalyptus*, *Leptospermum* and *Casuarina* (Andersen and New 1987, Andersen 1987, 1988a,b, 1989b), and *Banksia* (Abbott 1985, Zammit and Hood 1986, Wallace and O'Dowd 1989, Vaughton 1990). In California, experiments examined the effects of predispersal floral and seed predators on seed production and seedling establishment of two species of goldenbushes (Asteraceae) that replace each other along an environmental gradient from coast to inland mountains (Louda 1982a,c, 1983, 1988). The results of the latter study challenge the competitive replacement hypothesis usually used to explain displacement of similar plant species along such an environmental gradient. These results are discussed in more detail in the next section.

At least three general ideas arise from experimental exclusion of insects from developing inflorescences and seeds. First, accurate estimation of insect influence on seed production requires an exclusion test. Second, distinguishing between resource and insect limitation of seed also requires an insecticide-exclusion test. And, third, the magnitude of insect influence on recruitment by native plants is affected by the size and dynamics of the permanent seed banks.

First, the conventional method for estimating the amount of damage inflicted by insects needs to be replaced. Andersen (1988a) highlights this critical, but still relatively unappreciated, point. The usual approach for quantifying seed loss to insects is to collect ripe fruit or infructescences and inspect them for insect damage. But this method is inadequate. It drastically underestimates the total loss to insect-feeding because it does not take into account the effects of feeding by early insect instars on development of new inflorescences, presentation of flowers, or abortion of immature fruits (Louda 1982a,c, Andersen 1988c). Of the thirteen insect exclusion experiments of which I am aware, 85% clearly illustrate that the fate of these early stages in seed development are crucial to determining the total number of seed matured. Thus, the conventional method misses losses to an important fraction of the potential seed supply, that caused by early feeding damage by insects.

For example, the average rate of predispersal seed predation estimated for several myrtaceous species in Australia increased from 10%, using the conventional method, to 64% using the insecticide-exclusion method (Andersen 1988c). Similarly, examining developed follicles of *Banksia spinulosa* var. *neoanglica* showed that 13% were damaged by insects (Vaughton 1990). An insecticide exclusion test, however, showed that insects further decreased seed set by mining the rachis. This attack halved the number of inflorescences producing any mature follicles, and reduced developed follicles by 45% (Vaughton 1990). Thus, the estimated reduction in number of seed caused by insects feeding on the infructescences increased from 13%, based on the observational method, to about 75% using the experimental-exclusion method. Similar outcomes are observed in other insecticide-exclusion experiments on many species occurring in diverse regions globally (e.g., Waloff and Richards 1977, Louda 1982a,c, 1983, Kinsman and Platt 1984, Zammit and Hood 1986, Wallace and O'Dowd 1989, Louda et al. 1990b). These data suggest that insect feeding damage to developing infructescences often creates the most significant difference in seed production between plants, as evidenced by differences between plants protected from, or exposed to, these insects. This effect is missed entirely when insect damage is assessed by examining only fruits or infructescences that survive the early feeding damage and injury.

Hand removal of late instars of external, fruit-feeding insects is sometimes used as an alternative to insecticide exclusion (e.g., Breedlove and Ehrlich 1968, 1972, Jordano et al. 1990). It is better than no manipulation of insect densities, but hand removal is also likely to underestimate early, insect-caused damage. Because most phytophagous insects are either cryptic or have strong behavioral avoidance of predators, hand removal of even large, late-instar larvae can be difficult to do effectively, and removal of small early instars is next to impossible. As illustrated above, though, early instars often strongly influence survival and subsequent seed production of early inflorescences and floral buds. Thus, this method underestimates the potentially significant effects of feeding by cryptic, nocturnal, and early instar larvae. Such qualifications suggest that insecticide exclusion may be necessary to quantify the full effect of insects on seed production.

Second, insect limitation can have symptoms similar to, observationally indistinguishable from, those of resource limitation of fruit production. Both can cause massive inflorescence and fruit abortion (e.g., Stephenson 1981, Hendrix 1988, Louda 1989b). The confounding of resource and insect limitation is a big problem, making attribution of cause for fruit abortion difficult and generally ambiguous. When inflorescence-feeding insects are present, separating the relative contribution of resources and insects to seed limitation requires effective insect exclusion. As far as I can tell, no resource limitation has been found in any case where the roles of insects have been studied using mechanical or insecticide exclusion tests. For example, the goldenbush species that I studied occur in a relatively dry, winter-rain coastal scrub and chaparral in southern California. Flowering and seed production vary and could have been interpreted as limited by water availability. The number of seeds matured by both goldenbush species at all sites increased two or six times, however, when insect feeding on inflorescences was decreased with insecticide (Louda 1982a,c, 1983). These results suggest that insects rather than resources, such as water, generally limited fruit production.

Similarly, for several Australian species restriction of seed production by inflorescence feeders, rather than by resources, has also been observed in exclusion experiments. Auld and Myerscough (1986) bagged inflorescences of *Acacia suaveolens* to prevent weevil attack on developing seed. Bagging increased by sixfold on average the number of seeds that matured. Zammit and Hood (1986) used insecticide to exclude insects from inflorescences of *Banksia ericifolia* and *B. oblongifolia*. Insecticiding doubled the number of inflorescences of *B. ericifolia* that set seed, and increased by 40% the average number of seeds matured by each inflorescence that developed. No response was observed for *B. oblongifolia*, but Zammit and Hood (1986) suggest that their insecticide treatment was started too late to exclude the early damage to inflorescence development. This result would be consistent with the data suggesting that most damage done to eventual seed set by insects often occurs very early in flowering. A similar methodological factor may also explain the lack of a significant exclusion effect on seed maturation of *B. grandis* (Abbott 1985). In the latter experiment, insecticide was applied only once, and its timing relative to inflorescence development was not specified. Vaughton (1990) found that insecticide applied to developing inflorescences of *B. spinulosa* var. *neoanglica* increased the number of: inflorescences producing fruit, seeds per infructescence, and total seeds produced. And Andersen (1989a) evaluated the effects of inflorescence-feeding insects on seed production of *Eucalyptus baxteri* from floral bud initiation through dispersal; he found that seed was reduced by 70%. Insect feeding reduced flowering intensity, flower success, premature fruit abscission, seed/ovule ratio, seed viability, and exclusion increased the supply of mature, dispersed seed. These results are similar to those found for the two goldenbushes in California scrub (Louda 1982a,b,c, 1983). All these results lead to the conclusion that an insect exclusion test must be done prior to concluding that resources are limiting the number of seeds matured, at least for species that have inflorescence- and seed-feeding insect herbivores.

The third point is that the size of the permanent seed bank strongly influences the population consequences of inflorescence feeding and reduction of current seed. A large, persistent seed bank will buffer severe changes in seed production caused by oscillations in the magnitude of infloresence damage. Alternatively, a transient seed bank will cause establishment to mirror that variation. In Australia, changes in plant density caused by variation in current seed production appear to be buffered by permanent seed banks for both *Acacia suaveolens* (Auld and Myerscough 1986) and *Eucalyptus baxteri* (Andersen 1989b). Thus, even though the proportion of ovules to viable, dispersed seed for *E. baxteri* was only 0.2%, current seed production was not considered likely to be the main limiting factor for this long-lived species (Andersen 1989a,b). This result led Andersen (1989a) to question the importance of seed predation for stable populations of long-lived species. Because the probability of population limitation by seed should decrease with long adult longevity, small seed size, and a large permanent seed bank (Louda 1989b), Andersen's results are consistent both with the experimental data, showing insect effects on the demography of more volatile, larger-seeded species with transient seed banks (above), and with the model presented below.

Insect Seed Predation on Goldenbushes in California

The work on predispersal seed predation for two species of goldenbush in California (Louda 1982a,b,c, 1983, 1988) illustrates at least four important points about the effects of inflorescence-feeding insects on plant dynamics and distribution. The studies also substantiate the need to experimentally determine insect influence, both on local abundance and on the regional distribution of plant densities along an environmental gradient. Inflorescence-feeding insects greatly reduced both individual reproduction and seedling establishment in both the coastal sage scrub and the inland chaparral study sites in San Diego County, California. For one of the two goldenbushes, however, predispersal seed predation decreased significantly with distance from the coast along the gradient. The net result of the differential losses near the coast is a more inland distribution for this species than expected. For the other species, seedling herbivory increased with distance from the coast, and compressed the distribution of remaining plants toward the coast. Both experiments illustrate the importance of spatially predictable variation in herbivore pressure on variation in plant population abundance in space.

The goldenbushes, small composite subshrubs, reach their highest abundances three to five years after disturbances, such as fires, in the mediterranean scrub and chaparral in southern California (Munz and Keck 1970). In San Diego County, the two species replace each other along an 80- to 100-km gradient from coast to inland mountains. *Haplopappus* (*Isocoma*) *venetus* is most characteristic in the coastal area, whereas *Haplopappus* (*Hazardia*) *squarrosus* is much more common in inland chaparral (Fig. 13.4). Both species have relatively large seeds and no significant permanent seed bank (Louda 1978). Individual plant size and growth for both species are greatest near the coast (Louda 1982a, 1983). Adults appear to persist for years but recruitment is rare in closed vegetation (P. Zedler, pers. comm.). *Haplopappus squarrosus* often sprouts after spring and early summer fires and immediately produces seed, leading to substantial seedling establishment in one-year-old burns; seedling densities can be an order of magnitude higher than in long-unburned chaparral (P. Zedler, pers. comm.). This recruitment response could be explained by increased resource availability after fire. It could also be explained, though, by decreased seed predation after fire. The experimental results provide support for the latter interpretation.

Goldenbushes have insects feeding on all stages of development of their inflorescences, including inflorescence buds, floral buds, flowers, developing seed, and mature seed. A guild of about 25 phytophagous insects, primarily tephritid flies and microlepidopterans, lays eggs into the developing inflorescences, including very small, immature inflorescences (Louda 1978). The

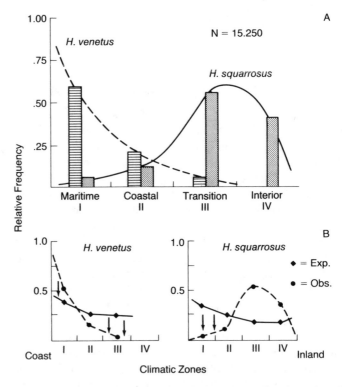

Figure 13.4. (A) The observed distribution of abundance of two species of golden-bushes (*Haplopappus squarrosus*, *Haplopappus venetus*, in coastal southern California. (B) The difference between observed and expected distribution of density of each species; for both species, insect feeding on inflorescences decreased the number of viable seeds released and led to proportionally lower seedling densities at all sites along the gradient. Because predispersal consumption by insects on *H. squarrosus* was much higher near the coast, it essentially eliminated major recruitment by this species near the coastal sites, and thus moved the modal class of its occurrence far inland. For *H. venetus*, subsequent herbivory on seedlings, which was especially high on new plants in the inland areas, augmented the significant reduction in seedling densities caused by insect feeding, and contributed to a coastal shift in the plant's distribution (From Louda 1989b).

consequences of inflorescence feeding by these insects on seed production and seedling establishment and juvenile recruitment were evaluated by comparing three treatments at four sites along the ocean-to-mountains gradient (Louda 1982a,c, 1983). These were: (1) insecticide spray (acephate [Isotox, Chevron Chemical Corp.]) applied in a water solution; (2) water-only spray as a control for adding water; and (3) no spray as a complete control. The basic prediction is that if resources or competition limit seed production or seedling density, then the reduction of insect feeding without resource augmentation should not increase either seed set or seedling numbers. Alternatively, if insect damage limits seed or seedlings, then reducing insect feeding should increase seed set and seedling numbers at the ambient levels of resources. The latter was observed in both species (Louda 1982a,b, 1983). Such results, reinforced by other insecticide-exclusion tests, lead me to suggest that short-lived perennials without permanent seed banks are particularly vulnerable to demographic effects of seed losses induced by inflorescence-feeding insects. Examples consistent with this suggestion include both polycarpic perennials in mediterranean scrub communities, such as *Haplopappus squarrosus* and *H. venetus*, and monocarpic perennials in grasslands, such as *Cirsium canescens* (Louda et al. 1990, 1992, Louda 1993, Louda and Potvin 1994).

Two proposed additional hypotheses contradict the significance of early flower and seed loss to insects. Early loss of an inflorescence may not reduce seed set as much as later losses, when more limiting resources have been invested. And early damage could lead to compensatory responses and additional flowers and seeds (Paige and Whitham 1987). The experimental results are directly applicable to an evaluation of the prediction and the two hypotheses.

First, the inflorescence feeding by insects significantly reduced seed production and input to the transient seed bank. The insecticide treatment, which did not exclude insects completely, increased seed 6.1 times, from 3.2% to 19.8% of flowers initiated, for *Haplopappus squarrosus* (Louda 1982b) and 2.2 times, from 22.0% to 49.1%, for *H. venetus* (Louda 1983). The earlier injury by oviposition occurred, the less likely it was that the inflorescence (head) developed further, eliminating the first hypothesis. Also, no compensation occurred in production of flowers or seed (Louda 1982a,c, 1983), eliminating the second hypothesis. Thus, I concluded that insects, not resources, limited the number of viable seeds of both goldenbush species at all my study sites along the coast-to-mountains gradient. No compensation occurred.

Second, the seed reduction caused by insects feeding on inflorescences, rather than safe sites, clearly determined densities of seedling establishment. The reduction in seed caused by insect feeding limited the number and density of seedlings for both *H. squarrosus* (Louda 1982a) and *H. venetus* (Louda 1983) at all my study sites along the coast-to-mountain gradient. The density of seedlings was directly proportional to the density of viable seed for both species. Because the rates of seedling establishment were similar among the three treatments, neither insecticide nor water nor safe sites could have determined seed germination. I concluded that seed availability, limited by insect feeding, rather than safe sites, determined seedling densities within sites all along the coast-to-inland gradient. Because resources did not limit seed production and safe sites did not limit seedling recruitment, the experimental results also suggest that a temporary escape from adapted seed predators is the most likely explanation for the seedling flush observed for *H. squarrosus* after fires.

Third, the spatial pattern of seedlings around each experimental parental plant of both species showed three important patterns. Seedling dispersion in both insecticide exclusion and controls in both years showed restricted seed dispersal, as expected for most plant species (Harper 1977). Also, the number of seedlings generally increased at all distances from the parental location when insects were excluded from the inflorescences, as expected by Janzen (1970). And finally, the predispersal seed predation lowered the density of the shadow but did not change the shape of the dispersion curve. Thus, the spatial pattern of recruitment observed was consistent with predictions made by Harper (1977) and Hubbell (1980) for rapidly, monotonically declining recruitment with distance from parent. The pattern is not consistent with Janzen's (1970) prediction that maximal seedling recruitment should occur at some intermediate distance from the parent because the highest levels of at least postdispersal seed predation should be near the parental plant.

Fourth, the differences established by manipulating attack of inflorescence-feeding insects persisted within each site for the next three years. Density-dependent seedling mortality may counter and compensate for the increased seedling density associated with predator removal (Harper 1977, Crawley 1990). In the studies of the goldenbushes, however, seedling densities were low and mortality was independent of density. I found no evidence of compensation for insect destruction of reproductive effort in number of flowers, seed vigor, seedling establishment, or juvenile survival. Insect feeding on developing seed did not affect the rate of survival of the seedlings that established (Louda 1982a,c, 1983). Survival rates of plants in subsequent life stages were similar between treatments (Louda 1982a,c, 1983). An interesting difference occurred, however, when seedling survival was compared among sites along the gradient. Different rates of seedling survival were observed between sites on the coast-to-mountains gradient. Mortality of seedlings of the inland goldenbush, *H. squarrosus*, was higher near the coast (Louda 1982a,b). Higher desiccation in the drier soils of the coast was the most likely explanation for differential survival near the coast and farther inland. Mortality of seedlings of *H. venetus*, the coastal goldenbush, was higher inland (Louda 1983). Desiccation is not a good explanation there because a high proportion (70%) of the seedlings in the inland area disappeared. A cage-exclosure experiment showed that vertebrate herbivory on new seedlings of this species was twice as high, and often lethal, in the inland area. Mills (1986) and Mills and Kummerow (1989) subsequently demonstrated experimentally that herbivory on seedlings could limit recruitment and survival of two other mediterranean scrub species.

Thus, in sum, inflorescence insects reduced seed and depressed seedling densities all along the gradient, and the difference between treatments persisted at each site. For the inland species (*H. squarrosus*), the seed predation effect was much more intense in the coastal region for the inland species. Higher seed predation coastally is sufficent in this case to explain a more inland population distribution than expected based on potential seed production (Louda 1982b). Also, seedling mortality caused by desiccation was also greater coastally for this species. Thus, the pattern of seedling mortality reinforced rather than compensated for the severe limitation imposed on plant density by inflorescence-feeding insects, especially near the coast. For the coastal species, the reduction in seedling densities caused by seed predation was significant, but equal, at all three sites along the gradient. But mortality of seedlings due to herbivory was much higher for this species inland. Thus, reproductive herbivory by insects limited seedling establishment densities, and vertebrate herbivory limited the survival of seedlings inland, moving the center of this species's distribution toward the coast.

Species replacement patterns, like that for the goldenbushes, are usually attributed to competitive displacement (e.g., MacArthur 1972, Cody and Mooney 1978, Hairston 1989). The data on the goldenbushes, though, provided the first definitive evidence for an alternative mechanism—differential predator effects—to explain a plant species replacement pattern along a gradient (Louda 1988). The experimental results clearly show that each species is limited by its consumers in the area of the other species' greatest abundance. Similar variation in intensity of interactions along gradients are documented in the rocky marine intertidal (Lubchenco 1978, Lubchenco and Cubit 1980, Lubchenco and Gaines 1981), in the tropical subtidal (Hay 1981, Hay et al. 1983), and at the marine–land interface in tropical areas (Louda and Zedler 1985, Smith 1987, Smith et al. 1989, Osborne and Smith 1990, O'Dowd and Lake 1990, 1991, Robertson 1991). We now have one clear-cut example of a gradient in the intensity of insect-herbivore damage and its effects on host–plant dynamics in terrestrial systems. These experiments and abundant anecdotal evidence suggest that this could be a general pattern (Louda et al. 1990b).

Recognizing such variation may help resolve seemingly disparate results among studies and put an end to unproductive conflict over the importance of insect herbivory. As I have argued before (Louda 1989b, Louda et al. 1990a), the question should no longer be: Is herbivory important? We have enough evidence now to ask more precisely: When, where and for which species is herbivory likely to be an important factor in determining plant population density and distribution? This more precise question cannot yet be answered. It begs for more experimental data on variation in herbivory and plant demography for colonizing versus for dominant plant species along environmental gradients within multiple terrestrial ecosystems, including the mediterranean-scrub communities.

Demographic Vulnerability to Granivory

Under what conditions should significant effects of seed predation be expected? The characteristics of the species showing significant effects of inflorescence-feeding insects in experiments provide clues for more general prediction of influence by floral- and seed-feeding consumers. Plant life history characteristics that have been shown to influence seed predation and seed availability for regeneration include: (1) short-lived perennial life history, and thus intermediate-length adult longevity; (2) relatively few, large inflorescences and seeds; (3) transient seed bank; (4) limited vegetative reproduction; (5) limited mean dispersal distance; and (6) dependence for establishment and persistence on coincidence of seed availability with small-scale disturbance (see Harper 1977, Fenner 1985). Both empirical and theoretical studies suggest that lack of a seed bank or vegetative reproduction increases the probability of extinction in a variable environment (Harper 1977, Brown and Venable 1985, Fenner 1985). This increase is associated with reproductive failure in one year. Thus, the higher the dependence on current seed production and the fewer the mechanisms to moderate effects of variation in seed reproductive failure, the greater the predicted influence of inflorescence- and seed-feeding insects. Predictability in the production of seed is the main strategy for dealing with unpredictability in the frequency and distribution of local disturbance.

Three traits seem particularly critical in plant vulnerability to insect-caused demographic effects. The first trait is a fugitive life history (e.g., Platt 1975, Louda 1982a, 1983, 1988), leading to a spatially dynamic metapopulation structure (Gilpin and Hanski 1991). A fugitive, colonizing life history strategy for a perennial species, especially one without buffering from a permanent seed bank, increases its dependence on current seed for regeneration. Recruitment requires high probability of coincidence of both openings in the vegetation and seed to colonize them. This requirement makes the availability of seed for recruitment, and anything influencing the number of seed, an increasingly important factor in population dynamics.

The second trait is a large, concentrated, energy-rich floral resource. Such floral resources are often associated with the fugitive life history strategy. Fugitive strategies are often most successful under harsh physical conditions or where frequent disturbance is common. These are situations under which large seeds should be an advantage in initial establishment. Alternatively, however, perennial plants with large seeds in conspicuous inflorescences provide a rich and trackable resource for inflorescence and seed-feeding insects. Thus, large seeds and predictable concentrations of floral resources facilitate specialization of an adapted suite of inflorescence feeders (Zwolfer 1965, 1988, Zwolfer and Romstock-Volkl 1991).

The third trait is a regeneration strategy that relies heavily on current seed rain, such as species without vegetative reproduction or a persistent seed bank (Louda and Potvin 1994). Thus, dependence on local disturbance for establishment and persistence, large and thus relatively vulnerable propagules, and short genertion time characterize species shown to be demographically vulnerable to effects of inflorescence-feeding insects. Other temperate species besides the goldenbushes for which experimental evidence shows that inflorescence-feeding insects reduce individual performance and fitness, and

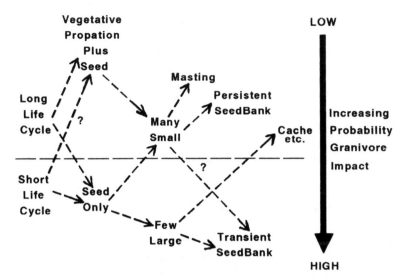

Figure 13.5. A conceptual model that represents an initial method for scaling the relative contribution of alternative plant life history traits to estimate the cumulative vulnerability of a plant species to increasing demographic consequences from its interaction with insect and vertebrate reproductive herbivores. The highest vulnerability is predicted for short-lived perennial plants that depend on a few large seeds that do not form a permanent seed bank in the soil, like fugitive species such as *Cirsium canescens*.

presumably population persistence, include *Cytisus (Sarothamnus) scoparius* (Waloff and Richards 1977), *Mirabilis hirsuta* (Kinsman and Platt 1984), and *Cirsium canescens* (Louda et al. 1990b, 1992, Louda and Potvin 1994).

These considerations lead me to suggest a conceptual model (Fig. 13.5) in which plant life history traits and their relationships combine to determine the probability of seed limitation, and thus influence of inflorescence- and seed-feeding consumers, on plant recruitment and persistence. Within this model, I assume that the effect of any factor that reduces availability of seed, including floral- and seed-feeding insects, increases as the importance of seed regeneration from current reproduction increases.

Alternative states of each characteristic, such as adult longevity, are contrasted in the model to illustrate the options (Fig. 13.5). Characters are arrayed as alternatives. The reproductive sequence proceeds from left to right. Each state of a character is related to its potential influence on seed regeneration by perennial plants in the face of flower and seed predator pressures. The lowest line, which includes short life cycle, regeneration solely by seed with a few large seeds and no permanent seed bank, is the combination of traits leading to the highest probability that population dynamics and persistence will be limited by seed consumers. Alternatively, characteristics that more or less buffer the population from dependence on current seed rain for regeneration and colonization, and reduce the probability of granivore limitation or regulation, are arrayed above. Life-history traits that decrease the tightness of the connection between current seed rain and persistence include vegetative reproduction, many small seeds, especially if they are masted, large, persistent seed banks, and animal-mediated seed dispersal (Grubb 1977, Harper 1977, Fenner 1985).

In sum, the probability of influence from reproductive herbivores, including classical granivores, is predicted to increase in its demographic importance as reliance on seed regeneration from current reproduction increases within the perennial plant life history. The greater this reliance, the greater the probability that any factor reducing the availability of seed, including floral- and seed-feeding insects, will limit recruitment, reduce density, and modify plant distribution. Thus, the cumulative effects of the plant's reproductive traits determine the severity of the seed loss demographically, and the probability that seed loss to consumers will limit plant population dynamics and influence plant community composition.

Acknowledgments

Many people have contributed to this work, and I want to thank them all. In particular, my research on *Haplopappus* could not have been done without the intellectual guidance and personal encouragement of Boyd D. Collier, Paul H. Zedler, and Robert F. Luck and the assistance of Gail Baker Gautier. This research was supported financially by both the Graduate Training Program in Ecology at San Diego State University and the National Science Foundation. My focus on plant population responses to insect consumers was stimulated by the work and insights of Richard D. Goeden, Daniel H. Janzen, and William W. Murdoch. My experimental approach to research on the population dynamics of plant–insect interactions bears the indelible mark of my opportunity to work with, and learn from: Joseph H. Connell, Paul K. Dayton, Bruce A. Menge, William W. Murdoch, Robert T. Paine, and Adrian Wenner. Finally, this chapter benefited from discussions with Nancy Huntly, V. Thomas Parker, and Paul H. Zedler.

References

Abbott I (1985) Reproductive ecology of *Banksia grandis* (Proteaceae). New Phytol 99:129–148

Abbott I, van Heruck P (1985) Comparison of insects and vertebrates as removers of seed and fruit in a Western Australian forest. Aust J Ecol 10:165–168

Andersen AN (1987) Effects of seed predation by ants on seedling densities at a woodland site in SE Australia. Oikos 48:171–174

Andersen AN (1988a). Dispersal distance as a benefit of myrmecochory. Oecologia 75:507–511

Andersen AN (1988b) Immediate and longer-term effects of fire on seed predation by ants in sclerophyllous vegetation in south-eastern Australia. Aust J Ecol 13:285–293.

Andersen AN (1988c) Insect seed predators may cause far greater losses than they appear to. Oikos 52:337–340

Andersen AN (1989a) How important is seed predation to recruitment in stable populations of long-lived perennials? Oecologia 81:310–315

Andersen AN (1989b) Impact of insect predation on ovule survivorship in *Eucalyptus baxteri*. J Ecol 77:62–69

Andersen AN, Ashton DH (1985) Rates of seed removal by ants at heath and woodland sites in southeastern Australia. Aust J Ecol 10:381–390

Andersen AN, New TR (1987) Insect inhabitants of fruits of *Eucalyptus, Leptospermum* and *Casuarina* in southeastern Australia. Aust J Zool 35:327–336

Ashton DW (1979) Seed harvesting by ants in forests of *Eucalyptus regrians* F. Muell. in central Victoria. Aust J Ecol 4:265–277

Auld TD (1983) Seed predation in native legumes of south-eastern Australia. Aust J Ecol 8:367–376

Auld TD (1986) Population dynamics of the shrub *Acacia suaveolens* (Sm.) Willd. Dispersal and the dynamics of the soil seed-bank. Aust J Ecol 11:235

Auld TD, Myerscough PJ (1986) Population dynamics of the shrub *Acacia suaveolens* (Sm.) Willd. Seed production and predispersal seed predation. Aust J Ecol 11: 219–234

Bartholomew B (1970) Bare zone between California shrub and grassland communities: The role of animals. Science 170:1210–1212

Belsky AJ (1986) Does herbivory benefit plants? A review of the evidence. Am Nat 127:870–892

Belsky AJ, Carson WP, Jensen CL, Fox GA (1993) Overcompensation by plants: Herbivore optimization or red herring? Evol Ecol 7:109–121

Breedlove DE, Ehrlich PR (1968) Plant-herbivore coevolution: lupines and lycaenids. Science 162:671–672

Breedlove DE, Ehrlich PR (1972) Coevolution: Patterns of legume predation by a lycaenid butterfly. Oecologia 10:99–104

Briese DT, Macauley BJ (1981) Food collection within an ant community in semiarid Australia, with special reference to seed harvesters. Aust J Ecol 6:1–19

Brown JH, Heske EJ (1990) Control of a desert–grassland transition by a keystone rodent guild. Science 250:1705–1707

Brown JH, Davidson DW, Reichman OJ (1979a) An experimental study of competition between seed-eating desert rodents and ants. Am Zool 19:1129–1143

Brown JH, Reichman OJ, Davidson DW (1979b) Granivory in desert ecosystems. Ann Rev Ecol System 10:201–227

Brown JH, Davidson DW, Munger JC, Inouye RS (1986) An experimental study of competition between seed-eating desert rodents and ants. Am Zool 19:1129–1143

Brown JS, Venable DL (1991) Life history evolution of seed-bank annuals in response to seed predation. Evol Ecol 5:12–29

Bullock SH (1974) Seed dispersal of *Dendromecon* by the seed predator *Pogonomyrmex*. Madroño 22:378–379

Bullock SH (1981) Aggregation of *Prunus ilicifolia* (Rosaceae) during dispersal and its effect on survival and growth. Madroño 28:94–95

Bullock SH (1989) Life history and seed dispersal of the short-lived chaparral shrub *Dendromecon rigida* (Papaveraceae). Am J Bot 76:1506–1517

Catling PC, Coordinator (1988) Vertebrates. In Specht RL (ed.) *Mediterranean-Type Ecosystems: A Data Source Book*. Kluwer Academic Publishers, Dordrecht, pp 171–194

Cody ML, Mooney HA (1978) Convergence vs. nonconvergence in Mediterranean-climate ecosystems. Ann Rev Ecol System 9:265–321

Cody ML, Fuentes ER, Glanz W, Hunt JH, Moldenke AR (1981) Convergent evolution in the consumer organisms of mediterranean Chile and California. In Mooney HA (ed) *Convergent Evolution in Chile and California*. Dowden, Hutchinson and Ross, Stroudsburg, PA, pp 144–192

Cowling RM, Lamont BB (1987) Post-fire recruitment of four co-occurring *Banksia* species. J Appl Ecol 24:645–658

Crawley MJ (1983) *Herbivory: The Dynamics of Animal–Plant Interactions.* U California Press, Berkeley

Crawley MJ (1989a) Insect herbivores and plant population dynamics. Ann Rev Entomol 34:531–564

Crawley MJ (1989b) The relative importance of vertebrate and invertebrate herbivores in plant population dynamics. In Bernays EA (ed) *Insect–Plant Interactions.* Vol I. CRC Press, Boca Raton, FL, pp 45–71

Crawley MJ (1990) The population dynamics of plants. Philos Trans Royal Soc London (B) 330:125–140

Crawley MJ (1992) Seed predators and plant population dynamics. In Fenner M (ed) *Seeds: The Ecology of Regeneration in Plant Communities.* C.A.B. International, London, pp 157–191

De Bach P (ed) (1964) *Biological Control of Insect Pests and Weeds.* Chapman and Hall, London

De Steven D (1991) Experiments on mechanisms of tree establishment in old-field succession: Seedling emergence. Ecology 72:1066–1075

di Castri F, Goodall DW, Specht RL (eds.) (1981) *Ecosystems of the World II: Mediterranean-Type Shrublands.* Elsevier, Amsterdam

Ehrlich PR, Birch LC (1967) The "balance of nature" and "population control." Am Nat 101:97–107

Fenner M (1985) *Seed Ecology.* Chapman and Hall, London

Fenner M (ed) (1992) *Seeds: The Ecology of Regeneration in Plant Communities.* C.A.B. International, London

Foster SA (1986) On the adaptive value of large seeds for tropical moist forest trees: A review and synthesis. Bot Rev 52:260–299

Fuentes ER, Le Boulenge P (1977) Prédation et compétition dans la structure d'une communauté herbacie secondaire du Chile central. La Terre et la Vie 31:312–326

Gange AC, Brown VK, Evans IM, Storr AL (1989). Variation in the impact of insect herbivort on *Trifolium pratense* through early plant succession. J Ecol 77:537–551

Gilpin ME, Hanski I (eds.) (1991) *Metapopulation Dynamics: Theory and Empirical Evidence.* Academic Press, New York

Goldberg DE, Werner PA (1982) The effects of size of opening in vegetation and litter cover on seedling establishment of goldenrods (*Solidago* spp.). Oecologia 60:149–155

Gross KL (1984) Effects of seed size and growth form on seedling establishment of six monocarpic perennial plants. J Ecol 72:369–387

Gross KL, Werner PA (1982) Colonizing abilities of "biennial" plant species in relation to ground cover: Implications for their distributions in a successional sere. Ecology 63:921–931

Gross RS, Werner PA (1983) Relationships among flowering phenology, insect visitors, and seed-set of individuals: Experimental studies on four co-occurring species of goldenrod (*Solidago*: Compositae). Ecol Mono 53:95–117

Grubb PJ (1977) The maintenance of species richness in plant communities: The importance of the regeneration niche. Bio Rev 52:107–145

Hairston NG (1989) *Ecological Experiments: Purpose, Design, and Execution.* Cambridge U Press, Cambridge

Halligan JP (1973) Bare areas associated with shrub stands in grassland: The case of *Artemisia californica.* BioSci 23:429–432

Halligan JP (1974) Relationship between animal activity and bare areas associated with California sagebrush in annual grassland. J Range Man 27:358–362

Harmon GD, Stamp NE (1992) Effects of postdispersal seed predation on spatial inequality and size variability in an annual plant, *Erodium cicutarium* (Geraniaceae). Am J Bot 79:300–305

Harper JL (1969) The role of predation in vegetational diversity. Brookhaven Symp 22:48–62

Harper JL (1977) *The Population Biology of Plants.* Academic Press, New York

Hay ME (1981) Herbivory, algal distribution, and the maintenance of between-habitat diversity on a tropical fringing reef. Am Nat 118:520–540

Hay ME, Colburn T, Downing D (1983) Spatial and temporal patterns in herbivory on a Caribbean fringing reef: The effects on plant distribution. Oecologia 58:299–308

Hendrix SD (1979) Compensatory reproduction in a biennial herb following insect defloration. Oecologia 42:107–118

Hendrix SD (1988) Herbivory and its impact on plant reproduction. In Lovett Doust J, Lovett Doust L (eds) *Plant Reproductive Ecology.* Oxford U Press, Oxford, pp 246–263

Hendrix SD, Trapp EJ (1989) Floral herbivory in *Pastinaca sativa*: Do compensatory responses affect offset reductions in fitness? Evolution 43:891–895

Horton JS, Wright JT (1944) The wood rat as an ecological factor in southern California watersheds. Ecology 36:244–262

Hubbell SP (1980) Seed predation and the coexistence of tree species in tropical forest. Oikos 35:214–229

Hughes L, Westoby M (1992) Fate of seeds adapted for dispersal by ants in Australian sclerophyll vegetation. Ecology 73:1285–1299

Huntly N (1991) Herbivores and the dynamics of communities and ecosystems. Ann Rev Ecol System 22:477–503

Jaksic FM, Fuentes ER (1980) Why are native herbs in the Chilean matorral more abundant beneath bushes: Microclimate or grazing? J Ecol 68:665–669

Janzen DH (1969) Seed eaters versus seed size, number, toxicity and dispersal. Evolution 23:1–27

Janzen DH (1970) Herbivores and the number of tree species in tropical forests. Am Nat 104:501–528

Janzen DH (1971) Seed predation by animals. Ann Rev Ecol System 2:465–492

Jordano D, Fernandez Haeger J, Rodriguez J (1990) The effect of seed predation by *Tomares ballus* (Lepidoptera: Lycaenidae) on *Astragalus lusitanicus* (Fabaceae): Determinants of differences among patches. Oikos 57:250–256

Keeley JE (1991) Seed germination and life history syndromes in the California chaparral. Bot Rev 57:81–116

Keeley JE, Hays RL (1976) Differential seed predation on two species of *Arctostaphylos* (Ericaceae). Oecologia 24:71–81

Kelly VR, Parker VT (1990) Seed bank survival and dynamics in sprouting and nonsprouting *Arctostaphylos* species. Am Mid Nat 124:114–123

Kinsman S, Platt WJ (1984) The impact of herbivores (*Heliodines nyctaginella*: Lepidoptera) upon *Mirabilis hirsuta*, a fugitive prairie plant. Oecologia 65:2–6

Kruger FJ, Mitchell DT, Jarvis JUM (eds) (1983) *Mediterranean-Type Ecosystems: The Role of Nutrients.* Ecological Studies 43. Springer-Verlag, Berlin 552 pp

Lamont BB, van Leeuwen SJ (1988) Seed production and mortality in a rare *Banksia* species. J Appl Ecol 25:551–559

Lloret F, Zedler PH (1991) Recruitment pattern of *Rhus integrifolia* populations in periods between fire in chaparral. J Veg Sci 2:217–230

Louda SM (1978) A test of predispersal seed predation in the population dynamics of *Haplopappus*. Ph.D. Thesis, San Diego State U, San Diego, and U California at Riverside, Riverside

Louda SM (1982a) Distribution ecology: Variation in plant recruitment in relation to insect seed predation. Ecol Mono 52:25–41

Louda SM (1982b) Inflorescence spiders: a cost/benefit analysis for the host plant, *Haplopappus venetus* Blake (Asteraceae). Oecologia 55:185–191

Louda SM (1982c) Limitation of the recruitment of the shrub *Haplopappus squarro-*

sus (Asteraceae) by flower- and seed-feeding insects. J Ecol 70:43–53

Louda SM (1983) Seed predation and seedling mortality in the recruitment of a shrub, *Haplopappus venetus* (Asteraceae), along a climatic gradient. Ecology 64: 511–521

Louda SM (1984) Herbivore effect on stature, fruiting and leaf dynamics of a native crucifer. Ecology 65:1379–1386

Louda SM (1988) Insect pests and plant stress as considerations for revegetation of disturbed ecosystems. In Cairns J (ed) *Rehabilitating Damaged Ecosystems.* Vol. II. CRC Press, Boca Raton, FL, pp 51–67

Louda SM (1989a) Differential predation pressure: A general mechanism for structuring plant communities along complex environmental gradients? Trends Ecol Evol 4:158–159

Louda SM (1989b) Predation in the dynamics of seed regeneration. In Leck MA, Parker VT, Simpson RL (eds) *Ecology of Soil Seed Banks.* Academic Press, New York, pp 25–51

Louda SM (1994) Experimental evidence for insect impact on populations of short-lived, perennial plants, and its application in restoration ecology. Pages 118–138 in Bowles M, Whelan C (eds) *Restoration Ecology: Concepts and Case Histories.* Oxford University Press, Oxford

Louda SM, Potvin MA (1994) Effect of inflorescence-feeding insects on seed production, cohort demography, and life time fitness of a native thistle Ecology 75: in press

Louda SM, Zedler PH (1985) Predation in insular plant dynamics: An experimental assessment of postdispersal fruit and seed survival, Enewetak Atoll, Marshall Islands. Am J Bot 72:438–445

Louda SM, Keeler KH, Holt RD (1990a) Herbivore influences on plant performance and competitive interactions. In Grace JB, Tilman D (eds) *Perspectives on Plant Competition.* Academic Press, New York, pp 413–444

Louda SM, Potvin MA, Collinge SK (1990b) Predispersal seed predation, postdispersal seed predation and competition in the recruitment of seedlings of a native thistle in sandhills prairie. Am Mid Nat 124:105–113

Louda SM, Potvin MA, Collinge SK (1992) Role of seed-feeding insects, vertebrates, and plant competition in the limitation of a native thistle population. In Menken SBJ, Visser JH, Harrewijn P (eds) *Insect–Plant Relations*, 8th ed. Kluwer Academic Publishers, B.V., Wageningen, pp 30–32

Lubchenco J (1978) Plant species diversity in a marine intertidal community: Importance of herbivore food preference and algal competitive abilities. Am Nat 112: 23–39

Lubchenco J, Cubit J (1980) Heteromorphic life histories of certain marine algae as adaptations to variations in herbivory. Ecology 61:676–687

Lubchenco J, Gaines SD (1981) A unified approach to marine plant-herbivore interactions. I. Populations and communities. Ann Rev Ecol System 12:405–437

MacArthur RH (1972) *Geographical Ecology.* Harper and Row, New York

Maschinski J, Whitham TG (1989) The continuum of plant responses to herbivory: The influence of plant association, nutrient availability, and timing. Am Nat 134: 1–19

McNaughton SJ (1979) Grazing as an optimization process: Grass–ungulate relationships in the Serengeti. Am Nat 113:691–703

Meyer GA, Root RB (1993) Effects of herbivorous insects and soil fertility on reproduction of goldenrod. Ecology 74:1117–1128

Mills JN (1986) Herbivores and early postfire succession in southern California chaparral. Ecology 67:1637–1649

Mills JN, Kummerow J (1989) Herbivores, seed predators, and chaparral succession. In Keeley SC (ed) *The California Chaparral: Paradigms Revisited.* Nat Hist Mus LA County, Los Angeles, CA, pp 49–55

Mooney HA (1977) Southern coastal scrub. In Barbour M, Major J (eds) *Terrestrial*

Vegetation of California. Wiley & Son, New York, pp 471–489

Morton SR (1985) Granivory in arid regions: Comparisons of Australia with North and South America. Ecology 66:1859–1866

Munz PA, Keck DC (1970) *A California Flora*. U California, Berkeley

Naylor REL (1984) Seed ecology. In Thomson JR (ed) *Advances in Research and Technology of Seeds*. Vol. 9. Centre for Agricultural Publication, Wageningen, pp 61–91

O'Dowd DJ, Gill AM (1984) Predator satiation and site alteration following fire: Mass reproduction of alpine ash *Eucalyptus delegatensis* in southeastern Australia. Ecology 65:1052–1066

O'Dowd DJ, Lake PS (1990) Red crabs in rain forest, Christmas Island (Indian Ocean): Differential herbivory on seedlings. Oikos 58:289–292

O'Dowd DJ, Lake PS (1991) Red crabs in rain forest, Christmas Island: Removal and fate of fruits and seeds. J Trop Ecol 7

Osborne K, Smith III TJ (1990) Differential predation on mangrove propagules in open and closed canopy forest habitats. Vegetatio 89:1–6

Ostfeld RS (1992) Small-mammal herbivores in a patchy environment: Individual strategies and population responses. In Hunter MD, Ohgushi T, Price PW (eds) *Effects of Resource Distribution on Animal-Plant Interactions*. Academic Press, San Diego, pp 43–74

Paige KN, Whitham TG (1987) Overcompensation in response to mammalian herbivory: The advantage of being eaten. Am Nat 129:407–416

Parker VT, Kelly VR (1989) Seed banks in California chaparral and other Mediterranean climate shrublands. In Leck MA, Parker VT, Simpson RL (eds) *Ecology of Soil Seed Banks*. Academic Press, San Diego, pp 231–255

Pierce SM, Cowling RM (1991) Dynamics of soil-stored seed banks of six shrubs in fire-prone dune fynbos. J Ecol 79:731–747

Platt WJ (1975) The colonization and formation of equilibrium plant species associations on badger disturbances in a tall-grass prairie. Ecol Mono 45:285–305

Price MV, Jenkins SH (1987) Rodents as seed consumers and dispersers. In Murray DR (ed) *Seed Dispersal*. Academic Press, Sydney, pp 191–235

Quinn RD (1990) Habitat preferences and distribution of mammals in California chaparral. U.S.D.A. Forest Service Pacific Southwest Research Station Research USDA, For. Ser, Pacific SW Research Paper PSW-202

Rabinowitz D, Rapp JK, Cairns S, Mayer M (1989) The persistence of rare prairie grasses in Missouri: Environmental variation buffered by reproductive output of sparse species. Am Nat 134:525–544

Robertson AI (1991) Plant–animal interactions and the structure and function of mangrove forest ecosystems. Aust J Ecol 16:433–444

Rockwood LL (1973) The effect of defoliation on seed production of six Costa Rican tree species. Ecology 54:1363–1369

Salisbury EJ (1942) *The Reproductive Capacity of Plants*. Bell, London

Scott JK (1982) The impact of destructive insects on reproduction in six species of *Banksia* L.F. (Proteaceae). Aust J Zool 30:901–921

Slobodkin LB, Smith FE, Hairston NG (1967) Regulation in terrestrial ecosystems, and the implied balance of nature. Am Nat 101:109–124

Smith TJ III (1987) Seed predation in relation to tree dominance and distribution in mangrove forests. Ecology 68:266–273

Smith TJ III, Chan H-T, McIvor CC, Robblee MB (1989) Inter-continental comparisons of seed predation in tropical, tidal forests. Ecology 70:146–151

Specht RL (ed) (1988) *Mediterranean-Type Ecosystems: A Data Source Book*. Kluwer Academic Publishers, Dordrecht

Stephenson AG (1981) Flower and fruit abortion: Proximate causes and ultimate functions. Ann Rev Ecol System 12:253–279

Swank SE, Oechel WC (1991) Interactions among the effects of herbivory, competition, and resource limitation on chaparral herbs. Ecology 72:104–115

Vaughton G (1990) Predation by insects limits seed production in *Banksia spinulosa* var. *neoanglica* (Proteaceae). Aust J Bot 38:335–340

Wallace DD, O'Dowd DJ (1989) The effect of nutrients and inflorescence damage by insects on fruit-set by *Banksia spinulosa*. Oecologia 79:482–488

Waloff N, Richards OW (1977) The effect of insect fauna on growth, mortality and natality of broom, *Sarothamnus scoparius*. J Appl Ecol 14:787–789

Weiner J (1985) Size hierarchies in experimental populations of annual plants. Ecology 66:743–752

Weis AE, Berenbaum MR (1989) Herbivorous insects and green plants. In Abrahamson WG (ed) *Plant–Animal Interactions*. McGraw-Hill, New York, pp 123–162

Wellington AB, Noble IR (1985) Seed dynamics and factors limiting recruitment of the mallee *Eucalyptus incrassata* in semi-arid, south-eastern Australia. J Ecol 73:657–666

Westman WE (1988) Vegetation, nutrition, and climate-data-tables. (3) Species richness. In Specht RL (ed) *Mediterranean-Type Ecosytems: A Data Source Book*. Kluwer Academic Publishers, Dordrecht

Westman W, O'Leary J (1986) Measures of resilience: The response of coastal sage scrub to fire. Vegetatio 65:179–189

Westoby M, Jurado E, Leishman M (1992) Comparative evolutionary ecology of seed size. Tr.E.E. 7:368–372

Whelan CJ, Willson MF, Tuma CA, Souza-Pinta I (1991) Spatial and temporal patterns of postdispersal seed predation. Can J Bot 69:428–436

Whitham TG, Maschinski J, Larson KC, Paige KN (1991) Plant responses to herbivory: The continuum from negative to positive and underlying physiological mechanisms. In Price PW, Lewinsohn TM, Wilson Fernandes G, Benson WW (eds) *Plant–Animal Interactions: Evolutionary Ecology in Tropical and Temperate Regions*. John Wiley & Sons, New York, pp 227–250

Whitham TG, Mopper S (1985) Chronic herbivory: Impacts on architecture and sex expression of pinyon pine. Science 228:1089–1091

Zammit C, Hood CW (1986) Impact of flower and seed predators on seed-set in two *Banksia* shrubs. Aust J Ecol 11:187–193

Zammit CA, Westoby MA (1988) Predispersal seed losses, and the survival of seeds and seedlings of two serotinous *Banksia* shrubs in burnt and unburnt heath. J Ecol 76:200–214

Zedler PH, Black C (1992) Seed dispersal by a generalized herbivore: Rabbits as dispersal vectors in a semiarid California vernal pool landscape. Am Mid Nat 128:1–10

Zwolfer H (1965) Preliminary list of phytophagous insects attacking wild Cynareae (Compositae) in Europe. CIBC Tech. Bull. No. 6:81–154

Zwolfer H (1988) Evolutionary and ecological relationships of the insect fauna of thistles. Ann Rev Entomol 33:103–229

Zwolfer H, Romstock-Volkl M (1991) Biotypes and the evolution of niches in phytophagous insects on Cardueae hosts. In Price PW, Lewinsohn TM, Wilson Fernandes G, Benson WW (eds) *Plant–Animal Interactions: Evolutionary Ecology in Tropical and Temperate Regions*. Wiley and Sons, New York, pp 487–507

V. Animal Physiology and Community Structure

14. Mediterranean Type of Climatic Adaptation in the Physiological Ecology of Rodent Species

Francisco Bozinovic, Mario Rosenmann, F. Fernando Novoa, and Rodrigo G. Medel

The balance between acquisition and expenditure of energy is critical to the survival and ecological success of vertebrates. This balance depends on the interplay among energy intake, processing, allocation, and expenditure (Karasov 1986, Kenagy 1987, Bozinovic 1992a).

Theoretically, the combination of organismal events, including structural features, the biochemical and physiological components of energy exchange between the organisms and their environment, and the thermodynamic efficiency of energy–matter transformations are under natural selection. An adaptive landscape that includes mechanisms and processes of energy, matter, and water interchange under specific biotic and abiotic environmental conditions, are some of the evident results of natural selection (MacMillen and Hinds 1992).

Some of the events associated with animal energetics are basal and maximal metabolic rate, thermal conductance, lower lethal and critical temperatures, body temperature, minimal and maximal temperature difference between animal and environment, limits of temperature tolerance, water economy, and evaporative water loss. For example, the maximal rate of energy metabolism may influence survival of mammals by setting the upper limits for sustained activity or for thermoregulatory thermogenesis—that is, escape from predators or exposure to cold (Hayes and Chappell 1986). Also, the maximal rate of metabolism affects rates of reproduction and limits of distribution (Karasov 1986, Weiner 1987, Bozinovic and Rosenmann 1989,

Peterson et al. 1990). On the other hand, the minimal or basal metabolic rate of thermoregulation, commonly used as an index of energy expenditure (but see Koteja 1991), exhibits a great amount of intra- and interspecific variation in mammals (McNab 1986). Some of this variation appears to be related to the specific traits of organisms, such as food habits, activity level, phylogenetic relationship, and reproduction; and to environmental characteristics, such as climate (e.g., Kleiber 1961, Hayssen and Lacy 1985, Elgar and Harvey 1987, McNab 1988, Derting 1989).

Energetics of endotherms and particularly of mammals are fundamentally determined by body mass (Kleiber 1961), although much of the residual variation in the allometry of mammalian energetics (secondary signals *sensu* Schmidt-Nielsen 1984) is associated with climate and habitat conditions (McNab 1988, Bozinovic and Rosenmann 1989). An animal's physiological response can range from immediate (in acute exposures), to acclimation (during chronic exposure ranging from several weeks to months), to an even longer ontogenetic (developmental) response. Finally, inherited responses over several generations could lead to evolutionary adaptation. A classic question is whether the physiological ecology of animals, particularly of rodents, is related to their similar genetic and developmental programs (phylogenetic constraints), independent of living under distinctive habitat conditions, or to evolutionary convergence.

The idea that convergence has occurred in organismic, populational, and community traits was derived from the theory of competition. According to expectations of niche theory, convergent evolution may occur in response to selective pressures on the realized and fundamental niche through two related, though distinct processes (Orians and Solbrig 1977). These are (1) interspecific interactions acting as selective pressures on phenotypes in such a way as to orient the evolution of species toward optimal values; and (2) physical environmental features affecting the mechanisms and processes of resource (= energy) acquisition, transformation and allocation to activity, growth, reproduction and metabolic maintenance, assuming negligible influence from the biotic environment. This last assumption is clearly simplistic, but it makes sense to conduct research in that avenue when we realize that most studies of convergence in animals have emphasized the realized niche of species but only incidentally the fundamental niche (e.g., maximal and basal rates of energy metabolism, thermal tolerance and capabilities, limits of urine concentration, and evaporative water loss).

In this chapter we study several physiological traits, including energetics and water economy, of organisms living in similar abiotic environments. The hypothesis of convergence establishes that under similar environmental conditions, species are very similar in attributes independent of their phylogenetic history (Cody and Mooney 1978). Lacking information about ancestral states of the species, we evaluate similarity in some physiological features rather than convergence. Schluter (1986) indicates that there is no straightforward relationship between similarity and convergence. Because comparisons may eventually shed light on convergence, we use an outgroup approach (Fuentes 1976, Blondel et al. 1984) by considering also species from a second ecosystem as a control for the mediterranean climate physiological comparisons. First, we compare the energetics of sympatric species of rodents from the mediterranean climate areas in central Chile (matorral) with their counterparts from California (chaparral). We use the sympatric species of rodents inhabiting the high Andean plateau in northern Chile as outgroup controls. Second, we analyze the water economy and balance of species from the Chilean matorral and conduct comparisons with chaparral and xeric Australian species.

If the physiological ecology of rodent species reflect convergent adaptation to a mediterranean climate, we expect that rodents from the matorral and the chaparral will show greater similarity in energetic and hydric parameters than will species from the control ecosystem. On the other hand, we may also expect to find lower tolerance to cold and reduced energetic capabilities in species inhabiting a warmer mediterranean climate (see Arroyo et al., this volume) than species living in the high Andean plateau, even though the three habitats are shared by species belonging to the same family (Cricetidae) and in some cases by the same genera (*Phyllotis* and *Abrothrix*, see Table 14.1). Also, we should expect more similar water economy patterns in species from the two mediterranean regions than in those inhabiting xeric and semixeric areas in Australia. To eliminate potential methodological distortions, we restrict our analysis to original data and information published by investigators who used methods similar to our own.

Energetics and Climate

Body Mass and Energy

As mentioned, one of the primary factors affecting basal rate of metabolism (BMR) and thermal conductance (C) is body mass (m_b). The ratio between BMR and C produces a temperature differential (ΔT_m) equal to the minimal difference between body temperature (T_b) and ambient temperature (T_a) at the lower limit of thermoneutrality (T_{LC}). This relationship is expressed:

$$\Delta T_m = T_b - T_{LC} = BMR/C \, [°C] \tag{1}$$

in which BMR is given in mLO_2/g h and C in mLO_2/g h °C. In allometric form:

$$\Delta T_m = T_b - T_{LC} = 3.56 m_b^{-0.284}/1.00 m_b^{-0.50}$$
$$= 3.56 m_b^{0.216} \, [°C] \tag{2}$$

Allometric equations are obtained from McNab (1988) and McNab and Morrison (1963) for BMR and C, respectively.

Theoretically, equation (2) can be used to estimate climate effects and to infer the evolutionary and ecological consequences of this factor and vice versa. As Bozinovic and Rosenmann (1989) point out, however, this relationship shows an acceptable resolving capacity only at interordinal level, because of the narrow and overlapping range of the difference $(T_b - T_{LC})$ shown by less-inclusive taxonomic groups. These authors also claim that by extending the temperature differential far beyond $(T_b - T_{LC})$, and taking as ΔT_M the maximum temperature difference between body and ambient temperature, a better correlation of this index with the range of climatic conditions prevailing in a given habitat was obtained. In this chapter we also determine and analyze the maximum metabolic effort in different habitats, using the maximum resting metabolic rate of thermoregulation (MMR). Under such conditions, small mammals should exercise maximal capacities. Thus, by analogy to equation (1):

$$\Delta T_M = T_b - T_{LL} = \text{MMR}/C \,[°C] \tag{3}$$

or, in allometric form:

$$\Delta T_M = T_b - T_{LL} = 27.14 m_b^{-0.32}/1.00 m_b^{-0.50}$$
$$= 27.14 m_b^{0.18} \,[°C] \tag{4}$$

The allometric equation relating MMR and m_b is from Bozinovic (1992a), and T_{LL} (lower lethal temperature) is the ambient temperature at which small mammals attain MMR; $T_b - T_{LL}$ is the maximum thermal differential (ΔT_M) between an endotherm and its environment. ΔT_M implies maximum metabolic effort whereas ΔT_m is measured under basal conditions. T_{LL} can also be predicted using this equation:

$$12.59 - 21.64 \log m_b \,[°C] \tag{5}$$

proposed by Bozinovic & Rosenmann (1989).

Here we hypothesize that species from mediterranean climates will experience more similar metabolic efforts than those from different climatic regions, independent of phylogenetic inertia or historical constraints.

Comparative Energetics of Rodent Species from Mediterranean Climates: Matorral-Chaparral

The original values and sources for the basal and maximum resting metabolic rate of thermoregulation, as well as body mass, thermal conductance, and body temperatures of twenty-one species of rodents are given in Table 14.1. Analyses were conducted using a single average point for each species. Data were compared by a multifactorial analysis of variance (MANOVA) with m_b as a covariate, and considering the climatic region as the main factor.

Table 14.1. Body mass (m_b in g), basal (BMR) and maximum (MMR) resting metabolic rate (in mLO_2/g h), conductance (C in mLO_2/g h °C), and body temperature (T_b in °C) of rodent species from different habitats

Habitat/Family/Species	m_b	BMR	MMR	C	T_b	Reference
High Andean Plateau (northern Chile)						
Caviidae						
Microcavia niata	255	0.69	7.03	0.689	37.4	(1, 2, 6)
Cricetidae						
Abrothrix andinus dolichonyx	23	1.92	12.71	0.139	37.5	(1, 2)
Akodon albiventer	31	1.50	11.80	0.155	36.9	(2)
Auliscomys boliviensis	77	1.44	7.61	0.102	36.3	(4, 5)
Calomys ducilla	16	1.80	14.00	0.254	38.0	(5)
Eligmodontia puerulus	21	1.71	11.50	0.152	36.9	(1, 2)
Phyllotis d. chilensis	49	1.34	7.17	0.116	36.4	(1, 2)
Matorral (central Chile)						
Octodontidae						
Octodon degus	195	0.93	5.50	0.072	37.0	(7)
Cricetidae						
Abrothrix longipilis	42	1.87	8.70	0.135	37.4	(3, 4)
Abrothrix olivaceus	27	1.83	9.20	0.152	37.2	(3, 4)
Oryzomys longicaudatus	28	1.81	9.80	0.159	37.3	(3, 4)
Phyllotis darwini	59	1.21	6.90	0.139	36.2	(3, 4)
Chaparral (California)						
Heteromyidae						
Dipodomys deserti	33	0.82	6.05	0.065	36.8	(8, 9)
Dipodomys merriami	34	1.45	7.62	0.136	34.1	(8, 9, 10)
Dipodomys heermanni	63	1.15	8.05	0.126[a]	36.5	(8, 9)
Dipodomys panamintinus	72	1.15	8.69	0.108	36.5	(8, 9)
Perognathus fallax	20	1.29	8.07	0.186	32.6	(8, 9, 10)
Microdipodops megacephalus	12	2.15	13.20	0.126[a]	36.0	(8)
Cricetidae						
Peromyscus californicus	41	1.38	5.09	0.105	35.9	(8, 10, 11)
Peromyscus eremicus	19	1.42	7.60	0.160	36.1	(8, 10, 11)
Peromyscus maniculatus	13	2.35	9.83	0.185	36.5	(8, 10, 11)

[a] Calculated value according to $C = 1.0\, m_b^{-0.5}$ (McNab and Morrison 1963).
References: (1) This study; (2) Bozinovic (1992a); (3) Bozinovic and Rosenmann (1988); (4) Bozinovic and Rosenmann (1989); (5) Rosenmann and Morrison (1974); (6) Bozinovic (1992b); (7) Rosenmann (1977); (8) Hinds and Rice-Warner (1992); (9) McNab (1979); (10) Hulbert et al. (1985); (11) Bradley and Deavers (1980).

Comparisons between BMR, MMR, and m_b for the species revealed non-significant differences between regions (Table 14.2; $F = 1.08$, $p > 0.05$ for BMR, and $F = 3.70$, p > 0.05 for MMR). Both BMR and MMR are similar to those predicted for m_b (Figs. 14.1, 14.2, and Table 14.2) regardless of the climatic region inhabited by the species of rodents. The aerobic scope, how-ever, (Fig. 14.3) was not significantly different among species from the medi-terranean climate (LSD test, $p > 0.05$), but it was significantly dissimilar between chaparral–high-Andean plateau, and matorral–high-Andean pla-

Table 14.2. Comparative energetic parameters (mean $\pm$ 1SE) of rodents from different habitats. Statistical significance after a multivariate ANOVA (MANOVA) is shown. Same letters represent nonsignificant differences between groups as revealed by the *a posteriori* LSD test when the overall MANOVA was significant

Parameter	High Andean Plateau	Matorral	Chaparral	P
Number of species	7	5	9	
Body mass (g)	67.00 ± 32.37	70.20 ± 31.74	41.22 ± 9.89	$> .05$
Metabolic rate (mLO$_2$/g h)				
BMR	1.48 ± 0.15	1.53 ± 0.19	1.46 ± 0.16	$> .05$
$BMR/3.56m_b^{-0.284}$	1.37	1.43	1.17	
MMR	10.26 ± 1.10	8.02 ± 0.79	8.25 ± 0.77	$> .05$
$MMR/27.14m_b^{-0.32}$	1.45	1.15	1.00	
Aerobic scope				
MMR/BMR	7.12 ± 0.64	5.34 ± 0.23^a	5.87 ± 0.45^a	$< .05$
$[MMR/BMR]/7.62m_b^{-0.04}$	1.10	0.83	0.89	
Conductance (mLO$_2$/g h °C)				
C	0.14 ± 0.02	0.15 ± 0.02	0.15 ± 0.02	$> .05$
$C/1.00m_b^{-0.5}$	1.45	1.26	0.96	
Temperature (°C)				
T_{LL}	-43.98 ± 9.60	-25.46 ± 4.17^b	-23.34 ± 5.51^b	$< .05$
$T_{LL}/[12.59 - 21.64 \, Log \, m_b]$	1.63	1.07	0.96	
ΔT_m	11.51 ± 0.94	11.76 ± 0.87	10.22 ± 0.77	$> .05$
$\Delta T_m/3.56m_b^{0.216}$	0.77	0.76	0.77	
ΔT_M	81.04 ± 9.64	62.48 ± 4.29^c	59.01 ± 5.73^c	$< .05$
$\Delta T_M/27.14m_b^{0.18}$	0.72	0.93	0.90	

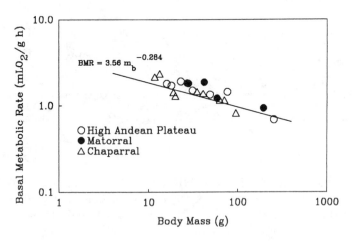

Figure 14.1. Relationship between basal metabolic rate and body mass in species of rodents from different geographic regions. The line represents the expected basal metabolism. No statistically significant differences were obtained between groups.

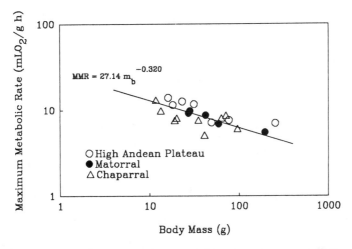

Figure 14.2. Relationship between thermoregulatory maximum metabolic rate and body mass in species of rodents from different geographic regions. The line represents the expected maximum metabolic rate. No statistically significant differences were obtained between groups.

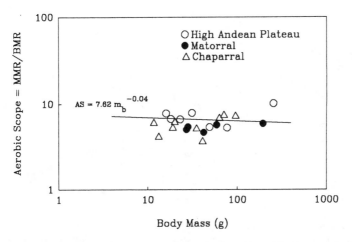

Figure 14.3. Relationship between aerobic scope ($= AS$) and body mass in the same species of rodents as that plotted in Fig. 14.1. The curve represents the standard relationship. See text for explanations and statistically significant differences and similarities in AS between regions.

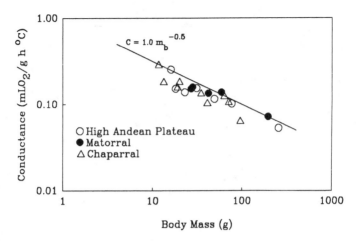

Figure 14.4. Relationship between conductance and body mass for the same species as in Fig. 14.1. The expected standard curve is shown. No significant differences were obtained between groups.

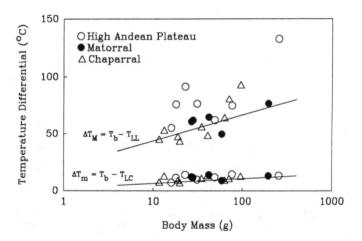

Figure 14.5. Relationship between minimum and maximum thermal differences and body mass in the same species as that shown in Fig. 14.1. The expected standard curves are shown.

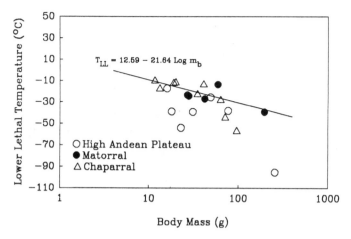

Figure 14.6. Relationship between lower lethal tolerance and body mass for the same species as in Fig. 14.1. The expected standard curve is shown. Significant differences were obtained between groups, but not between species from mediterranean climates.

teau (LSD test, $p < 0.05$, Table 14.2), in spite of the closer evolutionary relationship of the species from the matorral and the high Andean plateau.

In that increasing hair length and/or density may be a design constraint for small mammals, and may lead to negative effects on their locomotion (but see Bozinovic and Merritt 1992), lower conductance in species inhabiting the high Andean plateau was not expected. In fact, thermal conductance in the range of expected values predicted for body mass (Fig. 14.4) was not significantly different between regions ($F = 0.02$, $p = 0.05$, Table 14.2). Interestingly, and contrary to ΔT_m ($F = 0.57$, $p < 0.05$, Table 14.2, Fig. 14.6), when maximum metabolic and thermoregulatory capabilities were considered and analyzed—that is, T_{LL} and ΔT_M (Figs. 14.5 and 14.6 respectively), species from the matorral in central Chile and chaparral in California did not show significant differences (LSD test, $p > 0.05$) but significant differences do appear when comparisons are conducted with species from the high Andean plateau (see Table 14.2).

Although Bozinovic et al. (1990) and Rosenmann et al. (1975) documented seasonal changes in metabolic capabilities and aerobic scope in the Andean mouse *Abrothrix andinus* and in the Arctic red-backed vole *Clethrionomys rutilus* respectively, the reported mass-specific MMR/BMR ratios in this chapter did not vary significantly between the California dwelling species and the ones from central Chile, but they are different from those of high altitude inhabitants in the Andean-range. This pattern suggests that the factorial aerobic scope, lower thermal tolerance, and thermal differential respond to the geographic regions of species.

On the other hand, when foraging, endotherms must dissipate excess heat. In fact, in central Chile, with warm-dry summers and cold-wet winters, some species of rodents appear to be limited in their use of space by biotic factors (but see Bozinovic et al. 1988, Bozinovic and Simonetti 1992). An example is the caviomorph, Octodontid rodent *Octodon degus*. Intolerance to ambient temperatures higher than 32°C because of low evaporative water loss (Rosenmann 1977), appears to constrain patterns of space and microhabitat use in this species. During summer, activity is restricted at midday when heat loads apparently exceed the capacity to dissipate heat by evaporation (V.O. Lagos pers. comm). This case exemplifies the link between energy and water budget, and the importance of analyzing regional comparisons of similarities and differences in both physiological mechanisms and processes (see below).

Water Economy of Rodents

Renal Performance and Maximal Urine Concentration

Studies of the physiological ecology of small endotherms to climatic conditions in semixeric (and xeric) regions of Chile, the southwestern United States and Australia should include assessments of renal performance and pulmocutaneus water loss. These parameters, when evaluated together with analyses of energy metabolism and thermoregulation, will facilitate reasonable interpretations of the means by which these species cope with abiotic habitat conditions (MacMillen and Lee 1970).

The capacity to concentrate urine has traditionally been viewed as an indicator of the efficiency of water regulation (Schmidt-Nielsen 1964, Abbott 1971, Cortés et al. 1988), and according to MacMillen (1983), this view is probably correct for rodents that lose body mass when deprived of water and are kept on a dry diet. Under such conditions rodents should exercise maximal capabilities of water conservation, reflected in maximal urine concentrations. Nevertheless, and as we shall see, this one physiological appraisal may not be (by itself) a thorough estimate of the organismic response to specific climatic conditions.

In Table 14.3 we show urine concentrations of water-dependent and water-independent species of rodents, while deprived of water or on minimal water rations. Xeric species from Australia are taken as a reference group of maximal renal performance for three water-independent murid rodents. In analyzing urine concentration (Table 14.3), it should be realized that these reported values were obtained under somewhat different conditions. It is well known that urine concentrations may increase with increasing experimental or field temperatures, and that these may also change with dietary composition (MacMillen and Hinds 1983). Values shown in Table 14.3 should be taken as approximate representative examples of renal performance. They are all we have at the moment, however, to characterize the three groups of rodents. For the five matorral species, the maximal capacity to concentrate urine ranged from about 3300 mOsm/kg in the fossorial *Spalacopus cyanus*

Table 14.3. Maximal urine concentration in matorral, in chaparral, and in xeric Australian rodents

Species/Habitat	Urine Concentration (mOsm/kg)		Reference
	Mean values	Maximal values	
Matorral (central Chile)			
Octodon degus	850	4443	(1, 2)
Abrothrix olivaceus	1305	4468	(1, 2)
Oryzomys longicaudatus	1537	4168	(1, 2)
Phyllotis darwini	1305	4468	(1, 2)
Spalacopus cyanus	—	3272	(1, 2)
Chaparral (California)			
Perognathus fallax	502	4617	(3)
Dipodomys merriami	3990	4650	(4)
Dipodomys spectabilis	3780	4090	(4)
Citellus leucurus	3730	3900	(4)
Neotoma albigula	2100	2670	(4)
Peromyscus maniculatus gambelli	—	3600	(5)
Peromyscus maniculatus sonoriensis	—	3650	(5)
Peromyscus crinitus	—	4000	(5, 6)
Peromyscus truei	—	4200	(7)
Xeric—Semixeric (Australia)			
Notomys alexis	6550	9370	(4)
Notomys cervinus	3720	4920	(4)
Leggadina hermannsburgensis	4710	8970	(4)

References: (1) Cortés (1985); (2) Cortés et al. (1990); (3) MacMillen and Hinds (1983); (4) in MacMillen and Lee (1967); (5) in MacMillen (1983); (6) Abbott (1971); (7) Bradford (1974).

to close to 4500 mOsm/kg in *Phyllotis darwini* and *Abrothrix olivaceus*. The mean value for the species inhabiting matorral was 4138 ± 498 mOsm/kg, a value similar (3930 ± 564 mOsm/kg) to that found for nine western North American species, including the cricetid wood rat *Neotoma albigula*. The average urine concentration for the Australian species was 7753 mOsm/kg. Based on a one-way ANOVA ($F = 15.264$, $P < 0.05$), and multiple LSD analysis, the difference between the matorral and the chaparral species was nonsignificant, whereas both groups were statistically different from the Australian murids.

In spite of the apparent physiological similarity of the matorral and chaparral rodents in their ability to concentrate urine, none of the species from the matorral seems to be water-independent (Cortés et al. 1988). In contrast, some heteromyids from the chaparral such as *Dipodomys merriami*, *Perognathus fallax*, and the cricetid *Peromyscus crinitus* are known to be water-independent. Similarly, the three murid species from Australia can survive and may even gain body mass on a diet of dry seeds without drinking water (see reviews of MacMillen and Lee 1967, MacMillen 1983).

During water deprivation, survival time in water-dependent Chilean matorral species varied from four days in *Oryzomys longicaudatus* to 13.4 days in *Octodon degus*. Conversely, chaparral species such as *Peromyscus californicus*, *P. eremicus*, and a few subspecies of *P. maniculatus* are able to survive for longer periods. Moreover, after 14 days on a dry diet, the loss of body mass ranged from 15% in *P. maniculatus gambelli* to 32% in *P. eremicus* (MacMillen 1983). During the same period, and based on its specific mass loss kinetics, a loss of 45% in mass should be expected in *O. degus* (Cortés et al. 1988). Notice that *O. degus* has been described as the matorral species most resistant to dehydration (Cortés et al. 1988).

Metabolic Water Production and Evaporative Water Loss

The early work of Schmidt-Nielsen and Schmidt-Nielsen (1951) demonstrated that some *Perognathus* spp. and *Dipodomys* spp. can maintain a positive water balance without water intake. These authors also demonstrated for a few xeric species of rodents that the main mechanisms of water input and output are metabolic water production (MWP) and evaporative water loss (EWL), respectively. Furthermore, the ratio between MWP and EWL was not altered by activity if the environmental temperature and relative humidity remained constant (Raab and Schmidt-Nielsen 1972). These relationships here provided the basis for comparing states of water balance in rodents from the Chilean matorral and the chaparral in western North America.

In Table 14.4, we show evaporative water loss for four matorral species together with the expected values for a similar-sized water-independent species from the Californian chaparral. Evaporative water loss in the matorral rodents ranges from 13% to more than 100% higher than in *Dipodomys*. Moreover, if the MWP/EWL ratio is used as a better estimation of the capability for water conservation, we find that this index is 1/10 to 1/3 the value of 1.0 described for 13 heteromyid species measured between 14° and 26°C. These values are shown in Table 14.5, and demonstrate that unlike the species inhabiting the matorral, the water economy of heteromyid species is balanced by equal rates of metabolic water production and evaporative water loss. Furthermore, MWP can be higher than EWL in several species of

Table 14.4. Evaporative water loss (*EWL*) in four matorral species and the expected values for a water-independent rodent of similar body mass from California

| Species | Evaporative water loss (mg H_2O/g h) | | |
	Measured[a]	Predicted[b]	Measured/predicted
Octodon degus	0.97	0.72	1.35
Abrothrix olivaceus	1.66	1.47	1.13
Oryzomys longicaudatus	3.59	1.60	2.24
Phyllotis darwini	2.37	1.59	1.49

[a] *EWL* data from Cortés et al. (1988, 1990).
[b] Predicted values from the allometric relationship for *Dipodomys*: $EWL = 8.993\ m_b^{-0.498}$ (Hinds and MacMillen 1985).

Table 14.5. Ratios between metabolic water production (MWP) and evaporative water loss (EWL) in rodents from the matorral in central Chile and in heteromyid species

Species	MWP/EWL	
	Matorral[a]	Heteromyids[b]
Octodon degus	0.311	1.0
Abrothrix olivaceus	0.171	1.0
Oryzomys longicaudatus	0.139	1.0
Phyllotis darwini	0.096	1.0

[a] Measured at 21°C (Cortés et al. 1988, 1990)
[b] Reported values for 13 heteromyid species measured between 14° and 26 °C (Hinds and MacMillen 1985).

Perognathus, *Dipodomys*, *Microdipodops*, and *Heteromys* when measured at 15°C. These results may explain the differences in survival time during water shortage and water dependence between the matorral and chaparral species, in spite of their physiological similarity in ability to concentrate urine.

Conclusions

The search for similarities and differences among biological systems is one of the basic avenues of inquiry for physiologists, ecologists, and evolutionary biologists, and convergence has been one of the most influential ideas guiding comparative research in physiology, ecology, and evolution. Although many studies of ecological convergence have been carried out from the 1970s onward, the hypothesis of convergence has by no means been exhausted.

The evidence for convergence in animals has mostly been sought from studies at the level of community structure (cf. Fuentes 1976, Mooney 1977, Orians and Solbrig 1977, Cody and Mooney 1978, Ricklefs and Travis 1980, Blondel et al. 1984, Morton and Davidson 1988). Few studies seem to have considered animal physiology as a test of convergence.

Despite its apparent simplicity, the hypothesis of convergence is extremely difficult to test and rarely permits all-or-nothing conclusions. Reasons lie in both conceptual and methodological difficulties. First, it is rarely indicated if comparisons are made to detect convergence or similarity (Schluter 1986; Wiens 1991). Recognizing similarity in physiological traits does not guarantee that they have converged from the past. To be sure, we must know their ancestral states. Because this kind of information rarely is available and is very difficult to obtain for physiological characteristics, we remain ignorant about the prevalence of physiological convergence. A potential way to solve this problem is to infer the body size (= mass) of species from the fossil record (e.g., Martin 1986). As we show in this chapter, body mass of species is related to several physiological parameters (metabolic rate, temperature differential, evaporative water loss; see also Peters 1983, Damuth and MacFadden 1990). Knowing the body size of ancestral species it should be possible to test the hypothesis of physiological convergence.

Another way to distinguish similarity from convergence is to use an outgroup approach, as we have done here. Although that method can serve as a first step, however, it does not resolve the arbitrary choice of the control. Second, the hypothesis of physiological convergence is a one-way hypothesis (see also Peet 1978). It can explain the similarity of physiological traits (e.g., maximal urine concentration) in organisms in different geographic regions, but says nothing about why they differ. In fact, we have shown here cases of physiological similarity between mediterranean regions but also physiological differences. An argument frequently used to explain the absence of convergence is that the environments where the organisms originally evolved were different.

Despite the many shortcomings in testing the hypothesis of convergence, it should be possible to devise new methods for testing convergence at the physiological level. It is our feeling that methods and techniques designed to test convergence at one level of biological resolution may not be applicable at other levels. Careful laboratory and field studies on the physiological ecology of species, complemented with good knowledge of body-size records of ancestral species, promise to be a valuable starting point in this direction.

Acknowledgments

We thank Mary Kalin Arroyo for her invitation and valuable criticisms. F.F. Novoa acknowledges a Fundación Andes doctoral fellowship. This article was financed by a grant from the International Foundation for Science—IFS No. B/2030-1 and FONDECYT, Chile Grant 1930866 to F. Bozinovic.

References

Abbott KD (1971) Water economy of the canyon mouse *Peromyscus crinitus stephensi*. Compar Biochem Physiol 38A:37–52

Blondel J, Vuilleumier P, Marcus LE, and Terouanne E (1984) Is there ecomorphological convergence among mediterranean bird communities of Chile, California and France? Evol Bio 18:141–213

Bozinovic F (1992a) Scaling of basal and maximum metabolic rate in rodents and the aerobic capacity model for the evolution of endothermy. Physiol Zool 65:921–932

Bozinovic F (1992b) Rate of basal metabolism of grazing rodents from different habitats. J Mam 73:379–384

Bozinovic F, Merritt JF (1992) Summer and winter thermal conductance of *Blarina brevicauda* (Mammalia: Insectivora: Soricidae) inhabiting the Appalachian mountains. Ann Carnegie Mus 61:33–37

Bozinovic F, Rosenmann M (1988) Comparative energetics of South American cricetid rodents. Comp Biochem Physiol 91A:195–202

Bozinovic F, Rosenmann M (1989) Maximum metabolic rate of rodents: physiological and ecological consequences on distributional limits. Funct Ecol 3:173–181

Bozinovic F, Simonetti JA (1992) Thermoregulatory constraints on microhabitat use by cricetid rodents in central Chile. Mammalia 56:364–369

Bozinovic F, Rosenmann M, Veloso C (1988) Termoregulación conductual en *Phyllotis darwini* (Rodentia: Cricetidae): efecto de la temperatura ambiente, uso de nido y agrupamiento social sobre el gasto de energía. Rev Chil Hist Nat 61:81–86

Bozinovic F, Novoa FF, Veloso C (1990) Seasonal changes in energy expenditure and digestive tract of *Abrothrix andinus* (Cricetidae) in the Andes range. Physiol Zool 63:1216–1231

Bradford DG (1974) Water stress of free-living *Peromyscus truei*. Ecology 55:1407–1414

Bradley SR, Deavers DR (1980) A re-examination of the relationship between thermal conductance and body weight in mammals. Compar Biochem Physiol 65A: 465–476

Cody ML, Mooney HA (1978) Convergence versus nonconvergence in mediterranean-climate ecosystems. Ann Rev Ecol System 9:265–321

Cortés A (1985) Adaptaciones fisiológicas y morfológicas de pequeños mamíferos de ambientes semiáridos. M.S. Thesis, Universidad de Chile

Cortés A, Zuleta C, Rosenmann M (1988) Comparative water economy of sympatric rodents in a Chilean semi-arid habitat. Comp Biochem Physiol 91A:711–714

Cortés A, Rosenmann M, Báez C (1990) Función del riñón y del pasaje nasal en la conservación de agua corporal en roedores simpátridos de Chile central. Rev Chil Hist Nat 63:279–291

Damuth J, MacFadden BJ (1990) *Body Size in Mammalian Paleobiology: Estimation and Biological Implications*. Cambridge U Press, Cambridge

Derting TL (1989) Metabolism and food availability as regulators of production in juvenile cotton rats. Ecology 70:587–595

Elgar MA, Harvey PH (1987) Basal metabolic rates in mammals: allometry, phylogeny and ecology. Funct Ecol 1:25–44

Fuentes ER (1976) Ecological convergence of lizard communities in Chile and California. Ecology 57:3–17

Hayes JP, Chappell MA (1986) Effects of cold acclimation on maximum oxygen consumption during cold exposure and treadmill exercise in deer mice, *Peromyscus maniculatus*. Physiol Zool 59:473–481

Hayssen V, Lacy RC (1985) Basal metabolic rates in mammals: taxonomic differences in the allometry of BMR and body mass. Comp Biochem Physiol 81A:741–754

Hinds DS, MacMillen RE (1985) Scaling of energy metabolism and evaporative water loss in heteromyid rodents. Physiol Zool 58:282–298

Hinds DS, Rice-Warner CN (1992) Maximum metabolism and aerobic capacity in heteromyid and other rodents. Physiol Zool 65:188–214

Hulbert AJ, Hinds DS, MacMillen RE (1985) Minimal metabolism, summit metabolism and plasma thyroxine in rodents from different environments. Comp Biochem Physiol 81A:687–693

Karasov WH (1986) Energetics, physiology and vertebrate ecology. Trends Ecol Evol 1:101–104

Kenagy GJ (1987) Energy allocation for reproduction in the golden-mantled ground squirrel. Sym Zool Soc London 57:259–273

Kleiber M (1961) *The Fire of Life*. John Wiley & Sons, New York

Koteja P (1991) On the relation between basal and field metabolic rates in birds and mammals. Funct Ecol 5:56–64

MacMillen RE (1983) Water regulation in *Peromyscus*. J Mam 64:38–47

MacMillen RE, Hinds DS (1983) Water regulatory efficiency in heteromyid rodents: a model and its application. Ecology 64:152–164

MacMillen RE, Hinds DS (1992) Standard, cold-induced, and exercise-induced metabolism of rodents. In Tomasi TE, Horton TH (eds) *Mammalian Energetics: Interdisciplinary Views of Metabolism and Reproduction.* Comstock Publishing, Ithaca, NY, pp 16–33

MacMillen RE, Lee AK (1967) Australian desert mice: independence of exogenous water. Science 158:383–385

MacMillen RE, Lee AK (1970) Energy metabolism and pulmocutaneous water loss of Australian hopping mice. Comp Biochem Physiol 35:355–369

Martin RA (1986) Energy, ecology and cotton rat evolution. Paleobiology 12:370–382

McNab BK (1979) Climatic adaptation in the energetics of heteromyid rodents. Comp Biochem Physiol 62A:813–820

McNab BK (1986) The influence of food habits on the energetics of eutherian mammals. Ecol Mono 56:1–19

McNab BK (1988) Complications inherent in scaling the basal rate of metabolism in mammals. Quar Rev Bio 63:25–54

McNab BK, Morrison PR (1963) Body temperature and metabolism in subspecies of *Peromyscus* from arid and mesic environments. Ecol Mono 33:63–82

Mooney HA (ed) (1977) *Convergent Evolution in Chile and California.* Dowden, Hutchinson and Ross, Stroudsburg, PA

Morton SR, Davidson DW (1988) Comparative structure of harvester ant communities in arid Australia and North America. Ecol Mono 58:19–38

Orians GH, Solbrig OT (eds) (1977) *Convergent Evolution in Warm Deserts.* Dowden, Hutchinson and Ross, Stroudsburg, PA

Peet RK (1978) Ecosystem convergence. Am Nat 112:441–444

Peters RH (1983) *The Ecological Implications of Body Size.* Cambridge U Press, Cambridge

Peterson CC, Nagy KA, Diamond J (1990) Sustained metabolic scope. Proc Nat Acad Sci 87:2324–2328

Raab JL, Schmidt-Nielsen K (1972) Effect of running on water balance of the kangaroo rat. Am J Physiol 222:1230–1235

Ricklefs RE, Travis J (1980) A morphological approach to the study of avian community organization. Auk 97:321–338

Rosenmann M (1977) Regulación térmica en *Octodon degus.* Med Amb (Chile) 3: 127–131

Rosenmann M, Morrison PR (1974) Maximum oxygen consumption and heat loss facilitation in small homeotherms by He-O$_2$. Am J Physiol 226:490–495

Rosenmann M, Morrison PR, Feist D (1975) Seasonal changes in the metabolic capacity of red-backed voles. Physiol Zool 48:303–310

Schluter DR (1986) Test for similarity and convergence of finch communities. Ecology 67:1073–1085

Schmidt-Nielsen K (1964) *Desert Animals: Physiological Problems of Heat and Water.* Clarendon Press, London

Schmidt-Nielsen K (1984) *Scaling: Why Is Animal Size So Important?* Cambridge U Press, Cambridge

Schmidt-Nielsen B, Schmidt-Nielsen K (1951) A complete account of the water metabolism in kangaroo rats and an experimental verification. J Cell Comp Physiol 38:165–182

Weiner J (1987) Limits of energy budgets and tactics in energy investments during reproduction in the Djungarian hamster (*Phodopus sungorus sungorus* Pallas 1770). Sym Zool Soc London 57:167–197

Wiens JA (1991) Ecological similarity of shrub-desert avifaunas of Australia and North America. Ecology 72:479–495

15. Multivariate Comparisons of the Small-Mammal Faunas in Australian, Californian, and Chilean Shrublands

Barry J. Fox

Comparisons between California and Chile have formed one of the major foci for studies of convergent evolution, and the volumes edited by Thrower and Bradbury (1977) and Mooney (1977) have greatly influenced studies of convergence. The study by Glanz (1977) in the former volume provides the starting point for the present analysis, which considerably broadens the scope of the comparisons he made, and builds on another comparative study of small mammals in shrublands from Australia, California, and South Africa (Fox et al. 1985). Cody et al. (1977) drew their conclusions on the comparative study of small mammals in mediterranean shrublands from the work of Glanz (1977) and his earlier doctoral studies.

All three regions in the Pacific Basin with mediterranean-climate ecosystems are considered in this comparison. The comparisons are not, as in the IBP study, limited to climatically matched sites; rather, the physiognomy of scrubs or shrublands has been used to form the "matching" criteria. I take this path because matching sites on variables such as latitude, rainfall, aspect, soil type, as in the IBP studies, can lead one to compare types of vegetation that differ substantially in structure. It is my view that although animal species' perception of habitat will often differ from that of ecologists (Fox 1984), it is safe to conclude that the structure of the vegetation is most important in selection of habitat by animal species. This view also governed an earlier study of the three mediterranean regions, Australia, California and South Africa, that compared response by the small-mammal communities in these shrublands to disturbance by fire (Fox et al. 1985). The approach used here is, for each of the regions with shrubland as their major vegetation type, to choose many sites and test the null hypotheses of no significant differences among them for the specific variables investigated. Although accepting the null hypothesis would be necessary, but not sufficient, to demonstrate convergence, rejecting that hypothesis would support lack of convergence on the variable or set of variables being tested.

Study Sites

To compare the three regions, sites were chosen from those already available in the literature, for which the appropriate data including biomass information were presented. For the initial analyses of community characteristics I have used Glanz (1977) for comparative data for both California and Chile; Fox et al. (1985) for comparative data from the eastern Australian sites, and Wirtz et al. (in press) for some additional Californian sites. Sources for the original data are listed in Table 15.1. One constraint on site selection was to ensure that all sites used in the analyses had not burned for at least four years. Table 15.1 lists variables in each mammal community for the eight Australian sites, eight Californian sites, and five Chilean sites.

Four of the Australian sites are in Myall Lakes (32°30′ S; 152°30′ E) on the northeastern coast of New South Wales with coastal vegetation comprising sclerophyll scrub and woodland, referred to as heath and mallee heath. These data are from a 7-ha site reported in Fox (1982), here referred to as total, as well as data from separate wet heath, dry heath, and mallee heath habitats that are separated along a moisture gradient produced by a 4-m rise in elevation from swale to low dune, on a Pleistocene sand substrate (Fox et al. 1985). Additional sites were situated to the north in coastal heath and mallee heath at Cooloola in southeastern Queensland (Dwyer et al. 1979), in coastal scrub to the south at Nadgee (Newsome and Catling 1983), a coastal site on the NSW–Victorian border, and farther south in coastal heath at Anglesea on the southern Victorian coast west of Melbourne (Wilson et al. 1990).

The main Californian site was at Echo Valley (32°54′ N; 116°39′ W), with sclerophyll scrub or chaparral as the vegetation type. Other IBP sites were a montane forest site located at Mount Laguna (32°50′ N; 116°26′ W), a coastal-scrub site at Camp Pendleton (33°21′ N; 117°31′ W), and a xeric site with coastal succulent scrub at Punta Banda (31°49′ N; 116°44′ W) (Glanz 1977). Additional chaparral sites were farther north, at Santa Margarita (Quinn 1979) and in the San Dimas Experimental Forest, 45 km east of Los Angeles at San Dimas. The latter two were a five-year-old burned site (Bell fire) and a long-unburned site (Bell control) (Wirtz et al. in press).

The main Chilean site was sclerophyll scrub or matorral at Fundo Santa Laura in central Chile (33°04′ S; 71°00′ W), with a montane-forest site at

Table 15.1. Community variables for each site, including number of common species, species pool, source of data, and type of vegetation. Each variable explained in text

Site	Vegetation	Source	Richness	Diversity	Equitability	Biomass Kg ha⁻¹	Density ha⁻¹	Spp. pool	Common spp.
Australia									
Myall Lakes	wet heath	(1)	4	2.96	0.74	0.408	7.1	6	3
Myall Lakes	dry heath	(1)	3	2.91	0.97	0.080	5.1	7	3
Myall Lakes	mallee	(1)	3	2.46	0.82	0.119	6.7	7	3
Myall Lakes	total	(1)	7	5.36	0.77	0.184	4.7	7	6
Nadgee	scrub	(2)	5	3.20	0.64	0.765	14.5	5	4
Cooloola	heath	(3)	7	4.58	0.65	1.560	6.5	9	4
Cooloola	mallee	(3)	5	2.22	0.44	1.150	11.1	10	3
Anglesea	coastal heath	(7)	7	4.51	0.64	0.770	16.2	9	5
California									
Santa Margarita	chaparral	(4)	8	3.21	0.42	0.460	10.5	9	6
Camp Pendleton	coastal sage	(5)	7	3.63	0.52	2.729	19.1	12	7
Echo Valley	chaparral	(5)	5	2.34	0.47	1.068	15.2	10	5
Punta Banda	dry coastal scrub	(5)	4	2.14	0.54	3.766	29.0	6	3
Mt. Laguna	montane	(5)	3	2.80	0.93	2.998	28.0	6	3
Bell Control	chaparral	(6)	5	3.15	0.63	7.000	57.0	9	3
Bell Fire	chaparral	(6)	6	3.21	0.54	3.500	72.0	9	5
San Dimas	chaparral	(6)	7	3.59	0.51	4.500	57.0	9	4
Chile									
Fundo St. Laura	matorral	(5)	7	5.97	0.82	0.345	7.5	9	4
Papudo/Zapallar	matorral	(5)	4	2.57	0.64	1.224	30.0	7	4
Cerro del Roble	montane	(5)	2	1.80	0.90	0.300	3.0	2	1
Cerro Poterillo	dry coastal scrub	(5)	5	2.20	0.44	3.893	62.0	5	3
Los Molles	coastal scrub	(5)	5	1.80	0.36	2.299	20.0	7	5

Sources: (1) Fox et al. 1985; (2) Newsome and Catling 1983; (3) Dwyer et al. 1979; (4) Quinn 1979; (5) Glanz 1977; (6) Wirtz et al. 1988; (7) Wilson et al. 1990

Cerro del Roble (32°58′ S; 71°30′ W), coastal scrub at Papudo/Zapallar (32°25′ S; 71°28′ W), and coastal-succulent scrub at Los Molles (32°13′ S; 71°32′ W) and, the most northerly location, at Cerro Potrerillo (30°20′ S; 70°43′ W), which is a xeric site.

For analyses on "mammal types," a species list only was required for each site, hence it was possible to expand the number of sites by incorporating a xeric heath site from northwestern Victoria (Cockburn 1981) and three nearby sites in the Victorian mallee with chenopod, *Triodia*, and open-heath vegetation (Bennett et al. 1989). An additional coastal sage site at Irvine could be included for California (Meserve 1976) and an additional matorral site from Fundo San Carlos de Apoquindo for Chile (Iriarte et al. 1989). Meserve and Glanz (1978) provide information on a coastal-scrub site at Los Vilos and two thorn-scrub sites at Fray Jorge and Guanoqueros in north-central Chile. These additional sites bring the total to 30: 12 Australian sites, 9 Californian sites, and 9 Chilean sites.

Community Variables and Analysis of Data

Six of the seven measures used by Fox et al. (1985) have been applied in this study. Trap success was not included because those data were not available for the sites used by Glanz (1977). One new variable (common species) has been included, however, to take into account that some species may be very common at a site and contribute substantially to the mammal community and biomass at the site, and others may be very rare and hence contribute little to either the biomass or the community's functioning. Where possible, data were used directly from the literature, but in some cases it was necessary to calculate values from the raw data provided (e.g., for species diversity and equitability).

The seven variables used from the literature are defined as: *species richness* (S)—the number of species encountered at the site in trapping sessions during the surveys reported in the literature used here; *species diversity* (D)—the number of individuals from each species in these surveys was used to calculate the inverse of Simpson's (1949) measure of concentration for a site $D = \sum_i p_i^{-2}$, where p_i is the number of individuals of the ith species as a proportion of the total number of individuals. This expression takes into account different patterns in relative abundance and estimates the number of equally abundant species that would give the same diversity (Δ_3, Hurlbert 1971); *equitability* (E)—a measure of the evenness of abundances for all species in an assemblage ($E = D/S$); *total biomass*—the summed value for biomass measured over all species present (Kg ha^{-1}); *total density*—the number of individuals of each species summed for all species present (ha^{-1}); *total species pool*—the maximum number of species previously recorded in the area of the site, but not necessarily captured during the surveys reported in this study; *number of common species*—the most abundant, most commonly trapped species at each site, excluding rare and infrequently trapped species that make small contributions to biomass at each site. The values for this variable are subjective—they represent my own assessment of the data presented, in conjunction with authors' comments, and are meant to establish similar criteria overall.

Univariate analyses of these variables, and other derived variables resulting from the multivariate analyses below, were subjected to one-way analysis of variance (ANOVA), followed by *a posteriori* tests for differences in each of the regions, using Scheffé's test. These analyses were conducted on a Macintosh IIcx microcomputer using the SuperAnova statistical package (Abacus Concepts 1989).

Multivariate Statistical Analyses

Discriminant function analysis was used to answer this question: can significant separation among the three regional groupings be achieved with a mathematical combination of the variables in the analysis? Such an *a priori* grouping, imposed on the sites by their geography, is a necessary requirement for using discriminant function analysis. This technique is useful because it provides a means for statistically testing for significant differences among the three groups using the variance ratio test of the variance among groups to the variance within groups in Mahalanobis distance between sites in Euclidian space. A stepwise procedure was used on a Vax mainframe computer (SSPS-X 1983) to determine the optimum function, that is, the one with the fewest variables relative to proportion of the variance explained.

Factor analysis was used in analyzing the "mammal types" data set, to address this question: Are there underlying strong relationships between sets of variables that reflect the different evolutionary and biogeographic histories of the "mammal types," and the different relative frequencies, found in each of the three regions? This analysis was conducted on a Macintosh IIcx microcomputer using the StatView statistical package (Abacus Concepts 1992), as were the regression and correlation analyses.

Community Variables

The simplest form of analysis examined the data matrix using a univariate one-way analysis of variance across the three groups. From this analysis, just two variables demonstrate significant differences across the groups: total biomass and total density. Biomass shows a significant difference between regions ($F = 6.3, p = 0.0085$) and *a posteriori* tests demonstrate that biomass from Australian sites (0.63 ± 0.19 kg ha^{-1}) is significantly less (Scheffé's $F = 6.2, p < 0.05$) than from Californian sites (3.25 ± 0.72 kg ha^{-1}) but that Chilean sites (1.61 ± 0.68 kg ha^{-1}) do not differ from either of the other

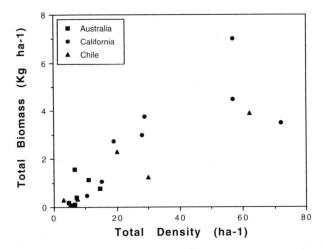

Figure 15.1. Scatterplot of total biomass as a function of total density for sites listed in Table 15.1.

areas. Density shows a similar pattern ($F = 4.4$, $p = 0.028$) with significant differences (Scheffé's $F = 5.0$, $p < 0.05$) between Australian (9.0 ± 1.6 ha^{-1}) and Californian sites (36.0 ± 8.1 ha^{-1}), neither of which differs from the Chilean sites (24.5 ± 10.5 ha^{-1}).

When biomass is plotted as a function of density for all sites a clear pattern appears (Fig. 15.1). As expected, the two are strongly correlated ($r = 0.841$, $p < 0.001$). It is also clear that Australian sites have both low density and low biomass, but there is no significant correlation between these variables ($r = 0.142$). Some Chilean sites exhibit low density and biomass, some have high density but only medium levels of biomass. Biomass appears to increase more slowly with density in Chile than in California where biomass as well as density can be high. These two variables show strong correlations within the Californian ($r = 0.732$) and Chilean ($r = 0.915$) regions.

A question of greater interest, however, is what significant differences there are among the groups when all variables are considered at once, better assessing the degree of convergence in the evolution of these faunas. The question is best answered by a multivariate analysis, and discriminant function analysis on the three a priori groups is the most appropriate technique. Two functions were selected that best discriminate among the three groups with a stepwise procedure using two variables in the equations: first, total biomass, and then total species pool, with which 80% of the sites could be classified into their correct groups.

The functions thus constructed can be correlated with the original variables to determine how they relate to each function. Function 1 shows strong positive correlation with biomass and density and function 2 shows strong positive correlation with total species pool, number of common species, and

Table 15.2. Variables entered into stepwise discriminant function, overall significance at each step, correlation between variables and each function, and coefficient used to calculate each function, based on the 20 sites from Australia (7), California (8), and Chile (5) in Table 15.1

Variable (entry order)	Significance (each step)	Correlation coefficient		Equation coefficient	
		Function 1	Function 2	Function 1	Function 2
Biomass (1)	0.013	0.842	−0.540	0.6526	−0.0044
Species Pool (2)	0.014	0.514	0.858	0.0181	0.4827
Intercept	—	—	—	−1.3878	−3.6359

species richness, and a minor correlation with diversity. Equitability shows a minor negative correlation with both axes. The discriminant functions chosen by the stepwise procedure use total biomass and total species pool to form the equations, and these results are summarised in Table 15.2. The discriminant function scores are plotted and identified in Figure 15.2, including the centroids for each region. The Anglesea data were not available for incorporation in the original analysis, hence discriminant function scores for Anglesea were calculated from the equations for the discriminant functions, and lie close to the other values from the Australian region.

The Californian sites have high positive values for both functions but Australian and Chilean sites have generally low or negative values on both axes. The Bell control site—chaparral that had not burned in twenty-eight

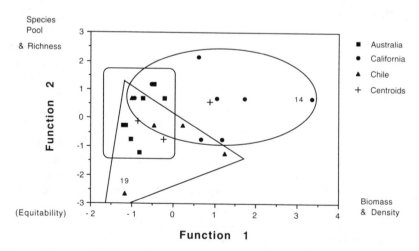

Figure 15.2. Scatterplot of sites on first two discriminant function axes derived from analysis of community variables at each site from Table 15.1 (symbols as for Fig. 15.1, with the centroid for each region also shown). See Tables 15.2 and 15.3 for variable contributions to the functions and the significance of comparisons. Some outliers specifically mentioned in the text are labeled.

Table 15.3. Significance tests, F-values (probability), for paired comparisons among the three regions, based on multivariate discriminant functions

Region	Australia	California
California	6.2	
	(0.01)	
Chile	1.1	4.3
	(0.36)	(0.03)

years, appeared as an outlier with substantially higher density and biomass than the other sites. This result contrasted markedly with the Chilean montane site, which also appeared to be an outlying site because of its low species richness.

The values for the F-statistic (df $= 2,16$) and its probability are shown for each of the three paired comparisons in Table 15.3. This table demonstrates that Australian and Californian sites are most different ($p = 0.001$); Chilean and Californian sites are also different ($p = 0.03$); but Australian and Chilean sites show no significant difference ($p = 0.36$).

"Mammal Types"

An alternative form of analysis would be to consider a comparison of "mammal types" as first proposed by Glanz (1977) in his study comparing California and Chile. To accommodate the mammal species encountered in Australian sites, two additional types are used here: an "insectivore" type and a "small possum" type. The results of this comparative exercise are set out in Table 15.4. The need for categories such as "small possum" for Australia and "squirrel" for California, which are present in one region only, emphasizes differences in the biogeography of these mammal faunas. Although Glanz (1977) did not recognize an "insectivore" type, in subsequent work reviewing insectivory in South American small mammals, he does indicate that some species of *Akodon* in Chile may have up to 56% animal material in their diet (Glanz 1982). This figure, though, falls well below his estimate of 92% for the marsupial *Marmosa elegans*, which is why I have included only this one species in the "insectivore" type for Chile.

The next obvious step is to characterize each site according to the mammal types that may be found there. Because the list of species present is all the information required for such an exercise it was possible to include thirty sites in Table 15.5.

The data from Table 15.5 were subjected to a factor analysis of the complete set of 9 "type" variables over the 30 sites, which produced 3 significant factors to explain a total of 67.7% of the variance in the data set. These factors were then correlated against the original 9 variables to determine the contribution each made to the 3 factors (Table 15.6). The first factor correlates most strongly with the "pocket-mouse" type 7 (-0.834), although the correlation is negative so that positive values for this factor would mean absence or low representation of this type. "Vole" type 5 (0.799) and "insectivore" type 8 (0.550) also had important positive correlation with factor 1, along with minor contributions from "ground squirrel" type 2 ($+0.404$) and "mouse" type 4 (-0.542). Factor 2 was positively correlated with "squirrel" type 1 (0.848), "gopher" type 6 (0.792) and "ground squirrel" type 2 ($+0.634$), and factor 3 had its strongest correlations with "rat" type 3 (0.836), "mouse" type 4 (0.726) and "small-possum" type 9 (-0.595), with a minor contribution from "ground-squirrel" type 2 ($+0.407$).

A score on each factor was then calculated for each site and scatterplots of these values are shown for the first three factors in Figure 15.3. The sites from the three regions are clearly separated and the individual sites are labeled as in Table 15.5. Californian and Chilean sites can be separated by their scores on factor 1. The Mt. Laguna montane site (17) is again an outlier with the highest value on factor 2 and the Echo Valley site (15) has the second-highest value, and with the Camp Pendleton sage site (14) these are the only sites separated from the otherwise very tightly clustered Californian sites (Fig 15.3B). Values for Australian sites are intermediate on factor 1 and low on factor 2 (Fig. 15.3B) and form a tight, easily identifiable group, the xeric site Big Desert (9) being the only one to show some slight separation toward the Californian sites. The Chilean sites also cluster tightly with the montane site (24 Cerro del Roble) being the only site at all removed from the main group (Fig. 15.3B). When factor 3 is plotted against factor 1 the Californian sites form a fairly tight group (again except for site 17) toward the positive end of the factor 3 axis and large negative values for factor 1, and the Chilean sites are grouped toward the the positive end of both factor axes, again except for site 24 (Fig. 15.3A). The Californian and Chilean outlying sites nearest to the origin are the montane forests (17 and 24) and these appear to be closer to the Australian xeric sites at Big Desert and in the Victorian mallee. The Fundo Santa Laura site (22) also appears to be something of an outlier here (Fig. 15.3A), mainly because of its high value for factor 3, which reflects strong representation from types 3 and 4, with no type 9.

One-way analysis of variance on the factor scores for each site over the three regions demonstrates significant differences for factor 1 ($F = 45.2$, $p = 0.0001$) with mean values of 0.951 for Chile, 0.210 for Australia, and -1.232 for California. Although no significant difference appears across regions for factor 2 ($F = 3.17$, $p = 0.058$), the differences for factor 3 are significant ($F = 6.46$, $p = 0.0051$), with mean values of 0.559 for California, 0.345 for Chile, and -0.678 for Australia. The results of a posteriori tests on these mean values using Scheffé's test are given in Table 15.7 for factors 1 and 3. All three regions are significantly different for factor 1 at the 5% level, and for

Table 15.4. Comparison of Australian, Californian, and Chilean small-mammal types found in shrublands (adapted from Glanz 1977)

Small-mammal type	Australia	California	Chile
1. "Squirrel" type: medium to large, diurnal, arboreal or semiarboreal, seed & fruit specialist	None	*Sciurus griseus* *Eutamius merriami*	None
2. "Ground-squirrel" type: large, diurnal, eats herbs, open/successional habitat, social	None 600–700 g, individual burrows	*Citellus beecheyi* 200–250 g, communal burrows	*Octodon degus*
3. "Rat" type: large, nocturnal, leaves & fruit, thick or spiny cover	*Rattus fuscipes* More omnivorous, prefers dense cover up to 1.5 m	*Neotoma fuscipes* *Neotoma lepida* Stick house & cactus patch habitat	*Abrocoma bennetti* *Octodon lunatus* Eats more herbs, prefers burrows in thorny/thick scrub
4. "Mouse" type: small to medium, large ears, long tail, small rear feet, omnivore, nocturnal, rocky or shrubby habitat	*Pseudomys novaehollandiae* small, eats seeds, plants, fungi, insects	*Peromyscus californicus* *P. boylii, eremicus, maniculatus* *Reithrodontomys megalotis* small, eats more seeds & fruits	*Phyllotis darwini* *Oryzomys longicaudatus* Larger, eats more herbs (Phd) & insects (Orl)
5. "Vole" type: small to large, small ears, short tail, small rear feet, terrestrial, crepuscular, some diurnal activity, thick ground cover, grasses, sedges, or shrubby litter habitat	*Pseudomys gracilicaudatus* *Rattus lutreolus* (Rl) Large; eats sedge stems and rhizomes; morphology like type 3 (Pgc) medium; eats seeds, insects, fungi, and plants, longer tail	*Microtus californicus* Eats grass and seeds, only in grassy habitats, morphology like type 6	*Akodon longipilis* *Akodon olivaceus* Eats many more insects and other invertebrates, in both shrubby & grassy habitats, morphology like type 4
6. "Gopher" type: medium, fossorial, very small ears, very short tail, long front claws, massive skull & forearms	None	*Thomomys bottae* Solitary, eats roots and leaves	*Spalacopus cyanus* Colonial, eats more bulbs

7. "Pocket Mouse" type: small to medium, short ears, long tail, large rear feet, saltatorial, open, dry habitats, eats seeds, nocturnal	*Notomys mitchelli* Drier areas only	*Dipodomys agilis* *Perognathus californicus* *Perognathus fallax*	None
8. "Insectivore" type: small, nocturnal, high metabolic rate, totally insectivorous diet	*Antechinus stuartii, A. flavipes* *Sminthopsis murina*	*Notiosorex crawfordi* *Sorex ornatus*	*Marmosa elegans*
9[a]. "Small Possum" type: very small, nocturnal, prehensile tail, nectarivore, some insects	*Cercatetus nanus*	None	None

[a] Type 9 has one highly specialized species, *Tarsipes rostratus*, which is found in Western Australia only, has reduced teeth and a featherlike tongue.

Table 15.5. Number of species of each small-mammal type (see Table 15.4) for each of the sites analyzed

No.	Site	Vegetation	Source	Small-mammal type								
				1	2	3	4	5	6	7	8	9
Australia												
1.	Myall Lakes	wet heath	1	0	0	0	1	2	0	0	1	0
2.	Myall Lakes	dry heath	1	0	0	1	2	0	0	0	1	0
3.	Myall Lakes	mallee	1	0	0	1	2	0	0	0	2	0
4.	Myall Lakes	total	1	0	0	2	2	2	0	0	2	0
5.	Nadgee	scrub	2	0	0	1	1	1	0	0	2	0
6.	Cooloola	heath	3	0	0	1	2	1	0	0	1	0
7.	Cooloola	mallee	3	0	0	1	2	1	0	0	2	0
8.	Anglesea	heath	7	0	0	2	2	1	0	0	3	0
9.	Big Desert	heath	8	0	0	0	1	0	0	1	1	1
10.	Victorian mallee	chenopod	9	0	0	0	1	0	0	0	1	0
11.	Victorian mallee	triodia	9	0	0	0	1	0	0	1	2	0
12.	Victorian mallee	open-heath	9	0	0	0	0	2	0	1	2	2
California												
13.	Santa Margarita	chaparral	4	0	0	2	4	0	0	3	0	0
14.	Camp Pendelton	coastal sage	5	0	1	2	5	1	1	2	1	0
15.	Echo Valley	chaparral	5	1	1	2	5	1	1	2	0	0
16.	Punta Banda	dry coastal scrub	5	0	0	1	3	0	0	2	0	0
17.	Mt. Laguna	montane	5	2	1	1	3	0	1	0	1	0
18.	Bell Control	chaparral	6	0	0	1	2	0	0	2	0	0
19.	Bell Fire	chaparral	6	0	0	1	3	0	0	2	0	0
20.	San Dimas	chaparral	6	0	0	1	3	1	0	2	0	0
21.	Irvine	coastal sage	10	0	0	2	5	1	0	3	0	0
Chile												
22.	F. Santa Laura	matorral	5	0	1	3	2	2	1	0	1	0
23.	Papudo, Zappallar	matorral	5	0	1	0	2	2	1	0	1	0
24.	Cerro del Roble	montane	5	0	0	0	1	1	1	0	0	0
25.	Cerro Poterillo	dry coastal scrub	5	0	1	1	1	2	0	0	0	0
26.	Los Molles	coastal scrub	5	0	0	1	2	2	1	0	1	0
27.	F.S.C. Apoquindo	matorral	11	0	1	1	2	2	0	0	1	0
28.	Guanoqueros	thorn scrub	12	0	1	1	2	2	0	0	1	0
29.	Fray Jorge	thorn scrub	12	0	1	1	2	2	0	0	1	0
30.	Los Vilos	coastal scrub	12	0	1	1	2	2	0	0	1	0

Sources (1) to (6) as for Table 15.1, (7) Wilson et al. (1990), (8) Cockburn (1981), (9) Bennett et al. (1989), (10) Meserve (1976), (11) Iriarte et al. (1989), (12) Meserve and Glanz (1978).

factor 3 California and Chile differ from Australia but not from each other. It is clear that in the three-dimensional factor space a significant separation will appear between sites from the three regions, as can be seen by considering Figs. 15.3A and 15.3B as the projection of these-sites onto an elevation and plan respectively for this three-dimensional space. California and Chile are well separated in the plan view (Fig. 15.3B), with the values for the Australian region lying between them. The elevation, however, (Fig. 15.3A) shows that the values for sites from the Australian region tend to lie below

Table 15.6. Factor loadings (correlations between factors and variables) demonstrating the contribution by each of the nine mammal types to three significant factors that explain a cumulative 67.7% of the variance. Correlations above 0.5 are shown in **bold typeface**, and below 0.2 are omitted for greater clarity

	Factor 1	Factor 2	Factor 3
Eigenvalue	2.808	2.071	1.218
Variance explained (%)	31.2	23.0	13.5
Variables			
Type 7	**−0.834**		0.315
Type 5	**0.799**		0.272
Type 8	**0.550**		−0.315
Type 1		**0.848**	
Type 6		**0.792**	
Type 2	0.404	**0.634**	0.407
Type 3			**0.836**
Type 4	−0.542	0.261	**0.726**
Type 9			**−0.595**

those from the other two regions because the Australian sites have fewer species from types 3 and 4, which produce the lower factor 3 scores.

Biomass and Productivity

The only similar comparative studies of mammals in mediterranean shrublands are those of Cody et al. (1977), Glanz (1982, 1977), and Fox et al. (1985), and data from the latter two were the basis for this study. Cody et al. (1977) based their comments on the work of Glanz (1977).

The differences between regions illustrated in Figure 15.1 appear to be related to productivity, particularly the low values observed from the Australian sites. Inspection of the soil-nutrient status in each of the regions gives some support to this view. Typical values of total nitrogen content and total phosphorus content of soil from shrublands in mediterranean climates are provided by di Castri (1981) and also by Read and Mitchell (1983); averages of these values together with mean values for the density and biomass of small mammals in each region are given in Table 15.8. Using these data and density values from Table 15.1 for all sites, it is possible to demonstrate a statistically significant regression of small-mammal density on nitrogen content ($F = 9.2$, $n = 21$, $p = 0.007$, see Fig. 15.4A). Biomass and nitrogen content show a similar relationship ($F = 11.9$, $p = 0.0027$ see Fig. 15.4B). These relationships, however, should be regarded as indicating biological significance only, because the soil-nutrient values are typical for the region and the small-mammal data apply to the sites examined. This relationship cannot be unequivocally established until soil-nutrient values are available from the

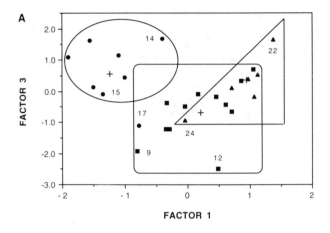

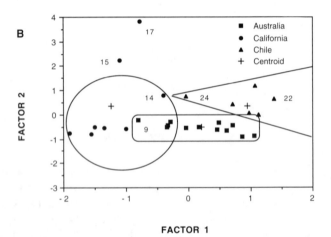

Figure 15.3. Scatterplot of the 30 sites on (A) the first and third factor axes, and (B) the first two factor axes, derived from analysis of mammal types at each site (see Tables 15.4, 15.5, 15.6). Symbols for each region are as for Fig. 15.1 and some outliers specifically mentioned in the text are labeled as in Table 15.5.

Table 15.7. Significance tests, Scheffe's F-values, for paired comparisons among the three regions, based on scores for factor 1 in the upper triangle and scores for factor 2 in the lower triangle.

Region	Australia	California	Chile
Australia	—	21.63[a]	5.72[a]
California	5.41[a]	—	43.40[a]
Chile	3.70[a]	0.14	—

[a] Value significant at 5% level.

Table 15.8. Typical values of total nitrogen content and total phosphorus content of soil from shrublands in Mediterranean climates are presented as % of dry mass values averaged from di Castri (1981) and from Read and Mitchell (1983): Also presented are mean values for the density and biomass of small mammals in each region (taken from Fig. 15.1). The regression of density on nitrogen content is significant.

$$(r = 0.998, \text{ adjusted } r^2 = 0.994, P = 0.035)$$
$$\{Mean\ density\ (\text{ha}^{-1}) = 2.238 + 154.3\ Nitrogen\ (\%)\}$$

	Australia	California	Chile
Nitrogen (%)	0.350	0.215	0.150
Phosphorus (%)	0.006	0.060	0.090
Mean Density (ha^{-1})	7.96	35.98	24.50
Mean Biomass (Kg ha^{-1})	0.609	3.253	1.612

same sites from which the small-mammal data are collected, to adequately assess the within-region variation.

The productivity difference among the three regions is most likely to be a large contributor to the significant separation observed between California and the other two regions (Fig. 15.2). It seems clear that productivity can influence the first discriminant function where density and biomass both contribute to the function. Any proposed link between the second discriminant function and productivity differences, though, is less clear, for the main contribution to this function comes from species richness and the species pool. Although some have suggested that increases in productivity produce increases in species richness, as well as density of individuals, others have suggested that the productivity–richness relationships may not be linear, and richness may increase and then decrease with increasing productivity (Tilman 1982, Abramsky and Rosenzweig 1984). I prefer not to link this discriminant function directly to productivity until further evidence is available.

The considerable overlap in values for function 2 for the three regions is responsible for the lack of significant differences, but though we may say that statistically there is no difference between regions we can be no means say that the community properties associated with species pool and species richness are equivalent in all regions, as might be expected from strong convergent evolution. Cody et al. (1977) found only a few one-to-one species comparisons for mammals, and they found reasonable evidence for convergent effects up to community-level processes and also that the convergence was more marked where communities were small, and more or less equal, so that the effect was most strongly exhibited at the level of α-diversity, the number of species coexisting within small patches of uniform habitat. The results in this study are not inconsistent with the conclusions of Cody et al. (1977) at this species richness level as the anovar across regions for richness ($p = 0.56$) and diversity ($p = 0.54$) were far from significantly different. The productivity differences demonstrated here, however, are not consistent with the

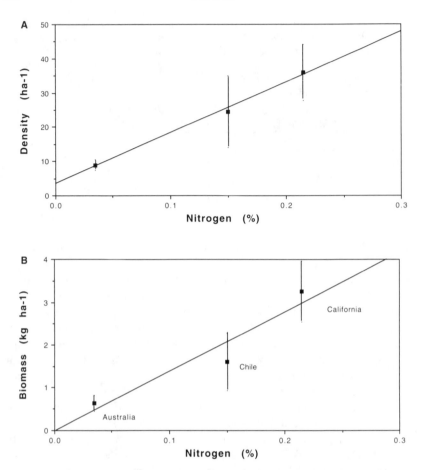

Figure 15.4. (A) density, and (B) biomass, as mean value (±standard error) for all sites (from Table 15.1, $n = 21$) as a function of total nitrogen content of soil (%) taken from Table 15.8. There is a significant regression of density on nitrogen content: Density (ha^{-1}) = 3.49 + 148.68 Nitrogen (%) ($r = 0.57$, adjusted $r^2 = 0.29$, $F = 9.2, p = 0.007$). There is a significant regression of biomass on nitrogen content: Biomass (Kg ha^{-1}) = 0.02 + 14.07 Nitrogen (%) ($r = 0.62$, adjusted $r^2 = 0.385$, $F = 11.89, p = 0.0027$).

same level of convergence described by Cody et al. (1977), with two reservations: the expansion to a three-way rather than a two-way comparison would tend to increase the likelihood of detecting differences between regions; and the Australian sites were not matched as were the sites used by Glanz (1977) and Cody et al. (1977). Fox et al. (1985) concluded that the mammal communities inhabiting Australian heathlands four to five years after fire differed significantly from those in shrublands from California and South Africa. They also found that density and biomass were the variables contributing most to the differences, that richness and diversity were not significantly different, and that productivity was implicated as the most likely contributing factor.

Summarizing the analysis of the community variables, it would seem that although differences occur in the data set for mammals in shrublands from these three continents, there are also similarities. These resemblances support the notion that current and recent climatic events, productivity, and other ecological factors influence variables such as density and biomass of individuals and number of species that may be found at any site.

Community Structure

Using of the number of species in each of nine "mammal types" (Table 15.4) to investigate the structure of the small-mammal communities for the three regions provided an alternative method for examining convergence. Although the community's structure can reflect the climatic and ecological factors currently influencing each region, one must also consider the potential for influence by events in the regions' evolutionary and biogeographic history. Unlike the analysis based on variables in the entire community (Fig. 15.2), when the greater detail of the "mammal types" was included it was clear that all three regions are distinct (Fig. 15.3).

Such distinct separation may indicate that the evolutionary and biogeographic history are important in determining structure in each region's mammal community and may outweigh the climatic and ecological influences currently acting on them. This view gains some support from the two groups represented in just one region: (1) the squirrel type in California, and (9) the small possum type in Australia. Neither of these types has analogues in either of the other regions. Three groups missing from one of the regions and from Australia are the ground squirrel type (2) and the gopher type (6), and the pocket-mouse type (7) is absent from Chile. Such omissions (from a total of nine "types") are a substantial flaw in the convergence argument because they would not be expected where climatic and ecological influences were the dominant factors influencing convergence. It appears that the level of "mammal types" is the appropriate one for such analyses. Progression through even greater levels of detail to individual species may provide interesting examples of convergence, with few one-to-one correspondences, but Cody et al. (1977) indicated that these examples were few. Although the specialist species, such as the gopher *Thomomys* in California and its analogue *Spalacopus* in Chile, provide the most convincing parallels demonstrating convergent evolution, the argument loses much of its force when they are found to be absent from another region that the convergence argument would require them to occupy. Demonstrating convergence at the community level is a necessary component of any argument for convergence, but one should not be distracted by having to determine which level of analysis to use, rather than addressing the initial question, ascertaining the degree of convergence. The mammal faunas inhabiting structurally similar shrubland vegetation from the three regions, with broadly similar climates, appear to be more strongly influenced by their evolutionary and biogeographic history than the climatic and ecological factors usually associated with evolutionary convergence.

Conclusions

Detailed comparisons of the small-mammal communities of shrublands in California and Chile were made as part of a major study of ecosystem convergence during the 1970s (Glanz 1977). These data and additional material published since were used as a basis for comparison with small-mammal communities inhabiting shrublands and woodlands in eastern Australia. A range of community variables such as species richness, diversity, equitability, density, biomass, and number of common species were used in a multivariate analysis showing that California differed significantly from both Australia and Chile, which displayed no significant differences between themselves. Biomass and density were the variables that contributed most to these differences in productivity, possibly because of differences in the soil-nutrient status of shrublands on the three continents. Community variables may reflect some climatic and ecological influences, most particularly productivity, as well as the regions' evolutionary and biogeographic history, although the latter effects appear to be much stronger.

The three regions were also investigated using the structure of the small-mammal communities at each site, determined by the number of species found in each of nine categories of "mammal types" that were identified. It was clear that the structure of the small-mammal communities were distinct in all three regions. Thus the evolutionary and biogeographic history determining the structure of the mammal community in each region clearly outweighs the climatic and ecological influences currently acting on it.

Acknowledgments

I first thank Mary Kalin Arroyo for the invitation and inspiration to write this chapter, and Eduardo Fuentes for the time and effort he spent showing me some of the mediterranean-climate ecosystems in Chile. Of course my thanks also go to those who published the data upon which such review articles depend, and these are listed in the text. I especially mention Bill Glanz, for without his studies pioneering the comparison between Californian and Chilean communities of mammals this study would not have been possible. This work was supported in part by grants from the Australian Reseach Council (D18616065 and A18930372) to study the structure of animal communities.

References

Abacus Concepts (1989) SuperAnova. Abacus Concepts Inc., Berkeley, CA

Abacus Concepts (1992) StatView. Abacus Concepts, Berkeley, CA

Abramsky Z, Rosenzweig ML (1984) Tilman's predicted productivity-diversity relationship shown by desert rodents. Nature 309:150–151

Bennett AF, Lumsden LF, Menkhorst PW (1989) Mammals of the mallee, southeastern Australia. In Noble JC and Bradstock RA (eds) *Mediterranean Landscapes in Australia: Mallee Ecosystems and Their Management*. CSIRO Publications, Melbourne, Australia, pp 191–220

Cockburn A (1981) Diet and habitat preference of the Silky Desert Mouse, *Pseudomys apodomoides* (Rodentia), in mature heathlands. Aust Wild Res 8:499–514

Cody ML, Fuentes ER, Glanz W, Hunt JH, Moldenke AR (1977) Convergent evolution in the consumer organisms of mediterranean Chile and California. In Mooney HA (ed) *Convergent Evolution in Chile and California Mediterranean Climate Ecosystems*. Dowden, Hutchinson and Ross, Stroudsville, PA, pp 144–192

di Castri F (1981) Mediterranean-type shrublands of the world. In di Castri F, Goodall DW, Specht RL (eds) *Mediterranean-Type Shrublands of the World*. Elsevier, Amsterdam, pp 1–52

Dwyer P, Hockings M, Wilmer J (1979) *Mammals of Cooloola and Beerwah*. Proc Royal Soc Queensland 90:65–84

Fox BJ (1982) Fire and mammalian secondary succession in an Australian coastal heath. Ecology 63:1332–1341

Fox BJ (1984) Small scale patchiness and its influence on our perception of animal species' habitat requirements. In Myers K, Margules C, Musto I (eds) Proc of a workshop on survey methods for nature conservation. CSIRO, Canberra, Australia, pp 162–178

Fox BJ, Quinn RD, Breytenbach GJ (1985) *A Comparison of Small-Mammal Succession Following Fire in Shrublands of Australia, California and South Africa*. Proc Ecol Soc Aust 14:179–197

Glanz W (1977) Small mammals. In Thrower NJW, Bradbury DE (eds) *Chile–California Mediterranean Scrub Atlas: A Comparative Analysis*. Dowden, Hutchison and Ross, Stroudsburg, PA, pp 232–237

Glanz W (1982) Adaptive zones of neotropical mammals: a comparison of some temperate and tropical patterns. In Mares MA, Genoways HH (eds) *Mammalian Biology in South America*. Vol 6, Spec Publ Ser, Pymatuning Lab Ecol, U Pittsburgh, Linesville, PA, pp 95–110

Hurlbert SH (1971) The non-concept of species diversity: A critique and alternative parameters. Ecology 52:577–586

Iriarte JA, Contreras LC, Jaksic FM (1989) A long-term study of a small-mammal assemblage in central Chilean matorral. J Mam 70:79–87

Meserve PL, Glanz WE (1978) Geographical ecology of small mammals in the northern Chilean arid zone. J Biogeog 5:135–148

Meserve PL (1976) Food relationships of a rodent fauna in a California coastal sage scrub community. J Mam 57:300–319

Mooney HA (1977) *Convergent Evolution in Chile and California: Mediterranean Climate Ecosystems*. Dowden, Hutchinson and Ross, Stroudsburg, PA

Newsome AE, Catling PC (1983) Animal demography in relation to fire and shortage of food: some indicative models. In Kruger FJ, Mitchell DT, Jarvis JUM (eds) *Mediterranean-Type Ecosystems: The Role of Nutrients*. Springer-Verlag, Berlin, pp 490–505

Quinn RD (1979) Effects of fire on small mammals in the chaparral. Cal-Neva Wild Trans 1979:125–133

Read DJ, Mitchell DT (1983) Decomposition and mineralization processes in mediterranean-type ecosystems and in heathlands of similar structure. In Kruger FJ, Mitchell DT, Jarvis JUM (eds) *Mediterranean-Type Ecosystems: The Role of Nutrients*. Springer-Verlag, Berlin, pp 208–232

Simpson EH (1949) Measurement of diversity. Nature 163:688

SPSS/X (1983) *A Complete Guide to SPSSX Language and Operations*. McGraw-Hill, New York

Thrower NJW, Bradbury DE (eds) (1977) *Chile-California Mediterranean Scrub Atlas: A Comparative Analysis*. Dowden, Hutchison and Ross, Stroudsburg, PA

Tilman D (1982) *Resource Competition and Community Structure*. Princeton U Press, Princeton, NJ

Wilson BA, Rogerson DJ, Moloney DJ., Newell GR, Laidlaw WS (1990) Factors affecting small mammal distribution and abundance in the Eastern Otway Ranges. Proc Ecol Soc Aust, 16:379–396

Wirtz WO, II, Hoekman D, Muhm JR, Souza SL (1988). Postfire rodent succession following prescribed fire in southern California chaparral. In Proc Sym Man amphibians, reptiles, small mammals in North America. USDA, For Ser, Rocky Mt For Range Exp Sta General Technical Report RM-166, pp 333–339

16. Role of Fossorial Animals in Community Structure and Energetics of Pacific Mediterranean Ecosystems

George W. Cox, Luis C. Contreras, and A.V. Milewski

Recent reviews reveal that fossorial animals (those both foraging and sheltering underground) have influential, or even keystone, roles in the dynamics of grassland, savanna, and mediterranean scrub ecosystems (Andersen 1987; Huntly and Inouye 1988; Contreras and McNab 1990). Fossorial animals include termites and other insects, insectivorous mammals such as moles, and members of several families of herbivorous rodents, including the Geomyidae of North America and the Octodontidae of Chile (Nevo 1979; Andersen 1987).

Fossorial rodents may be the major herbivores in mediterranean ecosystems. Though they usually forage for a diverse array of species, sometimes they concentrate on plants of a specific structure or life form, thus influencing vegetational structure and composition. Examples of this influence include restricting geophytes (i.e., herbaceous plants that die back each year to underground tubers, bulbs, or corms) to microhabitats that cannot be exploited by pocket gophers (Cox and Allen 1987a) and excluding woody plants from montane meadows occupied by these animals (Cantor and Whitham 1989).

Fossorial animals extensively disturb the soil by tunneling and depositing mined earth on the surface. These activities affect the establishment and survival of herbaceous plants. In time, mining and translocation of soil may also modify landscape topography, creating large-scale patterns of mounds with level or depressed intermound areas (Cox and Scheffer 1991).

In this chapter we examine the different patterns of fossorial animal activity in California, Chile, and Australia. Our first objective is to synthesize information on the influence of fossorial animals on vegetational structure and energetics in mediterranean ecosystems in these three regions, where plant and animal taxa are independently derived and differ markedly in degree of convergence. Second, we examine the biogeographic and evolutionary history of relationships involving fossorial animals, and interpret the degree of coevolutionary relationship of these species with plants in their respective ecosystems.

California: Herbivory by Solitary, Territorial Pocket Gophers of the Genus *Thomomys*

The genus *Thomomys* (Geomyidae) comprises six species of pocket gophers that range from central Mexico to southern Canada and from the Pacific Coast to the western portion of the Great Plains. In California, the valley pocket gopher, *T. bottae*, is widespread at low and intermediate elevations and the similar but less studied *T. monticola* and *T. mazama* occur in the central and northern mountains. We focus here on the valley pocket gopher, which occupies grasslands, shrublands, and woodlands with the most strongly developed mediterranean climate.

Adult valley pocket gophers weigh 40 to 200 g, males being larger than females. The valley pocket gopher is a solitary, territorial, herbivorous, fossorial rodent of arid to semiarid low and middle elevations in the southwestern United States and northwestern Mexico. The species inhabits grasslands, deserts, open shrubland and woodland, and the borders of salt marshes, occupying soils ranging from nearly pure sands to heavy clays. Adults defend individual burrow systems against other individuals; the animals thus are uniformly spaced when populations are dense (Howard and Childs 1959; Hansen and Remmenga 1961). Males and females share burrows for short periods only during the mating season. Densities of the species range up to 75 per hectare.

Pocket gophers dig and maintain extensive tunnel systems for foraging and for seeking mates. Valley pocket gophers are able to vary their foraging mode depending on the cost of digging and the availability of various food materials. On the coastal San Diego County mesas, the two principal foraging modes, both highly seasonal, are extended tunneling and surface-access tunneling (Cox and Hunt 1992) (Figure 16.1). Extended tunnels are long and horizontal and vary in depth, and are marked at intervals by surface heaps consisting of excavated soil. These tunnels lead from the nest area to outer portions of the territory or to the tunnels of potential mates. Extended tunneling is carried out primarily during winter and early spring, when the soil is moist and soft, and the energy cost of excavation is low. The foods harvested by extended tunneling consist of forb and grass roots and

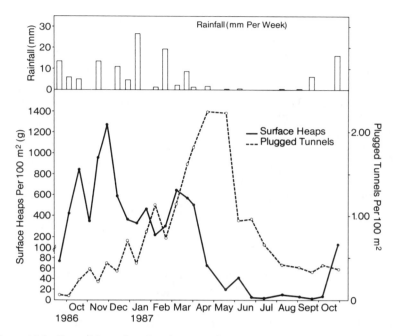

Figure 16.1. Deposition of surface heaps and construction of surface-access tunnels by pocket gophers on a southern California coastal mesa site in relation to rainfall, 1986–1987.

the corms of herbaceous species such as *Brodiaea* and *Muilla*. Shoot materials too are harvested around and openings dug to permit surface deposition of heaps.

Unlike extended tunnels, surface-access tunnels are short, dug laterally and upward from segments or ends of extended tunnels to openings at the surface. These tunnels often are not associated with surface heaps, indicating that the soil excavated is redeposited below ground. This mode of foraging permits the harvest of living or dead plant material near the surface opening, typically no farther than 10 cm. The animal probably does not completely leave the tunnel when engaged in this type of foraging. In winter and spring, leaves of herbaceous plants such as *Erodium*, *Hypochoeris*, and *Avena* are harvested in this manner. Surface-access tunneling is the primary mode of foraging in summer when dry and hard soils make the cost of excavation high (Figure 16.1). At this time materials such as *Erodium* and *Avena* seeds and dried grass shoots are taken.

The bulk of the valley pocket gopher's diet consists of the shoots of forbs and grasses, along with roots, corms, and other below-ground plant parts (Hunt 1989, 1992). Shoots and roots of shrubs and cacti are also eaten in some locations.

Pocket gophers store foods such as seeds, roots, and corms in underground caches. Cache formation may be opportunistic, and may occur when large quantities of storable foods can be acquired by expending little energy. On the San Diego mesas, extended tunneling is probably the primary foraging mode that yields corms and other storable below-ground plant materials in winter and spring. Caches may hold a few hundred grams of roots and corms. In summer, surface-access tunneling enables pocket gophers to form caches of 50 to 100 grams of *Erodium* seeds.

That caching is opportunistic is suggested by the many caches that are in backfilled sections of the tunnel system (G.W. Cox, pers. obs.). Indeed, the major *Erodium* caches are in surface-access tunnel complexes at the ends of extended tunnels that have been extensively backfilled with loose soil. These caches appear to be untended, and are near the surface where summer dryness impedes spoilage. Caches of this type may be exploited only in late summer and autumn when hard soil is combined with unavailability of surface foods. The loosely backfilled tunnels then may be reexcavated easily, allowing animals to utilize the cached foods.

Herbivory of below-ground parts of native perennials, including woody plants, can limit their abundance and distribution in grasslands (Cox and Allen 1987a; Cox 1989). On the San Diego coastal mesas, geophytes such as *Brodiaea* and *Chlorogalum* are frequently restricted to stony vernal pool basins or other shallow, stony soils that pocket gophers can penetrate only with difficulty. In northern Arizona, root herbivory by pocket gophers appears to keep trembling aspen, *Populus tremuloides*, from colonizing mountain meadows (Cantor and Whitham 1989).

Surface heaps deposited by pocket gophers influence composition of the plant community by acting as colonization sites for plants (Laycock 1958; Foster and Stubbendieck 1980; Grant and McBrayer 1981; Hobbs and Mooney 1985; Peart 1989). On the San Diego coastal mesas, sites extensively disturbed by tunneling and deposition of heaps are dominated by introduced annual grasses and forbs, suggesting that pocket gophers help maintain dominance by these species. Deposition of heaps by pocket gophers is a major way in which diversity is maintained in annual grasslands (Hobbs and Hobbs 1987; Peart 1989).

Individual surface heaps are generally a fraction of a square meter in area, so that they often influence plant-community structure by creating fine-grained patchiness. Sometimes, however, pocket gophers intensively and repeatedly exploit areas several to many square meters in extent. In these locations, repeated disturbance maintains a low-diversity community of annuals. This phenomenon is similar to that of grazing lawns maintained by large above-ground herbivores.

Estimates of the quantity of soil annually mined and deposited in surface heaps by species of *Thomomys* range from 11.2 to 104.9 Mg/ha, with the highest values coming from animals living in irrigated agricultural fields (Cox 1990). Surface deposits are only a portion of the soil mined (Andersen 1987; Thorne and Andersen 1989). On the San Diego coastal marine terraces, Cox (1990) estimates that subsurface redeposition by valley pocket gophers is nearly 2.5 times that of surface deposition, and that total annual mining equals about 28.5 Mg/ha or about 21.8 m^3/ha, assuming a bulk density of 1.31 g/cm^3. If most of this mining occurs in the surface 20-cm zone, it is equivalent to an annual turnover of slightly more than 1% of the soil.

Many areas where pocket gophers are abundant also have Mima mounds. These are earth mounds that range up to about 2 m in height, 40 m in diameter, and 100 per ha in density. Strong evidence suggests that these mounds were formed by the activity of pocket gophers (Cox 1984; Cox and Gakahu 1986, 1987; Cox et al. 1987, 1989; Cox and Scheffer 1991). Mima mounds occur where poorly drained or shallow unmounded soil would have severely limited the availability of sites suitable for pocket-gopher nests. Any sites that were favorable are therefore hypothesized to have been occupied generation after generation, with outward tunneling from these centers causing backward displacement of soil that gradually built mounds. Evidence for this activity comes from quantitative and experimental study of soil mining and translocation by pocket gophers (Cox and Allen 1987b; Cox 1990) and from analysis of mound-field geometry western North America (Cox and Gakahu 1986; Cox et al. 1987).

Recent studies in California quantitatively show how soil mined by pocket gophers is translocated moundward. The average distance that a unit of mined soil is moved toward the mound center increases outward from the center but decreases as the mound grows in height (Cox and Allen 1987b). When these findings are coupled with estimates of the volume of soil mined at various distances from mound centers, most mound growth can be accounted for by soil mining and translocation by pocket gophers (Cox 1990).

Mima mounds greatly influence plant community structure because of their deep, loamy soils and favorable moisture relations. In coastal southern California, these mounds, without frequent fire, are foci for establishment by various suffrutescent and woody shrubs (Cox 1986). They also permit upland plants to extend their distribution into coastal salt marshes (Cox and Zedler 1986).

Chile: Herbivory by *Spalacopus cyanus*, the Colonial Coruro

The coruro, *Spalacopus cyanus* (Octodontidae), is a fossorial rodent restricted to central Chile, where it occurs in coastal regions from Caldera (27°03′ S), near the southern edge of the Atacama Desert, to Quirihue (36°17′ S), north of Concepción, and in the Andes east of Santiago from Alicahue (32°19′ S) to Los Cipreses (34°01′ S), as well as in scattered localities on slopes of the intervening central-valley region (Contreras et al. 1987). Adults weigh 80 to 120 g, those from lower altitudes being smaller and showing more sexual dimorphism than those from high Andean localities

(Contreras 1986). The coruro occupies habitats ranging from alpine grass-land in the Andes to *Acacia* savanna in the central valley and coastal stabi-lized dunes and sandy grasslands. In all these communities woody plants cover less than about 60% of the ground surface (Contreras et al. 1987). When coruros live in habitats with high shrub cover, their burrows are lo-cated in open intershrub areas covered by herbaceous plants.

The coruro is colonial, living in groups of about 6 to 15 individuals (Reig 1970; Torres-Mura 1990). Coruros dig extensive foraging tunnels like those of pocket gophers. Reig (1970) considers the colonies "nomadic," in the sense that they quickly exploit the food resources at one site and then shift to a new one. Continuing studies of colonies in an arid mediterranean envi-ronment, however, indicate that they occupy the same general area for long periods, but shift their foraging locations extensively within this area.

Coruros feed extensively on corms of geophytes such as huilli, *Leuco-coryne ixioides* (Alliaceae), and *Rhodophialia* spp. (Amaryllidaceae) (Reig 1970; Torres-Mura 1990). Geophytes are abundant in arid and semiarid regions in Chile (Hoffmann and Kummerow 1978). Tunneling in open areas where geophytes grow is not random, but may be guided by visual inspection of the ground surface, for coruros often survey their surroundings from the mouths of tunnels in daylight (Contreras and Gutierrez 1991). Older litera-ture anecdotally indicates that coruros cache large quantities of bulbs or corms (Mann 1945; Housse 1953). In a few recent excavations of coruro tunnel systems, however, only small quantities of bulbs (4 to 6) have been found in chambers (Torres-Mura 1990). Coruros also eat shoots of grasses and forbs, when these are available around the openings of their tunnels (L.C. Contreras pers. obs.).

Soil mining by a coruro colony was investigated over nine months at Fray Jorge National Park (30°38′ S, 71°40′ W). This study suggested that tunnel-ing can be extensive for much of the year. The cumulative volume of soil deposited in surface heaps increased progressively from October 1990 through June 1991 (Figure 16.2). Tunneling and harvesting of the geophyte foods encountered appears to be the primary mode of foraging at this loca-tion. The colony members excavate enough soil that they encounter subter-ranean plant foods sufficient to supply their energetic needs. Preliminary studies indicate that in the dry season (January–February) and in dry years, huilli corms form the bulk of the diet at Fray Jorge National Park.

In a semiarid coastal site in north-central Chile (30°06′ S, 71°21′ W), ex-ploitation of geophytes by coruros appeared to stimulate reproduction in at least some of the plants that are primary coruro foods (Contreras and Gutierrez 1991). Harvesting depleted the abundance of large huilli corms, but stimulated germination and perhaps growth of young plants, thus lead-ing to an increase in density of small corms. The interval between harvests in an area may depend on the time required for geophytes to mature (Torres-Mura 1990). Soil mining and deposition by coruros also promoted the growth of annuals such as *Mesembryanthemum crystallinum*, a prostrate,

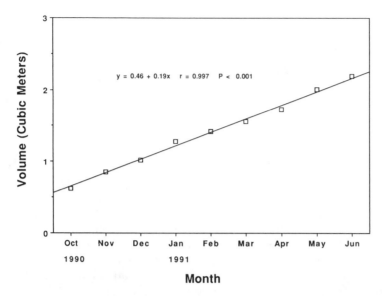

Figure 16.2. Cumulative volume of soil deposited in surface heaps by members of a coruro colony at Fray Jorge National Park, Chile, from October 1990 through June 1991.

introduced succulent that occurs on soils disturbed by various processes, but not on undisturbed soils. The abundance of this species in areas with coruro burrows was more than four times that in areas free of burrows, contributing to a total dry biomass of herbaceous plants more than 60% greater than in burrow-free areas (Contreras and Gutierrez 1991).

Only preliminary estimates of soil mining, translocation, and redeposition have been obtained for coruros. The abundance of coruro colonies varies greatly. In some locations, colonies and their areas of disturbance are hundreds of meters apart; in others, closely packed colonies create continuous disturbance for many kilometers. At Fray Jorge National Park, surface-heap deposition by members of one colony was 60 to 100 heaps per month over a 9-month period, corresponding to an average soil volume of about 2.5 m³/ year. The amount of subsurface redeposition of mined soil is unknown. For a density of three colonies per hectare, the estimate above for mining by one colony suggests a total of about 7.5 m³/ha or 9.83 Mg/ha, assuming a bulk density of 1.31 g/cm³ for mined soil. Thus, soil mining by coruros in extensive and probably comparable to that of most North American pocket gophers.

No evidence is reported for the formation mounds of Mima type by coruros in low-elevation sites in Chile, even over shallow hardpan comparable to that of coastal mesa soils in southern California. At low elevations, where their habitat is extensive, tunneling by coruro colonies appears to be spread over wide areas and may shift greatly from month to month. As a result, coruros in most locations do not show patterns of soil translocation toward activity centers that are occupied for many generations. In narrow, streamside meadows at Farellones, 3000 m high in the Andes, however, we have observed structures that may be low Mima mounds measuring 11.25 to 13.50 m in diameter and 0.69 to 0.77 m in height. Thus, Mima mounds may be formed where habitat is suitable, but they are not extensive.

Australia: Absence of Fossorial Mammalian Herbivores

Fossorial rodents are lacking and native rodents scarce in Australia (Watts and Aslin 1981). The only small fossorial mammal is the desert-inhabiting, insectivorous marsupial mole *Notoryctes* (Notoryctidae). No fossil record is known for *Notoryctes* (Nevo 1979; Strahan 1983). The taxonomic relationship of *Notoryctes* to other marsupials is uncertain. *Notoryctes* evidently represents an ancient divergence in marsupial evolution (Archer 1981). Australian rodents all belong to the family Muridae, which probably reached the continent only about 4 to 5 million years B.P. (Lee et al. 1981).

Not only are fully fossorial herbivorous mammals absent in mediterranean Australia, but also small terrestrial grazers and small seed-eaters are poorly represented (Milewski and Cowling 1985). The smallest grazing or browsing mammals are hare-wallabies (*Lagostrophus* spp.) and small wallabies of the genus *Macropus*, which are 2 to 5 kg in body mass. The vertebrates that eat geophytes are omnivores or generalized feeders of low abundance or wide-ranging behavior. These include several species of rat-kangaroos (Potoridae, 1 to 2 kg in body mass) and bandicoots (Peramelidae, 0.5 to 1.6 kg in body mass). The large, parrotlike corellas (*Cactua*, 0.4 to 0.9 kg in body mass) are omnivorous, and habitually excavate shallow fungi and geophytes (Strahan 1983; Forshaw 1989; Seebeck et al. 1989). The swamp rat, *Rattus lutreolus* (120 to 140 g in body mass) of eastern and southeastern Australia and Tasmania feeds mainly on stem bases and rhizomes of sedges (Braithwaite and Lee 1979). The bush rat, *Rattus fuscipes greyi* (ca. 80 g.) is moderately abundant in some mediterranean areas in South Australia, and is omnivorous, consuming some shallowly buried tubers (Watts and Aslin 1981). Australian consumers of geophytes are not known to cache their food.

The fossorial herbivore niche in Australia is partially filled by harvester termites, at least along the dry tropical fringe in the mediterranean zone. Species of *Amitermes* and *Drepanotermes*, which harvest the dead aboveground parts of grasses and other herbaceous plants, excluding geophytes, are present in greater diversity and biomass than in mediterranean regions of Chile and California (Watson 1982).

Some portion of the herbaceous productivity of mediterranean vegetation is harvested by large marsupials (*Macropus*, *Vombatus*), although these animals were never abundant. Under aboriginal conditions, much of the sparse herbaceous growth was burned.

Role of Fossorial Animals in Community Structure

Observations of pocket gophers and coruros indicate that soil disturbance by fossorial rodents promotes dominance by annual plants. At present in California, these plants are primarily introduced mediterranean annuals and are among the animals' primary foods. In Chile, although moderate disturbance by coruros may increase diversity in the plant community, heavy disturbance may promote dominance by aggressive annuals and reduce species diversity (Contreras and Gutierrez 1991). Thus, lack of fossorial mammals may have contributed to the poor representation of annuals in the native flora and vegetation in mediterranean regions in Australia.

Pocket-gopher herbivory limits local distribution of some perennial plants in North America. This effect is particularly evident on the Columbia Plateau in eastern Washington and Oregon, where some highly preferred geophytes are limited to stony habitats that are difficult for pocket gophers to enter (Cox and Allen 1987a; Cox 1989). As we have mentioned, in California some geophytes are also preferred by pocket gophers. Conversely, in Chile exploitation by coruros apparently promotes increased recruitment of some geophytes, as it has been reported that bathyergid mole rates do in South Africa (Lovegrove and Jarvis 1986).

In Australia, vertebrates neither restrict the distribution of geophytes nor aid in their propagation. Except for *Dioscorea* and *Drosera*, which are particularly palatable and nutritious, Australian geophytes lack bulbils or cormlets that similar plants in California and Chile often possess (Pate and Dixon 1982). These small organs may be shed or dispersed when a bulb or corm is harvested by an animal. But many Australian geophytes, particularly orchids, have root tubers that are easily killed by exposure or fragmentation.

Role of Fossorial Animals in Ecosystem Energetics

Populations of fossorial animals may be the principal herbivores in mediterranean ecosystems with an open canopy and a well-developed herbaceous stratum. Pocket gophers occur in moderately dense populations in mediterranean North America, where their energy consumption may equal or exceed that of above-ground grazing animals. Gettinger (1984), for example, estimated that a population of about 45 valley pocket gophers/ha in a southern California mountain area would harvest about 470 kg/ha of plant material annually, consuming about 2200 MJ/ha. This estimate is higher than that obtained for Utah mountain meadows, where Andersen and MacMahon (1981) estimated annual energy consumption at 636 MJ/ha for a population ranging in density between 4.4 and 62.5 animals/ha. The significance of coruro populations in energy flow through Chilean ecosystems remains unevaluated. Reig (1970), however, recorded a density of about three colonies (about 18 to 45 total animals) per hectare, in a coastal locality ($32°35'$ S, $71°26'$ W) between Valparaíso and Papudo. This density of animals suggests that the role of coruros in ecosystem energetics is similar to that of pocket gophers.

Fossorial mammals also influence the seasonal pattern of primary production by promoting annuals in the vegetation. Human beings have accentuated the influence of fossorial rodents by reducing populations of their predators and by introducing European annuals that are highly responsive to disturbance and, at least in California, are major foods for fossorial rodents.

Mima mounds intensify heterogeneity of primary productivity across the landscape by creating patches of deep-mound soils and thin intermound soils. In drier parts of western North America, for example, the maximum herbaceous biomass achieved on mounds is 1.5 to 3.1 times that of intermound areas (McGinnies 1960; Allgood and Gray 1974).

Biogeographic and Evolutionary Origins

The mediterranean areas in California, Chile, and Australia differ greatly in their geological background and in the evolutionary histories of their vegetation and herbivores. Fossorial mammalian herbivores evolved in the Americas, but not in Australia.

In California, fossorial mammals are in the genus *Thomomys*, which has a broad geographic and ecological range. Although fossorial mammals, especially the Geomyidae, are notorious for their variability in morphological characters of many sorts (Patton and Smith 1990), much of this variability may represent adaptation to local conditions of habitat and food. Members of the genus *Thomomys* appear to be adaptable generalists that can switch modes of food exploitation and can adjust patterns of life history to varied habitat conditions with minor genetic changes. Patton and Brylski (1987), for example, found that adjacent genetically similar populations of the valley pocket gopher in desert and irrigated alfalfa fields differed in body size, sexual dimorphism, sex ratio, fecundity, and population density.

The evolutionary history of geomyid rodents in North America extends back to the Miocene. Unfortunately, the affinity of many fossil taxa to modern forms remains uncertain. The genera *Geomys*, *Pappogeomys*, and *Thomomys* (Subfamily Geomyinae) data back only to early Blancan time (early Pliocene, about 3.5 million years B.P.), however, and appear in the North American fossil record with complete suites of derived characteristics (Patton and Smith 1990; J.H. Wehlert, pers. comm.). This finding suggests that the major evolutionary steps in the origin of these modern forms occurred elsewhere. Mexico and Central America are the current centers of diversity for the Geomyidae, and possibly for geomyid evolution (Wehlert and Souza 1988). The forms that have reached temperate North America may thus be adaptable colonizing species that are the most recent expansions from this center. This description may be particularly applicable to *Thomomys*, which includes species that have invaded the most arid habitats occupied by members of the family Geomyidae. Retention of generalist characteristics of colonizing species also may have been favored by Pleistocene glacial cycles that alternately opened and closed vast areas in western North America to *Thomomys* occupancy.

Chile, on the other hand, may exemplify a situation of long evolution by fossorial rodents and geophytes within the region itself, resulting in some mutual evolutionary adjustment between coruros and their food plants. The flora of this region is rich in geophytes (Hoffmann 1989) and herbaceous plants, and the soils rich in nutrients (Miller and Hajek 1981). Early octodontid rodents were present in southern South America in the Oligocene, about 35 million years B.P., prior to the Andean orogeny (Contreras et al. 1987). It appears likely that *Spalacopus* evolved in the Chilean region, contemporaneously with development of grassland and open mediterranean scrub vegetation along the western slopes of the Andes during the Pliocene and Pleistocene. Thus, *Spalacopus* may have become more specialized for exploiting geophytes than has *Thomomys* in North America, and Chilean geophytes, in turn, may have evolved adjustments to intensive herbivory by *Spalacopus*.

Understanding the ecological history of the mediterranean regions in Australia requires that one consider the conditions and history of the continent as a whole (Braithwaite 1990). A major feature of the continent is ancient, heavily weathered, nutrient-deficient soils (Lindsay 1985). This characteristic has evidently placed a high evolutionary premium on mechanism for acquiring and retaining nutrients. Australian herbaceous plants, including geophytes, have frugal nutrient economies, and show many specialized mechanisms for phosphorus and nitrogen uptake, such as microbial symbioses and carnivory (Specht 1973; Beadle 1981; Beard 1983).

Oligotrophy in Australia is associated with low availability of palatable geophytes and grasses. Most families of geophytes are absent from the mediterranean regions in Australia (Table 16.1). In particular, Australia lacks members of the family Alliaceae, which are abundant in both California and Chile and are among the geophytes most palatable to mammals. The cover, biomass, and productivity of geophytes are less in Australia than in any other mediterranean region, whether in mature or recently burned stands (Gardner 1949; Levyns 1961; Gill et al. 1981). On the arid fringe of the mediterranean zone, geophytes make up less than 1% of species (Beadle 1981). The treeless Nullarbor Plain has only one geophyte genus, with two species, and is probably the poorest winter-rainfall area in the world for geophytes. The Mojave Desert in California, comparable in area and cli-

Table 16.1. Genera of geophytes with large (exceeding 2 cm in maximum diameter) bulbs (B), corms (C), or rhizome tubers (T) in mediterranean areas in Chile, California, South Australia, and Western Australia (Hoffmann 1989; Munz 1959; Pate and Dixon 1982)

Family	Chile	California	South Australia	Western Australia
Alliaceae	Leucocoryne (B) Brodiaea (B) Gilliesia (B) Solaria (B) Nothoscordum (B) Tristagma (B) Ipheion (B) Pabellonia (B) Zoellnerallium (B) Gethyum (B) Miersia (B)	Allium (B) Brodiaea (C) Erythronium (C) Muilla (C)		
Alstroemeriaceae	Alstroemeria (T) Bomarea (T)			
Calochortaceae		Calochortus (B)		
Amaryllidaceae	Placea (B) Phycella (B) Rhodophialia (B) Traubia (B)		Calostemma (B)	
Dioscoreaceae	Dioscorea (T)		Dioscorea (T)	Dioscorea (T)
Haemodoraceae				Haemodorum (B)
Hyacinthaceae	Camassia (B)	Camassia (B) Chlorogalum (B)		
Iridaceae	Calydorea (C) Herbertia (B)			
Liliaceae		Lilium (B) Fritillaria (B) Zigadenus (B)		
Melanthiaceae		Odontostomum (C)		
Tecophilaeaceae	Conanthera (C) Tecophilaea (C) Zephyra (C)			

mate, is rich in geophytes. Mediterranean Western Australia lacks Amaryllidaceae, widespread elsewhere. Native grasses are scarce, both in number of species and in biomass. The perennial grasses that do occur are unpalatable and poorly adapted to heavy grazing. Annual herbaceous plants are also scarce and many of them unpalatable (Specht 1973; Baird 1977).

The absence of fossorial mammalian herbivores thus reflects the scarcity of suitable food plants over the long history of mammalian evolution in Australia. The scarcity and nutritional inadequacy of herbaceous foods, both geophytes and nongeophytes, have not allowed fossorial mammals to specialize for exploiting them. Termites may have partially exploited this niche through their ability, aided by gut symbionts, to consume plant materials with a high C : N ratio and to fix N. The large marsupial grazers consume a portion of the production of herbaceous plants, also aided by gut symbionts. Other portions of herbaceous productivity are consumed by vertebrate omnivores, terrestrial insects such as grasshoppers, and fire.

Conclusions

The mediterranean regions in California, Chile, and Australia have presented different conditions for evolution of fossorial herbivores. The vegetational characteristics of California and Chile have permitted evolutionary specialization by small mammals for this niche. This specialization has not appeared in Australia, probably because geophyte foods are scarce, even though an insectivorous mammal highly adapted for fossorial life did evolve. In North America, perhaps favored by repeated climatic and vegetational shifts associated with continental glaciation, geomyid pocket gophers have remained generalist feeders on many herbaceous and woody plants. In Chile, perhaps favored by long biogeographic isolation, coruros have evolved specialized dependence on geophyte foods, and geophyte species in turn have made some evolutionary adjustment to coruro herbivory.

Acknowledgments

Collaborative studies of the fossorial mammal's role in mediterranean ecosystems in California and Chile are being supported by grant INT-8919473 from the U.S. National Science Foundation. In Chile, this work has also been funded by grants FONDECYT 90/930, DIULS 120-2-67, and U.S. National Science Foundation BSR 90-20047. We thank CONAF (Corporación Nacional Forestal) IV Region for permission to work and use facilities in Fray Jorge National Park. We are grateful to J.R. Gutierrez, V. Valverde, and O. Contreras for permitting us to use unpublished data on coruros, and M. Muñoz for providing information on the form and size of Chilean geophytes. We also thank Robert Black, Richard Braithwaite, A.R. Main, John H. Wahlert, and Kathy S. Williams for critical comments on earlier drafts of this manuscript.

396   G.W. Cox, L.C. Contreras, and A.V. Milewski

References

Allgood FP, Gray F (1974) An ecological interpretation for the small mounds in landscapes of eastern Oklahoma. J Env Qual 3:37–41.

Andersen DC (1987) Below-ground herbivory in natural communities: A review emphasizing fossorial animals. Quart Rev Bio 62:261–286

Andersen DC, MacMahon JA (1981) Population dynamics and bioenergetics of a fossorial herbivore, *Thomomys talpoides* (Rodentia: Geomyidae), in a spruce-fir sere. Ecol Mono 51:179–202

Archer M (1981) A review of the origins and radiations of Australian mammals. In Keast A (ed.) *Ecological Biogeography of Australia.* Vol. 3. W. Junk, The Hague, pp 1435–1488

Baird AM (1977) Regeneration after bushfire in King's Park, Perth, Western Australia, J Royal Soc W. Australia 60:1–22

Beard JS (1983) Ecological control of the vegetation of southwestern Australia: moisture versus nutrients. In Kruger FJ, Mitchell DT, Jarvis JUM (eds) *Mediterranean-Type Ecosystems: The Role of Nutrients.* Springer-Verlag, New York, pp 66–73

Beadle NCW (1981) The vegetation of the arid zone. In Keast A (ed.) *Ecological Biogeography of Australia.* W. Junk, The Hague, pp 695–716

Braithwaite RW (1990) Australia's unique biota: Implications for ecological processes. J Biogeog 17:347–354

Braithwaite RW, Lee AK (1979) The ecology of *Rattus lutreolus.* I. A Victoria heathland population. Aust J Wild Res 6:173–189

Cantor LF, Whitham TG (1989) Importance of belowground herbivory: Pocket gophers may limit aspen to rock outcrop refugia. Ecology 70:962–970

Contreras LC (1986) Bioenergetics and distribution of fossorial *Spalacopus cyanus* (Rodentia): Thermal stress or cost of burrowing? Physiol Zool 59:20–28

Contreras LC, Gutierrez JR (1991) Effects of the subterranean herbivorous rodents *Spalacopus cyanus* on herbaceous vegetation in arid coastal Chile. Oecologia 87:106–109

Contreras LC, McNab BK (1990) Thermoregulation and energetics of subterranean mammals. In Nevo E, Reig O (eds) *Evolution of Subterranean Mammals at the Molecular and Individual Level.* Wiley-Liss, New York, pp 231–250

Contreras LC, Torres-Mura JC, Yañez JL (1987) Biogeography of octodontid rodents: An eco-evolutionary hypothesis. Fieldiana (New Series) 39:401–411

Cox GW (1984) The distribution and origin of Mima mound grasslands in San Diego County, California. Ecology 65:1397–1405

Cox GW (1986) Mima mounds as an indicator of the presettlement grassland–chaparral boundary in San Diego County, California. Am Mid Nat 116:64–77

Cox GW (1989) Early summer diet and food preferences of northern pocket gophers in north central Oregon. Northwest Sci 63:77–82

Cox GW (1990) Soil mining by pocket gophers along topographic gradients in a Mima moundfield. Ecology 71:837–843

Cox GW, Allen DW (1987a) Sorted stone nets and circles of the Columbia Plateau: A hypothesis. Northwest Sci 61:179–185

Cox GW, Allen DW (1987b) Soil translocation by pocket gophers in a Mima moundfield. Oecologia 72:207–210

Cox GW, Gakahu CG (1986) A latitudinal test of the fossorial rodent hypothesis of Mima mound origin. Zeit Geomorph 30:485–501

Cox GW, Gakahu CG (1987) Biogeographical relationships of mole rats with Mima mound terrain in the Kenya highlands. Pedobiologia 30:263–275

Cox GW, Hunt J (1992) Relation of seasonal activity patterns of valley pocket gophers to temperature, rainfall, and food availability. J Mam 73:123–134

Cox GW, Scheffer VB (1991) Pocket gophers and Mima terrain in North America. Nat Areas J 11:193–198

Cox GW, Zedler JB (1986) The influence of Mima mounds on vegetation pattern in the Tijuana Estuary salt marsh, San Diego County, California. Bull S Calif Acad Sci 85:158–172

Cox GW, Gakahu CG, Allen DW (1987) The small stone content of Mima mounds of the Columbia Plateau and Rocky Mountain regions: Implications for mound origin. Great Basin Nat 47:609–619

Cox GW, Gakahu CG, Waithaka JM (1989) The form and small stone content of large earth mounds constructed by mole rats and termites in Kenya. Pedobiologia 33:307–314

Forshaw JM (1989) *Parrots of the World*, 3rd ed. Blandford Press, London

Foster MA, Stubbendieck J (1980) Effects of the plains pocket gopher (*Geomys bursarius*) on rangeland. J Wild Man 33:74–78

Gardner CA (1949) The vegetation of Western Australia. J Royal Soc W Aust 28:11–87

Gettinger RD (1984) Energy and water metabolism of free-ranging pocket gophers, *Thomomys bottae*. Ecology 65:740–751

Gill AM, Groves RH, Noble IR (1981) *Fire and the Australian Biota*. Aust Acad Sci, Canberra

Grant WE, McBrayer JF (1981) Effects of mound formation by pocket gophers (*Geomys bursarius*) on old field ecosystems. Pedobiologia 22:21–28

Hansen RM, Remmenga EE (1961) Nearest neighbor concept applied to pocket gopher populations. Ecology 42:812–814

Hobbs RJ, Hobbs VJ (1987) Gophers and grassland: A model of vegetation response to patchy soil disturbance. Vegetatio 69:141–146

Hobbs RJ, Mooney HA (1985) Community and population dynamics of serpentine grassland annuals in relation to gopher disturbance. Oecologia 67:342–351

Hoffmann AE (1989) Chilean geophyte monocotyledons: taxonomic synopsis and conservation status. In Benoit IL (ed) *Red Book of Chilean Terrestrial Flora*. CONAF, Santiago, pp 141–151

Hoffmann A, Kummerow J (1978) Root studies in the Chilean matorral. Oecologia 32:57–69

Housse PR (1953) *Animales Salvajes de Chile en su Clasificación Moderna*. Ediciones U Chile, Santiago

Howard WE, Childs HE (1959) Ecology of pocket gophers with emphasis on *Thomomys bottae mewa*. Hilgardia 29:277–358

Hunt J (1989) *Feeding Ecology of* Thomomys bottae *in a Mima Mound–Grassland Habitat in San Diego County, California*. M.S. Thesis, San Diego State U, San Diego, CA

Hunt J (1992) Feeding ecology of valley pocket gophers (*Thomomys bottae sanctidiegi*) on a California coastal grassland. Am Mid Nat 127:41–51

Huntly NH, Inouye RS (1988) Effects of gophers (*Geomys bursarius*) on ecosystems. BioScience 38:786–793

Laycock WA (1958) The initial pattern of revegetation of the pocket gopher mounds. Ecology 39:346–351

Lee AK, Baverstock PR, Watts CHS (1981) Rodents: The late invaders. In Keast A (eds) *Ecological Biogeography of Australia*. Vol. 3. W. Junk, The Hague, pp 1521–1553

Levyns MR (1961) Some impressions of a South African botanist in temperate Western Australia. J S African Bot 27:87–97

Lindsay AM (1985) Are Australian Soils Different? Proc Ecol Soc Aust 14:83–97

Lovegrove BG, Jarvis JUM (1986) Coevolution between mole-rats (Bathyergidae) and a geophyte, *Micranthus* (Iridaceae). Cimbebasia 8:79–85

McGinnies WJ (1960) Effect of Mima-type relief on herbage production of five seeded grasses in western Colorado. J Range Man 13:231–234

Mann G (1945) Mamíferos de Tarapacá. Observaciones realizadas durante una expedición al alto norte de Chile. Biologica 2:23–134

Milewski AV, Cowling RM (1985) Anomalies in the plant and animal communities in similar environments at the Barrens, Western Australia, and the Caledon Coast, South Africa. Proc Ecol Soc Aust 14:199–212

Miller PC, Hajek E (1981) Resource availability and environmental characteristics of mediterranean type ecosystems. In Miller PC (ed) *Resource Use by Chaparral and Matorral*. Springer-Verlag, New York, pp 17–41

Munz PA (1959) *A California Flora*. California Press, Berkeley

Nevo E (1979) Adaptive convergence and divergence of subterranean mammals. Ann Rev Ecol System 10:269–308

Pate JS, Dixon KW (1982) *Tuberous, Cormous, and Bulbous Plants: Biology of an Adaptive Strategy in Western Australia*. U Western Australia Press, Perth

Patton JL, Brylski PV (1987) Pocket gophers in alfalfa fields: causes and consequences of habitat-related body size variation. Am Nat 130:493–506

Patton JL, Smith MF (1990) The evolutionary dynamics of the pocket gopher *Thomomys bottae*, with emphasis on California populations. U Calif Pub Zool 123:1–161

Peart RR (1989) Species interactions in a successional grassland. III. Effects of canopy gaps, gopher mounds and grazing on colonization. J Ecol 77:267–289

Reig OA (1970) Ecological notes on the fossorial octodont rodent *Spalacopus cyanus* (Molina). J Mam 51:592–601

Seebeck JH, Bennett AF, Scotts DJ (1989) Ecology of the Potoroidae—A review. In Grigg G, Jarman P, Hume I (eds) *Kangaroos, Wallabies, and Rat-kangaroos*. Surrey Beatty & Sons, New South Wales, pp 67–88

Specht RL (1973) Structure and functional response of ecosystems in the mediterranean climate region of Australia. In Di Castri F, Mooney HA (eds) *Mediterranean-Type Ecosystems: Origin and Structure*. Springer-Verlag, New York, pp 113–120

Strahan R (1983) *The Australian Museum Complete Book of Australian Mammals*. Angus and Robertson Publishers, North Ryde, New South Wales, Australia

Thorne DH, Andersen DC (1989) Long-term soil disturbance pattern by *Geomys bursarius*. J Mam 71:84–89

Torres-Mura JC (1990) *Uso del Espacio por el Roedor Fosorial* Spalacopus cyanus. M.S. Thesis, Faculty of Sciences, U Chile, Santiago

Watson JAL (1982) Distribution, biology and speciation in the Australian harvester termites, *Drepanotermes* (Isoptera: Termitinae). In Barker WR, Greenslade PMJ (eds) *Evolution of the Flora and Fauna of Arid Australia*. Peacock Publications, Frewville, South Australia, pp 263–265

Watts CHS, Aslin HJ (1981) *The Rodents of Australia*. Angus and Robertson, London

Wehlert JH, Souza RA (1988) Skull morphology of *Gregorymys* and relationships of the Entoptychinae (Rodentia, Geomyidae). Amer Mus Nov No. 2922:1–13

VI. Vegetation Change and Human Influences

17. The Human Role in Changing Landscapes in Central Chile: Implications for Intercontinental Comparisons

Eduardo R. Fuentes and Mauricio R. Muñoz

During the early 1970s (e.g., Mooney 1977) much effort was made to compare the Chilean matorral and Californian chaparral, two types of vegetation in similar mediterranean climates, but with the dominant woody species having different evolutionary histories. The underlying hypothesis was that if similar climates significantly affect ecosystem structure, then the matorral and chaparral should be more similar to each other than to the corresponding nearby ecosystems under different climatic constraints. The hypothesis did not specify identity under similar climatic conditions, but only relative similarity.

A basic assumption for the Chilean matorral and Californian chaparral was that individual "historical accidents" would not have had sufficient influence to mask the whole convergence phenomenon. Thus, if nonconvergence was found it would be a consequence of a general climatic effect acting on whole suites of species and not one species completely transforming the ecosystem. In other words, one would find no keystone species in the wide sense (Lawton and Brown 1993).

The overall results of early research efforts (Mooney 1977; Thrower and Bradbury 1977; Cody and Mooney 1978) showed strong similarities for some individual plant and animal species, but various groups of plants and animals exhibited different degrees of similarity when compared for their individual attributes (morphology, behavior, and physiology) and in their community arrangements. Thus, in a two-dimensional graph, placing individual similarity on the ordinate and community similarity on the abscissa, some groups, such as shrubs and trees, would occupy the extreme left upper corner of the graph. In other words, they were morphologically similar but dissimilar in their community arrangements. That is, "analogue" species did not have similar rankings in their relative abundance, nor did they have similar absolute abundance in the two communities. The Chilean matorral was shown to be more open (with Mediterranean annuals dominating in the open spaces) and structurally more diverse than the chaparral (Mooney 1977).

Conversely, some groups (e.g., lizards) occupied a position on the extreme right, at the lower corner of the plot, indicating high community similarity and low individual similarity among analogue species (Fuentes 1976). Finally, various groups showed intermediate similarities at the individual and community levels (Cody 1973), and some even showed low similarity along both axes (Cody and Mooney 1978; Jaksic 1982).

Subsequent research in these mediterranean ecosystems has been directed toward explaining these patterns. It has been possible to show, for example, that current differences in distribution of herbs and grasses in California and central Chile can be explained at least partially by the influence exerted by recently introduced rabbits in Chile (Jaksic and Fuentes 1980).

A completely different explanation for the dissimilarities between the two areas is based on possible differences in the resource base (Miller 1981). According to this hypothesis, organisms in the two regions may have been keying on different resource bases, and thus did not converge. If corrected for the resources available, a greater degree of convergence of the communities in the two continents would be expected. At least for plants, however, a series of simulation and experimental studies (Miller 1981) indicated that the discrepancies were not related to different resource supplies in California and Chile.

Miller and collaborators, like earlier authors (Mooney 1977), finally concluded that stronger human disturbance in Chile was probably the main explanation for the lack of convergence. The idea was that because Chile was more disturbed than California (see also Aschmann 1990), it would be in a different equilibrium mediated by humans and out of equilibrium with climate, hence not be strictly comparable with California. Humans would thus be a keystone species capable of masking convergence. At the time, however, no experimental data were presented to show the consequences that human activities can have.

The aim in this chapter is to briefly summarize the main findings obtained during the last fifteen years on the role of humans in changing the landscapes in central Chile, and discuss how this knowledge modifies our predictions about convergence among mediterranean ecosystems. We give evidence on how clearing, coppicing, fires, and grazing can change patches of vegetation and discuss how these changes and hunting could have affected the higher trophic levels in the matorral ecosystem.

Types of Human Alteration

Human beings have altered the landscapes in central Chile by changing the patch-types and the frequency of the different elements. Some areas such as the Intermediate Depression, which is geomorphologically and physiognomically analogous to the Central Valley in California, have experienced both types of changes (Fuentes 1990a; Fuentes et al. 1990a). Here evergreen sclerophyllous vegetation has been replaced either by winter-deciduous *Acacia caven- Prosopis chilensis* savannas or agricultural fields. The latter changes are human-induced and have completely transformed the original plant and animal communities in the Intermediate Depression.

In the Andes and in the Coast range surrounding the Intermediate Depression, a gradient of human disturbance frequently may be found. On the lower slopes, the original woodland vegetation was cleared and today, different types of agricultural fields and early successional plant communities dominated by such species as *Acacia caven, Baccharis* spp., or *Trevoa trinervis* can be seen. The mid- and high slopes have been less dramatically modified and show evidence of what can now be recognized as previous scattered clearings derived from rain-fed agriculture, woodcutting, grazing, fires, and hunting (Fuentes and Hajek 1979; Fuentes 1990b). As a consequence, the landscape now appears as irregularly patchy in shrub composition, with an understory dominated by annual Mediterranean herbs (Fuentes et al. 1990b). It was in these less altered sites that the convergence studies in the early 1970s were concentrated (Thrower and Bradbury 1977). In spite of all efforts made to study areas where human influence was minimal, Santa Laura, El Roble, and Zapallar, the main study sites, nevertheless had been substantially altered by human activity. At that time, however, the real role of human influence was incompletely understood.

Clearing

Clearing for agriculture has been important not only on flat areas, but also on the slopes surrounding such areas. Land clearing can significantly affect landscape physiognomy. On dry slopes, a scant cover of early successional *Baccharis* and *Trevoa* spp. or, if grazing is frequent enough, a sparse cover of *Acacia caven* and annual grasses dominates. On wetter slopes, a multispecies structure can establish (Fuentes et al. 1986). In the latter the clump structure closes to form a continuous canopy at rates that mainly depend on rainfall and exposure. On wet areas (more than 600 mm per year) canopy closure seems to occur after twenty to forty years, whereas on the drier slopes (less than 300 mm annual rainfall) it is not known yet if the canopy ever closes completely or if the clumpy structure remains as the asymptotic state. In such human-influenced vegetation, species alignments and spatial distribution are likely to differ from the original pristine vegetation.

Fires and Woodcutting

Fire has always been an important instrument for clearing land in central Chile. The limited evidence available suggests that fires in Chile have a somewhat different role from that documented in the other four MTE's (Mediterranean Type Ecosystems). Of the five areas in the world with a mediterranean climate, only central Chile does not have naturally occurring summer storms. In California, the Mediterranean Basin, Australia, and the Cape region in South Africa, lightning associated with summer storms is a frequent and "natural" source of fire. Other ignition sources such as vulcanism may have played some role, but the main natural cause of fires over evolutionary time has been the incursion of tropical storms during the hot and dry summer.

In Chile, the high Andes stabilize the position of the anticyclone and isolate the country from incursions of Atlantic summer storms. In central Chile, fire has traditionally been associated with human disturbance. Some evidence suggests, however, that natural fires could have preceded the arrival of human beings in Chile, about 11,000 years B.P. Pollen cores in central Chile show charcoal remains from about 40,000 B.P., well before people arrived (Montane 1968). Chile has a source of ignition that elsewhere has been a minor influence compared with lightning-namely, vulcanism. In central Chile recently active volcanoes are numerous and have a rich eruption record, covering very large areas with lava as well as ash. This record, as well as evidence on early development of lignotubers in plants, has been used (Fuentes and Espinoza 1986) to postulate that vulcanism could have produced sporadic natural fires during the recent evolutionary history of the Chilean matorral.

What consequences does this irregular history of fires have on the current composition of the Chilean matorral? One alternative is that as in California (Keeley 1987), some Chilean shrubs are induced to germinate by fires and that current increments in frequency of fire have shifted the community's composition toward these species.

Muñoz and Fuentes (1989) nevertheless, concluded that seeds of Chilean matorral shrubs are not induced to germinate by fire. When seeds of shrubs were treated under laboratory conditions (Muñoz and Fuentes 1989), they responded similarly to Californian species. But results in the field were quite different. In the Chilean matorral, neither field fires nor ash-soils induce germination in seeds. High seed mortality occurs with field fires as a consequence of the very high temperatures attained at the level of the soil seed bank. Muñoz and Fuentes (1989) concluded that if vulcanism has been a selective pressure for evolution in lignotubers and the resprouting behavior of shrubs, it has not been sufficient to select for fire-induced seed germination. At this level, these studies suggest that human-mediated fires should not significantly affect a community's composition.

Fuentes et al. (1993), however, also suggested that not all fires in the matorral are hot and uniform enough to consume all shrubs. They concluded that fires tend to have a patchy effect, calcinating some shrubs and merely burning the leaves, twigs, and branches of others. They showed that seeds of *Trevoa trinervis* and *Muehlenbeckia hastulata* found under lightly burned shrubs had statistically higher germination rates than the controls and those found in the soil under "ghosts" left by calcinated shrubs, suggesting that these two species could potentially be overrepresented today as a result of fire.

Many shrub and tree species in all five mediterranean regions respond to fire by resprouting (Keeley 1986). Because resprouting is a very general attribute of plants, however, it cannot be conclusively regarded as an adaptation to the conditions prevailing in these areas. Shrubs and trees in central Chile have been shown to resprout after fire, coppicing, and heavy grazing. This capacity is associated with the presence of an underground lignotuber (xylopodium) in many species, a structure rich in buds that is already visible at the seedling stage (Montenegro et al. 1978).

Different resprouting rates among species could change the species composition in a dense stand if fires or coppicing are frequent enough. Experimental results from an area near Santiago, confirm this hypothesis. After fires and woodcutting, a faithful copy of the previous species assemblage is not produced. Monitoring the resprouting response for the first year and a half after experimentally burning indicated that species differ significantly in their recovery and perhaps even in the final volume achieved (Fig. 17.1). In spite of vigorous initial responses, the recoveries of typical matorral species such as *Colliguaja odorifera*, *Lithrea caustica*, and *Peumus boldus* were not parallel after the fire, coppicing, and cutting plus litter-removal treatments. For example, *Colliguaja odorifera* was favored over *P. boldus*. Moreover, canopy growth exhibited evidence of some stagnation, suggesting very slow and unequal recuperation of species after these types of disturbance, especially in dryer locations. A consequence of these responses would be the rather open physiognomy exhibited by the matorral and perhaps somewhat altered floristic composition favoring species such as *C. odorifera*.

The paucity of the recovery must also be seen as influenced by the high soil-erosion rates that have been experimentally shown to ensue after shrub defoliation (Espinoza and Fuentes 1984). Resprouting, although present, is slow, and the ghosts are not covered in the first year, leading to soil erosion. This factor further reduces the possibility of recuperation of the original physiognomy because appropriate sites for establishing new genets are lacking.

Observations of seedling composition in recent ghosts left by shrubs after burning indicate deviations from the earlier species composition, favoring different plant species (Fig. 17.2). In addition, in ghosts left after burning, cutting, and cutting plus litter removal, the soil nutrient levels vary (Table

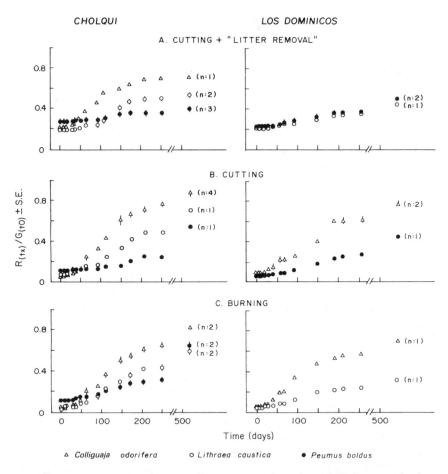

Figure 17.1. Resprouting rates in different matorral species: (△) *Colliguaja odorifera*; (○) *Lithrea caustica*; (●) *Peumus boldus*. The recovery of a plant is shown as the quotient between its instantaneous recovery diameter (R_{1+x}) and its initial (pretreatment) diameter (G_{t0}). The experimental treatments: (a) cutting plus litter removal, (b) cutting, and (c) burning were conducted at Cholqui (rainfall ca. 500 mm) and Los Dominicos sites (rainfall ca. 400 mm) near Santiago.

17.1), and species established after each of these treatments are also different (Table 17.2).

The subsequent medium-term changes after the establishment of these different suites of species is still unknown. Results, though, tend to reinforce our earlier finding about germination after cold fires, resprouting after hot fires, and coppicing—namely, that vegetation stands seem to be modified in composition following disturbance caused by human beings.

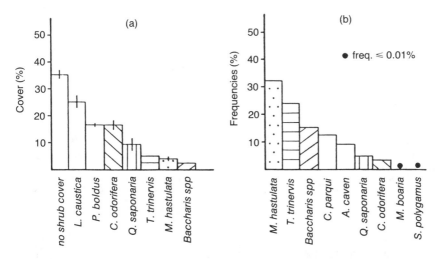

Figure 17.2. Changes in species composition after matorral burning. (a) native species composition in three linear transects of 50 m each, including one at an unburned site; (b) seedlings growing in "ghosts" left by shrubs after burning.

Table 17.1. Nutrients in the remaining soil after burning, cutting, and cutting plus litter removal

	Ghost origin		
	Cutting	Burning	Cutting-litter
Organic matter (%)	10.5–11.2	12.1–9.4	8.0–5.2
N (ppm)	71.0–13.0	74.0–68.0	18.0–22.0
P (ppm)	37.0–30.0	280.0–280.0	18.0–22.0
K (ppm)	440.0–625.0	2.851.0–1.615.0	508.0–299.0
pH	5.9–6.7	8.1–7.8	6.7–6.1
Salinity (mm ha/cm)	0.3–0.1	0.8–0.6	0.1–0.1

Grazing

Today the main grazers in the matorral are rabbits, hares (Jaksic and Fuentes 1980), and goats (Contreras et al. 1986; Fuentes 1990b). Here we provide evidence that these herbivores are another factor that leads to change in stand composition through their foraging patterns on adult plants and seedlings.

The closest native analogue species of the introduced goat (*Capra hircus*) is the guanaco (*Lama guanicoe*). For small mammals, *Octodon degus*, *Abrocoma bennetti*, and *Lagidium viscacia* were the prevalent vertebrate herbivores in the past. The abundance of these mammals has changed significantly, with *O. degus* now being more abundant and the other two probably less abundant than in the past (Miller 1980; Simonetti 1989).

Table 17.2. Number of seedlings established in shrub ghosts after burning, cutting, and cutting plus litter removal

	Cholqui $n = 35$			Los Dominicos $n = 20$		
	Cut	Burned	Cut-litter	Cut	Burned	Cut-litter
Muehlenbeckia hastulata	22	31	8	44	51	0
Trevoa trinervis	17	20	6	12	27	0
Baccharis spp.	3	4	18	0	2	10
Acacia caven	0	3	14	1	0	0
Quillaja saponaria	2	5	0	2	0	0
Colliguaja odorifera	2	2	1	0	3	0
Maytenus boaria	0	2	0	0	0	0
Schinus polygamus	1	0	0	0	0	0

One possibility is that goats, introduced rabbits (*Oryctolagus cuniculus*) (Jaksic and Fuentes 1990), and hares (*Lepus capensis*), simply are replacing the native herbivores, so that no shifts in vegetation would be expected. Observations and experimental results indicate, however, that the foraging patterns of these introduced herbivores differ markedly from those of the native species on adult shrubs and seedlings (Simonetti and Fuentes 1983). Whereas goats are browsers and grazers, guanacos tend to consume more annual herbs, even during the dry season (Table 17.3). Moreover, the species of shrubs eaten by the two herbivores also tend to differ (Table 17.4). Strictly then, on a trophic basis, guanacos and goats are not analogues. Therefore, the introduction of goats following the local extinction of guanaco does not constitute an ecological substitution of functionally equivalent species.

The significance of different foraging behaviors in guanaco and goats is both direct and indirect. The direct effect is reduced vegetation cover near places where goats are kept. An indirect effect results from a change in the competitive equilibrium between shrub species (Fuentes and Gutiérrez 1981). As pointed out elsewhere (Fuentes and Etchegaray 1983), the significance of herbivores to plants must be measured not only by immediate re-

Table 17.3. Foraging pattern of guanacos and goats at the Peñuelas site, expressed as percentage (± 2 S.E.) of the time taken for guanacos and goats to consume a plant

	Guanacos		Goats	
	Wet season	Dry season	Wet season	Dry season
Herbs	95.4 ± 4.5	70.0 ± 16.8	33.4 ± 12.4	2.3 ± 2.5
Shrubs	4.6 ± 4.5	30.0 ± 16.8	66.6 ± 12.4	97.7 ± 2.6
n	2119	986	1412	566
p	13	15	13	7

[a] n is the number of times a plant was consumed in p periods of observation.
Source: Based on Simonetti and Fuentes 1983.

Table 17.4. Acceptance of shrubs and herbs by guanacos and goats in the Chilean matorral: Figures are the mean percentage of time (± 2 S.E.) during which guanacos (GU) and goats (GO) consume a specific shrub or herb in relation to the number of trials

		Season			
		Wet	N^a	Dry	N
Acacia caven	GU	33 ± 67	6	98 ± 3	108
	GO	77 ± 31	119	98 ± 4	212
Baccharis linearis	GU	19 ± 22	70	17 ± 21	16
	GO	15 ± 11	114	50 ± 50	5
Maytenus boaria	GU	100 ± 0	44	99 ± 2	96
	GO	99 ± 1	535	99 ± 1	270
Muehlenbeckia hastulata	GU	0	8	91 ± 16	56
	GO	74 ± 33	68	80 ± 40	21
Trevoa trinervis	GU	66 ± 66	53	0	[b]
	GO	84 ± 32	95	0	[b]

[a] N is the number of observations.
[b] *Trevoa trinervis* loses its leaves in early summer.

productive output, or immediate defoliation, but also by loss of competitive status and greater susceptibility to herbivore attack in the near future.

When both direct and indirect effects are considered, goats must be seen as a new source of disturbance whose effects are not comparable, qualitatively and quantitatively, to that produced by native herbivores, including phytophagous insects (Fuentes and Etchegaray 1983). Experimental results with goat preferences (Fuentes and Etchegaray 1983; Simonetti and Fuentes 1983) and shrub responses (Torres et al. 1980) indicate that species such as *Lithrea caustica* tend to increase in abundance with goat grazing, in relation to such species as *C. odorifera*. The complete picture is not yet clear, but the data strongly suggest that goats not only change plant cover, but also mediate shifts in species composition.

Introduced herbivores also affect seedling composition. Seedlings of woody species are especially sensitive to the long summer drought, with shrub recruitment occurring sporadically, especially in wet years (Fuentes et al. 1984; 1986). Herbivores are most likely to seek shrub seedlings in the summer season when the annual cover is reduced and brown, and the seedlings stand out as the only green material to be eaten. Experiments on artificially planted seedlings are revealing. Large (30 cm, about 2 years old) and small (less than 5 cm, current year) seedlings were planted under shrubs and in the open spaces between shrubs at the Los Dominicos site, on the Andean foothills east of Santiago and at Peñuelas, in the wetter Coast range near Santiago. It has been amply demonstrated that the prevailing microclimate is more benevolent under than between shrubs (del Pozo et al. 1989). Therefore, experiments were designed to test effects of climate and herbivores on

Table 17.5. Percentage of seedling mortality (after 140 days) at the Los Dominicos site: Seedlings were either protected by wire cages (Dessication) or exposed to rabbit grazing (Rabbits + Desiccation)

		Rabbits + desiccation	Desiccation
Small seedlings	Under shrubs	80	40
(3–5 cm)	Between shrubs	100	100
Large seedlings	Under shrubs	73	0
(30 cm)	Between shrubs	80	0

Source: Simonetti and Fuentes 1983.

seedlings of two sizes. Half of the seedlings were protected from rabbits and hares with wire-mesh cages.

The results in Table 17.5 are for percentage of deaths after 140 days, including the summer drought period. All small seedlings between shrubs died of desiccation, but large seedlings died of the combined leporid-desiccation effect (Table 17.5). That is, larger seedlings survive drought better than small ones, but rabbits and hares significantly reduce survival of the former in the open spaces. Later in the season all seedlings had died from action of the introduced leporid herbivores.

Under shrub clumps, survival through the drought period is relatively higher in general. Consequently the proportion of seedlings in the shrub clumps that were eventually consumed by leporids was higher. All this evidence suggests that introduced leporids modify stand structure by decreasing survival of seedlings in the open spaces, perhaps more drastically during very wet years when initial survival there, without rabbits, would have been higher. Leporids would also make recruitment under shrub clumps more chancy and more sporadic than if they were absent. Overall this sequence would lead to a structurally more open matorral than in the past. Rabbits and hares also seem to prefer the seedlings of species such as *Quillaja saponaria* over others such as *Schinus molle* (Fuentes and Jaksic unpub.), suggesting another source of shifts in composition of the current canopy.

An experiment to compare the effects of leporids and guanaco on small (5 cm) seedlings was conducted at the Peñuelas Station (Table 17.6). Leporids have important effects both under and between shrub clumps, whereas the guanaco has little effect in the open spaces and no effect under shrubs. Once again, the introduced herbivores differ qualitatively and quantitatively from the native ones.

Table 17.6. Seedling mortality: guanacos vs. rabbits (after 140 days) at Peñuelas

	Under shrubs	Between shrubs
Rabbits	90	85
Guanacos	0	10

Source: Simonetti and Fuentes 1983

Table 17.7. Seedling mortality: rabbits vs. goats (after 140 days) at Peñuelas site

	Under shrubs		Between shrubs	
	Rabbits	Goats	Rabbits	Goats
Near house (200 m)	60	20	100	50
Far (2000 m)	20	0	60	0

Source: Simonetti and Fuentes 1983

Table 17.7 compares the effect of goats and leporids on seedling (5 cm) mortality under and between shrubs, and at two distances from the enclosures where goats are kept at night. It can be seen that goats as well as rabbits have decreasing influence with distance. The distance effect (Simonetti 1989) is probably a particular case of the more general tendency for increased leporid abundance in the more open spaces in central Chile. In this case the general openness of the vegetation would reflect a combination of the land being cleared by shepherds and subsequent goat defoliation. These results suggest, on the one hand, that the place of human habitation (Fuentes 1990a,b) is also important regarding relative herbivore effects. On the other hand, the effects of introduced herbivores should be seen as the combined total, and not as a series of independent influences.

The importance of combined effects between removal treatments and herbivore responses can be evaluated by comparing the amounts of leporid

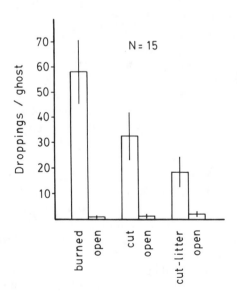

Figure 17.3. Amount of leporid droppings in ghosts left by shrubs after experimental burning, cutting, and cutting plus litter removal.

droppings in ghosts left by shrubs after experimental burning, cutting, and cutting plus litter removal (Fig. 17.3). In all three cases, visits by rabbits and hares were more frequent under ghosts than in nearby controls, indicating a strong interaction effect between these treatments and herbivore responses. These interactions, as well as that between lagomorphs and goats, suggest that in the field herbivores can significantly affect composition of vegetation and cover. Without an experimental approach, it would be difficult to separate the role of the different factors in changing the composition of the vegetation stands.

Upper Trophic Levels

Differences in the effect of native and introduced herbivores is further exaggerated when we consider their absolute abundance. The guanaco, a wild animal, was probably never as abundant in the past as goats are today. In the past the area had many more mountain lions (*Felis concolor*) than are there today. Today the latter are practically extinct from the area of mediterranean climate. Mountain lions hunt guanaco in the extreme south of the country (Raedecke 1978) and in the past they are also likely to have had a part in keeping guanaco densities low in central Chile.

Some evidence indicates that the whole fauna in central Chile has been selectively changed *via* hunting (Miller 1980) and therefore that other predators, in addition to mountain lions, have altered abundances. Changes in relative abundances cannot be evaluated at present, but almost certainly, hunting has been an important factor for current small-mammal abundances. Human-induced change of habitat availability has probably also modified the abundance of small mammals. Open stands of Chilean matorral tend to have more species and higher densities of small mammals than more densely covered ones (Fulk 1975; Glanz 1977; Jaksic et al. 1981). In relatively undisturbed shrublands, both composition and abundance of the small-mammal fauna change according to shrub cover (Table 17.8). The number of species ranges from three to six, with seven being present overall. Except for *Akodon* species, the remaining five occur in both sparse and dense patches. *Abrocoma bennetti*, *Marmosa elegans*, *Octodon degus*, and *Phyllotis darwini* are usually more abundant in the sparsely covered shrubland, the opposite being true for *Oryzomys longicaudatus*. *Akodon longipilis* was found only in the dense patches, and *A. olivaceus* only in sparse ones. Overall, small mammals were 2.2 times more abundant in the sparsely covered shrublands than in the densely covered areas. Total numbers of small mammals, however, are less variable in dense patches.

One may ask whether small mammal abundances are similar in patches of different origin, but with similar cover. That is, does small-mammal diversity differ in "old" and "new" patches, new ones being those for which human disturbance is evident and old patches being those in which it is not known if they are on the asymptote or still closing.

Table 17.8. Species composition and abundance of small mammals in undisturbed shrublands at the Los Dominicos site. Figures are the minimum of animals known to be alive. S = sparse, D = dense

| | 1980 | | | | | | 1981 | | | |
| | Winter | | Spring | | Summer | | Fall | | Spring | |
	S	D	S	D	S	D	S	D	S	D
Abrocoma bennetti	—	—	1	—	3	1	1	2	4	—
Akodon longipilis	—	—	—	—	—	1	—	1	—	2
Akodon olivaceus	—	—	—	—	3	—	5	—	1	—
Marmosa elegans	11	6	9	5	0	2	10	6	2	1
Octodon degus	6	—	5	—	13	5	5	2	4	1
Oryzomys longicaudatus	2	3	1	5	2	1	3	3	—	2
Phyllotis darwini	10	4	12	3	10	1	28	11	8	7
Total individuals	29	13	28	13	31	11	52	25	19	13
No. species	4	3	5	3	5	6	6	6	5	5

Source: Simonetti and Fuentes unpubl

We tested this idea by comparing the data in Table 17.8, representing old patches without known human disturbance for at least 40 years, with similar data from a nearby patch showing evidence of recent (about five to ten years) human disturbance (Jaksic et al. 1981). Small-mammal assemblages indeed differed between the similarly covered patches of different origin (Table 17.9). Although species composition is the same, total and relative abundances vary markedly. Small mammals, as assessed by trapping success, were more abundant in the older patch. *Octodon degus* dominated the recently disturbed patch and all other species were scarcely represented. In other words, not only plant-species cover but also time since disturbance is a factor affecting small-mammal species diversity. Clearly then, the effects of human disturbance on the vegetation are not compensated for at the herbivore level. On the contrary, disturbance has an important effect, modifying the abundance of small mammals.

Table 17.9. Density (per hectare) of small mammals in patches with different histories of human disturbance

| | Type of shrubland | | |
| | Undisturbed (old) | | Disturbed (recently) |
	Open	Dense	Open
Abrocoma bennetti	5.7	4.0	1.7
Akodon longipilis	0	5.3	0
Akodon olivaceus	5.7	0	9.3
Marmosa elegans	20.1	26.7	1.0
Octodon degus	20.8	10.7	79.3
Oryzomys longicaudatus	5.0	18.7	2.3
Phyllotis darwini	42.8	34.7	6.4
Trapping success (%)	12.3	5.9	3.4

Source: Jaksic 1981, Simonetti and Fuentes unpub

Do human-induced changes in the shrub and grass components have an effect only at the small-mammal level, or are there further consequences at higher trophic levels? Raptor species are generally opportunistic hunters, concentrating on the most abundant locally available prey within specific size ranges (Jaksic 1982). Hence it can be expected that species such as *Octodon degus*, overrepresented in human-disturbed communities, will also be well represented in the stomach contents of predators. In central Chile, *O. degus* is the "staple" food item for most predators and is a main element upon which the whole predator community is based (Jaksic 1982). This strong dependence on one diet item is one of the reasons no convergence was found in predator assemblages when Chile, California, and Spain were compared (Jaksic 1982). Simonetti (1988) claimed that human disturbance can be regarded as the ultimate factor governing guild structure of predatory assemblages in central Chile, and that lack of convergence at this level can at least partially be attributed to this distortion (see also Meserve 1988).

Conclusions

In Chile, people have introduced and eliminated species, increased the frequency and intensity of fire, modified the patterns of grazing and browsing, effected wholesale clearing and woodcutting, and used slopes for small-scale rain-fed agriculture (Fuentes and Hajek 1979). All these activities, but especially rain-fed agriculture, sometimes on slopes up to 30°, have drastically changed the physiognomy of the landscape. Species that had treelike habits, such as *Lithrea caustica*, now exhibit a shrubby multistemmed structure. Landscapes appear now as a patchwork in which different factors combine to produce alternative trajectories of vegetation change (Fuentes et al. 1986; Fuentes 1990a). We are far from understanding the effects of different combinations of these disturbance factors. It is clear, though, that in the past woodland and forest were more extensive than we see today. Much of today's shrublands (how much is not known) are anthropogenic.

We can now ask what all this background means for the convergence studies carried out in central Chile and California in the 1970s. In Chile, human beings have effectively played the role a keystone species, very significantly modifying the landscape and successional trajectories. Would convergence have been more extensive with other mediterranean landscapes if such landscapes had been compared before people arrived?

We know that forests were in the Mediterranean Basin in the past, and that, just as in Chile, the shrubby formations we see today are mainly a consequence of people as an additional driving force in the landscape (Thirgood 1981; Rother 1984). Consequently the past vegetation in the Mediterranean Basin and central Chile might have been more convergent without human beings. No one has compared in detail the mallee in western Australia and the evergreen sclerophyllous woodlands in central Chile, and for the moment it is not possible to comment on present or past similarities in these systems.

There is some evidence, however, that human disturbance could have also drastically and pervasively affected the physiognomy of Californian chaparral (Dodge 1975; Conrad and Oechel 1982; Aschmann 1990). In California, people have altered the frequency of fire and increased grazing pressure, and as a consequence native grasslands and woodlands have diminished and thick brushlands have increased in coverage (Dodge 1975). If Dodge's hypothesis is correct, pre-human California, may have had more woodlands and grasslands than we see today. Some of the extensive area of chaparral in California could also be a byproduct of human disturbance. Hence central Chile and California may have been more similar in the past.

Nevertheless, it is important to recall that some ecological and morpho-physiological (individual) convergences were found in spite of human effects in both regions. Morphophysiological similarities relate to similar climate constraints in both areas and are relatively easy to understand. The community-level similarities found are less easy to comprehend because human beings are such an important modifying factor. The organisms that showed greater similarity in community structure (lizards—Fuentes 1976, and some foliage-gleaning birds—Cody 1973) seem to respond to broad habitat attributes, which in turn are perhaps more difficult to modify. At present it is difficult to separate prehuman influence from recent human disturbances.

Acknowledgments

Javier Simonetti made useful suggestions that improved an earlier version of the manuscript. Mauricio R. Muñoz was financed by a Fundación Andes doctoral fellowship. Research reported has been financed through FONDECYT, Chile Grant 614-1989.

References

Aschmann H (1990) Human impact on the biota of mediterranean climate regions of Chile and California. In Groves RH, di Castri F (eds) *Biogeography of Mediterranean Invasions.* Cambridge U Press, Cambridge, pp 33–41

Cody ML (1973) Parallel evolution and bird niches. In di Castri F, Mooney HA (eds) *Mediterranean-Type Ecosystems: Origin and Structure.* Springer-Verlag, New York, pp 307–338

Cody ML, Mooney HA (1978) Convergence versus non-convergence in mediterranean-climate ecosystems. Ann Rev Ecol System 9:265–321

Conrad CE, Oechel WC (eds) (1982) *Dynamics and Management of Mediterranean-Type Ecosystems.* Berkeley, CA

Contreras D, Gasto J, Cosio F (eds) (1986) Ecosistemas pastorales de la zona mediterranea árida de Chile. I. Estudio de las comunidades agrícolas de Carquindaño y Yerba Loca del secano costero de la Región de Coquimbo. UNESCO-MAB, Montevideo, Uruguay

del Pozo AH, Fuentes ER, Hajek ER, Molina JD (1989) Zonación microclimática por efecto de los manchones de arbustos en el matorral de Chile central. Rev Chil Hist Nat 62:85–94

Dodge JM (1975) Vegetation changes associated with land use and fire history in San Diego County. Ph.D. Thesis, U California, Riverside

Espinoza G, Fuentes ER (1984) Medidas de erosión en los Andes centrales: Efectos de pastos y arbustos. Terra Aust 3:75–86

Fuentes ER (1976) Ecological convergence of lizard communities in Chile and California. Ecology 57:3–17

Fuentes ER (1990a) Landscape change in mediterranean-type habitats of Chile: Patterns and processes. In Zonneveld IS, Forman RTT (eds) *Changing Landscapes: An Ecological Perspective.* Springer-Verlag, New York, pp 165–190

Fuentes ER (1990b) Central Chile: How do introduced plants and animals fit into the landscape? In Groves RH, di Castri F (eds) *Biogeography of Mediterranean Invasions.* Cambridge U Press, Cambridge, pp 43–49

Fuentes ER, Espinoza G (1986) Resilience of central Chile shrublands: A vulcanism-related hypothesis. Interciencia 11:164–165

Fuentes ER, Etchégaray J (1983) Defoliation patterns in matorral ecosystems. In Kruger KJ, Mitchell DT, and Jarvis JVM (eds) *Mediterranean-Type Ecosystems. Ecological Studies,* Vol. 43. Springer-Verlag, New York, pp 525–542

Fuentes ER, Gutiérrez J (1981) Intra- and interspecific competition between matorral shrubs. Oecol Plant 2:283–289

Fuentes ER, Hajek ER (1979) Patterns of landscape modification in relation to agricultural practice in central Chile. Envir Cons 6:265–271

Fuentes ER, Otaiza RD, Alliende MC, Hoffmann A, Poiani A (1984) Shrub clumps of the Chilean matorral vegetation: Structure and possible maintenance mechanisms. Oecologia 62:405–411

Fuentes ER, Hoffmann AJ, Poiani A, Alliende MC (1986) Vegetation change in large clearings: Patterns in the Chilean matorral. Oecologia 68:358–366

Fuentes ER, Avilés R, Segura A (1990a) Landscape change under indirect effects of human use: The savanna of central Chile. Land Ecol 2:73–80

Fuentes ER, Avilés R, Segura A (1990b) The natural vegetation of a heavily man-transformed landscape: The savanna of central Chile. Interciencia 15:293–295

Fuentes ER, Segura AM, Holmgren M (1993) Are the responses of matorral shrubs different from those in an ecosystem with a reputed fire history? In Moreno J, Oechel W (eds) *Fire in Mediterranean-Type Ecosystems.* Springer-Verlag, New York: in press

Fulk G (1975) Population ecology of rodents in the semiarid shrubland of Chile. Occas. Paper Mus Texas Tech U 33:1–40

Glanz WE (1977) Comparative ecology of small mammal communities in California and Chile. Ph.D. Thesis, U California, Berkeley

Jaksic FM (1982) Predation upon vertebrates in Mediterranean habitats of Chile, Spain and California: A comparative analysis. Ph.D. Thesis, U California, Berkeley

Jaksic FM, Fuentes ER (1980) Why are herbs in the Chilean matorral more abundant beneath bushes: Microclimate or grazing? J Ecol 68:665–669

Jaksic FM, Fuentes ER (1990) Ecology of a successful invader: The European rabbit in central Chile. In Groves RH, di Castri F. (eds) *Biogeography of Mediterranean Invasions.* Cambridge U Press, Cambridge

Jaksic FM, Yañez JL, Fuentes ER (1981) Assessing a small mammal community in central Chile. J Mam 62:391–396

Keeley JE (1986) Resilience of mediterranean shrub communities to fires. In Dell B, Hopkins AJM, Lamont BB (eds) *Resilience of Mediterranean-Type Ecosystems.* W. Junk Publishers, Netherlands, pp 95–112

Keeley JE (1987) Role of fire in seed germination of woody taxa in California chaparral. Ecology 68:434–443

Lawton JH, Brown VK (1993) Redundancy in ecosystems. In Schulze E, Mooney

HA (eds) *Biodiversity and Ecosystem Function.* Springer-Verlag, Berlin, pp 255–270

Meserve P (1988) Are predator diets a consequence of human disturbance in central Chile? A reply to Simonetti. Rev Chil Hist Nat 61:159–161

Miller PC (ed) (1981) *Resource Use by Chaparral and Matorral.* Springer-Verlag, New York

Miller S (1980) Human influences on the distribution and abundance of wild Chilean mammals: Prehistoric–present. Ph.D. Thesis, U Washington

Montane HJ (1968) Paleo-Indian remains from Laguna de Tagua-Tagua, central Chile. Science 161:1137–1138

Montenegro G, Rivera O, Bas F (1978) Herbaceous vegetation in the Chilean matorral: Dynamics of growth and evaluation of allelopathic effects of some dominant shrubs. Oecologia 36:237–244

Mooney HA (ed) (1977) *Convergent Evolution in Chile and California: Mediterranean Climate Ecosystems.* Dowden, Hutchinson and Ross, Stroudsburg, PA

Muñoz MR, Fuentes ER (1989) Does fire induce shrub germination in the Chilean matorral? Oikos 56:177–181

Raedecke K (1978) El Guanaco de Magallanes, Chile. Su distribución y biología. Pub. Tech. No. 248. Corp Nac For, Santiago, Chile

Rother K (1984) *Mediterrane Subtropen.* Holler und Zwick, Verlags-GmbH, Braunschweig

Simonetti JA (1988) The carnivorous predatory guild of central Chile: A human-induced community trait? Rev Chil Hist Nat 61:23–25

Simonetti JA (1989) Microhabitat use by small mammals in central Chile. Oikos 56:309–318

Simonetti JA, Fuentes ER (1983) Shrub preference of native and introduced Chilean matorral herbivores. Ecol Appl 4:269–272

Thirgood JV (1981) *Man and the Mediterranean Forest: A History of Resource Depletion.* Academic Press London, England

Thrower NJW, Bradbury DE (eds) (1977) *Chile-California Mediterranean Shrub Atlas: A Comparative Analysis.* Dowden, Hutchinson and Ross, Stroudsburg, PA

Torres JC, Gutiérrez J, Fuentes ER (1980) Vegetative responses to defoliation of two Chilean matorral shrubs. Oecologia 46:161–163

18. Evaluating Causes and Mechanisms of Succession in the Mediterranean Regions in Chile and California

Juan J. Armesto, Patricia E. Vidiella, and Hector E. Jiménez

Succession is a central concept of ecology. Understanding the causes and mechanisms that change vegetation in ecological time can help answer practical (management) as well as theoretical (community-assemblage) questions. Approaches based on accepting one or a few broadly applicable mechanisms have proved disappointing (e.g., Connell and Slatyer 1977; Finnegan 1984). A modern view of succession requires that we consider the multiple forces that drive vegetational change and their interactions (Pickett et al. 1987; Walker and Chapin 1987; Burrows 1990). In this chapter we use a multi-factorial analysis to compare successional mechanisms in two regions distinguished by broad climatic and physiognomic resemblance: central Chile and California (Thrower and Bradbury 1977).

Earlier comparative studies (e.g., Mooney 1977) were focused mainly on convergent evolution of organisms and community structure in the two regions, but gave little attention to community dynamics in ecological time, such as the pathways communities followed in recovering from disturbance. Comparative analyses of succession have been limited in scope because they lacked a theory that would integrate the causes that promote change in vegetation (Pickett et al. 1987). We first introduce a hierarchic framework identifying both general causes and specific mechanisms of succession. We then apply this framework in examining the forces driving vegetation change, and predicting responses by local plant assemblages to these forces in Chile and California. The comparative approach (see Pickett and Armesto

1991) is considered useful for assessing how important are the factors that we suspect influence vegetational change in each region, as well as for testing how applicable successional theory may be.

Causes of Succession

Succession can be ascribed to three universal causes (Table 18.1): (1) A site becomes available for invasion, say when the former occupants die; (2) species invade the site at differing rates, because of greatly varying methods of dispersal, location of seed sources with respect to the opening, and propagule pool; and (3) species' performance (e.g. production of cover and biomass) at the open site varies with time because of interspecific differences in life cycle, growth rate, and so on. Understanding vegetation dynamics leads us to study these three primary causal phenomena, and wherever the three take place we should observe vegetational change (Pickett and McDonnell 1989). By analyzing the physical and biological systems and the species' attributes that underlie these three primary causes, we assemble a hierarchic framework (Table 18.1) with which we can examine specific mechanisms for succession operating in an ecological system (Pickett et al. 1987) and eventually propose empirical models that account for the observed successional pathways (e.g., Armesto and Pickett 1985). Using this general framework, we identify and compare the major forces driving secondary succession in the matorral in central Chile and the chaparral in California. Much of the data base is from published sources available for Chile and California. Few long-term studies of vegetation dynamics have been undertaken in central Chile, and so we build our conclusions mainly on our own limited field experience. In both geographic areas, we focus on evergreen sclerophyllous vegetation.

Table 18.1. Hierarchy of successional causes with special reference to chaparral and matorral

General causes	Contributing processes or attributes
1. Availability of sites	Disturbance regime[a]
2. Availability of species	1. Dispersal[a]
	2. Propagule pool[a]
3. Differential species' performance	1. Resource availability
	2. Life-history strategy[a]
	3. Ecophysiological tolerances[a]
	4. Competition
	5. Allelopathy[a]
	6. Herbivory, predation[a]

[a] Topics specifically addressed in this chapter.
Source: Pickett et al. (1987).

Availability of Open Sites

Sites become open to invasion because individuals die following physical or biological disturbance or both (White 1979). To understand early secondary succession, we must examine factors generating open sites, such as disturbance and its effects on plants and animals. In mediterranean regions in Chile and California, present disturbance regimes are superficially similar, and include fire and human influence. Underneath, however, lie important historical differences (Aschmann and Bahre 1977). Many fires in California originate from lightning during summer storms (Fig. 18.1). These storms sometimes coincide with another major climatic influence, the hot "Santa Ana" winds, blowing toward the coast, which are more frequent in fall (Keeley 1977). Coincident lightning and hot winds may be responsible for larger fires. These climatic factors have prevailed during the Quaternary (Axelrod 1989), so that the fire regime has been a major force driving vegetation dynamics in the chaparral over evolutionary time (Minnich 1989). Although no long-term trends appear in the number of fires initiated by lightning each year (Fig. 18.2), presently human beings cause more than 80% of the fires in the chaparral, and thus have altered the seasonal distribution of fires, formerly controlled by climate alone (Keeley 1982). The area burned by people is around 100 times that burned by lightning fires (Keeley 1982), mainly because large anthropogenic fires are set during the period of the Santa Ana winds.

In the Chilean matorral, almost all fires recorded since 1972 by the Chilean National Forest Service (CONAF) are caused directly or indirectly by people (Fig. 18.3). Among the most frequent causes of fire reported by CONAF are clearing of agricultural land (14%), motor vehicles on roadsides

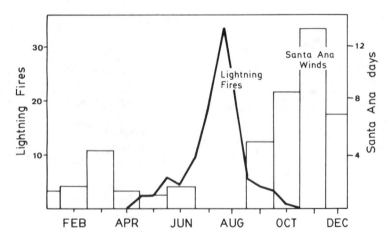

Figure 18.1. Frequency of fires attributed to lightning and the seasonal distribution of Santa Ana winds in California (after Keeley 1982).

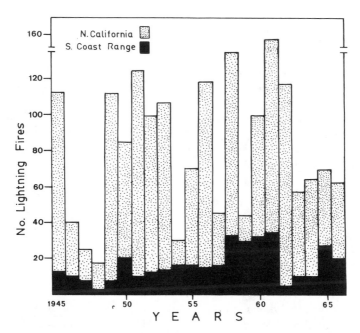

Figure 18.2. Number of fires caused by lightning in California since 1945 (after Keeley 1982).

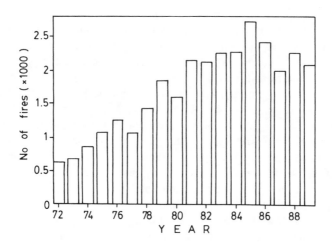

Figure 18.3. Number of fires per year in central Chile since 1972 (source CONAF 1986).

(24%), intentionally set fires (15%), and forest clearing (14%). Lightning storms are rare in central Chile, and are not listed as a common cause of fire (CONAF 1986). There is no evidence that recurrent natural fire was ever important in central Chile before people arrived, at least in the Coast range (Aschmann 1977; Aschmann and Bahre 1977; Heusser 1983), and winds of the Santa Ana type are unknown in the matorral. The CONAF statistics show a growing trend in number of fires per year (Fig. 18.3) and in area of matorral burned annually, suggesting a direct relationship between fire frequency and changes in use of land, including recent urbanization and associated growth in population.

In the past, other forms of anthropogenic disturbance have also been important in opening up the Chilean matorral. Domestic and industrial use of wood as fuel has been especially influential in central Chile, but less significant in California (Aschmann and Bahre 1977). Woodcutting contributes to opening up the matorral vegetation and changing the microclimate inside stands. The subsequent shading and death of the multiple stems and sprouts born from stumps after each cutting (Armesto and Smith, unpub.) have accumulated increasing amounts of dry biomass. Consequently, modern vegetation in central Chile has probably become more susceptible to extensive fires propagated once a fire is set.

Invasion and Establishment

Studies of succession in the chaparral demonstrate that areas disturbed by fire are rapidly recolonized (Hanes and Jones 1967; Hanes 1971; Vogl 1982). This recovery indicates that propagule availability does not limit revegetation in California. Conversely, arrival of propagules seems to be a substantial limitation in matorral succession (Fuentes et al. 1984; Armesto and Pickett 1985). Although both matorral and especially chaparral species are capable of resprouting, one study suggests that recruitment from seedlings may be limited by the propagule pool in the matorral (Jiménez and Armesto 1992). Some tree species, such as *Cryptocarya alba* (Lauraceae) are entirely absent from the seed pool in open areas (Table 18.2), presumably because their seeds are short-lived and sensitive to drought (Bustamante 1992). Another factor might be a strong tendency for obligate outbreeding in woody mediterranean species in Chile, where self-incompatibility is often associated with low seed production (Arroyo and Uslar 1993). For most chaparral shrub species, the propagule pool is several orders of magnitude larger than that of matorral shrubs (Table 18.2), even though many chaparral species are able to resprout. Seed availability may be the greatest limiting step for initiating succession in central Chile, particularly if seeds are killed by fire, as shown by Muñoz and Fuentes (1989).

The predominant dispersal agents for the matorral seeds are birds and wind (Fig. 18.4), suggesting a tendency for long-range dispersal. But these dispersal mechanisms are less frequent among chaparral shrubs (Fig. 18.4).

Table 18.2. Shrub seed banks in the California chaparral and the Chilean matorral

| | No. seeds/m^2 | |
	Woodland	Open scrub
Chile		
Baccharis sp.	0	87
Peumus boldus		
1988	183	98
1989	191	46
Colliguaja odorifera	10	31
Lithrea caustica	50	56
Cryptocarya alba	31	0
Quillaja saponaria	20	46
Muehlenbeckia hastulata		
1988	10	333
1989	0	46
Trevoa trinervis		
1988	61	132
1989	302	92
California		
Adenostoma fasciculatum		
Lake county		5385–320,000
Contra Costa county		450–3745
Arctostaphylos glauca		
1972		346
1982		298
A. glandulosa		
S. California, 1972		4116
S. California, 1982		3028
N. California, 1982		8422
A. canescens		4500
A. pechoensis		4490
A. crustacea		978
A. viscida		28,177
A. mewukka		6463
Ceanothus greggii		369
C. leucodermis		87

Source: Chile, Jiménez and Armesto (1992); California, Parker and Kelly (1989).

Autochory and gravichory, which are often associated with more limited seed dispersal ranges, are more abundant in the chaparral (see Hoffmann and Armesto, this volume). A significant consequence is that for species having bird-disseminated seeds in the matorral, recruitment is often restricted to the vicinity of previously established pioneer shrubs, which act as perches for birds (Armesto and Pickett 1985; Fuentes et al. 1986). Without shrub perches, the input of seeds to open areas may be very low, preventing recolonization.

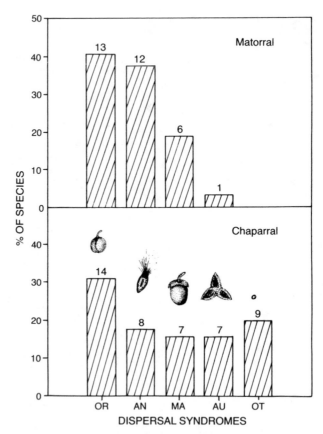

Figure 18.4. Dispersal syndromes in two comparable sites in Chilean matorral and Californian chaparral. OR = ornithochory; AN = anemochory; MA = mammalochory; AU = autochory; OT = other dispersal modes.

Species Performance in Open Sites

Nutrient Limitation

Burned sites in California are usually colonized by the same shrub species present before the fire (Hanes 1971; 1981), except for especially hot fires or sites where fires are extremely frequent (Vogl 1982; Zedler et al. 1983). Apparently the change in available soil nutrients associated with fire (DeBano and Conrad 1978) does not preclude reinvasion of the open site by a similar floristic assemblage. Nitrogen fixation in root nodules has been demonstrated for a number of chaparral shrubs (Rundel and Neel 1978; Ellis and Kummerow 1989), hence lowered nitrogen concentrations in burned soil are not a problem for recolonization. Although information on pre- and postdisturbance nutrient availability is lacking for the matorral, some colonizing shrubs of open sites, such as *Acacia caven* and *Trevoa trinervis*, are known to fix nitrogen (Rundel and Neel 1978; Silvester et al. 1985). Because few sclerophyllous shrubs are pioneers in recently opened matorral sites (Armesto and Pickett 1985), evergreen sclerophyllous shrubs that become established later in succession may be nitrogen limited. Limited seed input, however, (see above) will also result in delayed invasion. Field assays are needed to evaluate how colonization is influenced by limited seed input versus nutrient limitation.

Life History

A number of life-history strategies have been described for chaparral species, many of which can be related to response to fire (Keeley and Keeley 1981; Keeley et al. 1981; Zedler et al. 1983; Keeley et al. 1985; Zedler and Zammit 1989). These include several groups of annuals, perennial herbs, and shrubs (Table 18.3), but fire-responsive tree species are infrequent in the chaparral. In Chile, plant life histories cannot be readily associated with a major disturbance factor. Native annuals are less frequent (Arroyo et al. this volume) in the matorral, and annual grasslands mostly consist of European and Californian weeds. Geophytes are a major herbaceous component in both matorral and chaparral, being abundant under the canopy in old-growth stands, as well as in open sites. Geophytes with subterranean bulbs may have been favored by fire in both regions. Although some matorral shrubs survive fires and can resprout from lignotubers (e.g., *Lithrea caustica*), their seeds cannot survive intense fire, and germination is not induced by fire (Muñoz and Fuentes 1989). Shrubs found in early secondary succession in the matorral (*Baccharis* spp., *Colliguaja odorifera*, *Acacia caven*), are often replaced by tree species (*Cryptocarya alba*, *Quillaja saponaria*, *Maytenus boaria*). This succession suggests that without frequent anthropogenic fire, most areas of

Table 18.3. Major life-history strategies of plants represented in the mediterranean vegetation of central Chile and California

Life-history strategy	Chile	Example	California	Example
Annuals				
weeds	Yes	*Erodium* spp.	Yes	*Cryptantha* spp.
pyrophytes	No		Yes	*Phacelia brachyloba*
Perennial herbs				
geophytes	Yes	*Pasithaea coerulea*	Yes	*Brodiaea ixioides*
pyrophytes	No?		Yes	*Helianthemum scoparium*
Shrubs				
sprouter	Yes	*Lithrea caustica*	Yes	*Quercus dumosa*
obligate seeder	Yes	*Trevoa trinervis*[a]	Yes	*Ceanothus greggii*
sprouter seeder	No?		Yes	*Adenostoma fasciculatum*
Trees	Yes	*Cryptocarya alba*	Yes	*Quercus* spp.

[a] Iván Lazo (pers. commun.)

matorral would develop into woodland. A different sequence is documented for chaparral, where unburned old-growth stands may become senescent (Vogl 1982). This observation further suggests that fire has differently affected evolution for the matorral and chaparral floras, hence the dynamics of these communities.

We have no detailed comparisons of ecophysiological tolerances of chaparral and matorral shrubs to light and moisture conditions in open- versus closed-canopy environments. We suspect that tolerances of most chaparral species, at least for light and water stress, may be broader than those of matorral species. We base this hypothesis on the observation that many chaparral shrubs become established in open sites (Hanes 1971). Some sclerophyllous trees (e.g., *Cryptocarya alba*) of the Chilean matorral, though have very narrow tolerances, at least for water stress (Martínez and Armesto 1983), and their seedlings become established in open areas only under the shade of "nurse shrubs" (Fuentes et al. 1986; Bustamante 1992). In central Chile, several shrubs and trees are exclusively found in the shaded understory in mesic matorral woodlands (Armesto and Martínez 1978) and can be classified as intolerant to full light. We are not aware, however, of any shrub species being restricted to the understory in the chaparral (Hanes 1971).

Allelopathy

Allelopathy has been considered vital in preventing herbs and shrub seedlings from establishing and growing in the chaparral understory during intervals between disturbances (Keeley et al. 1985). Leachates accumulating from leaves in the soil under shrubs could inhibit germination and growth of some species. The evidence supporting this effect (McPherson and Muller 1969) remains controversial, however (Keeley et al. 1985). Recent experimental work has failed to demonstrate that the leachate has a generalized inhibitory effect (Keeley et al. 1981; 1985). In two separate assays, leachate of *Adenostoma fasciculatum* leaves negatively affected the germination of three herbaceous species of chaparral, affected not at all 19 species, and positively affected 16 species (Keeley et al. 1985). In Chile, a limited study on how leaf extracts from five sclerophyllous shrubs influenced herb germination (Montenegro et al. 1978) failed to show negative effects. Some inhibitory effects, though, have been attributed to leachate from leaves of the drought-deciduous shrub *Flourensia thurifera* (Fuentes et al. 1987), which is restricted to dry matorral sites, but its effects are not widespread in the matorral. Evidence that inhibitory toxic substances deter species invasion in either the matorral or in the chaparral remains doubtful.

Herbivory

Differential responses by plant species to herbivory and seed predation may account for some transitions during succession. The herbivores' function in maintaining open areas in the chaparral has been documented by Bartholomew (1970), and their influence in early postfire succession has been evalu-

Table 18.4. Major mammalian herbivores and seed predators in mediterranean regions of Chile and California

	Chile	California
Large native browsers	Locally extinct *Lama guanicoe*	Common *Odocoileus hemionus*
Specialized granivorous rodents	None	Common Heteromydae
Native grazers	Mainly *Octodon degus*	Many species *Lepus, Sylvilagus, Thomomys*
Introduced grazers	Many, widespread	Few, restricted

ated by Mills (1986) and Mills and Kummerow (1989). Several native mammals are important seed eaters, browsers, and grazers in the chaparral (Table 18.4), including mule deer, rodents, and lagomorphs. These species may kill or prevent growth of shrub seedlings (Mills and Kummerow 1989), influencing the rates at which shrubs reestablish following disturbance. Seed eaters may not always reduce successional rates significantly, however, because many species can resprout after disturbance (Keeley and Keeley 1981; Zedler et al. 1983). In Chile, several native and exotic mammals (Table 18.4) have been shown to kill shrub seedlings in open areas of matorral (Fuentes and Simonetti 1982; Fuentes et al. 1983). These herbivores may have significant influence in precluding reinvasion of open areas by shrub seedlings (Fuentes et al. 1984), especially considering that seed inputs and seedling establishment rates are very low for some species (Jiménez and Armesto 1992). Because of their large populations and intense activity in open areas, exotic mammals, particularly rabbits and livestock, seem to play a key role in maintaining open areas in the matorral (Fuentes and Simonetti 1982). Native herbivores, though, are much less abundant and are restricted to the vicinity of shrubs (Jaksic et al. 1979; Jaksic and Soriguer 1981). The most common native rodent, *Octodon degus*, consumes primarily grasses and forbs rather than seeds (Fuentes and Le Boulengé 1977). The lack of specialist seed eaters among Chilean matorral rodents (Glanz and Meserve 1982) is particularly striking and deserves further study, especially considering the seed bank of matorral shrubs, which is much smaller than that of their chaparral counterparts (Table 18.2). Another distinction between the two systems is the extensive influence of harvester ants on shrub seed banks in the chaparral (Mills and Kummerow 1989). Evidence of granivory by ants in the matorral is slight (Bustamante 1992).

A Convergent Model of Succession in Mesic Sites

Although chaparral succession generally seems to be driven by different forces and to follow pathways different from those of matorral succession, an example from the Berkeley hills (McBride 1974) is clearly similar to the

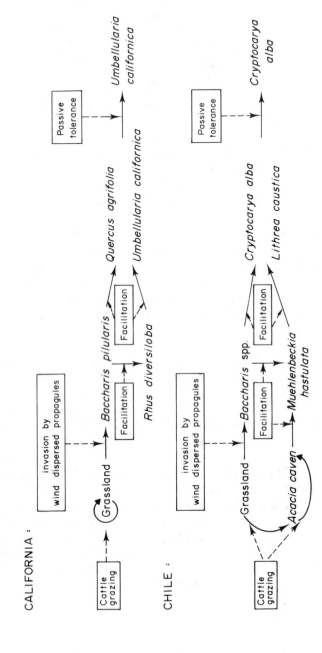

Figure 18.5. Models of succession in mesic Californian chaparral and Chilean matorral.

successional model proposed for mesic sites in the matorral by Armesto and Pickett (1985). After being eradicated by livestock, California grasslands turned slowly into a monospecific *Baccharis pilularis* brushland, which is eventually invaded by woodland species such as *Quercus agrifolia* and *Umbellularia californica* (Fig. 18.5). Sometimes mixed stands are formed by *Baccharis* and *Rhus diversiloba*. Seedling recruitment is closely associated with established *Baccharis* individuals (McBride 1974). Although *Baccharis* has windborne achenes, all other shrub species invading later in succession are dispersed by animals, presumably by birds. A similar successional model applies to post-agricultural succession in mesic matorral sites (Fig. 18.5). The pioneer shrubs in old fields are a number of *Baccharis* species, and many of the sclerophyllous shrubs and trees invading later have bird-disseminated propagules that become established only beneath pioneer shrubs (Fuentes et al. 1986). With no fire and grazing, both successions may lead to dominance by the longer-lived tree species. One large difference in succession between the two regions is that in Chile continuous grazing pressure leads from grassland to an *Acacia caven* savanna (Ovalle et al. 1990; Peralta et al. 1992), with sparse shrub cover and an understory of European weeds. Grazed areas in California remain as grassland dominated by native annuals.

Conclusions

The general successional framework presented here has proved useful for identifying and evaluating similarities and differences in vegetation in chaparral and matorral (Table 18.5). First, the disturbance regime, which determines the temporal and spatial distribution of successional sites, differs

Table 18.5. Relative importance of various processes contributing to vegetation change in matorral and chaparral

Process	Chile	California
Wildfires	0	b
Anthropogenic fire	c	b
Long-distance dispersal	c	a
Propagule pool	a	c
Tolerance to drought	c	b
Allelopathy	0	a
Native herbivores	a	c
Introduced herbivores	c	?

[a] = occasionally important
[b] = important
[c] = very important
0 = absent

greatly between the two regions. Lightning fires have been important at least since the Quaternary (Axelrod 1989) in promoting vegetational change in the chaparral but not in the matorral. Anthropogenic disturbances are significant in both regions, but appear to have different consequences for the organisms. In California, human beings have altered the geographic distribution, frequency, and intensity of fire (Keeley 1982; Minnich 1989), a preexisting form of disturbance. Because many species have life-history traits related to the fire cycle, often recovery by the original vegetation is rapid in open areas. In the Chilean matorral, man has introduced fire as a recent and unprecedented type of perturbation to organisms and to the system (Aschmann and Bahre 1977; Armesto and Gutiérrez 1978). Consequently, we might expect the increase in frequency and extent of anthropogenic fires to more strongly influence vegetation in the matorral than in the chaparral. Some burned areas in central Chile, previously dominated by evergreen sclerophyll shrubs, are occupied by succulents, drought-deciduous, and aphyllous shrubs characteristic of arid environments (Armesto and Gutiérrez 1978). Grazed areas are occupied by a sparse cover of drought-tolerant *Acacia* shrubs (Ovalle et al. 1990). Seed dispersal into open sites and seedling tolerance to drought seem to limit recovery of the matorral, but they are not so critical in most disturbed sites in California. Many chaparral species maintain large dormant seed banks which survive fire (Parker and Kelly 1989), hence recovery after fire includes seedling establishment as well as resprouting. Seed pools of shrubs in disturbed matorral sites are small and lack some important species (Jiménez and Armesto 1992). The ranges of tolerance to light and water stress appear to be narrower for matorral species and therefore more influential in determining the sequence of invasion, although this point deserves critical examination.

Competition does not appear to be significant in determining successional pathways in either chaparral or matorral, at least during early succession, perhaps because other influences (e.g., low seed input, low seedling survival, high herbivore pressure) limit species' abundance at this stage. Allelopathy's influence in preventing vegetational change in the chaparral seems to have been overemphasized, and does not appear to be important in the matorral. Finally, herbivory seems a major influence in both chaparral and matorral succession, but a significant difference appears in the evolutionary context of this interaction in the two systems. As we have seen, most mammalian herbivores presently influencing regeneration of the matorral are introduced—some of them, such as feral rabbits and hares—as early as the nineteenth century. These herbivores lack natural enemies and can maintain large populations (Jaksic and Soriguer 1981). But chaparral succession is influenced mainly by native herbivores, and some plant traits such as production of large seed crops may be a response to this pressure. Nevertheless, herbivore populations in the chaparral may be larger now than in the past because of increasing hunting pressure on their predators (Vogl 1982). Consequently, herbivore influence has probably grown in recent times in both chaparral and matorral, probably slowing recovery of vegetation in disturbed sites.

In summary, present matorral and chaparral vegetation dynamics are mainly controlled by anthropogenic influences. Human influence seems to have more drastically affected natural regeneration of the matorral because of the extensive devastation of shrublands and widespread introduction of novel forms of disturbance, such as fire and exotic grazers and browsers. The main consequences would be slower successional rates and persistence of anthropogenic plant formations, such as the *Acacia caven* savanna in central Chile. When fire is excluded, change in vegetation in mesic sites seems to follow similar successional trajectories in both regions.

Acknowledgments

We are grateful to S.T.A. Pickett and the editors of this book for helpful comments on earlier versions of the manuscript. We thank P. Rundel and J. Keeley for providing important references and advice. Partial funding for this work was provided by a grant from FONDECYT, Chile, to E. Fuentes. Final manuscript preparation was supported by FONDECYT, Grant 92/1135 to the first author.

References

Armesto JJ, Gutiérrez JR (1978) El efecto del fuego en la estructura de la vegetación de Chile central. Anal Mus Hist Nat, Valparaíso 11:43–48

Armesto JJ, Martínez, JA (1978) Relations between vegetation structure and slope aspect in the mediterranean region of Chile. J Ecol 66:881–889

Armesto JJ, Pickett STA (1985) A mechanistic approach to the study of succession in the Chilean matorral. Rev Chil Hist Nat 58:9–17

Arroyo MTK, Uslar P (1993) Breeding systems in a temperate mediterranean-type climate montane sclerophyllous forest in central Chile. Bot J Linn Soc 111:83–102

Aschmann, H (1977) Aboriginal use of fire. In Proc Sym Env Cons Fire and Fuel Man Medit Ecosys. Gen Tech Rep WO-3, USDA For Serv, Washington, DC, pp 132–140

Aschmann H, Bahre CJ (1977) Man's impact on the wild landscape. In Mooney HA (ed) *Convergent Evolution in Chile and California*. Dowden, Hutchinson & Ross, Stroudsburg, PA, pp 73–84

Axelrod DI (1989) Age and origin of chaparral. In Keeley SC (ed), *The California Chaparral: Paradigms Reexamined*. Sci Ser No. 34. Nat Hist Mus LA County, Los Angeles, CA, pp 7–19

Bartholomew B (1970) Bare zone between California shrub and grassland communities: The role of animals. Science 170:1210–1212

Burrows CJ (1990) *Processes of Vegetation Change*. Unwin Hyman, London

Bustamante R (1992) Granivoría y espaciamiento entre plántulas y sus plantas madres: El efecto de la distancia entre plantas madres. Ph.D. Thesis Facultad de Ciencias, U Chile

CONAF (1986) Estadísticas sobre plan de manejo del fuego 1964–1986. Centro de

Documentación Estadística, Departamento de Manejo del Fuego, Corporación Nacional Forestal, Santiago, Chile

Connell JH, Slatyer RO (1977) Mechanisms of succession in natural communities and their role in community stability and organization. Am Nat 111:1119–1144

DeBano LF, Conrad CE (1978) The effect of fire on nutrients in a chaparral ecosystem. Ecology 59:489–497

Ellis BA, Kummerow J (1989) Structure and function in chaparral shrubs. In Keeley SC (ed), *The California Chaparral: Paradigms Reexamined*. Sci Ser No. 34. Nat Hist Mus LA County. Los Angeles, CA, pp 140–150

Finnegan B (1984) Forest succession. Nature 312:109–114

Fuentes ER, Le Boulenge PY (1977) Prédation et compétition dans la dinamique d'une communauté herbacée secondaire du Chili central. Terre et Vie 31:313–326

Fuentes ER, Simonetti JA (1982) Plant patterning in the Chilean matorral: Are the roles of native and exotic mammals different? In Proc Sym Dyn Man Medit Ecosys. Gen Tech Rep PSW-58 USDA For Serv, Berkeley, CA, pp 227–233

Fuentes ER, Jaksic FM, Simonetti JA (1983) European rabbits vs. native mammals: Effects on shrub seedlings. Oecologia 58:411–414

Fuentes ER, Otaiza RD, Alliende MC, Hoffmann A, Poiani A (1984) Shrub clumps of the Chilean matorral vegetation: Structure and possible maintenance mechanisms. Oecologia 64:405–411

Fuentes ER, Hoffmann AJ, Poiani A, Alliende MC (1986) Vegetation change in large clearings: Patterns in the Chilean matorral. Oecologia 68:358–366

Fuentes ER, Espinoza GA, Gajardo G (1987) Allelopathic effects of the Chilean matorral shrub *Flourensia thurifera*. Rev Chil Hist Nat 60:57–62

Glanz WE, Meserve PL (1982) An ecological comparison of small mammal communities in California and Chile. In Proc Sym Dyn Man Medit Ecosys. Gen Tech Rep PSW-58 USDA For Serv, Berkeley

Hanes TL (1971) Succession after fire in the chaparral of southern California. Ecol Mono 41:27–52

Hanes TL (1981) California Chaparral. In di Castri F, Goodall DW, Specht RL (eds) *Ecosystems of the World 11: Mediterranean-Type Shrublands*. Elsevier, Amsterdam, pp 139–174

Hanes TL, Jones HW (1967) Postfire chaparral succession in southern California. Ecology 48:259–264

Heusser CJ (1983) Quaternary pollen record from Laguna de Tagua Tagua, Chile. Science 219:1429–1431

Jaksic FM, Soriguer RC (1981) Predation upon the European rabbit (*Oryctolagus cuniculus*) in mediterranean habitats of Chile and Spain: A comparative analysis. J An Ecol 50:269–281

Jaksic FM, Fuentes ER, Yañez JL (1979) Spatial distribution of the old world rabbit (*Oryctolagus cuniculus*) in central Chile. J Mam 60:206–209

Jiménez HE, Armesto JJ (1992) Importance of the soil seed bank of disturbed sites in Chilean matorral in early secondary succession. J Veg Sci 3:579–586

Keeley JE (1977) Fire-dependent reproductive strategies in *Arctostaphylos* and *Ceanothus*. In Proc Sym Envir Cons Fire Fuel Man Medit Ecosys. Gen Tech Rep WO-3, USDA For Serv, Washington DC, pp 391–396

Keeley JE (1982) Distribution of lightning and man-caused wildfires in California. In Proc Sym Dyn Man Medit Ecosys. Gen Tech Rep PSW-58 USDA For Serv, Berkeley, pp 431–437

Keeley JE, Keeley SC (1981) Post-fire regeneration of southern California chaparral. Am J Bot 68:524–530

Keeley SC, Keeley JE, Hutchinson SM, Johnson AW (1981) Postfire succession of the herbaceous flora in southern California chaparral. Ecology 62:1608–1621

Keeley JE, Morton BA, Pedrosa A, Troter P (1985) The role of allelopathy, heat and charred wood in the germination of chaparral herbs and suffrutescents. J Ecol 73:445–458

Martínez JA, Armesto JJ (1983) Ecophysiological plasticity and habitat distribution in three evergreen species of the Chilean matorral. Oecol Plant 4:211–219

McBride JR (1974) Plant succession in the Berkeley hills, California. Madroño 22: 317–329

McPherson JK, Muller CH (1969) Allelopathic effects of *Adenostoma fasciculatum*, Chamise, in the California chaparral. Ecol Mono 39:177–198

Mills JN (1986) Herbivores and early postfire succession in southern California chaparral. Ecology 67:1637–1649

Mills JN, Kummerow J (1989) Herbivores, seed predators, and chaparral succession. In Keeley SC (ed) *The California Chaparral: Paradigms Reexamined*. Sci Ser No. 34. Nat Hist Mus LA County, Los Angeles, CA, pp 49–55

Minnich RA (1989) Chaparral fire history in San Diego county and adjacent northern Baja California: An evaluation of natural fire regimes and the effects of suppression management. In Keeley SC (ed) *The California Chaparral: Paradigms Reexamined*. Sci Ser No. 34. Nat Hist Mus LA County. Los Angeles, CA, pp 37–47

Montenegro G, Rivera O, Bas F (1978) Herbaceous vegetation in the Chilean matorral. Dynamics of growth and evaluation of allelopathic effects of some dominant shrubs. Oecologia 36:237–244

Mooney HA, ed. (1977) *Convergent Evolution in Chile and California: Mediterranean Climate Ecosystems*. Dowden, Hutchinson & Ross, Stroudsburg, PA

Muñoz MR, Fuentes ER (1989) Does fire induce shrub germination in the Chilean matorral? Oikos 56:177–181

Ovalle C, Aronson J, Del Pozo A, Avendaño J (1990) The espinal: Agroforestry systems of the mediterranean-type climate region of Chile. Agrofor Sys 10:213–239

Parker VT, Kelly VR (1989) Seed banks in California chaparral and other mediterranean climate shrublands. In Leck MA, Parker VT, Simpson RL (eds) *Ecology of Soil Seed Banks*. Academic Press, New York, pp 231–255

Peralta I, Rodríguez J, Arroyo MTK (1992) Breeding systems and aspects of pollination in *Acacia caven* (Mol.) Mol. (Leguminosae: Mimosoideae) in the mediterranean type climate zone of central Chile. Botanische Jahrbucher 114:297–314

Pickett STA, Armesto JJ (1991) The theoretical motivation for ecological comparisons. Rev Chil Hist Nat 64:391–398

Pickett STA, McDonell MJ (1989) Changing perspectives in community dynamics: A theory of successional forces. Trends Ecol Evol 4:241–245

Pickett STA, Collins SL, Armesto JJ (1987) Models, mechanisms and pathways of succession. Bot Rev 53:335–371

Rundel PW, Neel JW (1978) Nitrogen fixation by *Trevoa trinervis* (Rhamnaceae) in the Chilean matorral. Flora 167:127–132

Silvester WB, Balboa O, Martínez JA (1985) Nodulation and nitrogen fixation in members of the Rhamnaceae (*Colletia, Retanilla, Talguenea* and *Trevoa*) growing in the Chilean matorral. Simbiosis 1:29–38

Thrower NJW, Bradbury DE (eds) (1977) *Chile–California Mediterranean Scrub Atlas: A Comparative Analysis*. Dowden, Hutchinson and Ross, Stroudsburg, PA

Vogl RJ (1982) Chaparral succession. In Conrad CE, Oechl WC (eds) Proc Sym Dyn Man Medit Ecosys. Gen Tech Rep PSW-58 USDA For Serv. Berkeley, pp 81–85

Walker LR, Chapin III FS (1987) Interactions among processes controlling successional change. Oikos 50:131–135

White PS (1979) Pattern, process and natural disturbance in vegetation. Bot Rev 45: 229–299

Zedler PH, Zammit CA (1989) A population-based critique of concepts of change in the chaparral. In Keeley SC (ed) *The California Chaparral: Paradigms Reexamined.* Sci Ser No. 34. Nat Hist Mus LA County. Los Angeles, CA, pp 73–83

Zedler PH, Gautier CR, McMaster GS (1983) Vegetation change in response to extreme events: The effect of a short interval between fires in California chaparral and coastal scrub. Ecology 64:809–818

Species Index

Italicized page numbers indicate figures and/or tables

Subject Index

Ecological Studies

Ecological Studies